T0189588

Lecture Notes in Computer Science 13390

More information about this series at https://link.springer.com/bookseries/558

Alberto Barrón-Cedeño ·
Giovanni Da San Martino · Mirko Degli Esposti ·
Fabrizio Sebastiani · Craig Macdonald ·
Gabriella Pasi · Allan Hanbury · Martin Potthast ·
Guglielmo Faggioli · Nicola Ferro (Eds.)

Experimental IR Meets Multilinguality, Multimodality, and Interaction

13th International Conference of the CLEF Association, CLEF 2022
Bologna, Italy, September 5–8, 2022
Proceedings

 Springer

Editors
Alberto Barrón-Cedeño (iD)
University of Bologna
Forlì, Italy

Mirko Degli Esposti (iD)
University of Bologna
Bologna, Italy

Craig Macdonald (iD)
University of Glasgow
Glasgow, UK

Allan Hanbury (iD)
TU Wien
Vienna, Austria

Guglielmo Faggioli (iD)
University of Padua
Padova, Italy

Giovanni Da San Martino (iD)
University of Padua
Padova, Italy

Fabrizio Sebastiani (iD)
Instituto di Scienza e Tecnologie dell'
Informazione "Alessandro Faedo"
Pisa, Italy

Gabriella Pasi (iD)
University Milano-Bicocca
Milan, Italy

Martin Potthast (iD)
Leipzig University
Leipzig, Germany

Nicola Ferro (iD)
University of Padua
Padova, Italy

ISSN 0302-9743   ISSN 1611-3349 (electronic)
Lecture Notes in Computer Science
ISBN 978-3-031-13642-9   ISBN 978-3-031-13643-6 (eBook)
https://doi.org/10.1007/978-3-031-13643-6

This Springer imprint is published by the registered company Springer Nature Switzerland AG
The registered company address is: Gewerbestrasse 11, 6330 Cham, Switzerland

Preface

Since 2000, the Conference and Labs of the Evaluation Forum (CLEF) has played a leading role in stimulating research and innovation in the domain of multimodal and multilingual information access. Initially founded as the Cross-Language Evaluation Forum and running in conjunction with the European Conference on Digital Libraries (ECDL/TPDL), CLEF became a standalone event in 2010 combining a peer-reviewed conference with a multi-track evaluation forum. The combination of the scientific program and the track-based evaluations at the CLEF conference creates a unique platform to explore information access from different perspectives, in any modality and language.

The CLEF conference has a clear focus on experimental information retrieval (IR) as seen in evaluation forums (like the CLEF Labs, TREC, NTCIR, FIRE, MediaEval, RomIP, TAC) with special attention paid to the challenges of multimodality, multilinguality, and interactive search, ranging from unstructured to semi-structured and structured data. The CLEF conference invites submissions on new insights demonstrated by the use of innovative IR evaluation tasks or in the analysis of IR test collections and evaluation measures, as well as on concrete proposals to push the boundaries of the Cranfield/TREC/CLEF paradigm.

CLEF 2022[1] was organized by the University of Bologna, Italy, and held during September 5–8, 2022. Despite the continued outbreak of the COVID-19 pandemic, the improvement of the overall situation allowed for organizing CLEF 2022 as an in-person event, after two editions—CLEF 2020 and 2021—were forced to be virtual only. The conference format remained the same as in past years and consisted of keynotes, contributed papers, lab sessions, and poster sessions, including reports from other benchmarking initiatives from around the world. All sessions were held in person but also allowed for remote participation for those who were not able to attend physically.

CLEF 2022 continued the initiative introduced in the 2019 edition, during which the European Conference for Information Retrieval (ECIR) and CLEF joined forces: ECIR 2022 hosted a special session dedicated to CLEF Labs where lab organizers presented the major outcomes of their labs and their plans for ongoing activities, which was followed by a poster session to encourage discussion during the conference. This was reflected in the ECIR 2022 proceedings, where CLEF Lab activities and results were reported as short papers. The goal was not only to engage the ECIR community in CLEF activities but also to disseminate the research results achieved during CLEF evaluation cycles through the submission of papers to ECIR.

The following scholars were invited to give a keynote talk at CLEF 2022: Rita Cucchiara (University of Modena and Reggio Emilia, Italy) and Benno Stein (Bauhaus-Universität Weimar, Germany).

CLEF 2022 received a total of 14 scientific submissions, of which a total of 10 papers (seven long and three short) were accepted. Each submission was reviewed by three Program Committee (PC) members, and the program chairs oversaw the reviewing

---

[1] https://clef2022.clef-initiative.eu/.

and follow-up discussions. Ten countries are represented in the accepted papers, as several of them were a product of international collaboration. This year, researchers addressed the following important challenges in the community: authorship attribution, fake news detection and news tracking, noise-detection in automatically transferred relevance judgments, the impact of online education on children's conversational search behavior, analysis of multi-modal social media content, knowledge graphs for sensitivity identification, a fusion of deep learning and logic rules for sentiment analysis, medical concept normalization, and domain-specific information extraction.

Like in previous editions, since 2015, CLEF 2022 invited CLEF lab organizers to nominate a "best of the labs" paper that was reviewed as a full paper submission to the CLEF 2022 conference, according to the same review criteria and PC. In total, seven full papers were accepted for this "best of the labs" section.

The conference integrated a series of workshops presenting the results of lab-based comparative evaluations. CLEF 2022 was the 13th edition of the CLEF conference and it was the 23rd year of the CLEF initiative as a forum for IR evaluation. A total of 15 lab proposals were received and evaluated in peer review based on their innovation potential and the quality of the resources created. The 14 selected labs represented scientific challenges based on new datasets and real-world problems in multimodal and multilingual information access. These datasets provide unique opportunities for scientists to explore collections, to develop solutions for these problems, to receive feedback on the performance of their solutions, and to discuss the challenges with peers at the workshops. In addition to these workshops, the labs reported results of their year long activities in overview talks and lab sessions. Overview papers describing each of the labs are provided in this volume. The full details for each lab are contained in a separate publication, the Working Notes[2].

The 14 labs running as part of CLEF 2022 comprised mainly labs that continued from previous editions at CLEF (ARQMath, BioASQ, CheckThat!, CheMU, eRisk, Image-CLEF, LifeCLEF, PAN, SimpleText, and Touché) and newer pilot/workshop activities (HIPE, iDPP, JOKER, and LeQUA). In the following we give a few details for each of the labs organized at CLEF 2022 (presented in alphabetical order):

ARQMath: Answer Retrieval for Mathematical Questions[3] aims to advance math-aware search and the semantic analysis of mathematical notation and texts. It offered the following tasks. Task 1: Answer Retrieval, given a math question post, return relevant answer posts. Task 2: Formula Retrieval; given a formula in a math question post, return relevant formulas from both question and answer posts. Task 3: Open Domain Question Answering, given a math question post, return an automatically generated answer that is comprised of excerpts from arbitrary sources and/or machine generated.

BioASQ: Large-scale Biomedical Semantic Indexing and Answering[4] aims to push the research frontier towards systems that use the diverse and voluminous information available online to respond directly to the information needs of biomedical scientists.

---

[2] Faggioli, G., Ferro, N., Hanbury, A., and Potthast, M. (Eds.). (2022). *CLEF 2022 Working Notes.* CEUR Workshop Proceedings (CEUR-WS.org), ISSN 1613-0073.

[3] https://www.cs.rit.edu/~dprl/ARQMath.

[4] http://www.bioasq.org/workshop2022.

It offered the following tasks. Task 1: Large-scale Online Biomedical Semantic Indexing involved classifying new PubMed documents before PubMed curators annotate (in effect, classify) them manually into classes from the MeSH hierarchy. Task 2: Biomedical Semantic Question Answering used benchmark datasets of biomedical questions, in English, along with gold standard (reference) answers constructed by a team of biomedical experts. The participants had to respond with relevant articles and snippets from designated resources, as well as exact and "ideal" answers. Task 3: DisTEMIST - Disease Text Mining and Indexing Shared Task focused on the recognition and indexing of diseases in medical documents in Spanish by posing subtasks on (1) indexing medical documents with controlled terminologies; (2) automatic detection indexing textual evidence (i.e., disease entity mentions in text); and (3) normalization of these disease entity mentions to terminologies. Task 4: Task Synergy - Question Answering for Developing Problems, biomedical experts posed unanswered questions for the developing problem of COVID-19, received the responses provided by the participating systems, and provided feedback, together with updated questions, in an iterative procedure that aimed to facilitate the incremental understanding of COVID-19.

CheckThat!: Lab on Fighting the COVID-19 Infodemic and Fake News Detection[5] aims at fighting misinformation and disinformation in social media, in political debates, and in the news. It offered the following tasks. Task 1: Identifying Relevant Claims in Tweets focused on disinformation related to the ongoing COVID-19 infodemic politics. It asked participants to identify which posts in a Twitter stream are worth fact-checking, contain a verifiable factual claim, are harmful to society, and why. This task was offered in Arabic, Bulgarian, Dutch, English, Spanish, and Turkish. In Task 2: Detecting Previously Fact-Checked Claims, given a check-worthy claim, and a set of previously-checked claims, participants had to determine whether the claim has been previously fact-checked with respect to a collection of fact-checked claims. The text could be a tweet or a sentence from a political debate. The task was offered in Arabic and English. In Task 3: Fake News Detection, given the text and the title of a news article, participants had to determine whether the main claim made in the article is true, partially true, false, or other (e.g., articles in dispute and unproven articles). This task was offered in English and German.

ChEMU: Cheminformatics Elsevier Melbourne University[6] focuses on information extraction in chemical patents, including tasks ranging from document- to expression-level. It offered the following tasks. Task 1a: Named Entity Recognition was aimed at identifying chemical compounds, their specific types, temperatures, reaction times, yields, and the label of the reaction. Task 1b: Event Extraction acknowledged that a chemical reaction leading to an end product often consists of a sequence of individual event steps. The task was to identify those steps involving the chemical entities recognized from Task 1a. Task 1c: Anaphora Resolution required requires the resolution of anaphoric dependencies between expressions in chemical patents. The participants were required to find five types of anaphoric relationships in chemical patents: coreference, reaction-associated, work-up, contained, and transform. In Task 2a: Chemical Reaction Reference Resolution, given a reaction description, participants had to identify

---

[5] https://sites.google.com/view/clef2022-checkthat.

[6] http://chemu2022.eng.unimelb.edu.au/.

references to other reactions that the reaction relates to, and to the general conditions that it depends on. Task 2b: Table Semantic Classification involved classifying tables in chemical patents into eight categories based on their contents.

eRisk: Early Risk Prediction on the Internet[7] explores the evaluation methodology, effectiveness metrics, and practical applications (particularly those related to health and safety) of early risk detection on the Internet. The main goal is to pioneer a new interdisciplinary research area that would be potentially applicable to a wide variety of situations and to many different personal profiles. Examples include potential pedophiles, stalkers, individuals that could fall into the hands of criminal organizations, people with suicidal inclinations, or people susceptible to depression. It offered the following tasks. Task 1: Early Detection of Signs of Pathological Gambling consisted of sequentially processing pieces of evidence and detecting early traces of pathological gambling (also known as compulsive gambling or disordered gambling), as soon as possible. Task 2: Early Detection of Depression consisted of sequentially processing pieces of evidence and detecting early traces of depression as soon as possible. Tasks 1 and 2 were mainly concerned with evaluating text mining solutions and thus concentrated on texts written in social media. Task 3: Measuring the Severity of the Signs of Eating Disorders consisted of estimating the level of features associated with a diagnosis of eating disorders from a thread of user submissions. For each user, the participants were given a history of postings and had to complete a standard eating disorder questionnaire (based on the evidence found in the history of postings).

HIPE: Named Entity Recognition and Linking in Multilingual Historical Documents[8] aims to assess and advance the development of robust, adaptable, and transferable named entity processing systems. Compared to the first HIPE edition in 2020, HIPE 2022 confronted systems with the challenges of dealing with more languages, learning domain-specific entities, and adapting to diverse annotation schemas. It offered the following tasks. Task 1: Named Entity Recognition and Classification (NERC) featured two subtasks—NERC-coarse on high-level entity types, for all languages, and NERC-fine on finer-grained entity types, for English, French, and German only. Task 2: Named Entity Linking (EL) involved the linking of named entity mentions to a unique referent in a knowledge base (Wikidata) or to a NIL node if the mention does not have a referent in the knowledge base.

iDPP: Intelligent Disease Progression Prediction[9] aims to design and develop an evaluation infrastructure for AI algorithms able to (1) better describe the mechanism of Amyotrophic Lateral Sclerosis (ALS) disease; (2) stratify patients according to their phenotype assessed throughout the disease evolution; and (3) predict ALS progression in a probabilistic, time dependent fashion. It offered the following tasks. Task 1: Ranking Risk of Impairment focused on ranking patients based on the risk of impairment in specific domains. We used the ALSFRS-R scale to monitor speech, swallowing, handwriting, dressing/hygiene, walking, and respiratory ability in time and asked participants to rank patients based on time to event risk of experiencing impairment in each specific domain. Task 2: Predicting Time of Impairment refined Task 1 by asking participants

---

[7] https://erisk.irlab.org/.

[8] https://hipe-eval.github.io/HIPE-2022/.

[9] https://brainteaser.health/open-evaluation-challenges/idpp-2022/.

to predict when specific impairments will occur (i.e., in the correct time window) by assessing model calibration in terms of the ability of the proposed algorithms to estimate a probability of an event close to the true probability within a specified time window. Task 3: Explainability of AI Algorithms called for position papers to start a discussion on AI explainability, including proposals on how the single patient data can be visualized in a multivariate fashion contextualizing its dynamic nature and the model predictions together with information on the predictive variables that most influence the prediction.

ImageCLEF: Multimedia Retrieval[10] promotes the evaluation of technologies for annotation, indexing, classification, and retrieval of multi-modal data, with the objective of providing information access to large collections of images in various usage scenarios and domains. It offered the following tasks. Task 1: ImageCLEFmedical focused on interpreting and summarizing the insights gained from radiology images, i.e. developing systems that are able to predict the UMLS concepts from visual image content, and implementing models to predict captions for given radiology images. The tuberculosis subtask called for systems that are able to detect lung cavern regions rather than simply provide a label for the CT images. Task 2: ImageCLEFcoral fostered tools for creating three-dimensional models of underwater coral environments. It required participants to label coral underwater images with types of benthic substrate, together with their bounding box, and to segment and parse each coral image into different image regions associated with benthic substrate types. Task 3: ImageCLEFaware concerned the online disclosure of personal data, which often has effects that go beyond the initial context in which data were shared. Participants were required to provide automatic rankings of photographic user profiles in a series of real-life situations, such as searching for a bank loan, accommodation, a waiter job, or a job in IT, with the ranking based on an automatic analysis of profile images and the aggregation of individual results. Task 4: ImageCLEFfusion involved system fusion—exploiting the complementary nature of individual systems to boost performance. Participants were tasked with creating novel ensembling methods that are able to significantly increase the performance of precomputed inducers in various use-case scenarios, such as visual interestingness and video memorability prediction.

JokeR: Automatic Wordplay and Humour Translation Workshop[11] aims to bring together translators and computer scientists to work on an evaluation framework for creative language, including data and metric development, and to foster work on automatic methods for wordplay translation. It offered the following tasks. Pilot Task 1: Classify and Interpret Wordplay involved, classifying single words containing wordplay according to a given typology, and providing lexical-semantic interpretations. Pilot Task 2: Translate Single Term Wordplay required the translation of single words containing wordplay. Pilot Task 3: Translate Phase Wordplay involved the translation of entire phrases that subsume or contain wordplay. In Task 4: Unshared Task, we welcomed submissions that use our data in other ways.

LeQua: Learning to Quantify[12] aims to allow the comparative evaluation of methods for "learning to quantify" in textual datasets; i.e. methods for training predictors of

---

[10] https://www.imageclef.org/2022.

[11] http://joker-project.com/.

[12] https://lequa2022.github.io/.

the relative frequencies of the classes of interest in sets of unlabeled textual documents. These predictors (called "quantifiers") are required to issue predictions for several such sets, some of them characterized by class frequencies radically different from the ones of the training set. This first edition of LeQua offered the following tasks. In Task 1 participants were provided with documents already converted into vector form; the task was thus suitable for participants who do not wish to engage in generating representations for the textual documents, but wanted instead to concentrate on optimizing the methods for learning to quantify. In Task 2 participants were provided with the raw text of the documents; the task was thus suitable for participants who also wished to engage in generating suitable representations for the textual documents, or to train end-to-end systems.

LifeCLEF: Biodiversity Identification and Prediction[13] aims to stimulate research in data science and machine learning for biodiversity monitoring. It offered the following tasks. Task 1: BirdCLEF involved bird species recognition in audio soundscapes. Task 2: PlantCLEF concerned image-based plant identification on a global scale (300K classes). Task 3: GeoLifeCLEF required location-based prediction of species based on environmental and occurrence data. Task 4: SnakeCLEF involved snake species identification in medically important scenarios. Task 5: FungiCLEF involved fungi recognition from image and metadata.

PAN: Digital Text Forensics and Stylometry[14] focuses on digital text forensics and stylometry, studying how to quantify writing style and improve authorship technology. It offered the following tasks. Task 1: Authorship Verification, given two texts, determine if they are written by the same author. Task 2: IROSTEREO, profiling Irony and Stereotype Spreaders on Twitter, given a Twitter feed, determine whether its author spreads Irony and Stereotypes. Task 3: Style Change Detection, given a document, determine the number of authors and at which positions the author changes. Task 4: Trigger Warning Prediction, given a document, determine whether its content warrants a warning of potential negative emotional responses in readers.

SimpleText: Automatic Simplification of Scientific Texts[15] addresses the challenges of text simplification approaches in the context of promoting scientific information access, by providing appropriate data and benchmarks, and creating a community of NLP and IR researchers working together to resolve one of the greatest challenges of today. It offered the following tasks. Task 1: What is in (or out)? Select passages to include in a simplified summary, given a query. Task 2: What is unclear? Given a passage and a query, rank terms/concepts that must be explained to understand this passage (definitions, context, applications, ...). Task 3: Rewrite this! Given a query, simplify passages from scientific abstracts. Task 4: Unshared task, we welcomed any submission that uses our data.

Touché: Argument Retrieval[16] focuses on decision making processes, be it at the societal or at the personal level, that often come to a point where one side challenges the other with a why-question, which is a prompt to justify some stance based on arguments.

---

[13] https://www.imageclef.org/LifeCLEF2022.
[14] http://pan.webis.de/.
[15] http://simpletext-project.com/.
[16] https://touche.webis.de/.

Since technologies for argument mining are maturing at a rapid pace, ad-hoc argument retrieval has also become within reach. Touché offered the following tasks. Task 1: Argument Retrieval for Controversial Questions, given a controversial topic and a collection of argumentative documents, participants had to retrieve and rank sentences (the main claim and its most important premise in the document) that convey key points pertinent to the controversial topic. Task 2: Argument Retrieval for Comparative Questions, given a comparative topic and a collection of documents, participants had to retrieve relevant argumentative passages for either the compared object or for both objects and detect the respective stances with regard to the object in question. Task 3: Image Retrieval for Arguments, given a controversial topic, participants had to retrieve images (from web pages) for each stance (pro/con) that show support for that stance.

The success of CLEF 2022 would not have been possible without the huge effort of numerous people and organizations, including the CLEF Association[17], the Program Committee, the Lab Organizing Committee, the reviewers, and the many students and volunteers who contributed.

Finally, we thank the University of Bologna (with special mention to the DIT, DIFA, and DISI departments), the Department of Mathematics of the University of Padua, and the AI4media H2020 project for their invaluable support. We thank the Friends of SIGIR program for covering the registration fees for a number of student delegates.

July 2022

Alberto Barrón-Cedeño
Giovanni Da San Martino
Mirko Degli Esposti
Fabrizio Sebastiani
Craig Macdonald
Gabriella Pasi
Allan Hanbury
Martin Potthast
Guglielmo Faggioli
Nicola Ferro

---

[17] http://www.clef-initiative.eu/association.

Organization

CLEF 2022, the Conference and Labs of the Evaluation Forum – Experimental IR meets Multilinguality, Multimodality, and Interaction, was hosted by University of Bologna, Italy.

General Chairs

Alberto Barrón-Cedeño	University of Bologna, Italy
Giovanni Da San Martino	University of Padua, Italy
Mirko Degli Esposti	University of Bologna, Italy
Fabrizio Sebastiani	ISTI-CNR, Italy

Program Chairs

Craig Macdonald	University of Glasgow, UK
Gabriella Pasi	University of Milano-Bicocca, Italy

Lab Chairs

Allan Hanbury	Vienna University of Technology, Austria
Martin Potthast	Leipzig University, Germany

Lab Mentorship Chair

Paolo Rosso	Universitat Politécnica de Valencia, Spain

Proceedings Chairs

Guglielmo Faggioli	University of Padua, Italy
Nicola Ferro	University of Padua, Italy

Program Committee

Martin Braschler	ZHAW Zurich University of Applied Sciences, Switzerland
Annalina Caputo	Dublin City University, Ireland
Fabio Crestani	USI, Switzerland
Giorgio Maria Di Nunzio	University of Padua, Italy

Guglielmo Faggioli	University of Padua, Italy
Jussi Karlgren	Spotify, USA
Liadh Kelly	Maynooth University, Ireland
Maria Maistro	University of Copenhagen, Denmark
Stefano Marchesin	University of Padua, Italy
Yashar Moshfeghi	University of Strathclyde, UK
Franco Maria Nardini	ISTI-CNR, Italy
Jian-Yun Nie	University of Montreal, Canada
Iadh Ounis	University of Glasgow, UK
Paolo Rosso	Universitat Politècnica de València, Spain
Eric Sanjuan	Laboratoire Informatique d'Avignon, Université d'Avignon, France
Ting Su	University of Glasgow, UK
Hanna Suominen	ANU, Australia
Theodora Tsirika	CERTH-ITI, Greece
Marco Viviani	Università degli Studi di Milano-Bicocca, Italy
Ellen Voorhees	National Institute for Standards and Technology (NIST), USA
Christa Womser-Hacker	University of Hildesheim, Germany

CLEF Steering Committee

Steering Committee Chair

| Nicola Ferro | University of Padua, Italy |

Deputy Steering Committee Chair for the Conference

| Paolo Rosso | Universitat Politècnica de València, Spain |

Deputy Steering Committee Chair for the Evaluation Labs

| Martin Braschler | Zurich University of Applied Sciences, Switzerland |

Members

Khalid Choukri	Evaluations and Language resources Distribution Agency (ELDA), France
Fabio Crestani	Università della Svizzera italiana, Switzerland
Carsten Eickhoff	Brown University, USA
Norbert Fuhr	University of Duisburg-Essen, Germany

Lorraine Goeuriot	Université Grenoble Alpes, France
Julio Gonzalo	National Distance Education University (UNED), Spain
Donna Harman	National Institute for Standards and Technology (NIST), USA
Bogdan Ionescu	Politehnica University of Bucharest, Romania
Evangelos Kanoulas	University of Amsterdam, The Netherlands
Birger Larsen	University of Aalborg, Denmark
David E. Losada	Universidade de Santiago de Compostela, Spain
Mihai Lupu	Vienna University of Technology, Austria
Maria Maistro	University of Copenhagen, Denmark
Josiane Mothe	IRIT, Université de Toulouse, France
Henning Müller	University of Applied Sciences Western Switzerland (HES-SO), Switzerland
Jian-Yun Nie	Université de Montréal, Canada
Eric SanJuan	University of Avignon, France
Giuseppe Santucci	Sapienza University of Rome, Italy
Jacques Savoy	University of Neuchâtel, Switzerland
Laure Soulier	Sorbonne University, France
Theodora Tsikrika	CERTH-ITI, Greece
Christa Womser-Hacker	University of Hildesheim, Germany

Past Members

Paul Clough	University of Sheffield, UK
Djoerd Hiemstra	Radboud University, The Netherlands
Jaana Kekäläinen	University of Tampere, Finland
Séamus Lawless	Trinity College Dublin, Ireland
Carol Peters (Steering Committee Chair 2000–2009)	ISTI-CNR, Italy
Emanuele Pianta	Centre for the Evaluation of Language and Communication Technologies (CELCT), Italy
Maarten de Rijke	University of Amsterdam, The Netherlands
Alan Smeaton	Dublin City University, Ireland

Supporters and Sponsors

ALMA MATER STUDIORUM
UNIVERSITÀ DI BOLOGNA

Contents

Best of 2021 Labs

Overviews of 2022 Labs

Full Papers

Full Papers

"Meanspo Please, I Want to Lose Weight": A Characterization Study of Meanspiration Content on Tumblr Based on Images and Texts

Linda Achilles[(✉)], Thomas Mandl, and Christa Womser-Hacker

Department of Information Science and Natural Language Processing,
University of Hildesheim, Hildesheim, Germany
{achilles,mandl,womser}@uni-hildesheim.de

Abstract. Past research has demonstrated a linkage between social media usage and disordered eating habits and body dissatisfaction. Trends relating to eating disorders develop around specific hashtags in communities in social networking sites such as Tumblr. One of these trends is #meanspiration, a tag that is used to request and give mean messages from/to social media users to inspire them to lose weight. In this study, images and texts of Meanspiration posts are automatically analyzed based on colorfulness, the images' emotional measures pleasure, arousal and dominance, whereas the textual information of the posts is evaluated based on sentiments, emotions and readability. These characteristics are used in a classification task to distinguish Meanspiration from regular content on Tumblr with 81% accuracy.

Keywords: Social Media · Image Analysis · Text Analysis · Emotion Analysis · Meanspiration · Tumblr

1 Introduction

Studies have shown that there is a link between the consumption of media (TV and magazines) and disordered eating behavior [14]. With the up-rise of social media, researchers investigated also the impact the use of these platforms has on dysfunctional eating patterns. Mabe et al. [22] found that a 20 min use of Facebook in their empirical study affects body weight and shape concerns in the study participants, compared to 20 min spent on other internet services. Tiggemann and Slater [29] demonstrated that the exposure to the Internet significantly correlated with body image concerns like the internalisation of beauty ideals, body surveillance and the drive for thinness. Also Holland and Tiggemann [16] confirmed in their review article the connection between social media usage and eating disorder symptoms. A study by Tiggemann and Anderberg [28] further explored the effect of the 'Instagram vs. reality' images trend on the body perception of women. The results show that watching natural pictures

A. Barrón-Cedeño et al. (Eds.): CLEF 2022, LNCS 13390, pp. 3–17, 2022.
https://doi.org/10.1007/978-3-031-13643-6_1

(in contrast to an idealized version of the same photo) on social media leads to less body dissatisfaction. Since these trends harbor great risks, early detection is particularly important with regard to the safety of social media users. The early detection of eating disorders in social media posts consequently became the focus of research groups and challenges, such as eRisk (early risk detection on the internet)[1] that is held in conjunction with the conferences of the CLEF initiative[2]. Besides the 'Instagram vs. reality' trend, also others arose, forming around specific hashtags.

One of these hashtags is #meanspiration or its abbreviation #meanspo. There is, at the time of writing this work, no study on this topic. A definition of Meanspiration therefore versions Urban Dictionary and can be defined as follows:

> "Meanspiration or Meanspo is a way of encouraging eating disorders by asking for and sending mean messages to social media users with the intention to inspire them to lose weight. The trend originated from Tumblr. Meanspiration is an artificial word consisting of *mean* and *inspiration*[3]".

Since studies about this phenomenon are underrepresented this work shall start filling this gap. Two research questions are formulated and are going to be answered in the course of this paper:

> RQ1: What are the characteristics of Meanspiration content in comparison to regular content on Tumblr in terms of image, text and a post's statistical attributes?
> RQ2: Can these characteristics be used to automatically distinguish between Meanspiration and regular content on social media?

To answer the research questions, posts and their corresponding images from the social media platform Tumblr are downloaded and analyzed. The extracted information is then used to perform a classification task on the social media posts.

2 Related Work

Besides Meanspiration, other artificial words arose in the eating disorder (ED) communities, like *Thinspiration* (combining 'thin' and 'inspiration'), a term that is used to share content intended to inspire social media users to become or remain very thin. Images depicting emaciated bodies or body parts, weight loss or diet plans and quotes as well as techniques to become thin, are shared using the hashtag Thinspiration on social media [1]. For that focus on the body Thinspiration content is interconnected with the pro anorexia (or pro-ED) community, a

---

[1] Website of the eRisk Lab: https://erisk.irlab.org/.
[2] Website of the CLEF initiative: http://www.clef-initiative.eu/.
[3] Taken from Urban Dictionary: https://www.urbandictionary.com/define.php?term =Meanspo.

movement that treats the illness of anorexia as an alternative lifestyle choice [2]. Thinspiration, just like Meanspiration, is a way of encouraging eating disorders. However, it was investigated in different areas of research, unlike the latter.

Automatic approaches were used to identify eating disorder related content on social networks. Chancellor et al. [5] developed and evaluated a supervised learning model that can distinguish between pro-ED and acceptable social media content by analyzing Tumblr photo posts. They showed that images in combination with texts play a big role in identifying deviant content in social media. One study [7] further analyzed the characteristics of posts in pro-ED communities that were removed from Instagram. They found that the removed content depicts potential harmful content like for instance self-harm behavior when compared to the posts that remain online. A different study [9] further investigated the lexical variations of hashtags that appeared after the banning of pro-ED tags on Instagram. The researchers could show that content moderation is not an efficient way to prevent the sharing of pro-ED content. Moreover, they could show that the communities, that used new lexical variants of banned tags, engaged stronger in their pro-ED communities. Researchers also addressed mental illness severity (MIS) in their work [6]. They observed a raise over time in MIS in social media authors who share pro-ED content on Instagram. De Choudhury [10] characterized anorexia on Tumblr. She showed that the pro-ED community focuses more on current life situations while they show avoidant behavior and attitudes. Furthermore, she showed that members of the pro-ED community have reduced levels of cognitive processing capabilities, which held the highest predictive power. Fettach and Benhiba [11] analyzed the main topics discussed in pro-ED and pro-recovery communities. In addition they performed a sentiment and a network analysis showing that that although the communities have common topics like for instance Thinspiration, they are discussed with different sentiments. Both communities show high connectivity networks. Topical trends were also investigated by Masood et al. [24]. They manually labeled anorexia related posts from Reddit and used them to differentiate pro-ED users and control group users. Chancellor, Mitra and De Choudhury [8] explored how Tumblr as a platform impacts the sustainable recovery from anorexia. They performed statistical analyses to calculate the chance of recovery in users. Their results show that only half of the studied cohort experiences recovery after a time period of four years. The likelihood of recovery remains low also after a period of six years (56%).

Since recovery from anorexia is difficult and after several years its likelihood does not change significantly anymore [8] the early detection of signs of anorexia is important. The early risk detection on the internet is the main objective of the eRisk Lab. The organizers of eRisk want to provide a platform for the evaluation methodologies, performance metrics, and building of test collections concerning issues of health and safety on the internet [21]. The early detection of signs of anorexia was part of a shared task in 2018 and 2019. A new measure, the early risk detection measure (ERDE), was introduced to punish systems that need a large number of social media posts to come to a correct classification and reward those that come to early decisions [19]. In 2018 best results for $ERDE_5$ (the value of the measure drops quickly after five processed posts) were achieved by implementing Flexible Temporal Variation of Terms (FTVT) and Sequential Incremental Classification (SIC). The team reached an $ERDE_5$ of 11.40%. The same team also reached best precision (p = 0.91) [12]. The most successful approaches for $ERDE_{50}$ (=5.96%) and F_1 (=0.85) were word embeddings and a linguistic metadata analysis [30]. In 2019 the use of neural networks in combination with a support vector machine achieved the highest F_1 (=0.71) [25]. The best precision (=0.77) was achieved by using a Random Forest Classifier [26]. The most successful $ERDE_5$ (=5.54%) and $ERDE_{50}$ (=2.97%) were achieved by implementing a SS3 text classifier [3]. In 2022 a new task, measuring the severity of the signs of eating disorders, is introduced.

Moreover, various qualitative studies addressed eating disorder related social media content. One approach, conducted by Wick and Harriger [34], focused on Thinspiration on Tumblr. They performed a content analysis of texts and images and showed that the majority of images showed thin bodies incorporating culturally based beauty ideals. The dominant themes in the posts' texts were food restraint as well as weight loss. Talbot et al. [27] also explored Thinspiration content, but distinguished it from the *Fitspiration* (inspiration to be fit) and *Bonespiration* (inspiration to have protruding bones), where they found that muscular bodies are only depicted in Fitspiration posts, but Thinspiration as well as Bonespiration feature very thin and objectified bodies. Ging and Garvey [13] analyzed Instagram images that were tagged with pro-ana hashtags. The researchers' approach revealed that large portions of the content were Thinspiration related, showing e.g. underweight bodies and thigh gaps. Another approach [4] focused in more detail on ED symptoms as they were expressed in Twitter tweets. However, an analysis of the used hashtags revealed that Thinspiration and its abbreviation *Thinspo* were among the three most common ones. The most discussed symptom was the concern about the body's shape. Branley and Covey [2] further investigated the communication about eating disorders on Twitter and Tumblr. They identified three predominately discussed topics: Pro-ana, the glorification of anorexia as a lifestyle choice, anti-ana, the explicit resistance against anorexia as a lifestyle and pro-recovery, a content that shares recovery experiences from eating disorders.

The studies mentioned above show that there is potential in identifying possibly harmful content on social media using automatic approaches. However, the phenomenon of Meanspiration is not addressed so far. Therefore, the following sections will present the research methodology of this paper and how the different measures can be used to distinguish Meanspiration posts from regular Tumblr content.

3   Research Methodology

This section describes the methodological approach. First the data collection is described, then the image analysis and the image measures are shown. Next the text analysis and the measures used for it are explained.

3.1 Data Collection

There are two data sets collected for this work: 1) A Meanspiration (referred to as 'Meanspo' for brevity) and 2) a control set.

Starting from the original tag #meanspiration more tags were manually searched for and included in the list by using the Tumblr search bar and collect more similar tags and derivatives of the original one. This way 22 Meanspiration hashtags were retrieved. Table 1 shows them in alphabetical order.

Table 1. Hashtags used to build the meanspo data set in alphabetical order

me4nsp0	meanso	meanspi	meanspiro	meanspox	
mean inspiration	meansp	meanspii	meanspo	meanspp	meanspx
mean inspo	meansp0	meanspir0	meanspocoach	meansppo	meanspø
meanpso	meanspa	meanspiration	meanspoo	meanspr0	

To characterize Meanspiration content on Tumblr, the posts tagged with one (or more) of the above mentioned hashtags were retrieved using the Tumblr application programming interface (API)[4]. The time period was between the years 2017 and 2021 in universal standard time (UTC).

The second data set was created to obtain a data set which represents standard Tumblr content to compare the Meanspiration posts to. To construct it a website that is used for social media marketing[5] was used to find the most common hashtag that co-occurred with #tumblr, which was #love. #Love posts of the year 2019 were downloaded while only keeping a subset of 15,000 posts as the control data set for this work.

Since the images used in the posts shall be analyzed besides the textual information, another algorithm was used to download all the images that were used in the posts. Table 2 summarizes the basic descriptive statistics of the two data sets.

3.2 Image Analysis

The qualitative study by Branley and Covey [2] described an impression, that pro-ED community content looks darker and 'gloomier' than content from other communities. Therefore, for this work, the colorfulness of images is calculated. The emotional measures pleasure, arousal and dominance were successfully used in order to characterize and detect self-harm in images shared on social media [33], so they were utilized for this study aswell.

Hasler and Süsstrunk [15] introduced a measure to calculate the colorfulness of an image. They investigated in an empirical approach how colorful participants perceived different images and found, through a set of experimental calculations, a measure that correlated with the empirical observation by 95.3%. The

---

[4] Link to the Tumblr API: https://www.tumblr.com/docs/en/api/v2.
[5] https://displaypurposes.com/hashtags/hashtag/tumblr.

Table 2. Basic descriptive statistics of the two data sets. Numbers are rounded to two decimal places.

	Meanspo	Control
Number of posts	14,925	15,000
Number of posts containing images	3,042 (20.38%)	10,532 (70.21%)
Number of images	5,059	11,988
Mean number of images per image post	1.66	1.12
Median of images per post	0	1
Number of notes on posts	906,216	932,622
Mean number of notes per post	60.72	62.17
Median of notes per post	7	1
Number of hashtags on posts	135,748	183,719
Mean number of hashtags per post	9.1	12.25
Median of hashtags per post	8	9
Number of unique users	7,092	9,183
Mean posts per user	2.1	1.63

colorfulness measure C is derived from the opposing colorspaces' mean and standard deviation. The higher the value, the more colorful the image is. Equation 1 shows how the colorfulness is calculated:

$$C = \sigma_{rgyb} + 0,3 \cdot \mu_{rgyb} \tag{1}$$

Valdez and Mehrabian [32] empirically researched the emotional impact of brightness (B) and saturation (S) of colors. They map the emotions to the dimensions pleasure, arousal and dominance. Equations 2, 3 and 4 depict how each measure is derived from the mean saturation and brightness of the images.

$$Pleasure = 0.69 \cdot B + 0.22 \cdot S \tag{2}$$

$$Arousal = -0.31 \cdot B + 0.60 \cdot S \tag{3}$$

$$Dominance = -0.76 \cdot B + 0.32 \cdot S \tag{4}$$

Previous work utilized these three measures in affective image analysis [23] as well as in exploring self-harm content in social media [33].

3.3   Text Analysis

For automatically processing the text of the Tumblr posts, three methods have been chosen: 1) A sentiment analysis, which was already implemented by Fettach and Benhiba [11] in the context of eating disorders. 2) A readability score was calculated to explore the assumption that Meanspiration content is more confusing compared to regular Tumblr content. 3) An emotion analysis was performed. The assumption here is that Meanspiration content contains more negative emotions (anger, fear, sadness) compared to regular Tumblr content.

For analyzing the sentiment of a post, the tool VADER [17] was used. This tool works rule-based and was especially developed to understand social media content, like for instance emojis, emoticons, the repetition of punctuation or capitalized words as a form of emphasizing the semantics. VADER calculates a compound score for each post, ranging between the extreme negative sentiment of -1, passing the neutral area between -0.05 and $+0.05$, to the extreme positive sentiment of $+1$[6]. Since Meanspiration is a trend to send and receive 'mean' messages, the assumption is that this content has more negative sentiments than the control data set.

To calculate the readability of a post, Flesch Reading Ease Score (FRES)[7] [18] was implemented. FRES makes use of the words, syllables and sentences of the post (see Eq. 5).

$$206.835 - 1.015 \left(\frac{\text{total words}}{\text{total sentences}} \right) - 84.6 \left(\frac{\text{total syllables}}{\text{total words}} \right) \tag{5}$$

Further, the emotions of each post were calculated[8]. The tool is compatible with five basic emotion categories: Happiness, anger, sadness, surprise and fear. It finds the emotion category from the single words of the post and calculates a value for each emotion and each post. The higher the score of one category, the more the post belongs to this emotion. This emotion analysis is more detailed than the sentiment analysis with VADER, where posts are only divided in positive, neutral and negative. This way a more detailed exploration of the posts' mood is possible than if only one tool would be used.

4   Results

4.1   RQ1: Visual and Textual Characteristics

In Fig. 1 the five images measures are represented as bars. For each, the normalized mean was calculated to compare the Meanspo and the control data sets. It becomes apparent, that indeed the Meanspo images are slightly less colorful than the ones from regular Tumblr content. Also in the emotional measures pleasure, arousal and dominance the Meanspiration pictures show slightly reduced values. Since the differences are so small, these measures seem to be no good features for the machine classification.

A radar chart (Fig. 2) was created to show how different the two groups are in terms of the emotions conveyed in the text. It becomes clear that the emotions fear, sadness and surprise occur much more frequently in the Meanspo

---

[6] VADER on Pypi.org: https://pypi.org/project/vader-sentiment/.

[7] FRES score as part of the textstat package on Pypi.org: https://pypi.org/project/textstat/.

[8] using text2emotion; On Pypi.org: https://pypi.org/project/text2emotion/.

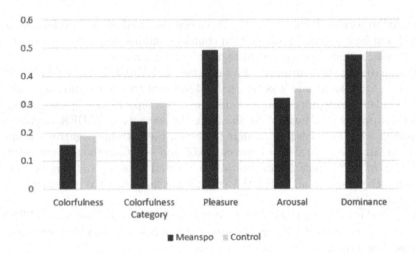

Fig. 1. Results of the visual analysis of the images. Bars represent normalized means of the measures.

group than in the control group. On the other hand, happy posts are about twice as common in the control group. Anger is rare for both. The assumption that negative emotions are more common in Meanspiration content is partly true. Fear and sadness are the emotions that appear more frequently, but anger does not. The usage of the texts' emotions can be a good indicator for automatic classification.

The assumption that Meanspo texts are more confusing than the ones of the control cannot be confirmed, since Table 3 shows that in the least readable category the control group has almost three times as many posts. However, it became also clear that the Control group has a lot more posts (43.49%) in the easiest readable category compared to Meanspiration (26.04%). From this point of view it is not possible to make a good distinction of the two groups based on their posts' readability.

Table 3. Results of the readability analysis. The total number of posts in both data sets differs from the numbers in Table 2, because some posts only contained images without any texts that could be analyzed here.

		Percentage (%)	Frequency	Percentage (%)	Frequency
Score	Reading Difficulty	**Meanspo (14,505)**		**Control (13,234)**	
90–100	Very easy	26.03	3,776	43.49	5,755
80–89	Easy	21.61	3,134	9.37	1,240
70–79	Fairly easy	19.32	2,803	12.33	1,632
60–69	Standard	12.92	1,874	7.03	931
50–59	Fairly difficult	8.31	1,206	6.12	810
30–49	Difficult	7.91	1,147	10.8	1,429
0–29	Very confusing	3.9	565	10.86	1,437

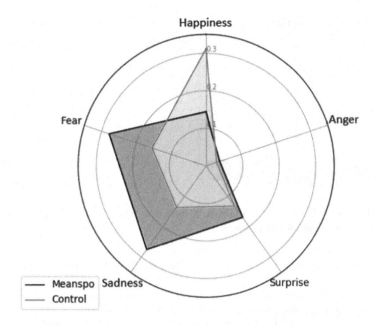

Fig. 2. Results of the emotion analysis. Radar chart shows the mean values for the five emotions.

The results of the sentiment analysis are shown as histograms for each group in Fig. 3. Besides the peak in the neutral sentiment area, that both groups have in common, it also becomes apparent that the control group has a bigger portion in the positive sentiments and only a few posts on the negative side. The Meanspo data set instead shows an overall smaller amount of neutral posts but more posts on the negative side, so the assumption that Meanspo content comprises overall more negative sentiment can be confirmed.

4.2 RQ2: Classification

Finally, supervised learning was utilized to construct a classifier to distinguish between the Meanspo and the control data set. A gradient boosting classifier was trained and the above mentioned measures were used as features for the model. Each Tumblr post is represented as one observation. Since not every post contains an image or image posts also do not necessarily contain texts the missing data was addressed by filling it with the mean value of each measure. A random search using 1,000 iterations was used to select the best performing model as the classifier. This model was validated using a standard 10-fold cross validation.

The features are split in different groups. For each group a new model was trained to better understand the performance of each group of measures. In addition, one approach combining all features was performed. For every model the average accuracy, precision, recall, macro F_1 and AUC (area under curve)

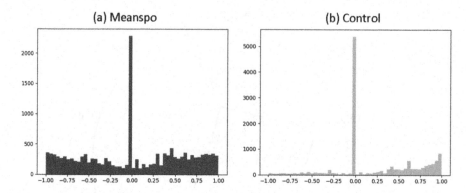

Fig. 3. Results of the sentiment analysis. X-axes show the compound value of VADER, Y-axes depict the number of posts.

are reported as metrics of performance. The groups of features are composed as follows:

- Images: colorfulness, pleasure, arousal, dominance
- Text: sentiment, readability***, happiness, sadness***, surprise, anger***, fear***
- Other: number of notes (rebloggs and likes), number of hashtags***, number of images

The asterisk represents statistical significance based on Mann-Whitney U tests with Holm-Bonferroni correction: * $p < .05$, ** $p < .01$, *** $p < .001$. The category *other* was derived from the basic descriptive statistic of the data sets and therefore included as features for this study. Table 4 summarizes the classification results.

Table 4. Performance of category of measures in classification

	acc	prec	rec	F1	AUC
Images	0.5	0.63	0.02	0.35	0.72
Text	0.73	0.7	0.82	0.73	0.83
Other	0.77	0.73	0.85	0.77	0.77
Images + Text	0.77	0.79	0.73	0.77	0.85
Images + Other	0.76	0.78	**0.86**	**0.81**	0.85
Text + Other	**0.81**	0.78	**0.86**	**0.81**	0.89
All	**0.81**	**0.81**	0.83	**0.81**	**0.9**

It becomes clear, that the best performing model, in terms of accuracy, precision, F_1 and AUC, is the one that combines all features in the classification task (accuracy = 0.81; F_1 = 0.81, precision = 0.81, AUC = 0.9). Among the single category classifications the image measures perform poorest (accuracy = 0.5; F_1 = 0.35), which confirms the assumption made above. The image measures do not show improvement over the baseline (baseline accuracy = 50%). The measures in the Other category perform best (accuracy = 0.77; F_1 = 0.77), but also the text measures show improvement over the baseline. The strong influence of the Other category measures becomes once more apparent when combining two categories. The combination of Text and Other as well as Images and Other have the same F_1 value like all measures combined (F_1 = 0.81).

The overall results show that Tumblr use to express Meanspiration bears distinctive markers, whose further exploration could help to better distinguish between normal content and potentially harmful one.

5 Discussion

The overall results show that there are differences between the regular content and the Meanspiration posts on Tumblr. Only small potential lies in the image features that were analyzed in this work. In order to address the small contribution of the selected image features to the classification accuracy, future approaches could analyze what the images actually depict. Qualitative as well as quantitative approaches are conceivable here. If the Meanspiration image content is better understood, then better measures can be selected for classification.

The trends of sharing Thinspiration content and its scientific research on it was described in the related work section of this paper. It was shown that Thinspiration is strongly related to the pro-ED community and disordered eating in general. Meanspiration is a trend related to Thinspiration, since its goal is also to inspire internet users to loose weight and the term is derived in the same way from two actual words. Since the nature of Meanspiration is to be mean to other social media users, it is likely that not only eating disorders are promoted by this trend. Other mental issues such as suicidal ideation, anxiety or self-harm tendencies could also be reinforced. In the pro-ED community 70% of patients reported the engagement of self-harm behavior [31]. The participation in Meanspiration could in addition negatively impact these kinds of behavior and worsen the overall condition of the affected. Therefore, it is important to distinguish positive social media trends from potentially harmful ones, especially because the related work shows how difficult and tedious the recovery process from illnesses like anorexia can be.

The relevance of the measures in the other category indicates that there is potential in investigating these features further. The number of notes is a measure to count likes and rebloggs, in other words, how much other users engage in the conversation about a specific topic. The big difference between the medians (see Table 2) indicates that in the Meanspo group the engagement is much higher, than in the control group. The lower number of tags can be traced back to thematically narrower content sharing and the number of images in a post as well as a ratio between textual and visual information per post can be an interesting future direction.

The missing image measures in text only posts were filled with the means of all observations for the classification task. The same method was applied to image only posts. When this approach would be used in a live monitoring of social media postings a good solution for the missing data should be developed. When the data is processed sequentially, like the shared tasks of eRisk propose it [20,21], the missing data could be predicted based on the single social media author, which could lead to better results.

It is also acknowledged that it is not clear to what extent the social media authors studied in this work, actually suffer from an eating disorder. All results are based on the usage of a series of hashtags referring to Meanspiration on Tumblr. It would be promising to repeat this methodological approach with data from another social network, like for instance Instagram, since it provides also a lot of visual data.

This study raises many more questions about Meanspiration on social media. Subsequent studies could address the language and the images of the posts more deeply, for instance by classifying what is actually depicted on the images. The employment of other models can be a promising addition as well. Furthermore, the analysis of streamed data to detect Meanspiration would be an interesting future step. More work can be done by diversifying the control data set by implementing more posts retrieved with different hashtags and by conducting the experiments with unbalanced training sets.

6   Conclusion

Meanspiration is a trend that emerged from Tumblr that encourages eating disorders by sending and receiving *mean* messages about one's body and/or weight to inspire weight loss. Eating disorders are a major health issue and social media usage can be linked to disordered eating. To address the automatic classification of Meanspiration and regular content, this study applied the analysis of image and text measures and utilized these as features in a machine learning classification. The results show that, Meanspiration is different from normal content mainly in the distribution of emotions and sentiments. Meanspiration posts more often comprise the emotions fear, sadness and surprise, while the control group's posts show more happy posts. In addition, Meanspiration posts have a greater share of negative sentiments in their texts. Concerning the image measures, both groups do not differ much. However, mean values of the measures are lower in Meanspiration images. These characteristics were leveraged as features in a classification task to distinguish Meanspiration from regular Tumblr content. An accuracy of 0.81 was achieved.

References

1. Boero, N., Pascoe, C.J.: Pro-anorexia communities and online interaction: bringing the pro-ana body online. Body Soc. **18**(2), 27–57 (2012)
2. Branley, D.B., Covey, J.: Pro-ana versus pro-recovery: a content analytic comparison of social media users' communication about eating disorders on Twitter and Tumblr. Front. Psychol. **8**, 1356 (2017)

3. Burdisso, S.G., Errecalde, M., Montes-y-Gómez, M.: UNSL at eRisk 2019: a unified approach for anorexia, self-harm and depression detection in social media. In: Cappellato, L., Ferro, N., Losada, D.E., Müller, H. (eds.) Working Notes of CLEF 2019 - Conference and Labs of the Evaluation Forum, Lugano, Switzerland, 9–12 September 2019. CEUR Workshop Proceedings, vol. 2380. CEUR-WS.org (2019). http://ceur-ws.org/Vol-2380/paper_103.pdf

4. Cavazos-Rehg, P.A., et al.: "I just want to be skinny": a content analysis of tweets expressing eating disorder symptoms. PloS One **14**(1), e0207506 (2019)

5. Chancellor, S., Kalantidis, Y., Pater, J.A., De Choudhury, M., Shamma, D.A.: Multimodal classification of moderated online pro-eating disorder content. In: Mark, G., et al. (eds.) Proceedings of the 2017 CHI Conference on Human Factors in Computing Systems, Denver, CO, USA, 06–11 May 2017, pp. 3213–3226. ACM (2017). https://doi.org/10.1145/3025453.3025985

6. Chancellor, S., Lin, Z., Goodman, E.L., Zerwas, S., De Choudhury, M.: Quantifying and predicting mental illness severity in online pro-eating disorder communities. In: Gergle, D., Morris, M.R., Bjørn, P., Konstan, J.A. (eds.) Proceedings of the 19th ACM Conference on Computer-Supported Cooperative Work & Social Computing, CSCW 2016, San Francisco, CA, USA, 27 February–2 March 2016, pp. 1169–1182. ACM (2016). https://doi.org/10.1145/2818048.2819973

7. Chancellor, S., Lin, Z.J., De Choudhury, M.: 'This post will just get taken down': characterizing removed pro-eating disorder social media content. In: Kaye, J., Druin, A., Lampe, C., Morris, D., Hourcade, J.P. (eds.) Proceedings of the 2016 CHI Conference on Human Factors in Computing Systems, San Jose, CA, USA, 7–12 May 2016, pp. 1157–1162. ACM (2016). https://doi.org/10.1145/2858036.2858248

8. Chancellor, S., Mitra, T., De Choudhury, M.: Recovery amid pro-anorexia: analysis of recovery in social media. In: Kaye, J., Druin, A., Lampe, C., Morris, D., Hourcade, J.P. (eds.) Proceedings of the 2016 CHI Conference on Human Factors in Computing Systems, San Jose, CA, USA, 7–12 May 2016, pp. 2111–2123. ACM (2016). https://doi.org/10.1145/2858036.2858246

9. Chancellor, S., Pater, J.A., Clear, T.A., Gilbert, E., De Choudhury, M.: #thyghgapp: Instagram content moderation and lexical variation in pro-eating disorder communities. In: Gergle, D., Morris, M.R., Bjørn, P., Konstan, J.A. (eds.) Proceedings of the 19th ACM Conference on Computer-Supported Cooperative Work & Social Computing, CSCW 2016, San Francisco, CA, USA, 27 February–2 March 2016, pp. 1199–1211. ACM (2016). https://doi.org/10.1145/2818048.2819963

10. De Choudhury, M.: Anorexia on Tumblr: a characterization study. In: Kostkova, P., Grasso, F. (eds.) Proceedings of the 5th International Conference on Digital Health 2015, Florence, Italy, 18–20 May 2015, pp. 43–50. ACM (2015). https://doi.org/10.1145/2750511.2750515

11. Fettach, Y., Benhiba, L.: Pro-eating disorders and pro-recovery communities on reddit: text and network comparative analyses. In: Proceedings of the 21st International Conference on Information Integration and Web-Based Applications & Services, iiWAS 2019, Munich, Germany, 2–4 December 2019, pp. 277–286. ACM (2019). https://doi.org/10.1145/3366030.3366058

12. Funez, D.G., et al.: UNSL's participation at eRisk 2018 lab. In: Cappellato, L., Ferro, N., Nie, J., Soulier, L. (eds.) Working Notes of CLEF 2018 - Conference and Labs of the Evaluation Forum, Avignon, France, 10–14 September 2018. CEUR Workshop Proceedings, vol. 2125. CEUR-WS.org (2018). http://ceur-ws.org/Vol-2125/paper_137.pdf

13. Ging, D., Garvey, S.: 'Written in these scars are the stories I can't explain': a content analysis of pro-ana and thinspiration image sharing on Instagram. New Media Soc. **20**(3), 1181–1200 (2018)
14. Grabe, S., Ward, L.M., Hyde, J.S.: The role of the media in body image concerns among women: a meta-analysis of experimental and correlational studies. Psychol. Bull. **134**(3), 460 (2008)
15. Hasler, D., Süsstrunk, S.: Measuring colorfulness in natural images. In: Rogowitz, B.E., Pappas, T.N. (eds.) Human Vision and Electronic Imaging VIII, Santa Clara, CA, USA, 20 January 2003. SPIE Proceedings, vol. 5007, pp. 87–95. SPIE (2003). https://doi.org/10.1117/12.477378
16. Holland, G., Tiggemann, M.: A systematic review of the impact of the use of social networking sites on body image and disordered eating outcomes. Body Image **17**, 100–110 (2016)
17. Hutto, C.J., Gilbert, E.: VADER: a parsimonious rule-based model for sentiment analysis of social media text. In: Adar, E., Resnick, P., De Choudhury, M., Hogan, B., Oh, A.H. (eds.) Proceedings of the 8th International Conference on Weblogs and Social Media, ICWSM 2014, Ann Arbor, Michigan, USA, 1–4 June 2014. The AAAI Press (2014). http://www.aaai.org/ocs/index.php/ICWSM/ICWSM14/paper/view/8109
18. Kincaid, J.P., Fishburne, R.P., Jr., Rogers, R.L., Chissom, B.S.: Derivation of new readability formulas (automated readability index, fog count and flesch reading ease formula) for navy enlisted personnel. Technical report, Naval Technical Training Command Millington TN Research Branch (1975)
19. Losada, D.E., Crestani, F.: A test collection for research on depression and language use. In: Fuhr, N., et al. (eds.) CLEF 2016. LNCS, vol. 9822, pp. 28–39. Springer, Cham (2016). https://doi.org/10.1007/978-3-319-44564-9_3
20. Losada, D.E., Crestani, F., Parapar, J.: Overview of eRisk: early risk prediction on the internet (extended lab overview). In: Cappellato, L., Ferro, N., Nie, J., Soulier, L. (eds.) Working Notes of CLEF 2018 - Conference and Labs of the Evaluation Forum, Avignon, France, 10–14 September 2018. CEUR Workshop Proceedings, vol. 2125. CEUR-WS.org (2018). http://ceur-ws.org/Vol-2125/invited_paper_1.pdf
21. Losada, D.E., Crestani, F., Parapar, J.: Overview of eRisk at CLEF 2019: early risk prediction on the internet (extended overview). In: Cappellato, L., Ferro, N., Losada, D.E., Müller, H. (eds.) Working Notes of CLEF 2019 - Conference and Labs of the Evaluation Forum, Lugano, Switzerland, 9–12 September 2019. CEUR Workshop Proceedings, vol. 2380. CEUR-WS.org (2019). http://ceur-ws.org/Vol-2380/paper_248.pdf
22. Mabe, A.G., Forney, K.J., Keel, P.K.: Do you "like" my photo? Facebook use maintains eating disorder risk. Int. J. Eat. Disord. **47**(5), 516–523 (2014)
23. Machajdik, J., Hanbury, A.: Affective image classification using features inspired by psychology and art theory. In: Bimbo, A.D., Chang, S., Smeulders, A.W.M. (eds.) Proceedings of the 18th International Conference on Multimedia 2010, Firenze, Italy, 25–29 October 2010, pp. 83–92. ACM (2010). https://doi.org/10.1145/1873951.1873965
24. Masood, R., Hu, M., Fabregat, H., Aker, A., Fuhr, N.: Anorexia topical trends in self-declared reddit users. In: Cantador, I., Chevalier, M., Melucci, M., Mothe, J. (eds.) Proceedings of the First Joint Conference of the Information Retrieval Communities in Europe (CIRCLE 2020), Samatan, Gers, France, 6–9 July 2020. CEUR Workshop Proceedings, vol. 2621. CEUR-WS.org (2020). http://ceur-ws.org/Vol-2621/CIRCLE20_14.pdf

25. Mohammadi, E., Amini, H., Kosseim, L.: Quick and (maybe not so) easy detection of anorexia in social media posts. In: Cappellato, L., Ferro, N., Losada, D.E., Müller, H. (eds.) Working Notes of CLEF 2019 - Conference and Labs of the Evaluation Forum, Lugano, Switzerland, 9–12 September 2019. CEUR Workshop Proceedings, vol. 2380. CEUR-WS.org (2019). http://ceur-ws.org/Vol-2380/paper_74.pdf

26. Ragheb, W., Azé, J., Bringay, S., Servajean, M.: Attentive multi-stage learning for early risk detection of signs of anorexia and self-harm on social media. In: Cappellato, L., Ferro, N., Losada, D.E., Müller, H. (eds.) Working Notes of CLEF 2019 - Conference and Labs of the Evaluation Forum, Lugano, Switzerland, 9–12 September 2019. CEUR Workshop Proceedings, vol. 2380. CEUR-WS.org (2019). http://ceur-ws.org/Vol-2380/paper_126.pdf

27. Talbot, C.V., Gavin, J., Van Steen, T., Morey, Y.: A content analysis of thinspiration, fitspiration, and bonespiration imagery on social media. J. Eat. Disord. 5(1), 1–8 (2017)

28. Tiggemann, M., Anderberg, I.: Social media is not real: the effect of 'Instagram vs reality' images on women's social comparison and body image. New Media Soc. 22(12), 2183–2199 (2020)

29. Tiggemann, M., Slater, A.: NetGirls: the Internet, Facebook, and body image concern in adolescent girls. Int. J. Eat. Disord. 46(6), 630–633 (2013)

30. Trotzek, M., Koitka, S., Friedrich, C.M.: Word embeddings and linguistic metadata at the CLEF 2018 tasks for early detection of depression and anorexia. In: Cappellato, L., Ferro, N., Nie, J., Soulier, L. (eds.) Working Notes of CLEF 2018 - Conference and Labs of the Evaluation Forum, Avignon, France, 10–14 September 2018. CEUR Workshop Proceedings, vol. 2125. CEUR-WS.org (2018). http://ceur-ws.org/Vol-2125/paper_68.pdf

31. Turner, B.J., Yiu, A., Layden, B.K., Claes, L., Zaitsoff, S., Chapman, A.L.: Temporal associations between disordered eating and nonsuicidal self-injury: examining symptom overlap over 1 year. Behav. Ther. 46(1), 125–138 (2015)

32. Valdez, P., Mehrabian, A.: Effects of color on emotions. J. Exp. Psychol. Gen. 123(4), 394 (1994)

33. Wang, Y., et al.: Understanding and discovering deliberate self-harm content in social media. In: Barrett, R., Cummings, R., Agichtein, E., Gabrilovich, E. (eds.) Proceedings of the 26th International Conference on World Wide Web, pp. 93–102. ACM (2017). https://doi.org/10.1145/3038912.3052555

34. Wick, M.R., Harriger, J.A.: A content analysis of thinspiration images and text posts on Tumblr. Body Image 24, 13–16 (2018). https://doi.org/10.1016/j.bodyim.2017.11.005

Tracking News Stories in Short Messages in the Era of Infodemic

Guillaume Bernard[1]([envelope]) [iD], Cyrille Suire[1], Cyril Faucher[1], Antoine Doucet[1] [iD],
and Paolo Rosso[2] [iD]

[1] Université de La Rochelle, Laboratoire L3i, 17000 La Rochelle, France
{guillaume.bernard,cyrille.suire,cyril.faucher,antoine.doucet}@univ-lr.fr
[2] Universitat Politècnica de València, València, Spain
prosso@dsic.upv.es
https://l3i.univ-larochelle.fr/

Abstract. Tracking news stories in documents is a way to deal with the large amount of information that surrounds us everyday, to reduce the noise and to detect emergent topics in news. Since the Covid-19 outbreak, the world has known a new problem: infodemic. News article titles are massively shared on social networks and the analysis of trends and growing topics is complex. Grouping documents in news stories lowers the number of topics to analyse and the information to ingest and/or evaluate. Our study proposes to analyse news tracking with little information provided by titles on social networks. In this paper, we take advantage of datasets of public news article titles to experiment news tracking algorithms on short messages. We evaluate the clustering performance with little amount of data per document. We deal with the document representation (sparse with TF-IDF and dense using Transformers [26]), its impact on the results and why it is key to this type of work. We used a supervised algorithm proposed by Miranda et al. [22] and K-Means to provide evaluations for different use cases. We found that TF-IDF vectors are not always the best ones to group documents, and that algorithms are sensitive to the type of representation. Knowing this, we recommend taking both aspects into account while tracking news stories in short messages. With this paper, we share all the source code and resources we handled.

Keywords: Text Classification and Clustering · News · Social data

1 Introduction

Tracking emergent topics from news is a long-standing task in natural language processing (NLP), investigated since the last century [2]. In many fields, from politics [25], IT services [22,28] to banking [20], the purpose of discovering and tracking related news stories is an important application. It helps taking better, faster decisions than one's competitors.

Furthermore, the emergence of news sources, from official agencies to institutional blog posts, including social networks, restructured the information sector.

A. Barrón-Cedeño et al. (Eds.): CLEF 2022, LNCS 13390, pp. 18–32, 2022.
https://doi.org/10.1007/978-3-031-13643-6_2

Since the 2010's, social networks have become a main source of information for a huge part of the population [30]. Consequently, official media relay their articles on Facebook or Twitter to draw audience to their websites. As a consequence, more and more people read social networks to inform themselves. On another hand, on-line social networks allowed non-institutional parties to publish and promote information and create communities.

Tracking news stories has been attempted with some success in recent years [20,22,32]. However, for most of experiments with short messages neither datasets nor implementations are shared [4,13,23,27,33]. With Covid-19, the world health organisation (WHO) introduced the concept of *'infodemic'* as *'too much information including false or misleading information in digital and physical environments during a disease outbreak'* [34]. We take benefit of this situation and of Covid-19 Twitter News datasets to conduct comparative experiments with different corpora.

In this paper, we address a few research issues and we propose a framework to track news stories in short messages. We focus on news article titles as they are shared on social networks, and aim to discover coherent clusters of events. We first address the problem of document representation and look into how algorithms interact with document features, formulating the hypothesis that they impact the results. Next, we experiment with news story tracking with article titles using two algorithms, one of them supervised and the other unsupervised. The latter is made relevant by the lack of annotated datasets in this research field. We also release the implementation of our tracking algorithms in Python Packages, as well as all the datasets and resources we used[1].

2 Related Work

The task of tracking news stories generally consists in ordering and clustering together documents reporting the same news story, written in identical or different languages [2]. A news story is an ordered collection of documents that relate a specific topic and all its subsequent developments [2]. The final football match of 2018 FIFA World cup is, for instance, the seminal event that is the root of a story. All the articles related to the preparation of the match, betting and editorials about the results are all related to the same event. It is part of a wider topic: 2018 FIFA World Cup, or more generally sports.

The Topic Detection and Tracking (TDT) project in 2002 [2] addressed the question of tracking news stories. Documents were grouped on macro topics: finance, sports, health, etc. In 2005, the Europe Media Monitor project [25] enriched the field of study with new results and strategies to identify emergent topics from press articles. They proposed an approach to track real-world events, not only macro-topics. The newsBrief system is still running[2] and gives a view of trending topics mentioned in news articles.

---

[1] Links to be added if the paper is accepted.

[2] https://emm.newsbrief.eu/.

Later on, after 2010, the Event Registry project [19,28] published a multi-lingual dataset of recent press articles. They used TF-IDF vectors with unsupervised algorithms to group articles related the same news events. The newsLens news tracking system [17] also benefits from TF-IDF vectors to cluster documents. The passage of time is materialised by time buckets: it assumes that articles close in time may relate to the same events [31]. Hence they discovered that buckets of 6 days, with 50% overlap between them are the most suitable parameters to treat group of news articles. Some time after, Miranda et al. [22] introduced and described a supervised and streaming algorithm able to cluster press articles into coherent mono-lingual and multi-lingual news stories. A more recent study [32] analysed the impact of vectorisation for news tracking algorithms and concluded that TF-IDF vectors provide competitive results and outperform dense vectors computed with doc2vec [18].

The code of implementations are rarely released but the algorithms are well described and datasets shared with the community to simplify the reproduction of experiments [17,22]. Miranda et al. [22] shared their implementation of their algorithm. In this paper, we will enhance this implementation with a new API and we will implement a baseline proposed in other research articles [21,28,32].

About the experiments performed on short messages, the propagation of the information on Twitter is studied since 2010 [13,23,27,33]. Researchers focused on tweets to track events discussed on the network. Tweets are short messages published on this social network, originally 140 characters long, 280 since 2017 [29]. Most of them used vectorisation to represent documents [24] while others used Twitter specific features, such as hashtags, internal links, followers or retweets to characterise tweets [4]. In this paper, we analyse two datasets from which we only keep news claims, that is to say, news article titles shared on the network, ignoring users reactions.

The scientific literature lacks news articles annotated in clusters of topics or stories. In the context of the pandemic, we took advantage of news article titles published on Twitter that are linked to fact checking services. It allows to know which articles are connected to the same event.

3   Datasets

Our task consists of building stories from documents written in natural language. We deal with article titles, which are short in size and contain a little amount of information. A suitable dataset for the task of tracking news stories has to provide events or clusters identifiers. In addition to generally used datasets [28] in this field, the emergence of Covid-19 datasets from Twitter with references to fact checking services such as PolitiFact[3] is the opportunity to carry out experiments with publicly available resources.

Our experiments focus on three available datasets, upon which we present relevant statistics in Table 1:

---

[3] https://www.politifact.com/.

Table 1. Statistics about the datasets chosen for the experiments.

Dataset	Language	Partition	Documents	Tokens in documents		Nb. of clusters	Cluster size	
				Avg.	Std.		Avg.	Std.
Event Registry	eng	Train	12,233	56	19	593	21	32
		Test	8,726	58	19	222	39	89
CoAID	eng	Train	72,045	179	82	375	192	146
		Test	32,100	214	79	125	257	163
FibVid	eng	Train	988	206	77	51	19	7
		Test	402	201	79	52	8	2

- **Event Registry** [10,28]: a widely known news tracking dataset that has been used in various recent researches [20,22,32] to tackle the issue of discovering news stories in press articles. It comprises events reported in multiple languages: English, German and Spanish and was collected in 2014 and 2015. It is composed of full article texts and titles. In our experiments, we only keep the title of each article.
- **CoAID** [9,14]: a Twitter dataset with Covid-19 related tweets written in English. It has been gathered during the first months of the Covid-19 pandemic, from January to May 2020. We keep only the 500 biggest clusters, as they capture 77% of all documents (104.145 tweets). It comprises news claims and user reactions. The first are, as authors describe, links to news websites and the tweet text is the title of the article. We ignore user reactions.
- **FibVid** [11,16]: another Twitter dataset with Covid-19 related tweets. It focuses more on users reactions, but similarly to CoAID, tweets reporting news are connected to news claims identifiers. It was built in 2020. Similarly to CoAID, news claims are retained, user reactions are ignored.

In the Event Registry dataset, each document is associated with a ground truth cluster identifier. Not the two others. For them, each tweet is connected to a news claim URI and we use it as a label for clustering analysis. This way, tweets connected to the same URI are considered within the same cluster, so within the same news story. The number of clusters given in Table 1 is the number of distinct clusters, so news stories, given by the labels in the datasets. The issue with Covid-19 outbreak and the struggle against the *infodemic* resides in detecting false or misleading information in news as they emerge. We twist the purpose of these corpora to apply them to news story tracking.

3.1 Represent Document with Multiple Features

Documents are represented with vectors of numbers in order to be compared and processed by computers. In most cases, vectors are computed using the TF-IDF weighting scheme. Sparse vectors of numbers are a strong baseline [32], compared to dense representations, to track documents that report similar stories. However, these conclusions are valid for full articles - not only titles - and use a doc2vec [18] dense vectorisation. Recent advances, with the introduction of the Transformer architectures provided new methods to represent documents, such as BERT [1, 15] or XL-NET [35]. On an attempt to focus on sentences rather than tokens and to capture sentence information, the Sentence-BERT representation has been introduced [26]. In this paper, we propose both to compare algorithms and the relative impact of document representation. To that extent, we use four representations to encode documents, two of them are sparse while the other two are Transformer-based dense vector representations.

Sparse Document Representation. Usually, TF-IDF weights are computed with the train part of the datasets. Here, we consider documents are handled in a stream and they are unknown before being processed. TF-IDF weighting models then have to be trained before processing data. We build them with huge sets of documents, independent from the data to weight. Different sizes of input documents used to feed the TF-IDF models give different weights. Hence, logically, they will produce different algorithm results.

To evaluate the impact of vectorisation, we propose to use different sets of documents to fit TF-IDF models and compute document vectors. We collected two: one with news articles, the other one with tweets [12]. News articles come from the Deutsche Welle (DW) website[4], which is scrapped to extract the title and body. DW is one of the only website that provides content in multiple languages and that is free to query and download. The other one is a collection of tweets, published in English from institutional press accounts. We manually chose press agencies or newspaper that publish on Twitter. There is an API limitation and we can only download 3200 tweets per source[5]. To overcome this problem, we selected a high number of press accounts. With this paper, we share the news articles and tweets identifiers with the code that weights the documents of each dataset listed in Sect. 3.

In addition to computing our own vectors, we use the pre-computed ones published by Miranda et al. [22] for the Event Registry dataset. We use them to have results comparable to previous works. In their paper, each document is characterised by several vectors of features associated to the text. This means for each document there are several TF-IDF vectors: one for the tokens, one for the lemmas and one for the named entities. For a news article which has a title and a body, there are at least six vectors: three for the title and three for the body. With titles we deal with only three vectors: the title tokens, lemmas and entities. To compute the respective weightings, we fit three different TF-IDF models. One will weight the tokens, another one the lemmas and the third one the entities. For all datasets of Sect. 3, we compute TF-IDF vectors using both sets of documents we collected: news and tweets.

To extract tokens, lemmas and entities from the titles, we use the spaCy software[6]. We only keep GPE, ORG and PER entities, as in Miranda et al. works

---

[4] https://www.dw.com.

[5] Twitter API v1: `get-statuses/user-timeline`.

[6] https://spaCy.io v3.2.1 with medium size models in English.

Table 2. Statistics on the content of the sets of documents used to fit TF-IDF weighting models.

Dataset	Language	Documents	Number of Unique Features in the Sets		
			Tokens	Lemmas	Entities
News	eng	79,856	13,135,162	12,205.181	881,298
Tweets	eng	55,792	546,625	544,538	49,540

[22]. To give an idea of the sizes of the TF-IDF datasets we used to weight the document features, some statistics are shown in Table 2. The software to compute weights is freely shared over the Internet [6].

Dense Document Representation. We use the Transformer architecture, especially the S-BERT [26] algorithms and models to encode the title texts into dense vectors. We select pre-trained models that focus on semantic similarity to compute title vectors. Among all the proposed models, we retained multilingual models with the highest scores in semantic search, available at the time of the experiments. They are `distiluse-base-multilingual-cased-v1` and `paraphrase-multilingual-mpnet-base-v2`. To simplify the remainder of the paper, we will respectively name them USE and MPNet.

With the two models, we encode the title texts into vectors of different sizes, 512 logits for USE, 768 for MPNet. While the cardinality of TF-IDF vectors is equal to the number of unique tokens, lemmas or entities found in the text, dense vector representation encodes documents in vectors of a fixed size. Contrary to the sparse TF-IDF document representation, we do not tokenize or extract entities from the text and encode the full sentence without any kind of pre-training. There remains a unique vector that encapsulates the whole text, instead of three with TF-IDF. Refer to [5] to encode texts with dense models.

4   Tracking Documents Reporting the Same Stories

To build news stories with short documents, we use the publicly available tracking algorithm proposed by Miranda et al. [22]. This is a supervised algorithm that dynamically creates clusters from incoming documents. In case there is no training data to create a clustering model, we propose another implementation of a news tracking algorithm based on K-Means.

4.1   Streaming Algorithm to Build News Stories

In this section, we provide more detailed explanations about the Miranda et al. algorithm we use in this article. This latter handles the documents of the dataset as a stream and each incoming document is compared to every existing cluster in the pool of already known clusters. Each existing cluster is a candidate in whom the document might be added if its similarity with the cluster is over a specific threshold T_1. If multiple candidates exist, the one with the highest similarity wins. On the contrary, if no candidate exists, a new cluster is created accordingly. The algorithm handles heterogeneous data with vectorised texts and with timestamps of documents and clusters. The similarity measure $sim(d, C)$ between a document d and a cluster C is detailed below.

First of all, for dates comparison, clusters keep track of two dates: a lower bound, with the oldest document date in the cluster, and a higher bound, with the most recent one. To compute time similarities, a Gaussian distribution is used with $\mu = 0$ and $\sigma = 3$. This latter parameter is to be seen as a number of days after which the similarity falls dramatically.

$$f(d_{date}, C_{date}) = \phi_{\mu,\sigma^2}(|d_{date} - C_{date}|) \tag{1}$$

After that, we compute text similarities. The cosine measure ($\theta(d^k, C^k)$) computes the similarity between the representative vectors (TF-IDF or dense) of the document and the cluster. The cluster features are the average of all the documents vectors it is composed of. $K = 3$ stands for the tokens, entities and lemmas vectors, as described in Subsect. 3.1 for TF-IDF. With dense representations, $K = 1$. There are two time similarities with the lower and upper bounds. In the Eq. 2, β acts as a logistic regression coefficient, α as the intercept. β_k balance the importance of features in the final similarity score.

$$g(d, C) = \sum_{k=0}^{K} \beta_k \times \theta(d^k, C^k) + \sum_{k=0}^{K=2} \beta_k \times f(d_i^k, C_i^k) + \alpha \tag{2}$$

To flatten the similarity scores within the $[0 : 1]$ interval, we put $g(d, C)$ into a sigmoid function, in compliance with the logistic regression. The final similarity is given by Eq. 3.

$$sim(d, C) = \frac{1}{1 + e^{-g(d,C)}} \tag{3}$$

Model coefficients β and threshold T_1 are trained with a logistic regression on the train part of the corpus. True label clustering is computed on the train part from which we keep all document - cluster similarity scores. To lower the number of negative examples, we keep, for each document - cluster comparison, the twenty highest negative similarity scores. A grid search gives the best model and the decision threshold T_1 is the one that maximizes F1 on the train set.

4.2   Unsupervised News Tracking with K-Means

In accordance with suggestions of previous authors [21,28,32] we use the K-Means algorithm as an unsupervised method to create news stories, with cosine similarity as the distance measure. We propose this algorithm to counter the lack of training data, which are rare in this field of research. To simulate the time that passes, the dataset is split into buckets of sliding windows [17,21]. We use the *newsLens* optimal parameters given in Sect. 2 with a window of 6 days.

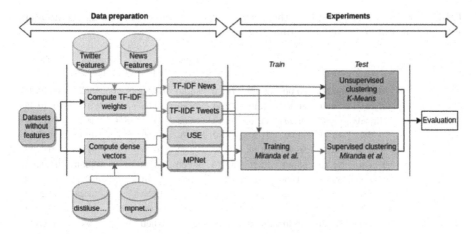

Fig. 1. Description of the whole process, from the documents in the datasets without features to the clustering evaluation. Two key parts are noticeable: dataset features computation and algorithm evaluation.

We try different configurations, as K-Means is an unsupervised algorithm. An *optimal* one for which we give the algorithm the true number of clusters for each window. The other one uses the Silhouette score to identify a coherent number of clusters. Early experiments implemented the *elbow method* that generated a too high number of cluster per window, providing unusable results. As it is necessary to compute numerous configurations in order to select the *optimal* number of clusters k, the time to compute clusterings may be very high for windows with lots of documents.

5   Experiments

With this paper, we release our implementation of the algorithms [7] and the training software [8] written in Python. It is a package with its own API. We intend to fulfil the lack of an end-to-end tool that builds news stories from documents. To the best of our knowledge, it does not exist yet.

The experiments consist of applying the algorithms mentioned in Sect. 4.1 with the datasets of Sect. 3, as described in Fig. 1. To evaluate the cluster results and coherence, standard and BCubed [3] evaluation metrics are computed. The latter is a more accurate evaluation method of clustering performance.

First, we run experiments with the algorithm developed by Miranda et al. [22] and the results are shown in Table 3. The three datasets are tested with document vectors computed with the sparse and dense models described in Subsect. 3.1. CoAID and FibVid are not multilingual, so we only focus on the English language in the Event Registry dataset. For the news story tracking results, the precision is good with Event Registry and CoAID. The low recall is correlated with a high number of clusters. By creating more clusters, the system focuses on precision,

Table 3. Experimental results of the Miranda et al. [22] algorithm on all the datasets described in Sect. 3. Sorted according to F1 BCubed score.

Corpus	Vectors	Standard			BCubed			Clusters		Time
		F1	P	R	F1	P	R	Real	Predicted	
Event Registry	MPNet	**54.50**	**90.00**	**39.00**	**74.30**	**85.60**	**65.70**	222	362	00:05:17
	News	74.80	86.10	66.10	72.30	72.10	72.60		206	00:01:45
	Miranda	61.80	98.20	45.10	73.00	95.90	59.00		902	00:02:33
	USE	46.50	91.00	31.20	68.80	89.70	55.90		644	00:08:10
	Tweets	52.90	96.90	36.30	65.50	93.30	50.50		1154	00:03:10
CoAID	Tweets	**54.70**	**65.90**	**46.70**	**61.70**	**80.60**	**50.00**	125	6,356	01:04:32
	News	50.50	64.00	41.70	57.60	77.90	45.70		6,621	01:09:38
	USE	12.50	23.50	8.50	20.90	64.00	12.50		13628	14:29:02
	MPNet	3.20	34.70	1.70	6.90	80.40	3.60		21826	22:57:34
FibVid	Tweets	**30.40**	**33.40**	**28.00**	**41.20**	**48.20**	**36.00**	52	98	00:00:02
	USE	29.10	31.50	27.10	41.20	49.30	35.90		116	00:00:04
	MPNet	24.70	20.20	31.80	39.60	39.40	39.80		85	00:00:03
	News	19.40	26.70	15.20	37.80	71.70	25.60		207	00:00:02

hence decreases the recall. Results are constantly bad with Fibvid, which seems not to be a very suitable dataset for this task: it comprises a low number of news claims over the number of user reactions (220 K) we eliminated. We also tracked the necessary time to process the datasets that depends on two factors: the number of documents and the number of clusters found by the algorithm. As we previously mentioned, each incoming document is compared to every existing cluster and as a consequence, a high number of clusters increases the processing time.

When analysing the results of Miranda et al. algorithm on the three datasets, we first notice the F1 scores are all below 75% and the precision is always very high compared to the recall. There is no clear trend in favour of a specific document representation. For Event Registry, MPNet representation is way higher than its competitors, the second one being the TF-IDF document representation based on the News TF-IDF corpus. With CoAID, we have the right opposite, MPNet is the worst one while the TF-IDF representation based on the Tweets TF-IDF corpus, then on the News one give close results. In addition to this clear distinction between Event Registry and CoAID vectors, the algorithm on FibVid behaves differently. The dense and Tweets TF-IDF representations give very close results in terms of precision and recall, while the News TF-IDF vectorisation is low.

On another hand, in Table 4 we report the results computed with the unsupervised algorithm. We notice the documents are clustered together with a high precision. The low recall, so low harmonic mean is explained by the high number of clusters found, a consequence of time windows. We encounter a similar but lesser phenomena with the other algorithm. With K-Means, the document vectors that produce the best clustering scores are not the same as in the other experiment. A noticeable point is that the unsupervised method is not able to

Table 4. Experimental results running the K-Means baseline on all the datasets described in Sect. 3. Method **T** stands for true number of clusters, **S** for Silhouette. Sorted according to F1 BCubed Silhouette score.

Corpus	Vectors	Method	Standard			BCubed			Clusters		Time
			F1	P	R	F1	P	R	Real	Predicted	
Event Registry	MPNet	T	**71.00**	**97.70**	**55.70**	**70.40**	**88.80**	**58.30**	**301**	**301**	**00:33:31**
		S	**62.40**	**78.10**	**51.90**	**74.00**	**80.00**	**68.80**	**301**	228	**18:18:53**
	USE	T	70.80	97.00	55.80	70.40	88.40	58.50	301	301	00:34:28
		S	54.40	68.30	45.20	71.40	78.80	65.20	301	196	10:45:14
	News	T	66.50	87.20	53.80	63.60	81.60	52.00	301	301	00:02:14
		S	31.40	57.30	21.60	57.90	76.40	46.70	301	259	02:22:49
	Miranda	T	66.70	86.90	54.20	63.50	81.30	52.10	301	301	00:02:13
		S	31.90	58.50	22.00	57.90	76.60	46.50	301	291	02:22:51
	Tweets	T	67.20	87.60	54.50	63.70	81.40	52.30	301	301	00:02:14
		S	32.10	55.90	22.50	57.40	75.00	46.50	301	234	02:21:46
CoAID	News	T	**34.10**	**49.70**	**26.00**	**35.10**	**66.40**	**23.80**	**965**	**965**	**00:31:49**
		S	**33.70**	**46.70**	**26.30**	**35.10**	**63.90**	**24.20**	**965**	655	**105:22:08**
	Tweets	T	34.00	45.40	27.20	36.60	68.80	24.90	965	965	00:28:46
		S	30.90	59.30	20.90	30.60	67.90	19.70	965	768	88:07:17
	MPNet	T	17.80	42.00	11.30	19.40	53.80	11.80	965	965	00:39:11
		S	12.70	24.90	8.60	16.00	43.40	9.80	965	688	01:37:36
	USE	T	22.10	48.70	14.30	23.20	60.10	14.40	965	965	00:32:52
		S	11.80	43.50	6.80	14.50	55.10	8.40	965	1.410	04:39:42
FibVid	MPNet	T	**24.70**	**62.20**	**15.40**	**38.40**	**79.70**	**25.30**	**224**	**224**	**00:05:46**
		S	**23.90**	**24.20**	**23.60**	**35.00**	**38.00**	**32.40**	**224**	65	**00:27:41**
	USE	T	25.30	61.00	16.00	38.70	78.80	25.60	224	244	00:05:32
		S	23.50	23.50	23.40	34.20	36.00	32.60	224	62	00:26:51
	News	T	21.70	54,20	13.60	36.10	76.40	23.60	224	224	00:00:54
		S	19.60	19.60	19.60	32.80	37.20	29.30	224	68	00:03:15
	Tweets	T	19.80	55.00	12.10	34.70	75.90	22.50	224	224	00:00:54
		S	19.30	19.00	19.60	32.80	37.20	29.40	224	68	00:03:13

cluster documents as well as the Miranda et al. [22] algorithm in all situations. Furthermore, the processing time with K-Means is incredibly high. The Silhouette coefficient process computes every possible clustering from 1 to the number of documents in the window.

With this algorithm and Event Registry, the dense MPNet representation over-performs the sparse TF-IDF vectorisations (with between 9 and 17 points less of F1 compared to News TF-IDF, between 4 and 11 points for Tweets TF-IDF). The difference between the gold clustering results and those given with Silhouette are closer with dense vectors. With CoAID, sparse TF-IDF vectors are better for clustering the dataset, in this case also, the results are close between the gold and silhouette method. With this dataset, dense vectorisation is not an option. Finally, with FibVid, the algorithm behaves similarly to the other one, and the results are close to each other with a F1 score of about 35%. With this last dataset, there is a huge gap between the precision obtained using the gold number of clusters and when discovering a number of clusters k with the Silhouette score.

In all cases, the Miranda et al. [22] algorithm proceeds better and faster than the baseline, even if we include the training time, that is, in this scenario, four times the testing duration.

6   The Issue of Short Message Similarities

Results published in Table 3 and Table 4 are low in comparison to other studies that processed the Event Registry dataset for the same task [20,22,32]. In these works, authors use the whole text and title of the article. Our approach focuses on news article titles, and we wonder how well it is possible to apply news tracking algorithms on this specific type of documents.

We compare the document - cluster similarities described in Subsect. 4.1 with two types of datasets: one with titles (the one we are using in this paper) and one that also includes article content text. We display in Fig. 2, the cosine similarities mean and standard deviation for document - cluster pairs belonging to the same news story (in green) or not (in red). This means for the feature f_entities_all in Fig. 2a, the mean similarity for this feature is around 0.65 for documents and clusters related to the same story, and almost 0 for documents and clusters that belong to different stories. The bigger the separation between green and red is, the more efficient the algorithm can be.

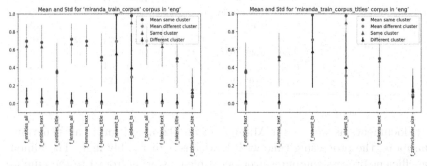

(a) With full articles, the article text gives a better separation of true and false clustering.

(b) With only titles, the dimensions that best discriminate documents and clusters are absent. There are less features.

Fig. 2. Document - cluster similarities on the training set with gold labels. (Color figure online)

Independently of the algorithm itself, whether Miranda et al. (Table 3) or K-Means (Table 4), the very nature of data is at stake. Considering only news article titles does not permit to separate well documents that are dissimilar; and to well cluster ones that report a same news story.

7   Discussion

As we showed in our experiments, it is possible to obtain rather good clustering results with a very little of information contained in article titles. On the other hand, we wanted to evaluate the impact of the document representation, with sparse TF-IDF weightings and dense vectors. It is impossible to conclude on a general trend that would allow us to give a recommendation on whether using one instead of the other. Our results may be considered as illogical, as it is reasonable to state that dense vectors should perform better in any circumstance as they better capture the context, over TF-IDF vectors that are necessarily shorter because of the limited number of tokens in article titles. Our suggestion, when tracking news stories in short messages, especially articles titles is to always test multiple document representations on the dataset in order to select the one that performs best. Even if the conclusion of a previous study [32] mentioned the pertinence of sparse vectors over dense ones, we assume this conclusion does not apply here. They handled full article content and similarly to the Transformer architecture, here only the 512 first tokens are used to compute the logits. In our case, the article title is always shorter than 512 tokens and the dense vectors represent the whole text. For full articles, longer than 512 tokens, it signifies deleting the rest of the text and removing pertinent information.

8   Conclusion

In this paper, we published an analysis of news propagation with tweets and articles titles coming from public sources. We tackled the issue of applying news tracking algorithms on short documents: article titles. We took advantage of the Covid-19 *infodemic* to twist the purpose of datasets dedicated to true and false news detection. We proposed to use the supervised algorithm released by Miranda et al. [22] to build stories from tweets when there are training data. For cases when they are missing, K-Means is a suitable unsupervised algorithm. We experimented the impact of document representation, with sparse TF-IDF vectorisations based on two corpora, and dense vectorisation with the Transformer architecture. We showed that the representation of a document is a major issue sometimes neglected in the literature. With this article, we release all resources: the code of the algorithms and the sets of data we collected to vectorise documents. We share with the community our implementations to let anyone reproduce our results and experiment with private datasets.

We showed one of the reason why clustering article titles works worse than when also taking the article content into account. Short messages do not contain enough discriminant data. We also lacked big datasets qualitatively annotated with events. By analysing Fibvid, we notice the quality of primary data could be the explanation of the rather bad results it produces.

On another hand, we computed the dense vectors with multilingual models [26]. These vectors are aligned in multiple languages, providing similar vectors for similar semantic in different languages. We will run new experiments on the

Event Registry dataset, which has a set of similar events reported in multiple languages. We expect to notice new outcomes in multilingualism for this type of task.

Acknowledgments. This work has been supported by the European Union's Horizon 2020 research and innovation program under grants 770299 (NewsEye) and by the ANNA project funded by the Nouvelle-Aquitaine Region. The research work by Paolo Rosso was partially funded by the Generalitat Valenciana under DeepPattern (PROMETEO/2019/121). The authors would like to thank the Polytechnic University Of València (UPV), Spain, which made this work possible, and its IT laboratory, DSIC.

References

1. Ai, M.: BERT for Russian news clustering. In: Proceedings of the International Conference "Dialogue 2021", p. 6. Moscow, Russia (2021)
2. Allan, J.: Introduction to topic detection and tracking. In: Allan, J. (ed.) Topic Detection and Tracking: Event-Based Information Organization, vol. 12, pp. 1–16. Springer, Boston (2002). https://doi.org/10.1007/978-1-4615-0933-2_1
3. Amigo, E., Gonzalo, J., Artiles, J., Verdejo, F.: A comparison of extrinsic clustering evaluation metrics based on formal constraints. Inf. Retrieval **12**(4), 461–486 (2009). https://doi.org/10.1007/s10791-008-9066-8
4. Atefeh, F., Khreich, W.: A survey of techniques for event detection in Twitter. Comput. Intell. **31**(1), 132–164 (2015). https://doi.org/10.1111/coin.12017
5. Bernard, G.: compute_dense_vectors. https://archive.softwareheritage.org/swh:1:dir:7b4552980670d658ab07e5458d8f3ee1956aae4b
6. Bernard, G.: compute_tf_idf_weights. https://archive.softwareheritage.org/swh:1:dir:76f1022d1380e5f1d39ba02924e9f8eb9906dd95
7. Bernard, G.: document_tracking. https://archive.softwareheritage.org/swh:1:dir:e51ab63fd7dcfa830773c8cdfe40979d64a63133
8. Bernard, G.: news_tracking. https://archive.softwareheritage.org/swh:1:dir:efa67f09d67b843a1a2a6f3cdac5aac96a46da9a
9. Bernard, G.: CoAID dataset with multiple extracted features (both sparse and dense) (2022)
10. Bernard, G.: Event registry dataset with multiple extracted features (both sparse and dense) (2022)
11. Bernard, G.: FibVID dataset with multiple extracted features (both sparse and dense) (2022)
12. Bernard, G.: Resources to compute TF-IDF weightings on press articles and tweets (2022). https://doi.org/10.5281/zenodo.6610406
13. Brigadir, I., Greene, D., Cunningham, P.: Adaptive representations for tracking breaking news on Twitter. arXiv:1403.2923 [cs], November 2014
14. Cui, L., Lee, D.: CoAID: COVID-19 healthcare misinformation dataset. arXiv:2006.00885 [cs] (2020)
15. Devlin, J., Chang, M.W., Lee, K., Toutanova, K.: BERT: pre-training of deep bidirectional transformers for language understanding. arXiv:1810.04805 [cs] (2019)
16. Kim, J., Aum, J., Lee, S., Jang, Y., Park, E., Choi, D.: FibVID: comprehensive fake news diffusion dataset during the COVID-19 period. Telemat. Inform. **64**, 101688 (2021). https://doi.org/10.1016/j.tele.2021.101688

17. Laban, P., Hearst, M.: newsLens: building and visualizing long-ranging news stories. In: Proceedings of the Events and Stories in the News Workshop, Vancouver, Canada, pp. 1–9 (2017). https://doi.org/10.18653/v1/W17-2701
18. Le, Q.V., Mikolov, T.: Distributed representations of sentences and documents. arXiv:1405.4053 [Cs] (2014)
19. Leban, G., Fortuna, B., Brank, J., Grobelnik, M.: Event registry: learning about world events from news. In: Proceedings of the 23rd International Conference on World Wide Web, WWW 2014 Companion, Seoul, Korea, pp. 107–110 (2014). https://doi.org/10.1145/2567948.2577024
20. Linger, M., Hajaiej, M.: Batch clustering for multilingual news streaming. In: Proceedings of Text2Story Co-Located with 42nd ECIR, Lisbon, Portugal, vol. 2593, pp. 55–61 (2020). http://ceur-ws.org/Vol-2593/paper7.pdf
21. Mele, I., Bahrainian, S.A., Crestani, F.: Event mining and timeliness analysis from heterogeneous news streams. Inf. Process. Manag. **56**(3), 969–993 (2019). https://doi.org/10.1016/j.ipm.2019.02.003
22. Miranda, S., Znotiņš, A., Cohen, S.B., Barzdins, G.: Multilingual clustering of streaming news. In: 2018 Conference on Empirical Methods in Natural Language Processing, Brussels, Belgium, pp. 4535–4544 (2018). https://www.aclweb.org/anthology/D18-1483/
23. Petrovic, S., Osborne, M., Lavrenko, V.: Streaming first story detection with application to Twitter. In: NACL 2010, Los Angeles, California, USA, pp. 181–189 (2010). https://dl.acm.org/citation.cfm?id=1858020
24. Phuvipadawat, S., Murata, T.: Breaking news detection and tracking in Twitter. In: 2010 IEEE/WIC/ACM International Conference on Web Intelligence and Intelligent Agent Technology, Toronto, AB, Canada, pp. 120–123 (2010). https://doi.org/10.1109/WI-IAT.2010.205
25. Pouliquen, B., Steinberger, R., Ignat, C., Käsper, E., Temnikova, I.: Multilingual and cross-lingual news topic tracking. In: Proceedings of the 20th International Conference on Computational Linguistics, COLING 2004, Geneva, Switzerland, p. 959-es (2004). https://doi.org/10.3115/1220355.1220493
26. Reimers, N., Gurevych, I.: Sentence-BERT: sentence embeddings using Siamese BERT-networks. In: Proceedings of the 2019 Conference on Empirical Methods in Natural Language Processing and the 9th International Joint Conference on Natural Language Processing (EMNLP-IJCNLP), Hong Kong, China, pp. 3982–3992 (2019). https://doi.org/10.18653/v1/D19-1410
27. Ritter, A., Mausam, Etzioni, O., Clark, S.: Open domain event extraction from Twitter. In: Proceedings of the 18th ACM SIGKDD International Conference on Knowledge Discovery and Data Mining, KDD 2012, Beijing, China, p. 1104 (2012). https://doi.org/10.1145/2339530.2339704
28. Rupnik, J., Muhic, A., Leban, G., Skraba, P., Fortuna, B., Grobelnik, M.: News across languages - cross-lingual document similarity and event tracking. J. Artif. Intell. Res. **55**, 283–316 (2016). https://doi.org/10.1613/jair.4780
29. Reddy, S.: Now on Twitter: 140 characters for your replies, March 2017. https://blog.twitter.com/en_us/topics/product/2017/now-on-twitter-140-characters-for-your-replies
30. Shearer, E., Mitchell, A.: News Use Across Social Media Platforms in 2020, January 2021. https://www.pewresearch.org/journalism/2021/01/12/news-use-across-social-media-platforms-in-2020/
31. Shinyama, Y., Sekine, S., Sudo, K.: Automatic paraphrase acquisition from news articles. In: Proceedings of the Second International Conference on Human Lan-

guage Technology Research, San Diego, California, pp. 313–318 (2002). https://doi.org/10.3115/1289189.1289218

32. Staykovski, T., Barron-Cedeno, A., da San Martino, G., Nakov, P.: Dense vs. sparse representations for news stream clustering. In: Proceedings of Text2Story Co-Located with the 41st ECIR, Cologne, Germany, vol. 2342, pp. 47–52 (2019). https://ceur-ws.org/Vol-2342/paper6.pdf

33. Weng, J., Lee, B.S.: Event detection in Twitter. In: Proceedings of the Fifth International Conference on Weblogs and Social Media, Barcelona, Catalonia, Spain, pp. 401–408 (2011). http://www.aaai.org/ocs/index.php/ICWSM/ICWSM11/paper/view/2767

34. World Health Organisation: Infodemic, January 2022. https://www.who.int/westernpacific/health-topics/infodemic

35. Yang, Z., Dai, Z., Yang, Y., Carbonell, J., Salakhutdinov, R., Le, Q.V.: XLNet: generalized autoregressive pretraining for language understanding. arXiv:1906.08237 [cs] (2020)

Leveraging Wikipedia Knowledge for Distant Supervision in Medical Concept Normalization

Annisa Maulida Ningtyas[1,2(✉)], Alaa El-Ebshihy[1,4], Guntur Budi Herwanto[2,3],
Florina Piroi[1], and Allan Hanbury[1]

[1] Technische Universität Wien, Vienna, Austria
{annisa.ningtyas,alaa.el-ebshihy,
florina.piroi,allan.hanbury}@tuwien.ac.at
[2] Universitas Gadjah Mada, Yogyakarta, Indonesia
gunturbudi@ugm.ac.id
[3] Universität Wien, Vienna, Austria
[4] Alexandria University, Alexandria, Egypt

Abstract. The majority of recent research has approached the Medical Concept Normalization (MCN) task as supervised text classification. However, combining all of the currently available training datasets for this task (CADEC, PsyTAR, COMETA) only covers a small fraction of the concepts contained in the Systematized Nomenclature of Medical-Clinical Terms (SNOMED-CT). In this work, we propose a distant supervision approach to broaden the training data coverage of the SNOMED-CT concepts by tapping into Wikipedia as a source of informal medical phrases. Based on our observations, components of Wikipedia articles (article summaries, Wikipedia's *redirect* pages, wikilinks data) contain informal medical terms that can be generalized to those used in social media posts. We extract the article summaries, Wikipedia's *redirect* pages, and wikilinks data from the Wikipedia articles relating to medical information. We pair this data with corresponding SNOMED-CT concepts. Our distant supervision approach was able to double the concept coverage from the public MCN data sets. Our experiments show that the proposed distant supervision data approach improved the model performance on the three publicly available MCN datasets.

Keywords: Distant Supervision · Medical Concept Normalization · Wikipedia

1 Introduction

Social media has become an important source of knowledge, with users searching for a broad range of information, from cooking recipes, to travel routes, to medical advice, to name a few. For the special case of medical information search, users are both looking for information, and for sharing experiences, or forming a community [22]. In this process new informal medical phrases are continuously

A. Barrón-Cedeño et al. (Eds.): CLEF 2022, LNCS 13390, pp. 33–47, 2022.
https://doi.org/10.1007/978-3-031-13643-6_3

Table 1. Example mappings between informal medical terms and medical terminologies in SNOMED-CT

Informal Medical Term	Normalized Medical Concept
Muscle pain	Myalgia (SNOMED-CT ID: 68962001)
Mellows me out	Feeling content (SNOMED-CT ID: 271599002)
Extreme pain	Severe pain (SNOMED-CT ID: 76948002)

emerging. Such phrases contain valuable information for medical workers, for pharmaceutical companies, for patients themselves.

These informal medical phrases contain valuable information that can be used to analyze various phenomena, such as Adverse Drug effects. This type of analysis, though, can be challenging since posts by social media users refer to medical terms in colloquial language, possibly with typographic errors [21]. In addition, there is a lexical and grammatical diversity in the language used in social media, depending on the user's level of background medical knowledge. In order to bridge this gap, Medical Concept Normalization (MCN) is used to standardise the informal medical phrases or terminology into formal medical terminology sourced from a medical knowledge base, such as the Systematized Nomenclature of Medical-Clinical Terms (SNOMED-CT). Table 1 provides a few examples of the MCN task output.

The current research approaches the MCN as a supervised text classification task [12,13,17,23,26]. The publicly available datasets for MCN provide mappings between informal medical terms and medical terminologies [3,10,29]. However, the combination of these datasets contains approximately 8,568 concepts out of the ~350,000 listed in SNOMED-CT. Therefore, deep learning models, which require a large amount of training data, are limited in their application to solving the MCN task. Extending the publicly available datasets to train deep learning models is costly, as annotations require expert knowledge.

One approach to addressing the data scarcity is to automatically generate labelled data with distant supervision methods using existing knowledge bases or dictionaries. Distant supervision in the MCN task is one of the approaches to overcome the low concept coverage of SNOMED-CT [21]. However, the current approach [21] is limited when the language gap between colloquial and formal medical terms is wide. For example, the informal term *'need to sleep constantly'* and *'Somnolence'* should be synonymous in the medical terminology. However, the proposed distant supervision approach [21] could not mapped between the two phrases due to their low linguistic similarity.

Wikipedia has a large number of articles related to the medical domain. According to Shafee et al. [25], the English Wikipedia contains approximately 30.000 medical articles, and Ngo et al. [19] estimate that around 80% of the SNOMED-CT concepts are covered by Wikipedia articles. Wikipedia's Manual of Style for medical-related articles[1] recommends that authors write in plain English and as simply as possible. For instance, when introducing new technical

---

[1] https://bit.ly/Wikipedia_Manual_of_Style_Medicine-related_articles.

terms, authors must provide an explanation in plain English, followed by the technical terms in parenthesis. It is advisable to use hyperlinks to direct readers to other pages for further information. Additionally, *redirect* pages are created to aid in the search process by providing alternate names. For example, *heart attack* redirects to *Myocardial Infarction*. Considering all these, we hypothesise that medical related Wikipedia articles incorporate colloquial medical terminology.

In this paper, we explore the suitability of the medical related Wikipedia articles as a source distant supervision dataset for the automatic collection of labelled data, to supplement the existing MCN datasets with additional informal phrases per concept. The MCN datasets we use are: (1) CADEC, an annotated dataset on Adverse Drug Effects (ADE) gathered from the "Ask a Patient" forum, consisting of 1,250 SNOMED-CT annotated text segments [10]; (2) Psy-TAR, a corpus with psychiatric medication ADEs, with 887 reviews from the "Ask a Patient" forum [29], structured similarly to CADEC; and (3) COMETA, a corpus of randomly collected reddit threads, with 18 disease categories [3].

Informal medical terminology is derived from Wikipedia elements such as abstract summaries, Wikipedia's redirect pages, wikilinks (i.e. hyperlinks to other Wikipedia articles), and medical term abbreviations. We map informal medical terms to SNOMED-CT concepts and determine the efficacy of the Wikipedia distant supervision data set thus extracted to MCN tasks.

2   Related Work

Distant supervision is an approach that uses information from knowledge bases to generate automatically labelled data [18]. The objective of this technique is to generate a labelled data set from an unlabelled data set where the labels are gathered from external sources using a semi-automated approach. Distant supervision is a frequently used technique for information extraction tasks such as Named Entity Recognition (NER) or Relation Extraction (RE), in which the labels come from a knowledge base or gazetteers [5,7,8,11,14,18].

Distant supervision has also been applied to a variety of tasks, including sentiment analysis using an emoticon dictionary [1,4] and identifying medical terms from Electronic Health Records (EHR) [6]. In the medical entity linking task, Vashishth et al. [27] create a training data set from PubMed abstracts to extract medical entity mentions automatically. A medical entity mention in a PubMed abstract is retained when text spans match an entity in the Medical Subject Heading (MeSH) thesaurus. Distant supervision has also been applied in MCN. Pattisapu et al. [21] employ distant supervision by extracting medical terms from medical forums and classifying them into SNOMED-CT concepts. To determine the corresponding medical concept, they compute the semantic similarity between medical phrases and medical concepts. This approach is limited by the language gap between informal and formal medical terms [21].

While automatically labeled data may have some drawbacks, such as noise and label error, it is easier to collect than the manually annotated data (often considered to be gold-standard), which requires careful consideration of prerequisites, annotation guidelines and settings. We aim to minimize the noise and label error by using the community driven knowledge which is Wikipedia.

3   Distant Supervision Approach

This section describes our distant supervision approach for the Medical Concept Normalization (MCN) task. Figure 1 shows the pipeline of our approach.

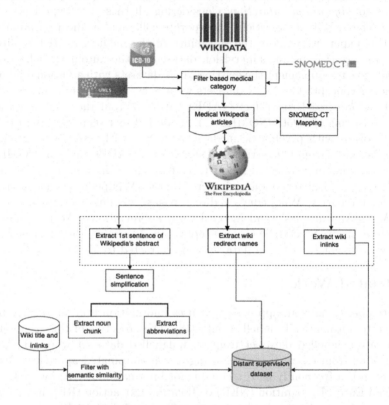

Fig. 1. The proposed Distant Supervision Approach

3.1   Data Description

Our distant supervision approach uses two main data sources, Wikipedia and Wikidata, where we use the January 2020 dump. The Wikipedia dump contains several features that we will be using to produce our distant supervision data[2]:

(1) The **article summary** is the first section of a Wikipedia article. We specifically consider the first sentence in this summary. A Wikipedia article's summary section contains several phrases that explain the article's concept. The majority of a concept's basic explanation and central idea are typically contained in the first sentence. The second sentence and its remainder are used to explain details that may be irrelevant to our MCN task. As a result, we limit our extraction of medical terms to the first sentence.

---

[2] *Distant supervision data* is defined in Sect. 4.

(2) The Wikipedia's *redirect* **pages** are alternative names, spellings, abbreviations, and common misspellings. Page redirects are frequently used to obtain the synonym from the Wikipedia article.

(3) The **wikilinks** are internal links between Wikipedia articles. Similar to redirects, they may point to synonyms and common phrases that refer to a concept.

In the typical MCN task, every informal medical phrase will be classified to one of the SNOMED-CT codes. Wikipedia articles do not directly store SNOMED-CT codes, therefore, we incorporate information from Wikidata[3], which contains several medical properties like SNOMED-CT, UMLs[4], and ICD-10[5] codes. ICD-10 is a more general classification system because it is oriented towards health statistics reporting. SNOMED-CT, on the other hand, gives a more detailed codification for the purpose of maintaining Electronic Health Records (EHR). UMLs combines many of the biomedical vocabularies, including SNOMED-CT, to facilitate interoperability between medical information systems and services, including EHR.

Wikidata is a knowledge graph generated by and from various Wikimedia projects, including Wikipedia [24]. Wikidata is a community driven knowledge base and can be freely edited by humans or machines. However, since it is a community driven knowledge, not all medical items in Wikidata are associated with a SNOMED-CT code, and they may be associated only with UMLs or ICD-10 codes. As a result, an additional mapping step from ICD-10 and UMLs to SNOMED-CT is required. In our work, we use the NLM's[6] SNOMED-CT US edition and its UMLs[7], and SNOMED International[8] knowledge base.

3.2 Identifying Medical Articles from Wikipedia

Since Wikipedia contains mostly non-medical articles, we need to identify those describing medical information to be used by our distant supervision approach. Even though Wikipedia makes use of several medical category labels, we tend to rely on Wikidata's medical properties to identify the medical articles, since we will be using these codes in our MCN. We identify three medical properties from Wikidata which are: (1) SNOMED-CT, (2) ICD-10, and (3) UMLs code. Whenever an article is labelled with a concept out of these three nomenclatures, we consider that article as one referring to medical information. To obtain this data, we issued a SPARQL query to the Wikidata SPARQL query endpoint[9].

Table 2 shows some query result examples. The Wikidata item *wd:Q121041* which is labelled with *appendicitis*, refers to Wikipedia article on Appendicitis[10].

---

[3] https://www.wikidata.org.
[4] https://www.nlm.nih.gov/research/umls/index.html.
[5] https://www.who.int/standards/classifications/classification-of-diseases.
[6] https://www.nlm.nih.gov/healthit/snomedct/us_edition.html.
[7] https://www.nlm.nih.gov/research/umls/index.html.
[8] https://www.nlm.nih.gov/healthit/snomedct/international.html.
[9] https://query.wikidata.org/.
[10] https://en.wikipedia.org/wiki/Appendicitis.

Table 2. Sample Wikidata Items with their related medical properties returned by the SPARQL query.

Item	ItemLabel	Wikipedia	SNOMED-CT	UMLs	ICD-10
wd:Q121041	appendicitis	Appendicitis	74400008	C0085693, C0003615	-
wd:Q68833	bone fracture	Bone fracture	-	C0016658	T14.2
wd:Q147362	ovarian cyst	Ovarian cyst	-	C0029513	-

Its SNOMED-CT property is *74400008*, and has two UMLs property values, *C0085693* and *C0003615*, and no ICD-10 code. As can be seen from the result, one Wikidata item may contain more than one UMLs.

3.3   Mapping Wikipedia Articles to SNOMED-CT

Each Wikipedia article retrieved is to be associated with a single SNOMED-CT concept that corresponds to the subject of the article. Other content in the article text that may be mapped to SNOMED-CT concepts are not considered in this work. As shown in Table 2, we can map the SNOMED-CT code directly from the Wikidata query result. We discovered that Wikidata entries which contain SNOMED-CT medical property, were only related with one SNOMED-CT concept per Wikipedia article. However, a Wikipedia article may be associated to one or more UMLs or ICD-10 codes. Therefore, we introduce a mapping step from ICD-10 and UMLs to SNOMED-CT concepts. We outline our mapping procedure in Algorithm 1.

The primary input to Algorithm 1 is the result of a Wikidata query. We examine each item in the query result that does not have a SNOMED-CT code. Then, using the ICD and UMLs mapping data sources, we determine the mapping of SNOMED-CT candidates. It is possible that the mapping generates multiple SNOMED-CT codes. Thus, to obtain the most closely matching SNOMED-CT code, we compute the Levenshtein distance between the SNOMED-CT label and the itemLabel from Wikidata. The candidate with the shortest distance would be assigned to the Wikidata item as the best match.

3.4   Lay Medical Terms Extraction

Once the medical Wikipedia articles have been mapped to SNOMED-CT concepts, we extract the medical phrases from the *article summary*, the Wikipedia's *redirect pages*, and the *wikilinks* data as explained in this section.

Extracting Medical Phrases from the Article Summary. According to our observations, the first sentence of a Wikipedia medical article may be too technical (i.e. already has the formal medical phrasing) to represent the MCN task. We first experimented with Wikipedia's Simple English article version[11].

---

[11] https://simple.wikipedia.org.

Algorithm 1: Mapping ICD-10 and UMLs CUI to SNOMED-CT

---

input : *query_result*
output: *best_match_snomed_code*

1 snomed_code ← List()
2 snomed_code.*append*(icdMapping (query_result.icd_code))
3 snomed_code.*append*(umlSmapping (query_result.umls_code))

4 **if** *length(snomed_code)==1* **then**
5 | best_match_snomed_code = snomed_code[0]
6 **else**
7 | best_match_snomed_code = GetBestSnomedCode (snomed_code, item_label)
8 **end**
9 return best_match_snomed_code

10 **Function** GetBestSnomedCode(*snomed_codes, item_label*):
11 | lev_distance ← List()
12 | **foreach** *scode ∈ snomed_code* **do**
13 | | distance ← Levenshtein_distance(scode, item_label)
14 | | lev_distance.append(distance)
15 | **end**
16 | return GetSnomedCodeWithMinimumDistance (lev_distance)

However, these simplified articles are not as complete in their information as the main Wikipedia articles. Therefore, we simplified the text of the main Wikipedia article using the MUSS sentence simplification model [16]. We decided to use MUSS sentence simplification as this model is trained, among others, also on Simple English Wikipedia. By using this simple sentence, the layman definition of a medical term (that is, the informal medical phrase) is extracted. Based on this sentence, we extract (1) noun phrases and (2) abbreviation:

(1) We extract medical terms using Noun Phrase Chunking. Since chunking noun phrases occasionally results in incomplete phrases, we extend the noun phrases by looking at the sentence dependency tree. We take the phrases composed of the tokens between the leftmost and rightmost syntactic descendants of an input token. Figure 2 shows an example where we use Spacy [9] to obtain the noun phrases and the token dependency.
(2) While examining the COMETA data set, we noticed that non-expert users frequently use abbreviations. As a result, we collect these abbreviations automatically and include them in our distant supervision data.

Extracting Medical Phrases from Wikipedia *Redirect* Pages and Wikilinks As explained in Sect. 3.1, the Wikipedia dump contains Wikipedia's *redirect* pages and wikilinks. A Wikipedia *redirect* is a page that directs the users to another page. For example, *'Heart Attack'* is redirected to the *'Myocardial Infarction'* article. Redirects involve Wikipedia article titles. A wikilink is an internal link to another Wikipedia page. For example, there are many hyperlinks in the Wikipedia article on *Chest Pain*, one of which is attached to the term

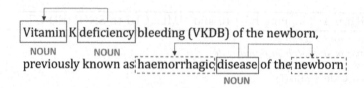

Fig. 2. Example of an extended noun phrases extraction. The arrows shows the leftmost and rightmost syntactic descendants of a token. The result of this extraction is: (1) Vitamin (2) Vitamin K deficiency and (3) haemorrhagic disease of the newborn.

heart attack. This term refers to the Wikipedia article on *Myocardial Infarction.* In this way we take the *heart attack* phrase as an informal medical term for *Myocardial Infarction.* We collected all of the redirects and wikilinks associated with a SNOMED-CT concept and removed duplicates as they appeared.

4   Experiment Setup

Our distant supervision approach produced a single dataset, which we refer to as *distant data.* In a set of three experiments, we evaluated the impact of combining our distant data with each of the three publicly available MCN datasets (CADEC, PsyTAR, and COMETA) in order to generate larger training data for our MCN model. Note that SNOMED-CT medical concepts are linked to informal medical terms in CADEC, PsyTAR, and COMETA.

It is important to note that our distant supervision dataset had more SNOMED-CT concepts than the publicly available MCN datasets. However, in this experiment, we did not exploit this fact, which we leave as future work. We only consider the SNOMED-CT concepts that found in the current MCN datasets. We created the training data for our experiments in the following way: (1) We computed the concepts overlap between our distant data and each of the original training splits of the public MCN datasets. This overlap is referred to as c_1 for CADEC, c_2 for PsyTAR, and c_3 for COMETA, which also represented the class label; (2) We only consider distant data and publicly available MCN datasets (CADEC, PsyTAR, and COMETA) where the concepts overlap (c_1, c_2, and c_3). The overlapped distant data is combined with each of the public MCN datasets based on the concepts overlap, which we refer as UD_1 for CADEC, UD_2 for PsyTAR, and UD_3 for COMETA; (3) We also trained MCN model with the subset of the original training set that based on concepts overlap for each of the publicly available MCN datasets. This is denoted by CD_1 for CADEC, CD_2 for PsyTAR, and CD_3 for COMETA; (4) Finally, we trained our MCN model using a subset of the distant data based on concept overlap, while retaining the validation and test set from each of the public MCN datasets. We refer this as DD_1 for CADEC, DD_2 for PsyTAR, and DD_3 for COMETA.

As a result, we trained our MCN model with 3 different training sets: (a) UD_1, UD_2, UD_3 (scenario (2)), which are the combination of the distant data with each of public MCN datasets based on the overlap concepts in scenario (1); (b) CD_1, CD_2, CD_3 (scenario (3)), which are the subset of the original training set based on the concept overlap in scenario (1); (c) DD_1, DD_2, DD_3 (scenario (4)), which are the subset of the distant data based on the concept overlap in scenario (1).

The validation is carried out by employing the dataset to address the MCN task, which is to classify informal medical phrases into one SNOMED-CT concepts (i.e. SNOMED-CT concepts represent the class labels) through multi-class classification. Our multi-class classification model is based on Gated Recurrent Units (GRU) algorithm [20], where we have one node per target SNOMED-CT concept in the final linear layer. Two contextual embeddings are used to vary the representation of our features. The first embedding, Hunflair [28], was trained on 23 biomedical datasets, while the second, the RoBERTa embedding [15], is a more general embedding trained on five datasets, namely BookCorpus, English Wikipedia, CC-News, OpenWebText, and Stories. The FlairNLP framework [2] is used for both the feature representation and text classification modules. The experiment based on the CADEC dataset was performed on the 5-fold dataset provided by Tutubalina et al. [26]. For PsyTAR [29] and COMETA [3], we repeat the experiment three times based on the original data, in order to see patterns of our model's performance. We evaluated our MCN model performance by using micro F1-score. In addition, statistical significance was computed using two-way ANOVA with ($p < 0.001$).

Table 3. Number of unique SNOMED-CT concepts in each dataset and the number of SNOMED-CT concepts that overlap with our distant dataset, DD.

Dataset	Unique concepts	Common concepts	DD only concepts
CADEC	1,029 [26]	58 (c_1)	9,701
PsyTAR	755 [29]	195 (c_2)	9,563
COMETA	7,648 [3]	1,247 (c_3)	8,512

5   Results

Table 3 shows the statistics of the overlap concepts (see, Common concepts). Our distant data set contains 9,759 unique concepts, out of which 8,546 concepts are not contained in any of the three publicly MCN datasets (CADEC, PsyTAR and COMETA). The average of F1-score for our MCN model trained on the evaluation scenarios described in Sect. 4 are shown in Table 4. It can be shown that the model trained on UD_i (1), performs better than (2) the DD_i, and (3) the CD_i, where i represents each of the public MCN datasets. In the experiment, we employed two different types of embeddings. According to the statistics provided in Fig. 3, RoBERTa performed better in CADEC and PsyTAR, which has more informal phrases. Meanwhile, Hunflair performed better in COMETA, which has more semi-formal phrases.

In general, we observe that our distant supervision data could boost the model performance on the UD_i scenario on all existing data (CADEC, PsyTAR, and COMETA). According to these findings, we can conclude that components

Table 4. F1-score comparison between our MCN model trained in overlap concepts with 3 different training set: (1) (UD_i) (2) the DD_i, and (3) the CD_i

Dataset	Hunflair			RoBERTa		
	CD_i	DD_i	UD_i	CD_i	DD_i	UD_i
CADEC ($i = 1$)	67.78	63.50	78.10	75.60	66.67	81.04
PsyTAR ($i = 2$)	87.93	79.28	90.92	89.18	79.44	91.03
COMETA ($i = 3$)	67.17	88.69	92.81	60.38	61.73	70.02

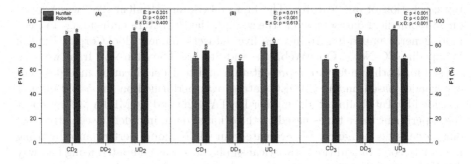

Fig. 3. Statistical results on (A) PsyTAR, (B) CADEC, (C) COMETA

of Wikipedia articles (article summary, Wikipedia's *redirect* pages, and wikilinks data) contain informal medical words that can be generalized to those stated in social media.

6   Discussion

We discovered that our distant supervision method increased the amount of medical terms in the available datasets (e.g., COMETA), as shown in Table 5. For instance, the medical concept *Crohn's disease of colon* is supported by one phrase (that is, there is one text segment mapped to it), which is *Crohn's colitis*. Our distant supervision method increased the number of medical phrase variations for *Crohn's disease of colon*, to *Crohn's disease, Lesniowski-Crohn disease, Crohn's disease of the esophagus, and granulomatous colitis*.

We must pay attention, though, to issues of topic shifting. We found the topic shifting issues in the previous example, the terms *Crohn's disease of the esophagus* is a different disease than *Crohn's disease*. We plan to address the topic topic shifting issue in our future work. Nevertheless, based on the result, we can argue that our distant data has a positive impact on model performance by increasing the number of informal medical phrases. This improvement can be observed in the UD_i column result in Table 4.

Moreover, we observe that the model trained on DD_1 and DD_2 set performed poorly in comparison to the model trained on CD_1 (CADEC) and CD_2 (Psy-TAR). However, the model trained on DD_3 (COMETA) produces the opposite result, indicating that DD_3 performs better than model trained in CD_3. We discovered that several medical phrases represented by DD_1 (CADEC) or DD_2 (PsyTAR) have different writing expression from the original test data. CADEC and PsyTAR appear to use more informal language which is not found in Wikipedia. Meanwhile, the COMETA dataset appears to be more consistent with what is written on Wikipedia. The examples of these data characteristics can be seen in Table 6. We can see that the concept *Alcohol intolerance* is supported with medical phrases *intolerance of alcohol* and *alcohol intolerance* in DD_2 (Psy-TAR) train set, meanwhile in the test set, it appears as *no longer to enjoy the occasional glass of wine or champagne b/c it makes me too drowsy* and *got really drunk really fast*. Due to the language gap between the medical phrases on the train and the test data, it is probable that the model was unable to accurately classify the phrase. Furthermore, we discovered that common terms in our distant supervision data set could correspond to one or more SNOMED-CT concepts. For instance, the term *"pain"* can refer to a number of SNOMED-CT concepts, including *Pain (SNOMED-CT ID:22253000)*, *Abdominal pain (SNOMED-CT ID:21522001)*, *Suffering (SNOMED-CT ID:706873003)*, and *Neuropathic pain (SNOMED-CT ID:247398009)*. Thus, we need to filter out ambiguous phrases to avoid injecting noise into the training data and models.

As stated previously, we found that the techniques we used to extract informal medical terms generated noise. We are aware of these noise issues and will address them in the future. On the other hand, we attempted to reduce noise when extracting formal medical terms. Wikidata has medical properties, such as SNOMED-CT, ICD-10, and UMLs, as indicated in Sect. 3. Thus, we contend that each Wikipedia article pertaining to one of these there medical concepts has been curated by humans. In compared to Pattisapu et al. [21], they used online discussion forum as a source of their distant supervision approach. Meanwhile, our research focused more on exploring Wikipedia articles as another source of distant supervision for MCN by reducing noise in the extracted formal medical phrases.

Table 5. Example for 3 sample medical concepts in CD_3 and DD_3 (COMETA)

Concept	Example of Medical Phrases in Human Annotated Data CD_3	Example of Medical Phrases in Automatically Generated Data DD_3
Backache	backaches, backpain, back pains, back ache	upper back pain, painful back, Back problems, bad back, Pain in the back
Myocardial infarction	myocardial infarction, heart attack, heart attacks	attack of heart failure, MI, heart infarction, severe heart attack, Cardial infarction
Crohn's disease of colon	Crohn' s colitis	Crohn's disease, Lesniowski-Crohn disease, Crohn's disease of the esophagus, granulomatous colitis

Table 6. Comparison of a sample of medical terms from the provided test set to our distant data

Dataset	Concept	Sample Medical phrases in DD_i Train set	Sample Medical phrases in CD_i Test Set
CADEC	Menorrhagia	Heavy menstruation, Heavy menstrual periods	heavy menstral bleeding with clots even though i had just finished my cycle a week before
PsyTAR	Alcohol intolerance	intolerance of alcohol, alcohol intolerance	no longer to enjoy the occasional glass of wine or champagne b/c it makes me too drowsy, got really drunk really fast
COMETA	Anhedonia	Social Anhedonia, lack of pleasure, decreased ability to feel pleasure, anhedonia	anhedonia

7 Conclusion and Future Work

In this paper, we proposed leveraging Wikipedia as a source of informal medical phrases to increase the SNOMED-CT concept coverage on three publicly available MCN datasets (CADEC, PsyTAR, and COMETA). We retrieved informal medical terms from Wikipedia articles' components (article summary, Wikipedia's redirect pages, and Wikilinks) and paired them with SNOMED-CT concepts.

Our distant supervision approach successfully mapped 9,759 SNOMED-CT concepts from the Wikipedia articles, of which 8,546 are SNOMED-CT concepts are not found in the public MCN datasets. The experimental results show that when we combine the data obtained by our distant supervision approach with each of the current MCN datasets, the model performance improves. Based on these findings, we conclude that the Wikipedia components that contain informal medical phrases can be used as training data to improve on the results for solving the MCN task.

Finally, it is important to note that Wikipedia is a community-driven knowledge source with the potential for data inaccuracy. Part of our future work is to perform noise filtering, applying specific filtering and disambiguation processes to reduce noise from unrelated or common phrases. For a more accurate MCN task, we also intend to combine our distant supervision approach with data augmentation.

Acknowledgements. The present work is part of PhD research funded by the Indonesia-Austria Scholarship Programme (IASP), a joint scholarship between the

Indonesian Ministry of Education and Culture (KEMDIKBUD) and Austrian Agency for International Cooperation in Education and Research (OeAD-GmbH). Reference number: ICM-2019-13880.

References

1. Abdul-Mageed, M., Ungar, L.: EmoNet: fine-grained emotion detection with gated recurrent neural networks. In: Proceedings of the 55th Annual Meeting of the Association for Computational Linguistics, vol. 1: Long Papers, pp. 718–728. Association for Computational Linguistics, Vancouver (2017). https://doi.org/10.18653/v1/P17-1067, https://www.aclweb.org/anthology/P17-1067

2. Akbik, A., Bergmann, T., Blythe, D., Rasul, K., Schweter, S., Vollgraf, R.: FLAIR: an easy-to-use framework for state-of-the-art NLP. In: NAACL 2019, 2019 Annual Conference of the North American Chapter of the Association for Computational Linguistics (Demonstrations), pp. 54–59 (2019)

3. Basaldella, M., Liu, F., Shareghi, E., Collier, N.: COMETA: a corpus for medical entity linking in the social media. In: Proceedings of the 2020 Conference on EMNLP, pp. 3122–3137. ACL, Online (2020)

4. Byrkjeland, M., Gørvell de Lichtenberg, F., Gambäck, B.: Ternary Twitter sentiment classification with distant supervision and sentiment-specific word embeddings. In: Proceedings of the 9th Workshop on Computational Approaches to Subjectivity, Sentiment and Social Media Analysis. pp. 97–106. Association for Computational Linguistics, Brussels (2018). https://doi.org/10.18653/v1/W18-6215, https://www.aclweb.org/anthology/W18-6215

5. Cao, Y., Hu, Z., Chua, T.S., Liu, Z., Ji, H.: Low-resource name tagging learned with weakly labeled data. In: Proceedings of the 2019 Conference on Empirical Methods in Natural Language Processing and the 9th International Joint Conference on Natural Language Processing (EMNLP-IJCNLP), pp. 261–270. Association for Computational Linguistics, Hong Kong (2019). https://doi.org/10.18653/v1/D19-1025, https://www.aclweb.org/anthology/D19-1025

6. Chen, J., et al.: A natural language processing system that links medical terms in electronic health record notes to lay definitions: system development using physician reviews. J. Med. Internet Res. **20**(1), e26 (2018). https://doi.org/10.2196/jmir.8669, https://www.jmir.org/2018/1/e26/

7. Dembowski, J., Wiegand, M., Klakow, D.: Language independent named entity recognition using distant supervision. In: Human Language Technologies as a Challenge for Computer Science and Linguistics. Proceedings of the 8th Language & Technology Conference, Poznań, Poland, 17–19 November 2017, pp. 68–72. Fundacja Uniwersytetu im. Adama Mickiewicza, Poznań (2019). http://nbn-resolving.de/urn:nbn:de:bsz:mh39-86198

8. Hedderich, M.A., Klakow, D.: Training a neural network in a low-resource setting on automatically annotated noisy data. In: Proceedings of the Workshop on Deep Learning Approaches for Low-Resource NLP, pp. 12–18. Association for Computational Linguistics, Melbourne (2018). https://doi.org/10.18653/v1/W18-3402, https://www.aclweb.org/anthology/W18-3402

9. Honnibal, M., Montani, I.: spaCy 2: natural language understanding with Bloom embeddings, convolutional neural networks and incremental parsing (2017). to appear

10. Karimi, S., Metke-Jimenez, A., Kemp, M., Wang, C.: Cadec: a corpus of adverse drug event annotations. J. Biomed. Inf. **55**, 73–81 (2015)
11. Lange, L., Adel, H., Strötgen, J.: NLNDE: enhancing neural sequence taggers with attention and noisy channel for robust pharmacological entity detection. In: Proceedings of The 5th Workshop on BioNLP Open Shared Tasks, pp. 26–32. Association for Computational Linguistics, Hong Kong (2019). https://doi.org/10.18653/v1/D19-5705, https://www.aclweb.org/anthology/D19-5705
12. Limsopatham, N., Collier, N.: Adapting phrase-based machine translation to normalise medical terms in social media messages. In: Proceedings of the 2015 Conference on EMNLP, pp. 1675–1680. ACL, Lisbon (2015)
13. Limsopatham, N., Collier, N.: Normalising medical concepts in social media texts by learning semantic representation. In: Proceedings of the 54th Annual Meeting of the ACL, pp. 1014–1023. ACL, Berlin (2016)
14. Lison, P., Barnes, J., Hubin, A., Touileb, S.: Named entity recognition without labelled data: A weak supervision approach. In: Proceedings of the 58th Annual Meeting of the Association for Computational Linguistics, pp. 1518–1533. Association for Computational Linguistics (2020). https://doi.org/10.18653/v1/2020.acl-main.139, https://www.aclweb.org/anthology/2020.acl-main.139
15. Liu, Y., et al.: Roberta: a robustly optimized bert pretraining approach. arXiv preprint arXiv:1907.11692 (2019)
16. Martin, L., Fan, A., de la Clergerie, É., Bordes, A., Sagot, B.: Multilingual unsupervised sentence simplification. CoRR abs/2005.00352 (2020). https://arxiv.org/abs/2005.00352
17. Miftahutdinov, Z., Tutubalina, E.: Deep neural models for medical concept normalization in user-generated texts. In: Proceedings of the 57th Annual Meeting of the ACL: Student Research Workshop, pp. 393–399. ACL, Florence (2019)
18. Mintz, M., Bills, S., Snow, R., Jurafsky, D.: Distant supervision for relation extraction without labeled data. In: Proceedings of the Joint Conference of the 47th Annual Meeting of the ACL and the 4th International Joint Conference on Natural Language Processing of the AFNLP, pp. 1003–1011 (2009)
19. Ngo, D.H., Truran, D., Kemp, M., Lawley, M., Metke-Jimenez, A.: Can wikipedia be used to derive an open clinical terminology? In: Digital Health: Changing the Way Healthcare is Conceptualised and Delivered: Selected Papers from the 27th Australian National Health Informatics Conference (HIC 2019), vol. 266, p. 136. IOS Press (2019)
20. Ningtyas, A.M., Hanbury, A., Piroi, F., Andersson, L.: Data augmentation for layperson's medical entity linking task. In: Forum for Information Retrieval Evaluation, pp. 99–106 (2021)
21. Pattisapu, N., Anand, V., Patil, S., Palshikar, G., Varma, V.: Distant supervision for medical concept normalization. J. Biomed. Inf. **109**, 103522 (2020). https://doi.org/10.1016/j.jbi.2020.103522, https://www.sciencedirect.com/science/article/pii/S1532046420301507
22. Pattisapu, N., Gupta, M., Kumaraguru, P., Varma, V.: Medical persona classification in social media. In: 2017 IEEE/ACM International Conference on Advances in Social Networks Analysis and Mining (ASONAM), pp. 377–384 (2017)
23. Pattisapu, N., Patil, S., Palshikar, G., Varma, V.: Medical concept normalization by encoding target knowledge. In: Proceedings of Machine Learning Research, pp. 246–259. PMLR (2020)

24. Piscopo, A., Vougiouklis, P., Kaffee, L.A., Phethean, C., Hare, J., Simperl, E.: What do wikidata and wikipedia have in common? an analysis of their use of external references. In: Proceedings of the 13th International Symposium on Open Collaboration. OpenSym 2017. Association for Computing Machinery, New York (2017). https://doi.org/10.1145/3125433.3125445, https://doi.org/10.1145/3125433.3125445

25. Shafee, T., Masukume, G., Kipersztok, L., Das, D., Häggström, M., Heilman, J.: Evolution of wikipedia's medical content: past, present and future. J. Epidemiol. Commun. Health **71**(11), 1122–1129 (2017)

26. Tutubalina, E., Miftahutdinov, Z., Nikolenko, S., Malykh, V.: Medical concept normalization in social media posts with recurrent neural networks. J. Biomed. Inf. **84**, 93–102 (2018)

27. Vashishth, S., Joshi, R., Newman-Griffis, D., Dutt, R., Rose, C.: MedType: improving medical entity linking with semantic type prediction. arXiv e-prints arXiv:2005.00460 (2020)

28. Weber, L., Sänger, M., Münchmeyer, J., Habibi, M., Leser, U., Akbik, A.: Hunflair: an easy-to-use tool for state-of-the-art biomedical named entity recognition. arXiv preprint arXiv:2008.07347 (2020)

29. Zolnoori, M., et al.: The psytar dataset: from patients generated narratives to a corpus of adverse drug events and effectiveness of psychiatric medications. Data Brief **24**, 103838 (2019)

Noise-Reduction for Automatically Transferred Relevance Judgments

Maik Fröbe[1]([✉]), Christopher Akiki[2], Martin Potthast[2], and Matthias Hagen[1]

[1] Martin-Luther-Universität Halle-Wittenberg, Halle, Germany
`maik.froebe@informatik.uni-halle.de`
[2] Leipzig University, Leipzig, Germany

Abstract. The TREC Deep Learning tracks used MS MARCO Version 1 as their official training data until 2020 and switched to Version 2 in 2021. For Version 2, all previously judged documents were re-crawled. Interestingly, in the track's 2021 edition, models trained on the new data were less effective than models trained on the old data. To investigate this phenomenon, we compare the predicted relevance probabilities of monoT5 for the two versions of the judged documents and find substantial differences. A further manual inspection reveals major content changes for some documents (e.g., the new version being off-topic). To analyze whether these changes may have contributed to the observed effectiveness drop, we conduct experiments with different document version selection strategies. Our results show that training a retrieval model on the "wrong" version can reduce the nDCG@10 by up to 75%.

Keywords: MS MARCO · monoT5 · Relevance transfer

1   Introduction

Retrieval models are usually trained and evaluated either (1) on datasets with up to several thousands of relevance judgments, carefully curated by expert annotators (e.g., for TREC tracks), or (2) on datasets with hundreds of thousands or more judgments inferred from user data. Particularly transformer-based retrieval models require many training instances to outperform traditional sparse models like BM25 [22]. One of the first datasets with a sufficiently large number of judgments was MS MARCO [7,23]. It was originally released with one positive passage-level judgment for each of 325,183 queries (i.e., only relevant instances are annotated) and later complemented by document-level judgments for Version 1. Despite some erroneous judgments [6,8,9], MS MARCO Version 1 has been the basis of training highly effective document retrieval models (e.g., monoT5 [24] which is the current state of the art[1] on Robust04). However, in 2021, the TREC Deep Learning tracks transitioned from using MS MARCO Version 1 as official training data to MS MARCO Version 2, a larger and improved version. It therefore came as a surprise that models trained on Version 2 were found to be less effective than models trained on Version 1 [9].

---

[1] https://paperswithcode.com/sota/ad-hoc-information-retrieval-on-trec-robust04

A. Barrón-Cedeño et al. (Eds.): CLEF 2022, LNCS 13390, pp. 48–61, 2022.
https://doi.org/10.1007/978-3-031-13643-6_4

Table 1. Examples of differences between versions of positive training instances from MS MARCO Version 1 (crawled in 2018) and Version 2 (2021). Text fragments highlighted in blue italics indicate relevance (erroneous versions have no blue italics).

Query	Relevant Document		Comment
	Version 1 (2018)	Version 2 (2021)	
what are deposit solutions banking	Oops! There was a problem! We had an unexpected problem processing your request.	*Deposit Solutions* Crunchbase *Company Profile* ...	Crawling error in V1
what are yellow roses mean	Meaning Of A Yellow Rose ... a yellow rose *stands for joy and happiness* ...	20 Best Knockout Roses To Make Your Garden Outstanding	Redirect in V2
how much magnesium in kidney beans	Kidney Beans ... *a cup of kidney beans contains 70 mg of magnesium* ...	Magnesium Grocery List. Bring this list to the store to ...	Content change in V2

The document-level relevance judgments for MS MARCO Version 1 were transferred 1:1 from the originally crowdsourced passage-level relevance judgments [8]. The transfer was based on a URL match, assuming that a document having the same URL as one that previously contained a relevant passage included in the original MS MARCO passage dataset is still relevant for the same query. The document-level judgment transfer from Version 1 to Version 2 relied on the same heuristic. However, the MS MARCO documents were crawled one year (Version 1) and four years (Version 2) after the original passage-level relevance judgments were obtained. Thus, some of the documents' content may have changed—possibly invalidating the passage-level judgments. A preliminary analysis of a sample of 50 instances showed that Version 2 has a comparable error rate to Version 1 [9]—whereas related work on a different web crawl found that re-crawling web pages after 3 years can yield quite substantial changes [15].

To analyze why retrieval models trained on MS MARCO Version 2 were less effective in the TREC 2021 Deep Learning track than models trained on Version 1, we compare the two versions of all 325,183 positive training instances using monoT5's estimated probability of a document being relevant to the respective query. Some cases with a substantially different probability are shown in Table 1. Overall, Version 1 contains about 3,800 such potential errors but Version 2 has about 13,100 (details in Sects. 3–5). Interestingly, snapshots from the Wayback Machine with an archival date closer to the date of the MS MARCO passage judgments only yield few additional cases. Finally, we compare the effectiveness of monoT5 models trained on the erroneous versions to models trained on the "correct" counterparts and observe that training on errors can reduce the nDCG@10 by up to 75% (Sect. 6). Our code and data are freely available[2].

---

[2] https://github.com/webis-de/CLEF-22

2   Related Work

The passage-level MS MARCO relevance judgments [8] (only positive instances included) enabled the training of data-hungry transformer-based retrieval models [22] and triggered research on identifying/sampling negative training instances [16,29,37,38]. For the two document-level MS MARCO versions created about one or four years later [6,8,9], the passage-level judgments were 1:1 transferred to the documents crawled for the same URL. Still, content changes may actually have invalidated some of the transferred judgments in the training data.

Evolution of Web Pages. Even though web pages change regularly, content and links usually remain highly similar within a couple of weeks [5,12,13,26,27]. But when more time has passed, two snapshots of a page can differ a lot. For example, a study by Fröbe et al. [15] showed that about 90% of the ClueWeb09 documents judged for some topic from the TREC Web tracks had a substantially different content in the ClueWeb12 crawled three years later—invalidating any URL-based judgment transfer. For the ClueWeb corpora, actually no judgments were transferred but a similar effect might have impacted the transition from MS MARCO Version 1 to 2. Besides the actual two MS MARCO document versions, we also study Wayback Machine snapshots that are close to the potential period of the MS MARCO passage-level judgments—inspired by recent studies that successfully enriched their datasets via the Wayback Machine [15,19].

Handling Training Data Errors. The two standard approaches to deal with errors in the training data of learning-to-rank algorithms [34] are (1) robust loss functions and (2) sample selection. While modifications of popular loss functions like adaptions of the cross-entropy loss [11] or generalizations of PeerLoss [34] aim to make them "robust" on noisy click data, sample selection aims to remove erroneous training instances [14]. Sample selection has been successfully applied to click logs [4,32] but also to MS MARCO [29,31]. For instance, Qu et al. [29] and Arabzadeh et al. [1] observed that unjudged MS MARCO passages (implicitly assumed to be non-relevant) can be more relevant to a query than the actually annotated positive instance. Taking this observation into account when sampling negative training instances, Qu et al. substantially increased the effectiveness of their final model [29]. Also Rudra et al. [31] applied sample selection and used only the most relevant passage of a relevant document as a positive instance during training. They assumed that the passage with the highest monoBERT score [25] is the most relevant to a query and removed the other passages of a positive document. We expand this idea to compare multiple versions of a document but use monoT5 [24] since it is more effective than monoBERT [36].

3   Identifying Potential Errors in the Training Instances

To study potential judgment "errors" in the MS MARCO document retrieval training data caused by the different crawling dates and possible content changes, we use monoT5 [24] to estimate the probability of a positive training document being relevant for the respective query. For each positive training instance, we compare the probabilities of the variants in Version 1 and Version 2 to identify discrepancies that may hint at errors on either side. We also use snapshots from the Wayback Machine to assess whether document versions close to the time of the MS MARCO passage-level judgments could "repair" some errors.

Overview of MS MARCO. Version 1 of the MS MARCO document-level dataset was crawled in 2018 and consists of 3,213,835 documents and 384,597 positive training instances (i.e., query–document pairs) released for the document retrieval tasks at the TREC 2019 and 2020 Deep Learning tracks [6,8]. Documents have a URL, a title field, and a body (HTML tags and boilerplate such as navigation elements removed by a proprietary approach [9]). The positive training instances were created by transferring the passage-level judgments obtained about one year earlier to the documents crawled for the same URL [8]. To somewhat assess the noise introduced by the different crawling date, Craswell et al. [9] used the NIST assessors' judgments on the 46 test queries of the 2020 track and found that for 11 of the 46 queries (i.e., 24%) the positive training instances were assessed as non-relevant—possibly hinting at crawling differences.

Version 2 of the MS MARCO document-level dataset was crawled in 2021 and consists of 11,959,635 documents and 331,956 positive training instances (i.e., query–document pairs) for the TREC 2021 Deep Learning track [9]. Documents now have a URL, title, body, and headings. The document pre-processing (i.e., identifying the body and headings) was different to Version 1, though. A proprietary query-independent approach identified the best non-overlapping passages in a document and concatenated them (mappings between the passage dataset and the document dataset were provided). The training instances were again created by transferring them on basis of the URLs.

Wayback Machine Snapshots. We use the Wayback Machine to compare the MS MARCO document versions with snapshots closer to the time of the passage-level judgments. For each training instance, we try to find one valid snapshot (i.e., successfully crawled with status code 200) from 2015, 2016, and 2017 using the Memento API[3]. If multiple snapshots are available for a year, we select the one closest to July 2nd since this day is the "middle" of the calendar year. We use the Resiliparse library [2] of ChatNoir [3] to extract the plain text and main content of the Wayback documents (this approach produces slightly different main content than the proprietary MS MARCO one, but we still deem the results as "good enough"). Overall, we found snapshots for 68,384 MS MARCO training instances (41,269 have a snapshot for all three years).

Preprocessing Steps. Since the monoT5 model that we use to identify potential errors cannot handle arbitrary input lengths, long documents need to be split into passages that are scored individually [24]. Since there is no explicit mapping to

---

[3] https://archive.readme.io/docs/memento

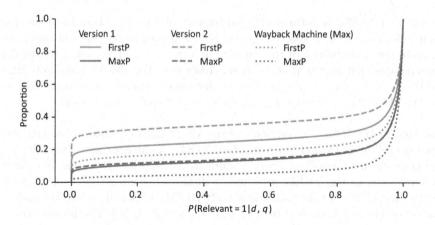

Fig. 1. Cumulative distribution of the monoT5 relevance probability estimates for the positive training instances in MS MARCO Version 1 and 2, and the "best" Wayback Machine snapshot using the first or highest scoring passage (FirstP or MaxP).

passages for Version 1 documents and our snapshots from the Wayback Machine, we use the TREC CAsT tools[4] to split all document versions into passages with the same pipeline. Following suggestions of Dai and Callan [10], we concatenate a document's title and body and split documents at the sentence level into fixed-length passages of approximately 250 terms—fixed-length passages were previously reported to be superior to variable-length passages [17].

Relevance Estimation with monoT5. We use the PyGaggle[5] implementation of monoT5 with its most effective pre-trained variant[6] to estimate the relevance of a document to a query. MonoT5 is based on the sequence-to-sequence model T5 [30] and ranks documents by the probability that, given the query and the document, the decoders' output is the literal "true" [28]:

$$P(\text{Relevant} = 1|d, q)$$

which estimates the probability that a document d is relevant to a query q. We apply monoT5 to all passages of a document and use two approaches that showed high effectiveness in previous work [24,31,35] to aggregate the passage level scores to document level scores: (1) FirstP where the probability of a document being relevant to a query is approximated by monoT5's prediction for the first passage, and (2) MaxP where the probability of a document being relevant to a query is approximated by monoT5's maximum prediction for any of its passages.

Results. Figure 1 shows the cumulative distribution of the monoT5 scores (i.e., the probabilities that a document is relevant to a query, as determined by

---

[4] https://github.com/grill-lab/trec-cast-tools
[5] https://github.com/castorini/pygaggle
[6] https://huggingface.co/castorini/monot5-3b-msmarco

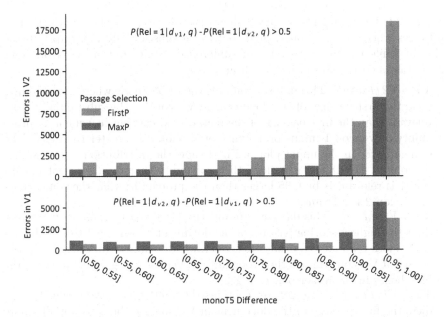

Fig. 2. Distribution of the monoT5 difference of error candidates in Version 2 (upper plot) and Version 1 (bottom plot). We report the difference in the monoT5 probabilities of the two positive training instances for all pairs with high discrepancies above 0.5 for two passage selection strategies (FirstP and MaxP).

monoT5) among the positive training instances for Version 1, Version 2, and snapshots from the Wayback Machine from 2015 to 2017 for FirstP and MaxP aggregation. The cumulative distributions for both Version 1 and 2 include all training instances, whereas we only retain those 41,269 instances into the Wayback Machine snapshots that were successfully crawled in all three years. For them, we select the maximum score of the three candidates as upper bound. Given only correct positive training instances (that are all relevant to its query), an ideal monoT5 model would assign probabilities of 1 to all of them. However, we observe that monoT5 predicts that a non-negligible number of documents is not relevant for all three corpora. Version 2 has the highest proportion of such potential errors (30.59% of positive instances have a probability below 10% for FirstP, respectively 10.52% for MaxP) while our upper bound using snapshots from the Wayback Machine has the smallest proportion (14.54% of positive instances have a probability below 10% for FirstP, respectively 3.07% for MaxP) showing that selecting a positive instance out of multiple versions is promising.

4   Training Datasets with Potential Errors

To assess the reliability of monoT5's relevance probabilities, we construct five datasets with potential errors by comparing the probabilities across different versions of the documents. Figure 2 shows the distribution of positive training instances where one version is predicted to be substantially more likely to be relevant while its counterpart is not (differences >0.5), indicating errors. Our five datasets cover cases of interest from this analysis:

- $V1 \gg V2$ *(FirstP)*. This dataset contains the 41,587 qrels where monoT5 predicts the first passage of the document in Version 1 to be substantially more relevant than the first passage of the recrawled document in Version 2 (probability of Version 1 minus probability of Version 2 is greater than 0.5). The orange boxes in the upper plot of Fig. 2 show the distribution of probability differences. For 18,382 instances, the probability that the document in Version 1 is relevant is by 0.95 larger than the probability that the counterpart in Version 2 is relevant.
- $V1 \gg V2$ *(MaxP)*. This dataset contains the 17,969 qrels where monoT5 predicts the highest-scoring passage of the document in Version 1 to be substantially more relevant than that of the recrawled document in Version 2 (probability of Version 1 minus the probability of Version 2 is greater than 0.5; the blue boxes in the upper plot of Fig. 2).
- $V2 \gg V1$ *(FirstP)*. This dataset contains the 9,991 qrels where monoT5 predicts the first passage of the document in Version 2 to be substantially more relevant than that of the recrawled document in Version 2 (the orange boxes in the lower plot of Fig. 2).
- $V2 \gg V1$ *(MaxP)*. This dataset contains the 15,817 qrels where monoT5 predicts the highest-scoring passage of the document in Version 2 to be substantially more relevant than that of the recrawled document in Version 2 (the blue boxes in the lower plot of Fig. 2).
- *Wayback Machine*. This dataset contains the 41,269 qrels where 5 versions of the positive documents are available: (1) Version 1, (2) Version 2, and (3) three snapshots from the Wayback Machine for 2015, 2016, and 2017. While the above datasets allow the assessment of the impact of errors, this one is used to assess if multiple versions are helpful on parts of MS MARCO without many errors.

Discussion. Our five datasets cover different parts of MS MARCO and are not representative of the complete corpus because they are intentionally focused on subsets of the training data that may contain many errors. Table 2 provides an overview of the datasets showing that all of them are rather dissimilar (the highest Jaccard Similarity with respect to included query IDs is 0.34 between the "V2 ≫ V1 (FirstP)" and the "V2 ≫ V1 (MaxP)" datasets). The first four have much fewer documents from Wikipedia and longer URLs with more parameters compared to "all" documents from MS MARCO. URL parameters are indicative of dynamic content and thus varying relevance, while Wikipedia articles remain topic-stable and thus relevant.

5   Review of Potential Errors in Positive Instances

We manually review 600 queries with their corresponding positive documents in Version 1 and 2 of MS MARCO, 100 each from our five datasets and a random

Table 2. Overview of characteristics of positive documents in our constructed training datasets using FirstP (F) respectively MaxP (M) aggregation. We report statistics on the URL, the most similar dataset measured as Jaccard similarity on the query IDs, and the most frequent domain.

Query Set	Relevant URLs		Most Similar		Most Frequent Domain	
	Len(Path)	Parameters	Name	Sim.	Domain	Percentage
V1 ≫ V2 (F)	38.04	0.06	V1 ≫ V2 (M)	0.31	wikipedia.org	8.06
V1 ≫ V2 (M)	36.90	0.07	V1 ≫ V2 (F)	0.31	wikipedia.org	4.37
V2 ≫ V1 (F)	33.57	0.04	V2 ≫ V1 (M)	0.34	wikipedia.org	9.01
V2 ≫ V1 (M)	35.28	0.04	V2 ≫ V1 (F)	0.34	wikipedia.org	9.09
Wayback M.	32.25	0.02	V1 ≫ V2 (F)	0.07	wikipedia.org	22.72
All	35.11	0.05	–	–	wikipedia.org	15.86

subset of 100 from all MS MARCO training queries. One annotator labeled the 600 instances in random order, and the annotations were randomly checked by two of the co-authors. For each instance, the annotator saw the query, the document in Version 1, and the document in Version 2 and labeled which of the two documents is more relevant to the query (if any). Table 3 provides an overview of the annotations for the five datasets and the random subset. Most of the document pairs of the random subset are equally relevant to the query (79 of the 100 labeled pairs have labels V1 = V2 = 1), but still, there are some errors (e.g., for 9 document pairs, Version 1 was relevant but Version 2 non-relevant, i.e., V1 = 1 > V2 = 0). The subsets that we constructed so that they contain many errors have, as expected, many errors (e.g., 73 of the labeled pairs from "V1 ≫ V2 (MaxP)" are indeed not relevant in Version 2 but in Version 1, i.e., V1 = 1 > V2 = 0). To estimate the overall number of errors in both versions, we use the MaxP variants (as FirstP has cases where the relevant passage comes later in the document) and find that the precision of monoT5 differs substantially for the two versions. Errors in Version 2 are detected with a precision of 0.73 (for 73 of the 100 reviewed "V1 ≫ V2 (MaxP)" pairs, the document in Version 1 is relevant but not relevant in Version 2, i.e., V1 = 1 > V2 = 0). Errors in Version 1 only have a precision of 0.25 (for 25 of the 100 reviewed "V2 ≫ V1 (MaxP)" pairs, the document in Version 2 is relevant but not relevant in Version 1, i.e., V2 = 1 > V1 = 0), resulting in a precision-oriented estimation that Version 2 has 13,117 errors while Version 1 has 3,954 errors.

6 Experiments

We fine-tune monoT5 on each of our datasets to assess if the potential errors identified in MS MARCO negatively affect their effectiveness. We therefore evaluate the effectiveness of these models on three benchmarks: (1) the 100 TREC Web track topics of the ClueWeb12, (2) the 88 topics of the document retrieval task of the TREC Deep Learning track from 2019 and 2020 (Voorhees et al. [33]

Table 3. Overview of our manual review of the relevance of positive documents for our datasets. For each dataset, we labeled 100 document pairs and report the absolute number of relevance preferences (e.g., V1 = V2 = 1: both relevant, V1 = 1, V2 = 0: V1 relevant, V2 non-relevant, etc.) and the precision and the estimated number of errors.

Query Set		Document Relevance							Prec.	Labels	Errors
Type	Selection	V1	0	0	1	1	2	1			
		V2	0	1	0	1	1	2			
$V1 \gg V2$	FirstP		1	4	48	37	6	4	0.48	41,587	19,962
	MaxP		5	3	73	11	7	1	0.73	17,969	13,117
$V2 \gg V1$	FirstP		0	21	7	55	2	15	0.21	9,991	2,098
	MaxP		0	25	5	51	0	19	0.25	15,817	3,954
Random	—		4	0	9	79	3	5	—	325,183	—
Wayback M.	—		0	1	3	89	7	0	—	41,269	—

recommend not to reuse the 2021 edition), and (3) all 250 topics of Robust04. Each training dataset has multiple versions of the positive document and we compare strategies to select the "best" version to demonstrate how the different versions impact effectiveness.

Trained Models. We conduct our experiments with the PyGaggle[7] implementation [18] of monoT5 as this model shows state-of-the-art effectiveness in a range of retrieval experiments [36]. Following Nogueira et al. [24], we use the base version of monoT5 and fine-tune it for one epoch on 10,000 randomly selected positive training instances from one of our five datasets, plus 10,000 randomly selected negative instances from the top-100 BM25 results on MS MARCO. This is repeated ten times using ten different seeds, thus obtaining ten fine-tuned monoT5 models per dataset. Independently of the passage aggregation strategy (FirstP or MaxP) used for the ground truth labels of each of our five datasets, five of the ten models per dataset use FirstP aggregation during training, and five use MaxP aggregation.

Using ir_datasets [20] for data-wrangling[8], we follow previously suggested training regimes [24,25,35], and pass relevant and non-relevant instances in alternating order within the same batch to a model during training. During inference, we rerank the top-100 BM25 results of PyTerrier [21] (default configuration) using the same passage aggregation used during training a given model.

Effectiveness of MonoT5 Trained on Erroneous Positive Instances. In our first experiments, we finetune monoT5 models on the four datasets which, according to the probabilities of the pretrained monoT5 model, contain errors in the positive training instances in one version of MS MARCO while the other version

---

[7] https://github.com/castorini/pygaggle
[8] https://github.com/allenai/ir_datasets

Table 4. Effectiveness of monoT5-base models trained on 20,000 instances from our constructed datasets. Positive instances are selected with one of three selection strategies: (1) BM25, (2) T5$_{Min}$, and (3) T5$_{Max}$. We report Precision@10 and nDCG@10 on the ClueWeb12 (2013 and 2014), the TREC Deep Learning document retrieval task (2019 and 2020), and Robust04 (all topics). Highest nDCG@10 in bold; † marks statistically significant differences to T5$_{Min}$ at $p = 0.05$, with Bonferroni correction.

Training Data		ClueWeb12		DL 19/20		Robust04	
Queries	Selection	P@10	nDCG@10	P@10	nDCG@10	P@10	nDCG@10
V1 ≫ V2 (FirstP)	BM25	0.517†	0.358†	0.580†	0.512†	0.359†	0.376†
	V1=T5$_{Max}$	**0.551†**	**0.385†**	**0.649†**	**0.586†**	**0.441†**	**0.448†**
	V2=T5$_{Min}$	0.425	0.282	0.450	0.388	0.294	0.297
V1 ≫ V2 (MaxP)	BM25	0.508†	0.352†	0.542†	0.474†	0.377†	0.380†
	V1=T5$_{Max}$	**0.557†**	**0.387†**	**0.620†**	**0.562†**	**0.436†**	**0.446†**
	V2=T5$_{Min}$	0.307	0.177	0.197	0.142	0.211	0.209
V2 ≫ V1 (FirstP)	BM25	0.455	0.308	0.547	0.466	0.384	0.383
	V1=T5$_{Min}$	0.468	0.314	0.534	0.452	0.345	0.349
	V2=T5$_{Max}$	**0.499**	**0.333**	**0.559**	**0.505†**	**0.386†**	**0.385†**
V2 ≫ V1 (MaxP)	BM25	0.422†	0.278†	0.449†	0.394†	0.324†	0.319†
	V1=T5$_{Min}$	0.367	0.238	0.385	0.316	0.287	0.279
	V2=T5$_{Max}$	**0.482†**	**0.318†**	**0.530†**	**0.476†**	**0.361†**	**0.367†**
Random	BM25	**0.546**	**0.371**	0.586	0.538	0.400†	0.404†
	T5$_{Min}$	0.532	0.369	0.591	0.531	0.376	0.384
	T5$_{Max}$	0.544	0.368	**0.616**	**0.570†**	**0.410†**	**0.412†**
BM25 (Baseline)		0.439	0.298	0.563	0.507	0.438	0.449

is correct: "V1 ≫ V2" selected by FirstP or MaxP, and "V2 ≫ V1" selected by FirstP or MaxP. We compare these datasets with random training queries.

In the datasets, two versions of each positive document (Version 1 and Version 2) are found. We compare three selection strategies to select which of the two is used for finetuning: (1) T5$_{Min}$ as baseline, which selects the document with the lower pretrained monoT5 score, (2) BM25, which selects the document with the higher BM25 score, and (3) T5$_{Max}$, which selects the document with the higher pretrained monoT5 score. It turns out that T5$_{Min}$ respectively T5$_{Max}$ almost unanimously select the document from Version 1 respectively Version 2 of MS MARCO, e.g., for "V1 ≫ V2", T5$_{Max}$ always selects Version 1, and consequently T5$_{Min}$ always selects Version 2.

Table 4 shows the effectiveness measured as Precision@10 and nDCG@10 for each combination of finetuning dataset, version selection strategy, and the three benchmarks ClueWeb12, TREC Deep Learning tracks 2019/2020, and Robust04. Each score reported results from applying each of the ten fine-tuned monoT5

Table 5. Precision@10 and nDCG@10 on three corpora for monoT5-base trained on 20,000 instances from the Wayback Machine data with 5 selection strategies: (1) BM25, (3) $T5_{Max}$, (2) $T5_{Min}$, (4) Version 1, and (5) Version 2. Highest nDCG@10 in bold; † marks statistical significance at $p = 0.05$ to $T5_{Min}$ with Bonferroni correction.

Training Data		ClueWeb12		DL 19/20		Robust04	
Queries	Selection	P@10	nDCG@10	P@10	nDCG@10	P@10	nDCG@10
	BM25	0.534	0.393	0.574	0.509	0.365	0.373
	$T5_{Max}$	0.543	**0.397**	0.620†	0.557†	0.396†	0.403†
Wayback M.	$T5_{Min}$	0.523	0.371	0.585	0.509	0.355	0.361
	V1	**0.562**	0.388	**0.641**†	**0.597**†	**0.439**†	**0.445**†
	V2	0.518	0.359	0.542†	0.472†	0.307†	0.316†
BM25 (Baseline)		0.439	0.298	0.563	0.507	0.438	0.449

models available for a dataset on a given benchmark to obtain ten runs, and then applying five-fold cross-validation over the benchmark's topics using the ten runs, as implemented by PyTerrier [21].

Many erroneous positive training instances can have a very dramatic impact on the effectiveness of ranking models. For the two "V1 ≫ V2" training datasets for which our monoT5 heuristic predicted that the positive document in Version 2 is not relevant while the positive document in Version 1 is relevant, we observe that BM25 and $T5_{Max}$ selection outperform the $T5_{Min}$ baseline statistically significant on all three benchmarks. The model trained on the positive instance selected with $T5_{Max}$ achieves an nDCG@10 of 0.562 on the TREC Deep Learning document retrieval task, while the model trained on positive instances selected with $T5_{Min}$ achieve only an nDCG@10 of 0.142. This behavior on the two "V1 ≫ V2" training data sets supports our manual review (cf. Sect. 5) that there is a substantial portion of positive training documents that were relevant to its query in Version 1 (selected by the $T5_{Max}$ strategy), which became non-relevant in Version 2 (selected by the $T5_{Min}$ strategy). Interestingly, many such cases can already be resolved by just using the version of the document with the higher BM25 score. Table 4 shows that training on erroneous positive instances from Version 2 of MS MARCO is very ineffective and that this effect is larger for the "V1 ≫ V2 MaxP" dataset than it is for the "V1 ≫ V2 FirstP" dataset. This is consistent with our manual review in Sect. 5, where the MaxP variant identified more errors in positive instances. Also the opposite direction, where the positive instance in Version 1 is not relevant to its query but the version of the document in Version 2 is more relevant can be confirmed by the effectiveness of models trained on the two "V2 ≫ V1" datasets: Selecting always the document from Version 2 for training achieves the most effective models, however, these effects are only significant for "V2 ≫ V1 (MaxP)", which is again consistent with our manual review from Sect. 5. The results for the random training queries show that selecting the better positive document out of Version 1 and Version 2 for training also increases model effectiveness, but only slightly because the random selection is less prone to noise compared to our other training datasets.

Using Snapshots from the Wayback Machine. To complement our experiments, we assess whether more versions of positive instances covering a wider time period may improve the effectiveness of finetuned models. We use our Wayback Machine dataset with 41,269 qrels having five versions of each positive document (Version 1, Version 2, 2015, 2016, and 2017, extracted from the Wayback Machine; cf. Sect. 4). We apply the same training procedure as above. We compare 5 strategies to select the positive instance out of the 5 versions of the positive document: (1) always using Version 1, (2) always using Version 2, and (3) $T5_{Min}$, (4) $T5_{Max}$, and (5) BM25.

Table 5 shows the effectiveness of monoT5 models trained on the Wayback Machine dataset for the five selection strategies on the three benchmarks. The overall picture is similar to the previous experiments: selecting the positive document with $T5_{Max}$ yields more effective models than the BM25 selection which is, in turn, again more effective than the $T5_{Min}$ selection. Interestingly, selecting always Version 1 is even more effective than $T5_{Max}$ and selecting always Version 2 is less effective than the $T5_{Min}$ strategy. The fact that the $T5_{Max}$ and $T5_{Min}$ selection strategies do not produce the most (respectively least) effective models shows that monoT5's probabilities are not suitable to distinguish among mostly correct positive documents and erroneous ones. The Wayback Machine dataset in Table 3 shows that only four out of 100 reviewed queries had incorrect positive documents, likely because "stable" domains like Wikipedia are overrepresented in the Wayback Machine dataset, as shown in Table 2. Hence, only substantial differences in the monoT5 relevance probabilities between versions are reliable. Switching to versions with a slightly higher monoT5 relevance probability does not improve the effectiveness of trained models.

7   Conclusion

Inspired by the effectiveness drop observed in the TREC 2021 Deep Learning track for models trained on MS MARCO Version 2 instead of Version 1, we have compared monoT5's estimated probabilities of judged documents being relevant for their queries in the two versions. Since the judgments were simply transferred after re-crawling documents for Version 2, larger differences in the probabilities might hint at major content changes. Our precision-oriented estimation predicts 13,100 such problems in Version 2—and only 3,800 in Version 1. In experiments, we show that models trained on the "wrong" document versions are highly ineffective. These cases thus probably contribute to the observed effectiveness drop.

Interesting directions for future work include a further investigation of other factors that may influence a model's effectiveness, such as the different preprocessing pipelines used for Versions 1 and 2, or the fact that Version 2 is larger than Version 1 (but same number of judgments). In addition, a more fine-grained classification of possible content changes might help to identify issues that can be neglected and issues that should be fixed during training dataset creation.

References

1. Arabzadeh, N., Vtyurina, A., Yan, X., Clarke, C.: Shallow pooling for sparse labels. CoRR abs/2109.00062 (2021)
2. Bevendorff, J., Potthast, M., Stein, B.: FastWARC: optimizing large-scale web archive analytics. In: Proceeding of OSSYM 2021. OSF (2021)
3. Bevendorff, J., Stein, B., Hagen, M., Potthast, M.: Elastic chatnoir: search engine for the clueweb and the common crawl. In: Pasi, G., Piwowarski, B., Azzopardi, L., Hanbury, A. (eds.) ECIR 2018. LNCS, vol. 10772, pp. 820–824. Springer, Cham (2018). https://doi.org/10.1007/978-3-319-76941-7_83
4. Cen, R., Liu, Y., Zhang, M., Zhou, B., Ru, L., Ma, S.: Exploring relevance for clicks. In: Proceeding of CIKM 2009, pp. 1847–1850. ACM (2009)
5. Cho, J., Garcia-Molina, H.: The evolution of the web and implications for an incremental crawler. In: Proceeding of VLDB 2000, pp. 200–209 (2000)
6. Craswell, N., Mitra, B., Yilmaz, E., Campos, D.: Overview of the TREC 2020 deep learning track. In: Proceeding of TREC 2020. NIST (2020)
7. Craswell, N., Mitra, B., Yilmaz, E., Campos, D., Lin, J.: MS MARCO: benchmarking ranking models in the large-data regime. In: Proceeding of SIGIR 2021, pp. 1566–1576. ACM (2021)
8. Craswell, N., Mitra, B., Yilmaz, E., Campos, D., Voorhees, E.: Overview of the TREC 2019 deep learning track. In: Proceeding of TREC 2019. NIST (2019)
9. Craswell, N., Mitra, B., Yilmaz, E., Campos, D.: Overview of the TREC 2021 deep learning track. In: Voorhees, E.M., Ellis, A. (eds.) Notebook. NIST (2021)
10. Dai, Z., Callan, J.: Context-aware document term weighting for ad-hoc search. In: Proceeding of WWW 2020, pp. 1897–1907. ACM (2020)
11. Feng, L., Shu, S., Lin, Z., Lv, F., Li, L., An, B.: Can cross entropy loss be robust to label noise? In: Proceeding of IJCAI 2020, pp. 2206–2212. IJCAI (2020)
12. Fetterly, D., Manasse, M., Najork, M.: On the evolution of clusters of near-duplicate web pages. In: Proceeding of LA-WEB 2003, pp. 37–45 (2003)
13. Fetterly, D., Manasse, M., Najork, M., Wiener, J.: A large-scale study of the evolution of web pages. In: Proceeding of WWW 2003, pp. 669–678 (2003)
14. Frénay, B., Verleysen, M.: Classification in the presence of label noise: a survey. IEEE Trans. Neural Netw. Learn. Syst. 25(5), 845–869 (2014)
15. Fröbe, M., et al.: CopyCat: near-duplicates within and between the clueweb and the common crawl. In: Proceeding of SIGIR 2021, pp. 2398–2404. ACM (2021)
16. Gao, L., Dai, Z., Fan, Z., Callan, J.: Complementing lexical retrieval with semantic residual embedding. CoRR abs/2004.13969 (2020)
17. Kaszkiel, M., Zobel, J.: Passage retrieval revisited. In: Proceeding of SIGIR 1997, pp. 178–185. ACM (1997)
18. Lin, J., Ma, X., Lin, S., Yang, J., Pradeep, R., Nogueira, R.: Pyserini: a python toolkit for reproducible information retrieval research with sparse and dense representations. In: Proceeding of SIGIR 2021, pp. 2356–2362. ACM (2021)
19. MacAvaney, S., Macdonald, C., Ounis, I.: Reproducing personalised session search over the AOL query log. In: Proceeding of ECIR 2022 (2022). https://doi.org/10.1007/978-3-030-99736-6_42

20. MacAvaney, S., Yates, A., Feldman, S., Downey, D., Cohan, A., Goharian, N.: Simplified data wrangling with ir_datasets. In: Proceeding of SIGIR 2021, pp. 2429–2436. ACM (2021)
21. Macdonald, C., Tonellotto, N., MacAvaney, S., Ounis, I.: PyTerrier: declarative experimentation in python from BM25 to dense retrieval. In: Proceeding of CIKM 2021, pp. 4526–4533. ACM (2021)
22. Mokrii, I., Boytsov, L., Braslavski, P.: A systematic evaluation of transfer learning and pseudo-labeling with BERT-based ranking models. In: Proceeding of SIGIR 2021, pp. 2081–2085. ACM (2021)
23. Nguyen, T., Rosenberg, M., Song, X., Gao, J., Tiwary, S., Majumder, R., Deng, L.: MS MARCO: a human generated machine reading comprehension dataset. In: Proceeding of CoCo@N(eur)IPS 2016. CEUR, vol. 1773. CEUR-WS.org (2016)
24. Nogueira, R., Jiang, Z., Pradeep, R., Lin, J.: Document ranking with a pretrained sequence-to-sequence model. In: Findings of EMNLP 2020, pp. 708–718. ACL (2020)
25. Nogueira, R., Yang, W., Cho, K., Lin, J.: Multi-stage document ranking with BERT, pp. 1–13. CoRR abs/1910.14424 (2019)
26. Ntoulas, A., Cho, J., Olston, C.: What's new on the web? the evolution of the web from a search engine perspective. In: Proceeding of WWW 2004, pp. 1–12. ACM (2004)
27. Olston, C., Pandey, S.: Recrawl scheduling based on information longevity. In: Proceeding of WWW 2008, pp. 437–446. ACM (2008)
28. Pradeep, R., Nogueira, R., Lin, J.: The expando-mono-duo design pattern for text ranking with pretrained sequence-to-sequence models, pp. 1–23. CoRR abs/2101.05667 (2021)
29. Qu, Y., et al.: Rocketqa: An optimized training approach to dense passage retrieval for open-domain question answering. In: Proceeding of NAACL 2021, pp. 5835–5847 (2021)
30. Raffel, C., et al.: Exploring the limits of transfer learning with a unified text-to-text transformer. J. Mach. Learn. Res. **21**, 140:1–140:67 (2020)
31. Rudra, K., Anand, A.: Distant supervision in bert-based adhoc document retrieval. In: Proceeding of CIKM 2020, pp. 2197–2200. ACM (2020)
32. Singla, A., White, R.: Sampling high-quality clicks from noisy click data. In: Proceeding of WWW 2010, pp. 1187–1188. ACM (2010)
33. Voorhees, E., Craswell, N., Lin, J.: Too many relevants: whither cranfield test collections? In: Proceeding of SIGIR 2022. ACM (2022)
34. Wu, X., Liu, Q., Qin, J., Yu, Y.: PeerRank: robust learning to rank with peer loss over noisy labels. IEEE Access **10**, 6830–6841 (2022)
35. Yates, A., Arora, S., Zhang, X., Yang, W., Jose, K., Lin, J.: Capreolus: a toolkit for end-to-end neural ad hoc retrieval. In: Proceeding of WSDM 2020, pp. 861–864. ACM (2020)
36. Yates, A., Nogueira, R., Lin, J.: Pretrained transformers for text ranking: BERT and beyond. In: Proceeding of SIGIR 2021, pp. 2666–2668. ACM (2021)
37. Zhan, J., Mao, J., Liu, Y., Guo, J., Zhang, M., Ma, S.: Optimizing dense retrieval model training with hard negatives. In: Proceeding of SIGIR 2021, pp. 1503–1512. ACM (2021)
38. Zhan, J., Mao, J., Liu, Y., Zhang, M., Ma, S.: Repbert: contextualized text embeddings for first-stage retrieval. CoRR abs/2006.15498 (2020)

The Effect of Prolonged Exposure to Online Education on a Classroom Search Companion

Mohammad Aliannejadi[1] , Theo Huibers[2] , Monica Landoni[3] ,
Emiliana Murgia[4] , and Maria Soledad Pera[5(✉)]

[1] University of Amsterdam, Amsterdam, The Netherlands
m.aliannejadi@uva.nl
[2] University of Twente, Enschede, The Netherlands
t.w.c.huibers@utwente.nl
[3] Università della Svizzera Italiana, Lugano, Switzerland
monica.landoni@usi.ch
[4] Università degli Studi di Milano-Bicocca, Milan, Italy
emiliana.murgia@unimib.it
[5] Boise State University, Boise, USA
solepera@boisestate.edu

Abstract. Exposure to technology impacts children's perception and conceptualisation of the way devices they regularly use work. This prompts us to study if almost two years of online teaching, enabled by a broad range of technologies, have influenced the way children imagine a search companion would look and behave when helping them perform school-related search tasks. We conducted a 2-stage study during which children ages 9 to 11 drew and described their imaginary search companion; they also chose a few desirable and non-necessary traits. By following the protocol of a study conducted pre-pandemic, we contextualise salient altered expectations that we attribute to exposure to technology prompted by the COVID-19 pandemic. We highlight and discuss emerging trends observed from the analysis of data gathered before and after the extensive online experience and how these will guide the design of functionality of a search companion for the classroom.

Keywords: children · classroom · search companion · COVID-19

1 Introduction

The design of technology to support children's education in and out of the classroom interests researchers and industry practitioners [4,8,19,36,37], as its intentional use can leave a lasting impact on students and teachers alike [16]. It is then imperative to carefully consider the complexities involved in designing and deploying technology for the classroom context, regardless of the instruction modality (i.e., in-person or remote) [9,26,40,41].

A. Barrón-Cedeño et al. (Eds.): CLEF 2022, LNCS 13390, pp. 62–78, 2022.
https://doi.org/10.1007/978-3-031-13643-6_5

A learning-related aspect sustained by technology is information gathering [16]. Children use mainstream search engines for locating resources that can formally (within structured assignments) or informally support knowledge acquisition [2,52]. Given the ubiquitous presence of voice assistants (VA), like Siri or Alexa, and the fact that even before they can read or write children can already interact with VA [29], it is not surprising for them to also turn to VA for inquiries concerning formal and informal learning settings [43,49]. VA, however, have not been designed with children in mind, instigating research to understand how children interact with VA, their perceptions and expectations, and the limitations faced [13,31,64]. Literature in this area aims to advance knowledge on child-VA interactions in the broad sense [64]. We instead seek to expand on foundational works focused on understanding how VA can aid the search process [24,64] to explicitly consider the classroom context.

We argue for the benefits of designing a *Search Companion for Children in the Classroom* (SCCC) to support learning [15,34], anchored on the search as a learning paradigm [11] and principles related to spoken conversational systems that help users navigate the information space, keep track of context, and seek a natural flow of conversation [50]. This SCCC could facilitate children's quests for curriculum-related information and offer necessary scaffolding while affording them the benefits of voice-based interactions they have grown accustomed to and minimising the barriers faced when using search engines (e.g., SERP navigation) or VA (e.g., query formulation via speech interfaces) [3,14,38,48,65]. Taking such a SCCC from theory to practice requires that we first understand what children expect from technology, how and why technology is used in a classroom setting, and which factors influence acceptance and success [32].

The COVID-19 pandemic has caused a shift from in-person to blended or virtual learning environments. This has translated into the integration of technologies into online lessons (e.g., search engines) as well as the adoption of technologies to support instruction delivery (e.g., Zoom) [17,18,47]. Screen time and interaction with VA among children have naturally increased over the past two years [57]. We wonder if the long-term exposure to and the broader adoption of technologies that directly or indirectly enable online teaching and learning has impacted what children expect from technology for the classroom–in particular from a SCCC that facilitates completion of online inquiries for learning purposes. This prompted us to replicate the study we ran [25] before the pandemic to compare trends and assess the potential impact extensive online learning has had on children and their perspective on technology in the classroom. In doing so, we explore changes in children's attitude towards technology in the classroom and their effects on the design of tools to scaffold their learning, specifically search as learning, both in terms of process and outcome. To control scope, we use a framework that establishes four pillars for the design and evaluation of information retrieval systems for children: (i) *strategy*, (ii) *user group*, (iii) *task*, and (iv) *context* [24]. Here, (i) personifying and empowering a SCCC, (ii) children[1] in

---

[1] From here on, whenever we say children we mean children aged 9 to 11.

primary five (ages 9 to 11), (iii) online inquiries about topics common among primary five curricula, and (iv) classroom setting.

Two questions drive our 2-stage exploration: **RQ1**: *What do children expect from a SCCC?* and **RQ2**: *Does prolonged exposure to online instruction impact children's expectations for a SCCC?* In Stage 1, we elicit children's needs and expectations for voice-driven technology that can ease information discovery in the classroom. In Stage 2, we examine data we collected pre-pandemic [25]. We then compare findings as data collection using the same protocol, where the only difference is in (ii). Neither user group of children in primary five was tech-savvy, i.e., children did not receive formal technology-related or search literacy instruction. Yet, children in Stage 1 have frequently used technology over the past two years given pandemic mandates.

The pandemic marks a turning point in attitude towards technology and its adoption. Reported outcomes are not meant as a rigid picture of the status quo but as insights into trends to help researchers and industry practitioners–in areas like Information Retrieval, Natural Language Understanding, Human-Computer Interaction, Spoken Dialogue, and Artificial Intelligence–better interpret the evolution of children's requirements for SCCCs. Findings also call for shifting a classic paradigm: Start by outlining requirements to design explicitly for a particular user group and context, as opposed to designing for average populations and then adapting to serve users with differing needs. In this way, we could better explore and define the dimensions impacting algorithmic and interface design.

2   Background and Related Work

Preference Elicitation. Drawing is a widely-used technique for eliciting feedback as it provides even young children with a convenient way to express themselves freely. The downside is in interpreting the artefacts produced by defining codes and procedures for assigning them. Besides the groundwork described in [28], we refer to [56] which describes how to involve primary school children in designing specific functions for a pedagogical agent. Combining an initial phase of free drawing with a follow-up "scaffolded ideation phase", made guideline extraction easier as children focused and elaborated on visual representations of good collaboration. In our study, we explore the influence that long-term usage of (educational) technologies to support learning can have on children's expectations of the look and feel of a SCCC. Obaid et al. [42] examined free drawings made by children and interaction designers. They contrasted the designs of educational robots while assessing the influence of knowledge of robotics on children's designs. We concentrate on children's designs but do not compare them with adults'. We engage children in structured activities to get a better understanding of their preferences and the reasons behind these preferences.

User Experience. Literature describing factors that foster successful interactions between assistants/agents and children are inconclusive. There is a consensus, however, on children preferring personification [64] much more than adults [28,45]. Hence, we explore children's preferences for personification for their SCCC. Because of reports comparing children and adult drawings, we use technical knowledge as another dimension that can influence users' perceptions [28]. We were inspired by findings of the FMBT design model [10], which stresses how the conversational Functions linked to content and the interaction process should be kept separated, the importance of different Modalities to support communication, the emphasis on Behaviour beyond functions and how Timing is an essential element of conversation. Luger and Sellen [32] linked users' perception of intelligence to the look and feel of the VA, how the system represented information and the quality of the interaction with users. Most unsatisfactory interactions were associated with a gap between expectation and reality; wider among users with lower expertise in technology. Examining (non-)human properties of VA to understand the expectations of children aged 3 to 6 and the impact on future development Xu and Warschauer [61] note that children are reticent to describe VA as living beings or artefacts; often attributing animate properties. Yang et al. [63] highlight the relevance of pragmatic (e.g., response content and interaction) and hedonic (e.g., comfort, pride, and fun) qualities. We adopt this categorisation for our analysis set on young users, not adults.

The context in which VA are used also influences users' perceptions. Matsui and Yamada [35] investigate user perception of humanoid VA and how "one's emotion infects the partner" in emotional contagion. They suggest that experience and familiarity are key to users; culture also plays a crucial role. Thus, for the school context, we turn to pedagogical agents, which are "interactive systems that teach by talking to students" [51]. As teachers, pedagogical agents help students learn a new topic or skill, as companions they provide emotional support, and as students they serve as a peers others can learn from. Similar findings are in [22,23] where primary school children created a teachable 3D Tutor, following activities like drawing sessions, devised in agreement with teachers. This resulted in a 3D tutor with the appearance and personality of a friendly Alien.

Learning. VA are seldom explored in a classroom setting. Lee et al. [28] study adults and children's perceptions of VA personas; Druga et al. [13] investigate instead question-asking behaviour when children (ages 5 and 6) turn to VA at home. Lovato et al. [31] aim to understand how children conceptualise the way VA work. Bhatti et al. [5] argue for the design of a VA that could act as a childcare assistant for parents of young children. Oranç and Ruggeri [43] looked at spontaneous interactions of children aged 3 to 10 with Alexa and concluded that the effects of familiarity with VA and technology, in general, had to be further studied in terms of possible influence on learning habits and success. Literature addressing the usefulness of VA supporting learning is often focused on fostering the development of a particular skill, such as science [55,60], literacy [59], reading [62], computer science [51] and history [33]. This evidences the need to allocate efforts to the design, development, assessment, and deployment of a SCCC which could instead support a broad range of inquiry tasks related to the primary school curriculum–something already known to be of interest among children aged 7 to 12 when it comes to VA supporting in-home learning [15]. Perhaps the closest related works are two that seek to understand what children expect for a VA [15] and a search companion [25] supporting learning. Still, both report findings are based on children's expectations before COVID-19.

3 Experimental Set-Up

Our study description, including rational procedure and protocol, was approved by the school ethics committee so that it could be administered by teachers as part of normal in-class teaching. Following the **protocol** and questions used to elicit children's responses in [24, 25], study participants completed three tasks.

- *Task 1.* Guided by their teacher, children engaged in a *drawing* activity during which they were asked to sketch their ideal SCCC, i.e., how it would look like. This allowed us to infer children's preferences in appearance.
- *Task 2.* Children wrote a brief *description* of how they expect a SCCC to look and behave. Task 2 enabled us to elicit attributes not captured in drawings, given children's disparate ability to draw and attention to detail.
- *Task 3.* From a pre-defined *trait* list inspired by [35, 63] and introduced in [25] (Table 1), children *identified* those desirable and non-necessary for the SCCC to be supportive of their classroom needs. Inspired by Yang et al. [63], we link traits to pragmatic (i.e., helpful and easy interactions with SCCCs) or hedonics (i.e., related to the fun, pride and comfort experienced when engaging with SCCCs) qualities. Task 3 enabled us to interpret Task 1 drawings and elucidate essential attributes to be considered when translating children's vision from theory to practice.

We rely on two sets of **data** collected using the same protocol at schools in Italy with the same program for instruction, which does not include formal training on (the use of) technology or search tools. We were permitted to use anonymised data (stored in a secure location) for research purposes.

- *PreCOVID-19Data* includes drawings, descriptions, and choices generated by 20 children in primary five; ages 9 to 11, gender uniformly distributed. This data–collected in late 2019–was made available by the authors in [25].
- *COVID-19Data* includes drawings, descriptions, and choices from 19 children (10 boys, 9 girls) in primary five; ages 9 to 11 (disjoint from PreCOVID-19Data) who engaged in a data-gathering session in January of 2022.

To examine data, we expand on the methodology proposed in [25]:

- *Code Analysis.* Inspired by [28], researchers and an education expert coded Task 1 drawings as *Animal, Device, Human, Robot,* and *Other.* Rare disagreements were solved using Task 2 descriptions.

Table 1. Task 3 traits. **H** indicate hedonic traits and **P** pragmatic ones.

Traits	
T1 (H). Behave a research fellow / peer	**T8 (P)**. Remember your previous requests
T2 (P). Be a research expert / librarian	**T9 (P)**. Keep the conversation fluidly without interruptions
T3 (H). Be like a human being	**T10 (P)**. Use external services when needed (e.g. open a video)
T4 (H). Take care of your privacy	**T11 (P)**. Respond promptly
T5 (H). Worry about not distracting you	**T12 (H)**. Make you feel safe
T6 (H). Learn your tastes and needs	**T13 (H)**. Make you feel proud because you use new technologies
T7 (P). Anticipate your requests for information	**T14 (H)**. Make you have fun

- *Description Analysis.* We examined term occurrence (stopword removal, lemmatization, and tokenization using Python's NLTK [7]) in Task 2 descriptions seeking common terminology used to describe a SCCC. We also considered description length as a proxy to assess engagement.
- *Trait Analysis.* We examined the frequency with which Task 3 traits were selected as desirable or non-necessary. We investigated how hedonic and pragmatic traits can inform different layers of design–the emotional, interactive, and internal architecture. Given children's preferences for personifications [64], we investigated connections between appearance and trait selection.

4   Results and Discussion

4.1   Stage 1: Children's Expectations

Appearance (Task 1) The 19 drawings in COVID-19Data were distributed as 4 Animals, 2 Humans, 3 Devices, 8 Robots, and 2 Others (Fig. 1). Among children's depictions, we saw a prevalence of robots and devices, much more so than human-like or animal-like personifications. Moving beyond the general appearance, an in-depth perusal of the intricacies captured in children's characterisations of a SCCC revealed that regardless of the code assigned, most drawings ($\sim$75%) exhibited 'technology-like' components. Animal-/human-like portrayals evinced a technology-like demeanour. This is evident, for example, in Figs. 1a and 1g, which depict a bear and a flower. Both with details that signal how the sketches depart from classic representations: 'I am Siri' linking to the well-known VA and an embedded audio device turning the flower into a gadget, respectively. Similarly, the human-like body in Fig. 1c mentions Alexa, another popular VA.

Descriptions (Task 2) Using brief descriptions in COVID-19Data, we built a word cloud emphasising prominent terms used to describe SCCCs (Fig. 2b). Analysing the word cloud and the descriptions, we infer that children envision a SCCC as always available (e.g., 'wifi' in Fig. 2b) and supportive of not just classroom-related concerns, but 'everything'. Clues in the descriptions (like 'voice', 'ask', and 'search') suggest that children see a SCCC as an extension of well-known VA or tools (see 'Alexa,' Siri,' and 'Google' on Fig. 2b). These are tools children already turn to for educational inquiries and beyond [6,30].

Traits (Task 3) One of the most thought-after traits was for the SCCC to 'learn your tastes and needs' (Fig. 5). This points to children expecting a SCCC to offer a personalised experience. Other salient traits included the ability for a SCCC to be mindful of children's privacy and one that behaves as a research fellow/peer. Regarding traits that could be overlooked, we saw less of a consensus. Except for a SCCC not needing to neither behave like a research expert nor look like a human being, the rest of the traits were often selected by at most a single child (Fig. 6d). The top desired and top non-necessary traits are a mix of hedonic and pragmatic choices. The desired ones, however, align with the behaviour of the SCCC, whereas non-necessary ones refer to appearance. We noted from Fig. 6b that except for pragmatic trait T9 (fluent conversation) which was often selected along with hedonic traits, the strongest co-occurrences were among T1, T4, and T6. From

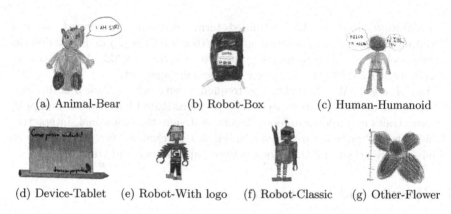

(a) Animal-Bear   (b) Robot-Box   (c) Human-Humanoid

(d) Device-Tablet   (e) Robot-With logo   (f) Robot-Classic   (g) Other-Flower

Fig. 1. Sample SCCCs included in COVID-19Data.

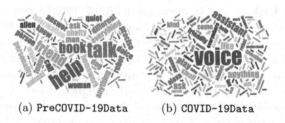

(a) PreCOVID-19Data   (b) COVID-19Data

Fig. 2. Word-cloud based on sentences collected in response to Task 2.

Fig. 7b we see children who designed personified SCCCs tended to choose a wider range of hedonic and pragmatic traits to attach to them than counterparts who went for non-personified depictions; similarly when considering non-necessary hedonic traits (Fig. 7d).

4.2 Stage 2: Unintended Consequences of the Pandemic

We gauge unintended pandemic consequences, as children worldwide were regularly exposed to and use technologies supporting remote teaching.

Appearance (Task 1). It emerges from Fig. 4a the increase in children envisioning SCCCs as robots. Drawings in PreCOVID-19Data were more evenly distributed across Animal, Human, Device, and Robot. Regardless of the category assigned to them, the illustrations in PreCOVID-19Data tended to be more friendly-looking and cute, with less of a 'technology' bias. This is visible in the sample illustrations in Fig. 3, all with friendly, happy faces, most with cute adornments (the bow in Fig. 3b); a sharp contrast with the sample sketches in Fig. 1, produced by children after prolonged exposure to technologies.

Descriptions (Task 2). Descriptions in COVID-19Data were much lengthier than those in PreCOVID-19Data–close to twice as much as shown in Fig. 4b. From the word clouds in Fig. 2, we see that descriptions in COVID-19Data included

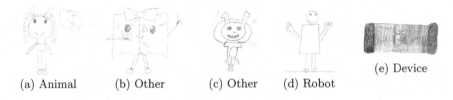

(a) Animal   (b) Other   (c) Other   (d) Robot

(e) Device

Fig. 3. Sample search companions in `PreCOVID-19Data`.

words commonly-attributed to tech-savvy individuals, e.g., 'voice', 'wifi', and 'understand'. Instead, descriptions in `PreCOVID-19Data` emphasised terms like 'help', 'friend', 'chatty', and 'funny'.

Traits (Task 3) There is evidence about expected traits a SCCC should exhibit. We begin by probing desired and non-necessary trait selections captured in Fig. 5. Based on `PreCOVID-19Data`, children gravitated towards traits that would result in a SCCC being fun (T14) but also able to prevent distractions when completing classroom assignments (T5) and be mindful of their privacy (T4). Instead, from `COVID-19Data` we noted that two years onwards, children voiced their longing for a SCCC capable of learning their tastes and needs (T6), behaving like a peer (T1), and enforcing privacy considerations (T4). Regardless of their (formal) exposure to technology, children are aware of privacy concerns that technologies they use must safeguard. Moreover, before COVID-19, children seemed to prioritise interaction characteristics (fun vs. distraction-trade-off), whereas children in `COVID-19Data` focused more on functional aspects of a SCCC. Overall, from the trait selection distribution showcased in Fig. 5, it is evident that there was a preference increase for T1, T4, T6, and T9 surfacing from `COVID-19Data` with respect to `PreCOVID-19Data`. There was less disagreement on requirements that could be overlooked. Except for T2 and T3, there is no majority agreement for non-necessary traits before or after COVID-19.

When examining the type of qualities expected in a SCCC, a mix of hedonic and pragmatic made it to the top based on `COVID-19Data`. From `PreCOVID-19Data`, it emerged that children desired traits were more prone to

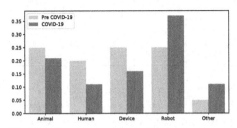

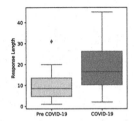

(a) Distributions of coded drawings, normalised by total count per data source.

(b) Length of Task 2 descriptions.

Fig. 4. Drawings and description analysis using Pre- and `COVID-19Data`.

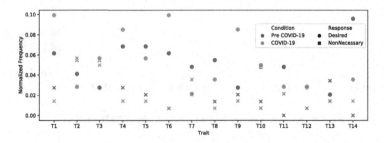

Fig. 5. Frequency distribution of traits (Table 1) that children find either desirable or non-necessary. Distribution normalised by total count per category.

be from a hedonic variety. The differing selection patterns are perhaps the most visible in Fig. 6, particularly for hedonic traits. Co-occurrences inferred from `PreCOVID-19Data` are visibly less strong than those inferred from `COVID-19Data`. This prompted us to investigate if chosen hedonic and pragmatic traits were conditioned by the appearance of an envisioned SCCC, both before and during the pandemic. We grouped children's drawings (excluding Others) into two groups: Personified (Human or Animal) and Non-personified (Robot or Device). As illustrated in Fig. 7, we saw variations in the required and non-necessary traits for a SCCC across trait type and appearance type. This is salient in hedonic traits, somewhat similar preferences during and before the pandemic on non-personified depictions, not so for personified depictions (Fig. 7a). The distribution of desired traits is consistent among children who did not personify their SCCC. In other words, choices inferred from `COVID-19Data` seem to converge on a smaller set of pragmatic requirements than those children who opted to personalise their SCCC. Children portraying a 'technology-like' SCCC were more likely to select a specific set of desired pragmatic traits, but like the children choosing to represent a personified SCCC, they too would go for a mixture of hedonic traits, possibly to suit their personality and make the search as a learning experience more engaging. Among non-necessary traits, pragmatic ones are similarly distributed regardless of children's level of exposure to technologies. We could then assume that, in general, drawings are good indicators of children's expectations and preferences.

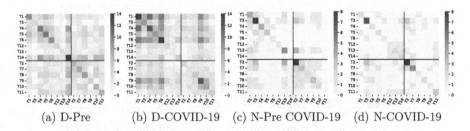

(a) D-Pre   (b) D-COVID-19   (c) N-Pre COVID-19   (d) N-COVID-19

Fig. 6. Heatmap of Desired (D) vs. Non-Necessary (N) trait co-occurrences. We divide the heatmaps into 4 sections based on trait type: Left top shows the co-occurrence of H-H traits, right top H-P, bottom left P-H, and bottom right P-P.

4.3 Discussion

To answer **RQ1**, we refer to Sect. 4.1. From sketches, we surmise children's awareness of technology and, indirectly, how this impacts their expectations for the look and functionality of SCCCs. Children demonstrated an affinity for technology-like facades. They favoured robots and device-style depictions more than human-/animal-like personifications. Children's tie to technology was also visible in their descriptions. They made sure the demeanour they envisioned for a SCCC came across; prominently employing words like 'voice', 'search', and 'assistant'. They also articulated characteristics of a SCCC which could be missed by simply focusing on surface appearance. Human-like appearance/behaviour is not a primary concern for children when describing their SCCCs.

Reoccurring themes refer to expectations for SCCC to be voice-driven and to be everywhere. Given that the study prompt explicitly asked children to *"Imagine using a Vocal Assistant like Alexa or Google to run your school searches, what would it look like?"*, we focus on the repetition in children's descriptions of words like 'Alexa' or 'voice'. We posit this choice reveals the importance for children to go beyond classic search systems and instead rely on conversational search systems. For the latter, we attribute this to children's normalising that inquiries on the classroom context no longer take place only in the physical classroom.

From trait selection, it emerged that children's requirements prioritised functional aspects of a SCCC. Their mixture of hedonic and pragmatic traits pointed to both personalization (modelling and responding to individual's interests and preferences) and privacy being key components to consider when outlining the architecture of a SCCC. This brings up a conundrum when designing and deploying technology for children, as personalization and privacy are often at odds [20]. We notice a shift from children trusting more authoritative figures [24] to expecting a SCCC to behave like a peer, rather than an expert. This could be interpreted as children having more realistic expectations of the SCCC, given their increased exposure to technology; also an expression of their need to find ways to share their search experience as learning online proved for most lonely.

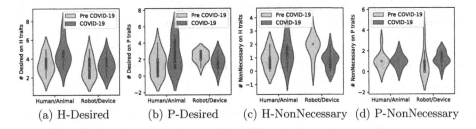

(a) H-Desired   (b) P-Desired   (c) H-NonNecessary   (d) P-NonNecessary

Fig. 7. Distribution of Desired (a & b) and Non-Necessary (c & d) selections on H (a & c) and P (b & d) traits, given by children who designed personified (Human/Animal) vs. non-personified (Robot/Device) SCCCs.

To answer **RQ2**, we revisit Sect. 4.2. It became apparent that children's expectations for a SCCC have altered across all tasks. We attribute this to the mere exposure effect [66] to what is now familiar–technology to support the classroom context. Children now seem able to identify specific functional requirements and express them. They are more likely to overlook the appearance of a SCCC in lieu of its ability to remember past actions, converse fluidly, and guard users' privacy. These findings are grounded on the type of sketches produced prior to COVID-19 and after two years of children experiencing the pandemic. Drawings exhibit a technology-like demeanour that before was not there. Findings are also informed by the fact that anthropomorphization (i.e., depicting SCCC as cute, friendly, and often human-/animal-like), no longer surfaces as a common expectation among children. Also contributing to these findings is the fact that children participating in Stage 1 appeared to be more engaged with the task, offering longer and more detailed descriptions than their counterparts before COVID-19. In the end, we posit that long-term exposure to technology to support remote instruction resulted in children acquiring a more technology-related vocabulary which they now can and want to use to articulate their requirements for a SCCC. The impact of long-term technology use is the easiest to spot when probing traits expected to be exhibited by a SCCC (Figs. 5, 6, and 7). Indeed, personalization and privacy remain desired traits even after two years of the COVID-19 pandemic. And yet, from juxtaposing selections, it is evident that children now prioritise functional aspects of a SCCC more than before the pandemic, at which time children seemed more focused on behavioural characteristics. Interestingly, the pandemic caused less of an impact on non-necessary requirements.

Our findings related to the differences in how children envision a SCCC echo those reported for the home and classroom context in [25, 28]. There, the authors emphasise that different personas are needed for different users. They attribute these users' background knowledge. Results on comparisons based on desired vs. non-necessary traits align with those we previously reported in [25]; in this study children known to be technically savvy tended to regard higher traits referring to the functionality of an ideal SCCC such as its ability to remember prior users' requests (T8). Our findings corroborate that background and informal exposure to technology often directly impact users' expectations.

5 Conclusions, Limitations, and Future Work

With our 2-stage study, we identified the initial requirements of a SCCC–a spoken conversational system meant to ease inquiry tasks related to the primary school curriculum. Outcomes disclose children's preference for a voice-driven companion, with 'technology-like' miens that can support classroom-related search needs regardless of the teaching modality. We glean from our preliminary exploration that informal guidance on the use of technology and persistent use of technology to support learning impact how children think about SCCCs. In view of these findings, which illustrate how rapidly children's preferences and attitudes towards technologies supporting inquiry tasks change, and indirectly influence their ever-evolving search behaviour, we discourage the research community from building on old assumptions and instead revisit explorations to capture current needs and open challenges. The roles children play in collaborative design must be revisited, as much as the power balance in inter-generational research teams. Lastly, the growth of familiarity with technology opens a window of opportunity for educational experts to use it in teaching more extensively and seamlessly.

One of the limitations of our work is the small *sample population*. Yet, this is a common sample size when conducting studies involving children [12,48]. Reported results reflect the preferences and expectations of *children aged 9 to 11* who are part of a specific school system. To best understand how online instruction may have inadvertently impacted children's view of technology, it will be necessary to repeat the exploration using the same protocol, but extending the age ranges of participants and including different school systems and countries. Given the focus on the impact of the COVID-19 pandemic, We did not analyse participants' gender, cultural background, socio-economic status, or stereotype biases. In the future, we aim to expand our study to consider how the aforementioned dimensions influence children's expectations [46,53]. Other immediate research direction includes extending our study to consider the role of emotions, given their impact on technology supporting learning [15]. Inspired by existing works discussing children's interactions with pedagogical agents [44,58], we also plan to account for traits that best align with the pedagogical aspects necessary to support the search as a learning paradigm as the foundation of a SCCC, such as their ability to provide the help needed to solve a task, as well as encouraging children to study and displaying emotional intelligence. Vtyurina et al. [54] studied how an intelligent VA could support complex search tasks for adults; they reported on the need for intelligent VA to recognise different types of tasks and user preferences and provide appropriate support accordingly. In line with our discoveries, it would be opportune to investigate whether children's requirements for a SCCC also differ when handling search tasks of varying levels of complexity.

Our work has implications for researchers and practitioners in broad areas. Children take for granted that a SCCC will be voice-driven; attributed to their early interactions with VA [29]. Yet, voice-driven technologies struggle to understand children [39]. Further, research on conversational user interfaces and their applicability to children and the classroom context is still preliminary [1]. We have focused on children, but it is important to note that the design and deployment of technologies to support the classroom must simultaneously account for the perspectives of multiple stakeholders (teachers, parents, industries, and the children themselves) if they are to be of practical use, and this is a nontrivial endeavour [27]. Educational and pedagogical implications related to the use of SCCC are not the only ones to consider; privacy and security are integral for safe deployment and use of technology among the sensitive population and context that are the focus of this work [21]. Given children's requirement for privacy

(Stage 1) it is pertinent to consider embedding within a SCCC opportunities for teaching about the safe use of digital technologies.

References

1. Allen, G.M., Gadiraju, U., Yang, J., Pera, M.S.: Using conversational artificial intelligence to support children's search in the classroom. In: CUI@ CSCW: Inclusive and Collaborative Child-Facing Voice Technologies: A Workshop at the Virtual ACM CSCW 2021 Conference (2021)
2. Anuyah, O., Milton, A., Green, M., Pera, M.S.: An empirical analysis of search engines' response to web search queries associated with the classroom setting. Aslib J. Inf. Manag. (2020)
3. Azpiazu, I.M., Dragovic, N., Pera, M.S., Fails, J.A.: Online searching and learning: YUM and other search tools for children and teachers. Inf. Retr. J. **20**(5), 524–545 (2017)
4. Barros, G., et al.: Learning interactions: robotics supporting the classroom. In: Stephanidis, C., Antona, M., Ntoa, S. (eds.) HCII 2021. CCIS, vol. 1421, pp. 3–10. Springer, Cham (2021). https://doi.org/10.1007/978-3-030-78645-8_1
5. Bhatti, N., Stelter, T.L., Scott McCrickard, D.: The interactive show: a conversational companion for young children and childcare assistant for parents. In: Proceedings of the 8th International Conference on Human-Agent Interaction, pp. 221–223 (2020)
6. Bilal, D., Gwizdka, J.: Children's query types and reformulations in google search. Inf. Process. Manag. **54**(6), 1022–1041 (2018)
7. Bird, S., Klein, E., Loper, E.: Natural Language Processing with Python: Analyzing Text with the Natural Language Toolkit. O'Reilly Media Inc., Sebastopol (2009)
8. Blikstad-Balas, M., Klette, K.: Still a long way to go: narrow and transmissive use of technology in the classroom. Nordic J. Digit. Lit. **15**(1), 55–68 (2020)
9. Burnett, C.: The Digital Age and Its Implications for Learning and Teaching in the Primary School. Cambridge Primary Review Trust York, Heslington (2016)
10. Cassell, J.: Embodied conversational agents: representation and intelligence in user interfaces. AI Mag. **22**(4), 67 (2001)
11. Collins-Thompson, K., Hansen, P., Hauff, C.: Search as learning (Dagstuhl seminar 17092). Dagstuhl Reports, vol. 7. Schloss Dagstuhl-Leibniz-Zentrum fuer Informatik (2017)
12. Downs, B., French, T., Wright, K.L., Pera, M.S., Kennington, C., Fails, J.A.: Searching for spellcheckers: what kids want, what kids need. In: Proceedings of the 18th ACM International Conference on Interaction Design and Children, pp. 568–573 (2019)
13. Druga, S., Williams, R., Breazeal, C., Resnick, M.: Hey google is it ok if I eat you?: Initial explorations in child-agent interaction. In: Proceedings of the 2017 Conference on Interaction Design and Children, pp. 595–600. ACM (2017)
14. Festerling, J., Siraj, I.: Alexa, what are you? Exploring primary school children's ontological perceptions of digital voice assistants in open interactions. Hum. Dev. **64**(1), 26–43 (2020)
15. Garg, R., Sengupta, S.: Conversational technologies for in-home learning: using co-design to understand children's and parents' perspectives. In: Proceedings of the 2020 CHI Conference on Human Factors in Computing Systems, pp. 1–13 (2020)

16. Hales, P.D., et al.: Alexa?: Possibilities of voice assistant technology and artificial intelligence in the classroom. Empower. Res. Educ. **3**(1), 4 (2019)

17. Hodges, C.B., Moore, S., Lockee, B.B., Trust, T., Aaron Bond, M.: The difference between emergency remote teaching and online learning (2020)

18. Hu, X., Chiu, M.M., Leung, W.M.V., Yelland, N.: Technology integration for young children during COVID-19: towards future online teaching. Br. J. Educ. Technol. **52**(4), 1513–1537 (2021)

19. Hyndman, B.: Ten reasons why teachers can struggle to use technology in the classroom. Sci. Educ. News **67**(4), 41–42 (2018)

20. Kucirkova, N., Toda, Y., Flewitt, R.: Young children's use of personalized technologies: insights from teachers and digital software designers in Japan. Technol. Knowl. Learn. **26**(3), 535–554 (2021)

21. Kumar, P.C., Chetty, M., Clegg, T.L., Vitak, J.: Privacy and security considerations for digital technology use in elementary schools. In: Proceedings of the 2019 CHI Conference on Human Factors in Computing Systems, pp. 1–13 (2019)

22. Landoni, M., Murgia, E., Gramuglio, F., Manfredi, G.: Fiction design of a 3d tutor for and with school children. In: Proceedings of the Eleventh International Conference on Advances in Computer-Human Interactions, ACHI, pp. 94–97 (2018)

23. Landoni, M., Murgia, E., Gramuglio, F., Manfredi, G.: Teaching an alien: children recommending what and how to learn. In: 2nd KidRec Workshop Colocated with ACM IDC (2018)

24. Landoni, M., Matteri, D., Murgia, E., Huibers, T., Pera, M.S.: Sonny, Cerca! Evaluating the impact of using a vocal assistant to search at school. In: Crestani, F., et al. (eds.) CLEF 2019. LNCS, vol. 11696, pp. 101–113. Springer, Cham (2019). https://doi.org/10.1007/978-3-030-28577-7_6

25. Landoni, M., Murgia, E., Huibers, T., Pera, M.S.: You've got a friend in me: children and search agents. In: Adjunct Publication of the 28th ACM Conference on User Modeling, Adaptation and Personalization, pp. 89–94 (2020)

26. Landoni, M., Huibers, T., Murgia, E., Pera, M.S.: Ethical implications for children's use of search tools in an educational setting. Int. J. Child Comput. Interact. **32**, 100386 (2021)

27. Landoni, M., Huibers, T., Pera, M.S., Fails, J.A.: 5th KidRec workshop: search and recommendation technology through the lens of a teacher. In: Interaction Design and Children, pp. 658–661 (2021)

28. Lee, S., Kim, S., Lee, S.: "What does your agent look like?" A drawing study to understand users' perceived persona of conversational agent. In: Extended Abstracts of the 2019 CHI Conference on Human Factors in Computing Systems, pp. 1–6 (2019)

29. Lovato, S., Piper, A.M.: "Siri, is this you?" Understanding young children's interactions with voice input systems. In: Proceedings of the 14th International Conference on Interaction Design and Children, pp. 335–338 (2015)

30. Lovato, S.B., Piper, A.M.: Young children and voice search: what we know from human-computer interaction research. Front. Psychol. **10**, 8 (2019)

31. Lovato, S.B., Piper, A.M., Wartella, E.A.: Hey google, do unicorns exist? Conversational agents as a path to answers to children's questions. In: Proceedings of the 18th ACM International Conference on Interaction Design and Children, pp. 301–313 (2019)

32. Luger, E., Sellen, A.: "Like having a really bad pa" the gulf between user expectation and experience of conversational agents. In: Proceedings of the 2016 CHI Conference on Human Factors in Computing Systems, pp. 5286–5297 (2016)

33. Mack, N.A., Moon Rembert, D.G., Cummings, R., Gilbert, J.E.: Co-designing an intelligent conversational history tutor with children. In: Proceedings of the 18th ACM International Conference on Interaction Design and Children, pp. 482–487 (2019)
34. Martha, A.S.D., Santoso, H.B.: The design and impact of the pedagogical agent: a systematic literature review. J. Educ. Online **16**(1), n1 (2019)
35. Matsui, T., Yamada, S.: Two-dimensional mind perception model of humanoid virtual agent. In: Proceedings of the 5th International Conference on Human Agent Interaction, pp. 311–316. ACM (2017)
36. Mercer, N., Hennessy, S., Warwick, P.: Dialogue, thinking together and digital technology in the classroom: some educational implications of a continuing line of inquiry. Int. J. Educ. Res. **97**, 187–199 (2019)
37. Meyer, M., et al.: How educational are "educational" apps for young children? App store content analysis using the four pillars of learning framework. J. Child. Media **15**(4), 526–548 (2021)
38. Milton, A., Anuya, O., Spear, L., Wright, K.L., Pera, M.S.: A ranking strategy to promote resources supporting the classroom environment. In: 2020 IEEE/WIC/ACM International Joint Conference on Web Intelligence and Intelligent Agent Technology (WI-IAT), pp. 121–128. IEEE (2020)
39. Monarca, I., Cibrian, F.L, Mendoza, A., Hayes, G., Tentori, M.: Why doesn't the conversational agent understand me? A language analysis of children speech. In: Adjunct Proceedings of the 2020 ACM International Joint Conference on Pervasive and Ubiquitous Computing and Proceedings of the 2020 ACM International Symposium on Wearable Computers, pp. 90–93 (2020)
40. Murgia, E., Landoni, M., Huibers, T.W.C., Fails, J.A., Pera, M.S.: The seven layers of complexity of recommender systems for children in educational contexts. In: Workshop on Recommendation in Complex Scenarios 2019, pp. 5–9 (2019)
41. Murgia, E., Landoni, M., Huibers, T., Pera, M.S.: All together now: teachers as research partners in the design of search technology for the classroom. In: KidRec 2021: 5th International and Interdisciplinary Perspectives on Children & Recommender and Information Retrieval Systems (KidRec) Search and Recommendation Technology through the Lens of a Teacher- Co-located with ACM IDC 2021, 26 June 2021, Online Event. arXiv preprint arXiv:2105.03708 (2021)
42. Obaid, M., Barendregt, W., Alves-Oliveira, P., Paiva, A., Fjeld, M.: Designing robotic teaching assistants: interaction design students' and children's views. In: ICSR 2015. LNCS (LNAI), vol. 9388, pp. 502–511. Springer, Cham (2015). https://doi.org/10.1007/978-3-319-25554-5_50
43. Oranç, C., Ruggeri, A.: "Alexa, let me ask you something different" children's adaptive information search with voice assistants. Hum. Behav. Emerging Technol. **3**(4), 595–605 (2021)
44. Pérez-Marín, D., Pascual-Nieto, I.: An exploratory study on how children interact with pedagogic conversational agents. Behav. Inf. Technol. **32**(9), 955–964 (2013)
45. Purington, A., Taft, J., Sannon, S., Bazarova, N., Taylor, S.H.: Alexa is my new BFF: social roles, user satisfaction, and personification of the amazon echo. In: Proceedings of the CHI Conference Extended Abstracts on Human Factors in Computing Systems, pp. 2853–2859. ACM (2017)
46. Rankin, Y.A., Henderson, K.K.: Resisting racism in tech design: centering the experiences of black youth. In: Proceedings of the ACM on Human-Computer Interaction, 5(CSCW1), pp. 1–32 (2021)

47. Sáiz-Manzanares, M.C., Marticorena-Sánchez, R., Ochoa-Orihuel, J.: Effectiveness of using voice assistants in learning: a study at the time of COVID-19. Int. J. Env. Res. Public Health **17**(15), 5618 (2020)

48. Szczuka, J.M., Strathmann, C., Szymczyk, N., Mavrina, L., Krämer, N.C.: How do children acquire knowledge about voice assistants? A longitudinal field study on children's knowledge about how voice assistants store and process data. Int. J. Child Comput. Interact. **33**, 100460 (2022)

49. Terzopoulos, G., Satratzemi, M.: Voice assistants and smart speakers in everyday life and in education. Inform. Educ. **19**(3), 473–490 (2020)

50. Trippa, J.R.: Spoken conversational search: audio-only interactive information retrieval. Ph.D. thesis, RMIT University (2019)

51. Urrutia, E.K.M., Ocaña, J.M., Pérez-Marín, D., Tamayo, S.: A first proposal of pedagogic conversational agents to develop computational thinking in children. In: Proceedings of the 5th International Conference on Technological Ecosystems for Enhancing Multiculturality, pp. 1–6 (2017)

52. Usta, A., Altingovde, I.S., Vidinli, I.B., Ozcan, R., Ulusoy, Ö.: How k-12 students search for learning? Analysis of an educational search engine log. In: Proceedings of the 37th International ACM SIGIR Conference on Research & Development in Information Retrieval, pp. 1151–1154 (2014)

53. Vanderlyn, L., Weber, G., Neumann, M., Väth, D., Meyer, S., Vu, N.T.: "it seemed like an annoying woman": on the perception and ethical considerations of affective language in text-based conversational agents. In: Proceedings of the 25th Conference on Computational Natural Language Learning, pp. 44–57 (2021)

54. Vtyurina, A., Savenkov, D., Agichtein, E., Clarke, C.L.A.: Exploring conversational search with humans, assistants, and wizards. In: Proceedings of the CHI Conference Extended Abstracts on Human Factors in Computing Systems, pp. 2187–2193. ACM (2017)

55. Ward, W., et al.: My science tutor: a conversational multimedia virtual tutor for elementary school science. ACM Trans. Speech Lang. Process. (TSLP) **7**(4), 1–29 (2011)

56. Wiggins, J.B., et al.: From doodles to designs: Participatory pedagogical agent design with elementary students. In: Proceedings of the 18th ACM International Conference on Interaction Design and Children, pp. 642–647. ACM (2019)

57. Wojcik, E.H., Prasad, A., Hutchinson, S.P., Shen, K.: Children prefer to learn from smart devices, but do not trust them more than humans. Int. J. Child Comput. Interact. **32**, 100406 (2021)

58. Woodward, J., McFadden, Z., Shiver, N., Ben-hayon, A., Yip, J.C., Anthony, L.: Using co-design to examine how children conceptualize intelligent interfaces. In: Proceedings of the CHI Conference on Human Factors in Computing Systems, pp. 575. ACM (2018)

59. Xu, Y., Warschauer, M.: A content analysis of voice-based apps on the market for early literacy development. In: Proceedings of the Interaction Design and Children Conference, pp. 361–371 (2020)

60. Xu, Y., Warschauer, M.: "Elinor is talking to me on the screen!" Integrating conversational agents into children's television programming. In: Extended Abstracts of the 2020 CHI Conference on Human Factors in Computing Systems, pp. 1–8 (2020)

61. Xu, Y., Warschauer, M.: What are you talking to?: Understanding children's perceptions of conversational agents. In: Proceedings of the 2020 CHI Conference on Human Factors in Computing Systems, pp. 1–13 (2020)

62. Xu, Y., Aubele, J., Vigil, V., Bustamante, A.S., Kim, Y.-S., Warschauer, M.: Dialogue with a conversational agent promotes children's story comprehension via enhancing engagement. Child Dev. **93**, e149–e167 (2021)

63. Yang, X., Aurisicchio, M., Baxter, W.: Understanding affective experiences with conversational agents. In: Proceedings of the CHI Conference on Human Factors in Computing Systems, p. 542. ACM (2019)

64. Yarosh, S., et al.: Children asking questions: speech interface reformulations and personification preferences. In: Proceedings of the 17th ACM Conference on Interaction Design and Children, pp. 300–312 (2018)

65. Yuan, Y., et al.: Speech interface reformulations and voice assistant personification preferences of children and parents. Int. J. Child Comput. Interact. **21**, 77–88 (2019)

66. Zajonc, R.B.: Attitudinal effects of mere exposure. J. Pers. Soc. Psychol. **9**(2p2), 1 (1968)

Rhythmic and Psycholinguistic Features for Authorship Tasks in the Spanish Parliament: Evaluation and Analysis

Silvia Corbara[1(✉)], Berta Chulvi[2,3], Paolo Rosso[2], and Alejandro Moreo[4]

[1] Scuola Normale Superiore, Pisa, Italy
`silvia.corbara@sns.it`
[2] Universitat Politècnica de València, Valencia, Spain
`berta.chulvi@upv.es, prosso@dsic.upv.es`
[3] Universitat de València, Valencia, Spain
[4] Istituto di Scienza e Tecnologie dell'Informazione, CNR, Pisa, Italy
`alejandro.moreo@isti.cnr.it`

Abstract. Among the many tasks of the authorship field, Authorship Identification aims at uncovering the author of a document, while Author Profiling focuses on the analysis of personal characteristics of the author(s), such as gender, age, etc. Methods devised for such tasks typically focus on the *style* of the writing, and are expected not to make inferences grounded on the *topics* that certain authors tend to write about. In this paper, we present a series of experiments evaluating the use of topic-agnostic feature sets for Authorship Identification and Author Profiling tasks in Spanish political language. In particular, we propose to employ features based on rhythmic and psycholinguistic patterns, obtained via different approaches of text masking that we use to actively mask the underlying topic. We feed these feature sets to a SVM learner, and show that they lead to results that are comparable to those obtained by a BETO transformer, when the latter is trained on the original text, i.e., potentially learning from topical information. Moreover, we further investigate the results for the different authors, showing that variations in performance are partially explainable in terms of the authors' political affiliation and communication style.

Keywords: Authorship Analysis · Text masking · Political speech

1 Introduction

In the authorship analysis field, Authorship Identification (AId) investigates the true identity of the author of a written document, and it is of special interest when the author is unknown or debated. Two of the main sub-tasks of AId are Authorship Attribution (AA) and Authorship Verification (AV): in the former, given a document d and a set of candidate authors $\{A_1, \ldots, A_m\}$, the goal is to identify the real author of d among the set of candidates; instead, AV is defined as a binary classification task, in which the goal is to infer whether A (the only

A. Barrón-Cedeño et al. (Eds.): CLEF 2022, LNCS 13390, pp. 79–92, 2022.
https://doi.org/10.1007/978-3-031-13643-6_6

candidate) is the real author of d or not. On the other hand, Author Profiling (AP) aims at distinguish among classes of authors, in order to investigate the authors' individual characteristics, which can span from their gender, to their nationality, to their mental state; these studies are especially important in order to find common traits among groups, or to reveal relevant information about a specific author. While tackling these classification problems, the researchers' goal is to devise methods able to discriminate the different styles of the authors under consideration, often relying on supervised machine learning.

In this article, we evaluate the use of topic-agnostic feature sets for AV, AA and various AP tasks (by gender, by age, and by political affiliation) for Spanish texts. Along with feature sets that are by now consolidated in the authorship field, we introduce rhythm- and psycholinguistics-based feature sets. Concretely, we propose to generate new masked versions of the original text extracting (i) the syllabic stress (i.e., strings of *stressed* and *unstressed* syllables), and (ii) the psycholinguistic categories of the words, as given by the LIWC dictionary (Sect. 3.2). The resulting representations are topic-agnostic strings from which we extract n-grams features. In order to asses the effect of our proposed feature sets on the performance, we carry out experiments of *ablation* (in which we remove one feature set from the model at a time) and experiments of *addition* (in which we add one single feature set to the model at a time). Our results seem to indicate that our topic-agnostic features bring to bear enough authorial information as to perform on-par with BETO, the Spanish equivalent to the popular BERT transformer, fine-tuned on the original (hence topic-aware) text. The code of the project can be found at GitHub.[1]

This work continues the preliminary experiments presented in the short paper by Corbara et al. [8]. With respect to the previous paper, we present a wider set of experiments, both in terms of size and diversity of the dataset, as well as in terms of the number of tasks. In the current paper, we consider a higher number of authors characterized by a finer-grained political spectrum, and we add experiments of AP. Moreover, we devise a new experimental protocol (Sect. 3.3) and and we take an initial step towards interpreting the impact that different feature sets bring to bear in the AV task (Sect. 5).

2   Related Work

The annual PAN[2] event presents various shared tasks both for AId and AP, often with challenging settings including the open-set or the cross-domain problems, and thus offers a very good overview of the most recent trends in this field. For example, in the recent 2021 edition [1], the participants tackled an AId task[3] and an AP task (based on the problem of identifying hate speech spreaders). In this

---

[1] https://github.com/silvia-cor/Topic-agnostic_ParlaMintES.
[2] https://pan.webis.de/.
[3] Precisely, the PAN2021 event presented a particular case of AV where the dataset contained pairs of documents, and the aim was to infer whether the two documents shared the same author; we call this task Same-Authorship Verification (SAV).

occasion is was observed that, although the presence of Neural Network (NN) methods increased with respect to past editions, and even though the best results in the AId and AP challenges were obtained by NN methods, simpler approaches based on (character or word) n-grams and traditional classification algorithms were still competitive; indeed, they were found to outperform NN methods in past editions [18]. In fact, the method by Weerasinghe et al. [27], based on the absolute difference among feature vectors fed to a logistic regression classifier, reached the third (with the large dataset) and fourth (with the small dataset) positions in the overall ranking for the AId task, while in the AP task only 7 out of 66 methods were able to surpass the baseline Support Vector Machine (SVM) fed with character n-grams.

In classical machine learning algorithms, the choice of the features to employ is crucial. In his survey, Stamatatos [24] discusses the features that are most commonly used in the authorship field; however, he also notes that features such as word and character n-grams might prompt methods to base their inferences on topic-related patterns rather than on stylometric patterns. In fact, an authorship classifier (even a seemingly good one) might end up unintentionally performing topic identification if domain-dependent features are used [2]. In order to avoid this, researchers might limit their scope to features that are clearly topic-agnostic, such as function words or syntactic features [15], or might actively mask topical content via a text-masking approach [14,25]. Continuing the first experimentation [8], in this project we focus our attention on features capturing the rhythmic and the psycholinguistic traits of the texts, by employing a text-masking technique based on syllabic stress and LIWC categories.

The idea of employing rhythmic features in the authorship field is not a new one. Their most natural use is in studies focused on poetry [19], although they have also been employed in authorship analysis of prose texts. In the work by Plecháč [23], the role of accent, or stress, is studied for AId problems in English; in the research by Corbara et al. [9], the documents are encoded as sequences of long and short syllables, from which the relevant features are extracted and used for AA in Latin prose texts, with promising results. Since Spanish derives from Latin, we aim to investigate the extent to which similar considerations apply to the Spanish language as well. In this case, we exploit the concept of *stress*, which gained relevance over the concept of *syllabic quantity* in Romance languages.

Linguistic Inquiry and Word Count (LIWC) [21] is a well-known software application for text analysis. LIWC is built around a word dictionary where each entry is associated with one or more categories related to grammar, emotions, or other cognitive processes and psychological concepts. Nowadays, LIWC has become a popular tool for the study of psychological aspects of textual documents, usually by employing the relative frequency of each LIWC category. It has been profitably used for the characterization of a "psychological profile" or a "mental profile mapping" for authorship studies [4,13], and also for the analysis of speeches regarding the Spanish political debate [11]. In a similar vein, García-Díaz et al. [12] designed UMUTextStats, a LIWC-inspired tool, and studied its application to AA and various AP tasks (gender, age range, and political spectrum) on a dataset of Spanish political tweets.

3   Experimental Setting

3.1   Dataset: ParlaMint

In this project, we employ the Spanish repository (covering the years 2015–2020) of the *Linguistically annotated multilingual comparable corpora of parliamentary debates ParlaMint.ana 2.1* by the digital infrastructure CLARIN,[4] which contains the annotated transcriptions of many sessions of various European Parliaments. Because of their declamatory nature, between the written text and the discourse, these speeches seem particularly suited for an investigation on rhythm and psycholinguistic traits. Apart from lowercasing the text, we did not apply any further pre-processing steps.

In order to have a balanced dataset, we select the parties with more than 300 speeches in the dataset and assign them to the Left, Right, Centre, or Regionalist[5] wing. In particular, we assign PSOE and UP to the Left, PP and PP-Foro to the Right, EAJ-PNV and JxCat-Junts to the Regionalist wing, and only the Ciudadanos (Cs) party to the Centre. We then delete all the speeches that have less than 50 words, and for each wing we select the 5 authors with most speeches in the dataset. The minimum number of samples per author is 70 (Bal Francés), while the maximum is 467 (Sánchez Pérez-Castejón). We randomly select 50 samples for each author to compose the training set, keeping all the remaining samples as test instances. We thus obtain 1000 training samples and 3048 test samples in total. Figure 1 reports the total number of words per author in the training set, divided by political wing, while Fig. 2 reports the distribution of authors by gender, age,[6] and political party.

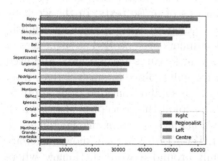

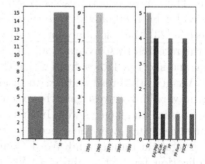

Fig. 1. Total number of words for each speaker in the training set, grouped by political wing.

Fig. 2. Number of speakers for each category in gender, age, and political party.

---

[4] https://www.clarin.si/repository/xmlui/handle/11356/1431.

[5] Regionalist parties aim for more political power for regional entities.

[6] Note that we use the decade of birth as representation of age group. We assign the closest decade label to each author's birth; for example, an author born in 1984 is assigned the label '1980', while an author born in 1987 is assigned the label '1990'.

3.2   Feature Extraction: BaseFeatures and Text Encodings

Our focus in this research is to evaluate the employment of rhythm- and psycho-linguistics-based features for AId and AP tasks. To this aim, we explore various combinations of feature sets, including other topic-agnostic feature sets commonly used in literature. In particular, we follow the same feature-extraction approach as in the preliminary experiments [8].

As a starting point, we employ a feature set comprised of features routinely used in the authorship field, including the relative frequencies of: function words (using the list provided by the NLTK library[7]), word lengths, and sentence lengths. We set the range of word (sentence) lengths to $[1, n]$, where n is the longest word (sentence) appearing at least 5 times in the training set. We call this feature set BASEFEATURES.

We also employ a text-masking approach, where we replace each word in the document with the respective Part-of-Speech tag (we exploit the POS annotation already available in the ParlaMint dataset). From the encoded text, we then extract the word n-grams in the range $[1, 3]$ and compute the TfIdf weights, which we use as features. We call this feature set POS.

We follow a similar approach to extract the rhythm of the discourse, i.e., we convert the document into a sequence of stressed and unstressed syllables, using the output of the RANTANPLAN library;[8] from this encoding, we extract the character n-grams in the range $[1, 7]$ and compute the TfIdf weights as features. We call this feature set STRESS.

Similarly, in order to encode the psycholinguistic dimension of the document, we employ the LIWC dictionary.[9] We define three macro-categories from a subset of the LIWC category tags, representing (i) grammatical information, (ii) cognitive processes or actions, and (iii) feelings and emotions.[10] For each macro-category, we perform a separate text masking by replacing each word with the

---

[7] https://www.nltk.org/.

[8] https://github.com/linhd-postdata/rantanplan.

[9] We employ the Spanish version of the dictionary, which is based on LIWC2007.

[10] We use the following categories: (i) YO, NOSOTRO, TUUTD, ELELLA, VOSUTDS, ELLOS, PASADO, PRESENT, FUTURO, SUBJUNTIV, NEGACIO, CUANTIF, NUMEROS, VERBYO, VERBTU, VERBNOS, VERBVOS, VERBOSEL, VERBELLOS, FORMAL, INFORMAL; (ii) MECCOG, INSIGHT, CAUSA, DISCREP, ASENTIR, TENTAT, CERTEZA, INHIB, INCL, EXCL, PERCEPT, VER, OIR, SENTIR, NOFLUEN, RELLENO, INGERIR, RELATIV, MOVIM; (iii) MALDEC, AFECT, EMOPOS, EMONEG, ANSIEDAD, ENFADO, TRISTE, PLACER. We avoid employing categories that would repeat information already captured by the POS tags, or topic-related categories (e.g., DINERO, FAMILIA).

Table 1. Example of the encodings employed in this project. Note there is not a one-to-one correspondence between syllables and stresses due to linguistic phenomenons across word boundaries (e.g., synalepha), which RANTANPLAN accounts for.

Original text:	Gracias		No	hay	que	restituir	lo	que	no	ha	existido	
POS:	NOUN	PUNCT	ADV	AUX	SCONJ	VERB	PRON	PRON	ADV	AUX	VERB	PUNCT
LIWC_GRAM:	w		NEGACIO	PRESENT	w	w	ELELLA	w	NEGACIO	PRESENTVERBOSEL	w	
LIWC_COG:	w		w	w	MECCOG	w	w	MECCOG	w	w	w	
LIWC_FEELS:	AFECTEMOPOS		w	w	w	w	w	w	w	w	w	
STRESS:	+−+−−−+−−+−+−											
English translation:	Thank you. There is no need to return what has not existed											

corresponding LIWC category tag.[11] From a single encoding, we extract the word n-grams in the range $[1, 3]$ and compute the TfIdf weights as features. We call these feature sets LIWC_GRAM, LIWC_COG, and LIWC_FEELS, respectively. We show an example of all the encodings in Table 1.

3.3 Experimental Protocol

We perform AId experiments in two settings: Authorship Verification (AV) for each author (where each test sample is labelled as belonging to that author, or not) and Authorship Attribution (AA) (where each sample is labelled as belonging to one of the 20 authors). We perform AP experiments by labelling each sample based on the gender, age group, political wing or political party of the author it belongs to. We assess the effect of the different feature sets by evaluating the performance of a classifier fed with them. As evaluation measure, for the AV task we use the well-known F_1 function, and for the AA and AP tasks we use the macro-averaged F_1 (F_1^M) and micro-averaged F_1 (F_1^μ) variants.

We employ SVM as learner;[12] the implementation we employ in this study is the SVC module from the `scikit-learn` package.[13] We perform the optimisation of various hyper-parameters: the parameter C, which sets the trade-off between the training error and the margin ($[0.001, 0.01, 0.1, 1, 10, 100, 1000]$), the kernel function (*linear*, *poly*, *rbf*, *sigmoid*), and whether the classes weights should be balanced or not. The optimization is computed in a grid-search fashion,

---

[11] Formally, LIWC can be seen as a map $m : w \rightarrow C$, where w is a word token and $C \subset \mathcal{C}$ is a subset of the psycholinguistic categories $\mathcal{C}$. Given a macro-category $M \subset \mathcal{C}$, we replace each word w in a document by the categories $m(w) \cap M$. If $|m(w) \cap M| > 1$, then a new token is created which consists of a concatenation of the category names (following a consistent ordering). If $m(w) \cap M = \emptyset$, then w is replaced with the 'w' symbol. (Note that some entries in LIWC have the suffix truncated and replaced with an asterisk '*', e.g., *president**; the asterisk is treated as a wildcard in the mapping function, and in case more than one match is possible, the match with the longest common prefix is returned).

[12] We also carried out preliminary experiments with Random Forest (RF) and Logistic Regression (LR). SVM showed a remarkably better performance than RF, while no significant differences were noticed between SVM and LR.

[13] https://scikit-learn.org/stable/modules/generated/sklearn.svm.SVC.html.

via 5-fold cross-validation on the training set. The best model is then retrained on the whole training set and is used to make predictions on the test set samples.

We apply a feature selection approach, since the number of different features generated using the LIWC encodings tends to be very high.[14] We keep only the 10% most important features (employing χ^2 as score measure of importance) for each feature set derived from the LIWC encodings.[15]

Finally, we also compare the results obtained with the aforementioned features with the results obtained by a method trained on the original text (hence, potentially mining topic-related patterns). To this aim, we employ the pre-trained transformer named 'BETO-cased', from the Hugginface library [6],[16] with the learning rate set to 10^{-6} and the other hyper-parameters set as default. We fine-tune the model for 50 epochs on the training set.

4   Results

We show the results of the AV experiments in Table 2. In the first batch of results, we show the performance of the feature sets in the experiments "by addition", using the BASEFEATURES set as a baseline. In the second batch of results, we report the experiments "by ablation", where we subtract each feature set from the combination of all the feature sets we are exploring (named ALL). These results are obtained using a SVM learner. Finally, we also report the results obtained using the BETO transformer. Moreover, we perform the non-parametric McNemar's paired test of statistical significance between the results obtained using our best SVM configuration and the results obtained using BETO [6], for each of the authorship tasks. We take 0.05 as the confidence level.

BETO obtains the best result in 10 out of 20 cases, 5 of which are statistically significant; conversely, the SVM classifier obtains the best performance 10 out of 20 cases, 7 of which are statistically significant. Thus, we might consider the performance of the two methods comparable, even though SVM does not exploit any topic-related information (as the BETO transformer instead could do). Focusing on the SVM results, we observe that the best-performing feature set is often (in 10 out of 20 cases) the one combining BASEFEATURES and POS, confirming that the syntactic encoding is a good indicator of style. The other best-resulting feature sets are mostly different "ablations" of the ALL set. In particular, it seems that the LIWC_FEELS features are rather detrimental, since the configuration ALL - LIWC_FEELS yields the best results in 5 cases. In line with our preliminary results [8], we observe severe fluctuations in performance across the authors, with the best result and relative merits of each of

---

[14] Indeed, LIWC_GRAM, LIWC_COG and LIWC_FEELS create the highest number of features in our experiments, ranging from 3000 to more than 20000.

[15] The selection is always carried out in the training set. During the 5-fold cross-validation optimization phase, feature selection is carried out in the corresponding 80% of the training set used as training.

[16] https://huggingface.co/dccuchile/bert-base-spanish-wwm-cased.   This   model obtained better results than the 'uncased' version in preliminary experiments.

Table 2. Results of the AV experiments. *

	Martínez	Sagastizabal	Rodríguez	Legarda	Agirretxea	Girauta	Esteban	Rajoy	Sánchez	Catalá	Montoro	Báñez	Iglesias	Rivera	Roldán	Bel	Bal	Calvo	GMarlaska	Montero
BaseFeatures	.616	.626	.272	.580	.277	.100	.510	.420	.528	.397	.468	.671	*.482*†	.526	.161	.599	.203	.227	.362	.479
+ POS	.742†	.761	.524	.672	.560	*.188*†	*.516*†	*.515*†	*.640*†	*.537*†	.540	.717	.444	*.654*†	.400	.609	*.449*	*.481*†	.425	.453
+ STRESS	.675	.618	.293	.586	.359	.086	.458	.414	.464	.373	.488	.645	.389	.557	.136	.662	.171	.204	.321	.381
+ LIWC_GRAM	.529	.517	.091	.621	.329	.060	.365	.165	.535	.277	.379	.539	.319	.443	.080	.460	.047	.189	.295	.431
+ LIWC_COG	.538	.503	.092	.640	.374	.070	.281	.259	.508	.367	.423	.603	.313	.323	.091	.506	.205	.167	.290	.456
+ LIWC_FEELS	.549	.408	.089	.521	.277	.051	.273	.229	.425	.214	.366	.543	.345	.402	.138	.483	.068	.131	.327	.359
ALL	.706	.646	.371	.650	.589	.081	.338	.492	.618	.503	.395	**.748**	.437	.622	.289	.677	.189	.383	.478	.401
- BaseFeatures	.724	**.781**	.372	**.734**	.524	.046	.362	.447	.514	.258	.328	.719	.307	.534	.349	.594	.206	.318	.423	.305
- POS	.599	.415	.229	.543	.403	.078	.291	.348	.504	.347	.356	.657	.403	.463	.237	.568	.090	.209	.409	.346
- STRESS	.723	.655	.441	.629	.545	.036	.379	.477	.602	.489	.409	.745	.437	.636	.310	.654	.189	.394	**.480**	.447
- LIWC_GRAM	.709	.568	.392	.552	.560	.131	.341	.449	.631	.518	.454	.689	.418	.604	.212	*.703*†	.200	.268	.420	.501
- LIWC_COG	.692	.568	.381	.708	.533	.088	.348	.469	.625	.441	.456	.707	.421	.587	.184	.652	.267	.329	.386	.384
- LIWC_FEELS	.710	.611	*.526*†	.572	*.674*	.056	.363	.513	*.640*†	.487	.435	.747	.454	.629	*.471*	.683	.175	.353	.408	*.570*†
Beto_base_cased	*.836*	*.798*	.314	*.771*	.632	*.247*	.352	*.757*	.388	*.729*	*.610*	*.800*	.437	.460	.468	.601	.381	*.494*	*.664*	.426

* The best result for SVM is in **bold**, while the best overall result is in *italic*; statistical significance is indicated with a † in the best SVM result.

the feature sets being strongly dependent on the author under consideration; for example, the feature set STRESS appears to be beneficial for authors like Agirretxea and Bel, while the same feature set seems to be detrimental for authors like Sagastizabal, Grande-Marlaska, and Montero.

We show the results of the AA experiments in Table 3. In these experiments, the ALL - STRESS and the ALL - LIWC_FEELS feature combinations employing the SVM learner obtain the best results, both outperforming BETO in a statistically significant sense. In fact, consistently with previous results [8], the feature sets LIWC_FEELS and STRESS exhibit a constant disturbance effect.

We show the results of the AP experiments in Table 4. While BETO and SVM do not show remarkable differences for *gender* prediction, SVM excels in the other tasks, with statistical significance in the case of *age* and *political wing*. Consistently with what observed for the AA and AV case, the POS feature set shows a clear proficiency, while the contrary can be said for LIWC_FEELS.

Overall, these experiments allow us to draw some interesting conclusions regarding the features we study for authorship analysis in the Spanish language: on the one hand, TfIdf-weighted n-grams computed on POS-tags encodings are effective for multiple tasks and settings; on the other hand, the feature sets STRESS and LIWC_FEELS tend to fare poorly. Interestingly enough, the combination of multiple topic-agnostic feature sets proved to fare comparably to, and to outperform in some cases, a state-of-the-art neural network that has full access to topic-related information.

5   Post-hoc Analysis of the AV Results

Given the differences in performance spotted in the AV results among authors, we further analyse the system's behaviour in order to outline a suitable explanation for such variances. To this aim, we resort to a series of tools for data analysis: the one-way ANOVA test (Sect. 5.1) and the Spearman test applied to the Analytic

Table 3. Results for the AA experiment. *

	F_1^M	F_1^μ
BaseFeatures	.401	.444
+ POS	.570	.620
+ STRESS	.392	.436
+ LIWC_GRAM	.430	.480
+ LIWC_COG	.446	.493
+ LIWC_FEELS	.348	.394
ALL	.580	.631
- BaseFeatures	.545	.599
- POS	.435	.485
- STRESS	*.585*	*.638*†
- LIWC_GRAM	.565	.615
- LIWC_COG	.562	.613
- LIWC_FEELS	*.585*	*.635*†
Beto_base_cased	.417	.471

Table 4. Results for the AP experiments. *

	Gender		Age		Wing		Party	
	F_1^M	F_1^μ	F_1^M	F_1^μ	F_1^M	F_1^μ	F_1^M	F_1^μ
BaseFeatures	.720	.802	.428	.478	.599	.631	.547	.563
+ POS	.751	.828	.545	.592	.685	.715	*.642*	*.681*
+ STRESS	.705	.803	.402	.459	.581	.613	.539	.561
+ LIWC_GRAM	.696	.788	.421	.469	.601	.636	.508	.536
+ LIWC_COG	.716	.812	.441	.511	.619	.653	.492	.502
+ LIWC_FEELS	.669	.777	.366	.434	.525	.554	.492	.531
ALL	.736	*.854*	.551	.589	.690	.719	.604	.648
- BaseFeatures	.709	.843	.485	.540	.650	.681	.552	.614
- POS	.700	.825	.411	.469	.640	.669	.498	.544
- STRESS	.732	.850	.553	.594	*.707*	.736†	.610	.658
- LIWC_GRAM	*.754*	*.854*	.522	.573	.677	.709	.611	.637
- LIWC_COG	.736	.838	.522	.564	.687	.717	.610	.637
- LIWC_FEELS	.725	.831	*.573*	*.609*†	.706	*.737*†	.613	.650
Beto_base_cased	.762	.847	.337	.420	.666	.698	.574	.662

* The best result for SVM is in **bold**, while the best overall result is in *italic*; statistical significance is indicated with a † in the best SVM result.

Thinking Index (ATI), the Categorical-versus-Narrative Index (CNI), and the Adversarial Style Index (ASI) (Sect. 5.2).

5.1   One-Way ANOVA Test for Political Groups

The one-way ANOVA is a parametric test used to check for statistically significant differences in any outcome among groups that are under one categorical variable. We employ this test in order to see if, by grouping the speakers by categories (wing, party, gender, or age), statistically significant differences in performance emerge from the adoption of certain sets of features in the AV task. We use 0.05 as confidence level, and we check that the assumptions for the test (independence, normality and homogeneity of variance) are met. We only consider groups with more than one member, e.g., when grouping by age, we do not consider the groups corresponding to decades 1950 and 1990, since each group would have only one member, Montoro and Rodríguez, respectively.

From this analysis, we have not found any significant difference employing the grouping by gender or by age. However, we have found that the F_1 results display significant differences for multiple feature sets and for BETO if grouped by political party, and especially so if grouped by political wing. The results are reported in Table 5. The Tukey's Honestly Significant Difference (HSD) [26], a statistical test that compares all possible pairs of means among various result groups, reveals that: i) when grouping by political wing, the significant difference always occurs between the Centre and the other wings, ii) when grouping by political party, the significant difference occurs between Cs and EAJ-PNV, and between Cs and PP. If we focus our attention to the features BASEFEATURES + LIWC_COG (the only SVM feature setting giving rise to statistically significant differences in performance when the groups are generated by wing, and also when the groups are generated by political party), it turns out that authors belonging to the Centre/Cs obtained F_1 scores significantly lower than authors from other groups (Fig. 3). In fact, the Cs is a relatively new political party (it was founded in 2006) and its members have been in various different parties before joining it; that could have lead to a certain difficulty in creating a specific personal style.

5.2   Spearman Coefficient Applied to Style Indices

The Analytic Thinking Index (ATI), introduced by Pennebaker et al. [22] and formerly named Categorical Dynamic Index, is a unit-weighted score computed from grammatical categories derived from LIWC. This measure is based on the observation that the use of articles and prepositions is associated with a more abstract thinking, while the use of pronouns, auxiliary verbs, conjunctions, adverbs, and negations is associated with a more intuitive and narrative style. The score is obtained by adding the percentages of the former, and subtracting the percentages of the latter. Thus, a positive score denotes a more analytic thinking, while a negative score denotes a more intuitive one. This score has been used to analyze long-term trends in political language in EEUU [17].

The Categorical-versus-Narrative Index (CNI), introduced by Chulvi et al. [7] and inspired by the study of Nisbett et al. [20], is similar in nature to ATI. The score is obtained by adding the percentages of nouns, adjectives, and prepositions, and subtracting the percentages of verbs, adverbs, and personal pronouns. Like ATI, a higher score of CNI denotes a language more focused on the exposition of abstract concepts, while a lower score denotes a language more prone to narration and storytelling.

The Adversarial Style Index (ASI), also proposed by Chulvi et al. [7], is a ratio representing how much an author refers directly to the political adversary in a confronting manner in political debates. The adversarial genre has been vastly studied in parliamentary and election debates, both in the English [5] and Spanish context [3]. The score is the ratio between the sum of the percentages of LIWC categories TUUTD and VOSUTDS (2nd person singular and plural Spanish pronouns), and the sum of the percentages of LIWC categories YO, NOSOTRO (1st person singular and plural Spanish pronouns), TUUTD and VOSUTDS.

We show the ATI, CNI and ASI scores computed for each author on the entire dataset in Fig. 4, 5 and 6, respectively.

We employ these measures to quantify the extent to which the AV performance correlates to certain styles of communication. To this aim, in Table 6 we show, for each of the indices, the Spearman correlation coefficient (r) between the classification scores and the authors' index scores. We see that BETO displays the strongest positive correlation with respect to ATI and CNI, followed closely by the BASEFEATURES + LIWC_COG feature set. This seems reasonable, since ATI and CNI hinge upon abstract explanations and concepts, which are captured by LIWC_COG, while a big portion of the training of BETO is comprised of sources like Wikipedia, legislative texts and talks. Interestingly, the correlations between F_1 scores and the psycholinguistic indexes obtained by each author is found to be statistically significant for more feature sets combinations for CNI (8) than for ATI (1). We hypothesize that this might be due to the fact that ATI, unlike CNI, captures a degree of formality that is rather

Table 5. ANOVA p-values on the F_1 results on the AV experiments when grouped by wing or by party. **

Table 6. Spearman r correlation between the F_1 values in the AV experiments and the indices values. **

	Wing	Party
BaseFeatures	**.026**	.085
+ POS	.078	.192
+ STRESS	**.014**	.060
+ LIWC_GRAM	**.023**	.054
+ LIWC_COG	**.008**	**.031**
+ LIWC_FEELS	**.042**	.117
ALL	**.045**	.136
- BaseFeatures	.065	.135
- POS	**.045**	.133
- STRESS	.071	.185
- LIWC_GRAM	.054	.159
- LIWC_COG	**.042**	.136
- LIWC_FEELS	.192	.399
Beto_base_cased	**.001**	**.008**

	ATI		CNI		ASI	
	r	p-value	r	p-value	r	p-value
BaseFeatures	.335	.148	.432	.057	-.444	.050
+ POS	.296	.205	**.460**	**.041**	**-.683**	**.001**
+ STRESS	.284	.225	.411	.072	**-.508**	**.022**
+ LIWC_GRAM	.403	.078	**.558**	**.011**	-.429	.059
+ LIWC_COG	**.489**	**.029**	**.621**	**.003**	**-.489**	**.029**
+ LIWC_FEELS	.438	.054	**.543**	**.013**	-.400	.081
ALL	.335	.148	**.483**	**.031**	**-.459**	**.042**
- BaseFeatures	.251	.286	.389	.090	**-.642**	**.002**
- POS	.360	.119	**.471**	**.036**	-.402	.079
- STRESS	.314	.177	**.463**	**.040**	**-.487**	**.029**
- LIWC_GRAM	.271	.248	.442	.051	-.435	.056
- LIWC_COG	.319	.171	**.453**	**.045**	**-.543**	**.013**
- LIWC_FEELS	.211	.373	.331	.154	**-.493**	**.027**
Beto_base_cased	**.544**	**.013**	**.675**	**.001**	**-.517**	**.020**

** Statistically significant values are in bold.

common in parliamentary speeches, hence preventing meaningful differences to emerge. Moreover, while 8 feature sets show a positive correlation with CNI, as many feature sets show a negative correlation with ASI, where BaseFeatures + POS holds the strongest correlation. For comparison, we show the plot of both CNI and ASI correlated with the results for the ALL feature set in Fig. 7 and Fig. 8 respectively. The opposite nature of the two indices is understandable, since a more adversarial style would naturally be less abstract and more focused on events and narration. Indeed, some studies, both regarding Question Time in English [10,16] and face-to-face Spanish political debates [3], noted that adversarial speeches in the parliamentary context present certain repeating oratory patterns. This could explain why the present features, and in particular the POS set, perform worse on speakers with higher ASI, who are likely to use common syntactic patterns. Conversely, it is easier to recognize speakers with a more abstract communication style.

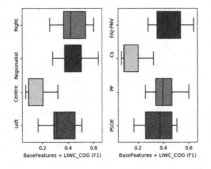

Fig. 3. AV results for the BaseFeatures + LIWC_COG feature set, divided by political wing and party.

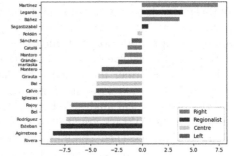

Fig. 4. Analytical Thinking Index values for each author.

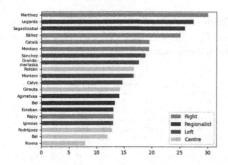

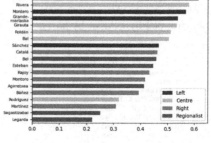

Fig. 5. Categorical-versus-Narrative Index values for each author.

Fig. 6. Adversarial Style Index values for each author.

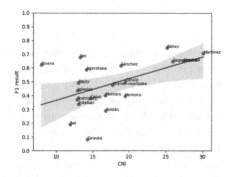

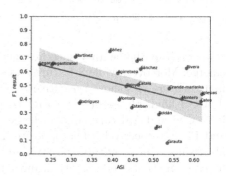

Fig. 7. Correlation among AV results for the ALL feature set and CNI values.

Fig. 8. Correlation among AV results for the ALL feature set and ASI values.

6   Conclusion and Future Work

In this research, we investigate the extent to which topic-agnostic features, and in particular rhythmic and psycholinguistic feature sets obtained via a text-masking approach, are useful for AId and AP tasks in the Spanish language, using a dataset of political speeches. We show that such features perform comparably to a BETO transformer fine-tuned with the non-masked texts (hence potentially learning from topic-related information) in all such tasks. Moreover, we conduct a series of statistical analysis, showing that the different results for the various authors in the AV task are at least partially linked with the political affiliation of the author and their communication style, with a positive correlation with abstract and categorical style, and a negative correlation with adversarial style.

In future work, we aim to extend this study to other forms of political communication (e.g., tweets [12]), and to further explore the relation between the author's profile and the classification performance with a more comprehensive statistical analysis.

Acknowledgment. The research work by Silvia Corbara was carried out during her visit at the Universitat Politècnica de València and was supported by the AI4MEDIA project, funded by the EU Commission (Grant 951911, H2020 Programme ICT-48-2020).

The research work by Paolo Rosso was partially funded by the Generalitat Valenciana under DeepPattern (PROMETEO/2019/121).

References

1. Bevendorff, J., et al.: Overview of PAN 2021: authorship verification, profiling hate speech spreaders on Twitter, and style change detection. In: Candan, K., et al. (eds.) CLEF 2021. LNCS, vol. 12880, pp. 419–431. Springer, Cham (2021). https://doi.org/10.1007/978-3-030-85251-1_26
2. Bischoff, S., et al.: The importance of suppressing domain style in authorship analysis. arXiv:2005.14714 (2020)
3. Blas-Arroyo, J.L.: 'Perdóneme que se lo diga, pero vuelve usted a faltar a la verdad, señor Gonzalez': form and function of politic verbal behaviour in face-to-face Spanish political debates. Discour. Soc. **14**(4), 395–423 (2003)
4. Boyd, R.L.: Mental profile mapping: a psychological single-candidate authorship attribution method. PLoS One **13**(7), e0200588 (2018)
5. Bull, P., Wells, P.: Adversarial discourse in Prime Minister's questions. J. Lang. Soc. Psychol. **31**(1), 30–48 (2012)
6. Cañete, J., Chaperon, G., Fuentes, R., Ho, J.H., Kang, H., Pérez, J.: Spanish pre-trained BERT model and evaluation data. In: PML4DC at ICLR 2020 (2020)
7. Chulvi, B., Rosso, P., Molpeceres, M.A., Sánchez-Junquera, J., Rodrigo, M.: Us and them: immigrant's stereotypes and language style on political parliamentary speeches (under revision) (2022)
8. Corbara, S., Chulvi, B., Rosso, P., Moreo, A.: Investigating topic-agnostic features for authorship tasks in Spanish political speeches. In: Rosso, P., Basile, V., Martínez, R., Mètais, E., Meziane, F. (eds.) NLDB 2022. LNCS, vol. 13286, pp. 394–402. Springer, Cham (2022). https://doi.org/10.1007/978-3-031-08473-7_36
9. Corbara, S., Moreo, A., Sebastiani, F.: Syllabic quantity patterns as rhythmic features for Latin authorship attribution. arXiv:2110.14203 (2021)
10. Fenton-Smith, B.: Discourse structure and political performance in adversarial parliamentary wuestioning. J. Lang. Polit. **7**(1), 97–118 (2008)
11. Fernández-Cabana, M., Rúas-Araújo, J., Alves-Pérez, M.T.: Psicología, lenguaje y comunicación: análisis con la herramienta LIWC de los discursos y tweets de los candidatos a las elecciones gallegas. Anuario Psicol. **44**(2), 169–184 (2014)
12. García-Díaz, J.A., Colomo-Palacios, R., Valencia-García, R.: Psychographic traits identification based on political ideology: an author analysis study on Spanish politicians' tweets posted in 2020. Futur. Gener. Comput. Syst. **130**, 59–74 (2022)
13. Gaston, J., et al.: Authorship attribution vs. adversarial authorship from a LIWC and sentiment analysis perspective. In: 2018 IEEE Symposium Series on Computational Intelligence (SSCI), pp. 920–927. IEEE (2018)
14. van der Goot, R., Ljubešić, N., Matroos, I., Nissim, M., Plank, B.: Bleaching text: abstract features for cross-lingual gender prediction. In: Proceedings of the 56th Annual Meeting of the Association for Computational Linguistics (ACL 2018), Volume 2: Short Papers, pp. 383–389 (2018)

15. Halvani, O., Graner, L., Regev, R.: TAVeer: an interpretable topic-agnostic author-ship verification method. In: Proceedings of the 15th International Conference on Availability, Reliability and Security (ARES 2020), pp. 1–10 (2020)

16. Harris, S.: Being politically impolite: extending politeness theory to adversarial political discourse. Discour. Soc. **12**(4), 451–472 (2001)

17. Jordan, K.N., Sterling, J., Pennebaker, J.W., Boyd, R.L.: Examining long-term trends in politics and culture through language of political leaders and cultural institutions. Proc. Natl. Acad. Sci. **116**(9), 3476–3481 (2019)

18. Kestemont, M., et al.: Overview of the author identification task at PAN-2018: cross-domain authorship attribution and style change detection. In: Cappellato, L., Ferro, N., Nie, J.Y., Soulier, L. (eds.) CLEF (Working Notes). CEUR Workshop Proceedings, vol. 2125. CEUR-WS.org (2018)

19. Neidorf, L., Krieger, M.S., Yakubek, M., Chaudhuri, P., Dexter, J.P.: Large-scale quantitative profiling of the old English verse tradition. Nat. Hum. Behav. **3**(6), 560–567 (2019)

20. Nisbett, R.E., Peng, K., Choi, I., Norenzayan, A.: Culture and systems of thought: holistic versus analytic cognition. Psychol. Rev. **108**(2), 291 (2001)

21. Pennebaker, J.W., Boyd, R.L., Jordan, K., Blackburn, K.: The development and psychometric properties of LIWC2015. Technical report (2015)

22. Pennebaker, J.W., Chung, C.K., Frazee, J., Lavergne, G.M., Beaver, D.I.: When small words foretell academic success: the case of college admissions essays. PLoS One **9**(12), e115844 (2014)

23. Plecháč, P.: Relative contributions of Shakespeare and Fletcher in Henry VIII: an analysis based on most frequent words and most frequent rhythmic patterns. Digit. Scholarsh. Humanit. **36**(2), 430–438 (2021)

24. Stamatatos, E.: A survey of modern authorship attribution methods. J. Am. Soc. Inform. Sci. Technol. **60**(3), 538–556 (2009)

25. Stamatatos, E.: Masking topic-related information to enhance authorship attribution. J. Am. Soc. Inf. Sci. **69**(3), 461–473 (2018)

26. Tukey, J.W.: Comparing individual means in the analysis of variance. Biometrics, pp. 99–114 (1949)

27. Weerasinghe, J., Singh, R., Greenstadt, R.: Feature vector difference based author-ship verification for open world settings. In: Working Notes of CLEF 2021 - Conference and Labs of the Evaluation Forum. CEUR-WS.org (2021)

The Impact of Pre-processing on the Performance of Automated Fake News Detection

Salar Mohtaj[1,2]([envelope]) and Sebastian Möller[1,2]

[1] Technische Universität Berlin, Berlin, Germany
{salar.mohtaj,sebastian.moeller}@tu-berlin.de
[2] German Research Centre for Artificial Intelligence (DFKI), Labor Berlin, Berlin, Germany

Abstract. Fake news spreading through social media has become a serious problem in recent years, especially after the United States presidential election in 2016. Accordingly, more attention has been paid to this issue by scientists to develop automated tools to combat those pieces of information that contain misinformation, using natural language processing methods. Although the performance of fake news detection models has increased by using more complex architectures and state-of-the-art models, less attention has been paid to the impact of pre-processing on the overall performance of such models. In this study, we focus on investigating the impact of pre-processing, especially removing URLs on the performance of fake news detection systems. We compared the performance of fake news detection in tweets as a text classification task, using support vector machine, long short-term memory networks, and BERT pre-trained model. In addition to URLs, we analyzed the impact of different approaches for dealing with emojis and Twitter handles on the performance of the models. Our results show URLs could be good clues for identifying fake news, despite the fact that they are usually removed in pre-processing step.

Keywords: Fake News Detection · Pre-processing · LSTMs · BERT

1 Introduction

Social media has played a crucial role in recent year, having a great impact on quite a few areas such as communication, entertainment, and politics. Beside their positive applications, social networks became an easily accessible medium to express fake claims and harmful contents in recent years. Disinformation and the potential approaches to combat fake news has been studied by academics, either from media or technology point of views. In this paper we focus on fake news and disinformation detection, using Natural Language Processing (NLP) models.

Although various state-of-the-art models have been proposed to better empower machines to fight untrue stories in social media, less attention has

A. Barrón-Cedeño et al. (Eds.): CLEF 2022, LNCS 13390, pp. 93–102, 2022.
https://doi.org/10.1007/978-3-031-13643-6_7

been paid to the impact of pre-processing on the overall performance of the models. Previous researches studied the impact of pre-processing on the performance of text classification task in terms of various aspects such as classification accuracy, text domain (e.g., e-mail and news), and text language [17]. Moreover, it has been shown that n-grams based on stop-words can be used to detect plagiarism, despite the fact that they are usually removed in the pre-processing step [16].

Since the text length is usually very short in social network content (especially in Twitter), the pre-processing has even more impact on the overall result of NLP task and can play a more important role. However, the URLs are usually either replaced by special token (e.g., replacing link with *"weblink"* string), or removed from ordinary text in the pre-processing step in quite a few researches on fake news and hate speech detection in social media [3,7,11,12]. On the other hand, more information could lead to better prediction of fake news. So, it is very important that how the URLs are dealt in the pre-processing phase. URLs can either simply be removed from ordinary text, or they can be replaced by the content of pages that they refer to. In this paper, we aim to analyze the impact (either positive or negative) of dealing with URLs, emojis and Twitter handle (i.e., username), as pieces of information that usually present in social media texts, on different fake news detection models.

For this purpose, we compared different scenarios in which different pre-processing steps have been applied on the input text, on the overall performance of Support Vector Machines (SVMs) [4], Long Short Term Memory Network (LSTM) [8] and the BERT transformer based pre-trained language model [6] approaches. We classified tweets into two categories, namely, *Real* and *Fake*, using the above mentioned models. Since these models cover different areas of Machine Learning (ML), from feature-based approaches to deep neural networks and transformers, we believe that the results can reveal the impact of pre-processing, regardless of complexity and the nature of the models.

Our experiments show that even though dealing with emojis and Twitter handles in tweets doesn't significantly affect the performance of a fake news detection system, URLs can remarkably impacts it. Our results show that rather than removing URLs in tweets or replacing them with a special token, the best practice to deal with URLs is replacing them by the content of web-page that they refer to.

The rest of the paper is organized as follow; In Sect. 2 we discuss a number of recent machine learning based models to detect fake stories in social media and news, as well as a few studies on the influence of pre-processing on different NLP tasks. Section 3 contains the detailed description about the corpus, models and experiments in this research. In Sect. 4 we present the obtained results by different classification models and finally in Sect. 5 we briefly conclude the paper and discuss some research questions that can be studied later to expand this research's findings.

2   Related Work

In this section, we review related works on fake news detection and also a number of papers that study the impact of pre-processing in different NLP tasks.

2.1   Fake News Detection

The approaches for detecting fake content on the web and especially on social media can be categorized into three categories; namely, a)feature-based, b) knowledge-based, and c)learning-based approaches [9]. In this section, we review some of the state-of-the-art supervised learning-based models. Deep neural networks and transformer based pre-trained language models (e.g., BERT) are commonly used by these methods to automatically categorize content into *"Fake"* and *"Real/True"* categories.

A text and image information based Convolutinal Neural Network (CNN) named as *TI-CNN* is proposed in [18]. The model try to automatically detect fake news, using textual and visual contents by projecting the explicit and latent features into a unified feature space. The authors used CNNs to extract latent features from texts and images of news articles and fed those features to a dense neural network to categorize news into two categories. The model is tested on a collection of data that focus on news about American presidential election.

A supervised ML technique to fake news detection is presented in [5]. They applied various classification models on a fake news dataset about US politics. The obtained results show Gradient Boosting Machine (GBM) could outperform the other models, including ensemble models.

Finally, an approach for automatic fake news detection based on BERT pre-trained model (FakeBERT) and three parallel blocks of 1d-CNN is proposed in [9]. They achieved an accuracy of 98.90% using BERT on a real-world fake news dataset from Kaggle, which outperforms the approaches based on feature-based ML.

2.2   The Impact of Pre-processing in NLP Tasks

In this section, we overview a few paper that studied the influence of different pre-processing steps on various NLP tasks.

The impact of pre-processing on the accuracy of ML techniques for the task of sentiment analysis has been studied in [1]. Regarding the pre-processing steps, they measured the impact of emoticons removal, stop-words removal, stemming and word vectorization on three different models, namely, Naive Bayes, Maximum entropy and SVMs. Their results show that Naive Bayes benefits a lot from applying pre-processing, while there is almost no improvement in case of maximum entropy by applying the mentioned pre-processing steps on the data. In addition, the results show that a slight improvement in accuracy of SVM has been seen after applying the pre-processing steps [1].

In another research, the influence of a set of different pre-processing steps on Twitter data has been studied [15]. They measured the impact of a number of pre-processing steps including lower-casing, removing of Twitter handles and hashtag, removing of Non-ASCII characters and white-spaces, removing of short and stop-words, and lemmatization on five different ML algorithms for the task of opinion mining in Twitter [15]. They measured the impact of pre-processing for TF-IDF and Bag of Word features. Their findings show that pre-processing leads to a substantial improvement of classifiers on BOW feature, while doesn't have a significant impact on the TF-IDF feature.

The effect of pre-processing on document categorization in Arabic text has been analyzed in [2]. In this research the impact of word normalization, stemming, and stop-word removal tested, using feature based classifier methods. The obtained results show that the combination of pre-processing tasks can play an important role in the overall performance of document categorization, while inappropriate combinations may degrade the classification accuracy.

To the best of our knowledge, there are no research on the impact of pre-processing, especially measuring the influence of URLs, on the task of fake news detection in social media content.

3   Experiments

In this part, we present the experiments, settings and the fake news detection corpus, which have been used to analyze the results.

3.1   Fake News Dataset

For measuring the impact of pre-processing, we used COVID-19 Fake News Dataset that contains *fake* and *real* news about *COVID-19* [13]. The data have been collected from different social media and fact checking web-sites. The data is designed to predict fake news as a binary classification task. It contains *10,700* posts which has been divided into train, validation and test parts[1]. However, these three parts are merged together in order to re-split the data for a 5-fold cross validation setting. Some statistics of the dataset are as follow (Table 1).

As it is highlighted in the table, more than half of the tweets in the dataset contain at least one URL in the text, and also there are quite a few tweets with Emojis and Twitter handles. It shows the importance of the approach that is used to pre-process social media content. URLs can particularly have a significant impact on the overall performance of the classification models to categorize tweets into real and fake classes.

Among the available datasets for this task, the *COVID-19 fake news detection* dataset has been chosen because of the following reasons:

– It covers social media, so the impact of replacing URLs with the content of web-pages is better measurable due to the short length of original texts.

---

[1] https://competitions.codalab.org/competitions/26655.

Table 1. *COVID-19* Fake News Detection Dataset Statistics

Attribute	#
Total number of documents	10700
Number of *real* documents	5600 (52%)
Number of *fake* documents	5100 (48%)
Average length of documents (in character)	181.8
The length of shortest document (in character)	18
The length of longest document (in character)	10170
Number of tweets that contain URL	5530
Number of tweets that contain Emoji	612
Number of tweets that contain Twitter handles	1989

- There are quite a few URLs in the dataset. Again, the existence of a large number of URLs in the data makes it easier to compare the impact of different approaches to deal with URLs.
- It a balanced and new data, that covers a trending topic.

3.2 Experiment Setting

As mentioned before, in this study we try to measure the impact of three pre-processing tasks, namely, replacing Twitter handles with a special token ["username"], replacing emojis with the expression they represent (e.g., a few terms that explain the emoji), and replacing URLs either with a special token ["weblink"] or with the text content of pages that URLs refer to. We elaborate on each of these tasks in this section.

Regarding the classification algorithms, we track the impact of the above mentioned pre-processing tasks on three models, namely, SVM, LSTMs, and BERT. These are among the best models that outperform the other approaches in related works.

Pre-processing Tasks. Before applying any of the above mentioned pre-processing steps in text, we converted the input text to lower-cases and also replaced multiple white-spaces by a single one as a simple normalization step. We named the normalized texts as "Ordinary Text" in the experiments.

For replacing Twitter handles and URLs with special tokens, we used regular expressions to find the targeted strings and replace them with the corresponding tokens. These are named "Replacing Twitter handles with a special token" and "Replacing URLs with a special token" in the experiments, respectively. Emojis are replaced with the expression they represent using Python demoji package[2]. As an example of this step, the package replaced the emoji for the flag of Germany with string "flag: Germany".

---

[2] https://pypi.org/project/demoji/.

For replacing a URL with the textual content of the page that it refers to, we first extracted all URLs from the whole dataset and got the content based on the domains. For URLs that refer to a Twitter post, tweepy package[3] is used to get the post bodies. For non-Twitter links, Python BeautifulSoup package[4] is used to extract all readable textual content from web-pages. There were a total number of 7,388 web links in the whole dataset. Table 2 presents the most frequent domains among the URLs in the data.

Table 2. Frequency of different domains among URLs in the data

Domain	Frequency
twitter.com	3696 (50%)
medscape.com	437 (7%)
news.sky.com	429 (6%)
thespoof.com	386 (6%)
politifact.com	168 (2%)
Others	2272 (29%)

Among the all links, we could successfully get the content of 3697 tweets/web-pages and replaced URLs with the corresponding textual content. The rest of the web-pages were not accessible because of various reasons, including suspension of some users in the platform, deletion of the post, to name but a few. We named those experiments based on the replaced URLs with texts as "Replacing URLs with the webpage's text" in Sect. 4.

ML Models. As mentioned before, we trained three models based on SVMs, LSTMs, and BERT for assigning documents into "real" and "fake" classes. In this section we briefly highlight the main setting and hyper-parameters that have been set for the experiments. All the experiments have been done on a 5-fold cross validation setting.

For the SVM model, we converted raw input texts into vectors of numbers, using *TF-IDF* vectorization model. We set *5,000* as the max number of tokens to convert. As a result less frequent tokens have been ignored in this process. We used *Sigmoid* kernels and set *1.0* as the L2 regularization factor.

For the LSTM model, we used a two layers LSTMs architecture and vectorized raw texts using *GloVe* [14] vectors with the embedding size of *300*. Moreover, we used a hidden size of *256* and set a dropout layer with the probability of *0.5*. The Adam optimizer [10] with learning rate of *1e-3* is used in the model and it is trained on batch sizes of *32* for *20* epochs.

---

[3] https://pypi.org/project/tweepy/.
[4] https://pypi.org/project/beautifulsoup4/.

Finally, for the BERT model we used the *bert-base-uncased* model from the Hugging Face library[5]. We fine tuned BERT model in three epochs with learning rate of *2e−5* and batch size of *32* [6].

The obtained results for the combinations of different models and pre-processing tasks are presented in the next section.

4 Results

For each of the classification models, we trained five different models using different versions of input text, as pre-processing point of view. We named these five experiments based on each version as follows:

- Ordinary Text, where none of the mentioned pre-processing steps are applied on the text
- Replacing URLs with a special token
- Replacing URLs with the webpage's text
- Replacing Twitter handles with a special token
- Replacing Emojis with the expression they represent

The obtained results from SVM, LSTM and BERT models are depicted in Figs. 1, 2 and 3, respectively.

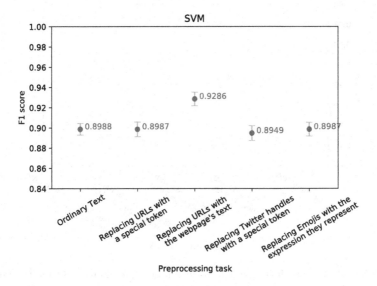

Fig. 1. The obtained results on 5-fold cross validation, based on SVM model. The blue points show the average of F1 in the whole folds and orange lines show the standard deviation. (Color figure online)

---

[5] https://huggingface.co/.

[6] More details about the implementation and parameters can be found in the GitHub repository of the project at https://github.com/salarmohtaj/FakeNews_Detection_Twitter.

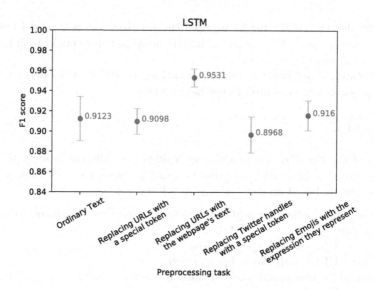

Fig. 2. The obtained results on 5-fold cross validation, based on LSTM model. The blue points show the average of F1 in the whole folds and orange lines show the standard deviation. (Color figure online)

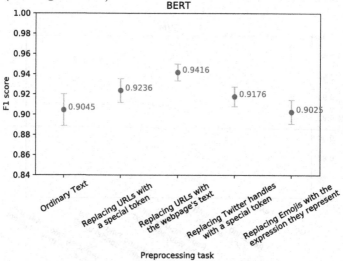

Fig. 3. The obtained results on 5-fold cross validation, based on BERT model. The blue points show the average of F1 in the whole folds and orange lines show the standard deviation. (Color figure online)

As it is presented in the figures, all three models could approximately achieve 90% of F1 score based on the "Ordinary Text" for "fake" as the target class. Also, in all experiments, replacing URLs with the textual content of web-pages that they refer to could significantly (approximately 3%) increase the f-score. The result seems robust comparing to the obtained results from the other pre-processing tasks, considering the standard deviation.

For the other pre-processing tasks, includes replacing URLs, and Twitter handles with special tokens ("weblink", and "username", respectively), and replacing emojis with the expression they represent, different models behave differently. In other words, a pre-processing may increase F-score in one model and degrade the score in another one. However, the rate of the change is not significant.

The obtained results show increasing the data from training a model for the task of fake news detection by replacing URLs with content of the target pages could make a remarkable improvement on the performance of different models.

5   Conclusion and Future Work

In this paper we studied the impact of different pre-processing tasks on the performance of different text classification models for the task of fake news detection in social media. Due to the nature of social media in which post texts are usually very short, the pre-processing and the approaches to deal with social media's special tokens (e.g., emojis, and handles) could play an important role in the overall performance of NLP systems.

Based on our findings, content of the web-page–URLs refer to–could significantly improve the quality of the fake content prediction. In addition, it has been shown that the other pre-processing tasks like replacing emojis with the expression they represent don't have remarkable impact on the performance of the models.

As future works, same experiments can be run on different fake news detection datasets, to better validate our initial findings in this paper. Moreover, the impact of other pre-processing steps, like stemming and stop-word removal can be measured on the performance of classification models for this task.

Acknowledgment. This research was funded in part by the German Federal Ministry of Education and Research (BMBF) under grant number 01IS17043 (project ILSFAS).

References

1. Alam, S., Yao, N.: The impact of preprocessing steps on the accuracy of machine learning algorithms in sentiment analysis. Comput. Math. Organ. Theory **25**(3), 319–335 (2019). https://doi.org/10.1007/s10588-018-9266-8
2. Ayedh, A., Tan, G., Alwesabi, K., Rajeh, H.: The effect of preprocessing on arabic document categorization. Algorithms **9**(2), 27 (2016). https://doi.org/10.3390/a9020027
3. Chen, B., et al.: Transformer-based language model fine-tuning methods for COVID-19 fake news detection. In: Chakraborty, T., Shu, K., Bernard, H.R., Liu, H., Akhtar, M.S. (eds.) CONSTRAINT 2021. CCIS, vol. 1402, pp. 83–92. Springer, Cham (2021). https://doi.org/10.1007/978-3-030-73696-5_9
4. Cortes, C., Vapnik, V.: Support-vector networks. Mach. Learn. **20**(3), 273–297 (1995). https://doi.org/10.1007/BF00994018

5. Datta, A., Si, S.: A supervised machine learning approach to fake news identification. In: Hemanth, D.J., Shakya, S., Baig, Z. (eds.) ICICI 2019. LNDECT,
vol. 38, pp. 197–204. Springer, Cham (2020). https://doi.org/10.1007/978-3-030-
34080-3_22
6. Devlin, J., Chang, M., Lee, K., Toutanova, K.: BERT: pre-training of deep bidirectional transformers for language understanding. In: Burstein, J., Doran, C., Solorio,
T. (eds.) Proceedings of the 2019 Conference of the North American Chapter of
the Association for Computational Linguistics: Human Language Technologies,
NAACL-HLT 2019, Minneapolis, MN, USA, 2–7 June 2019, Vol. 1 (Long and
Short Papers), pp. 4171–4186. Association for Computational Linguistics (2019).
https://doi.org/10.18653/v1/n19-1423
7. Gupta, A., Sukumaran, R., John, K., Teki, S.: Hostility detection and covid-19
fake news detection in social media. CoRR abs/2101.05953 (2021). https://arxiv.
org/abs/2101.05953
8. Hochreiter, S., Schmidhuber, J.: Long short-term memory. Neural Comput. **9**(8),
1735–1780 (1997). https://doi.org/10.1162/neco.1997.9.8.1735
9. Kaliyar, R.K., Goswami, A., Narang, P.: FakeBERT: fake news detection in social
media with a BERT-based deep learning approach. Multimedia Tools and Appl.
80(8), 11765–11788 (2021). https://doi.org/10.1007/s11042-020-10183-2
10. Kingma, D.P., Ba, J.: Adam: a method for stochastic optimization. In: Bengio, Y.,
LeCun, Y. (eds.) 3rd International Conference on Learning Representations, ICLR
2015, San Diego, CA, USA, 7–9 May 2015, Conference Track Proceedings (2015).
http://arxiv.org/abs/1412.6980
11. Mohtaj, S., Schmitt, V., Möller, S.: A feature extraction based model for hate
speech identification. CoRR abs/2201.04227 (2022). https://arxiv.org/abs/2201.
04227
12. Mohtaj, S., Woloszyn, V., Möller, S.: TUB at HASOC 2020: Character based LSTM
for hate speech detection in Indo-European languages. In: Mehta, P., Mandl, T.,
Majumder, P., Mitra, M. (eds.) Working Notes of FIRE 2020 - Forum for Information Retrieval Evaluation, Hyderabad, India, 16–20 December 2020. CEUR Workshop Proceedings, vol. 2826, pp. 298–303. CEUR-WS.org (2020). http://ceur-ws.
org/Vol-2826/T2-26.pdf
13. Patwa, P., et al.: Fighting an infodemic: COVID-19 fake news dataset. In:
Chakraborty, T., Shu, K., Bernard, H.R., Liu, H., Akhtar, M.S. (eds.) CON-
STRAINT 2021. CCIS, vol. 1402, pp. 21–29. Springer, Cham (2021). https://doi.
org/10.1007/978-3-030-73696-5_3
14. Pennington, J., Socher, R., Manning, C.D.: Glove: Global vectors for word representation. In: Moschitti, A., Pang, B., Daelemans, W. (eds.) Proceedings of the
2014 Conference on Empirical Methods in Natural Language Processing, EMNLP
2014, 25–29 October 2014, Doha, Qatar, A meeting of SIGDAT, a Special Interest
Group of the ACL, pp. 1532–1543. ACL (2014). https://doi.org/10.3115/v1/d14-
1162
15. Pimpalkar, A.P., Raj, R.J.R.: Influence of pre-processing strategies on the performance of ML classifiers exploiting TF-IDF and bow features. ADCAIJ: Adv.
Distrib. Comput. Artif. Intell. J. **9**(2), 49 (2020)
16. Stamatatos, E.: Plagiarism detection using stopword n-grams. J. Assoc. Inf. Sci.
Technol. **62**(12), 2512–2527 (2011). https://doi.org/10.1002/asi.21630
17. Uysal, A.K., Günal, S.: The impact of preprocessing on text classification. Inf.
Process. Manag. **50**(1), 104–112 (2014). https://doi.org/10.1016/j.ipm.2013.08.006
18. Yang, Y., Zheng, L., Zhang, J., Cui, Q., Li, Z., Yu, P.S.: TI-CNN: convolutional
neural networks for fake news detection. CoRR abs/1806.00749 (2018). http://
arxiv.org/abs/1806.00749

Short Papers

Short Papers

Business Document Information Extraction: Towards Practical Benchmarks

Matyáš Skalický[iD], Štěpán Šimsa[iD], Michal Uřičář[iD], and Milan Šulc[(✉)][iD]

Rossum.ai, Prague 8, Czech Republic
{matyas.skalicky,stepan.simsa,michal.uricar,milan.sulc}@rossum.ai

Abstract. Information extraction from semi-structured documents is crucial for frictionless business-to-business (B2B) communication. While machine learning problems related to *Document Information Extraction* (IE) have been studied for decades, many common problem definitions and benchmarks do not reflect domain-specific aspects and practical needs for automating B2B document communication. We review the landscape of Document IE problems, datasets and benchmarks. We highlight the practical aspects missing in the common definitions and define the *Key Information Localization and Extraction* (KILE) and *Line Item Recognition* (LIR) problems. There is a lack of relevant datasets and benchmarks for Document IE on semi-structured business documents as their content is typically legally protected or sensitive. We discuss potential sources of available documents including synthetic data.

Keywords: Document Understanding · Survey · Benchmarks · Datasets

1 Introduction

The majority of B2B communication takes place through the exchange of *semi-structured*[1] *business documents* such as invoices, purchase orders and delivery notes. Automating information extraction from such documents has a considerable potential to reduce repetitive manual work and to streamline business communication. There have been efforts to provide standards for electronic data interchange of business document metadata [5,7,52]. Despite, e.g., electronic invoices taking place rapidly [16], the standards did not get globally adopted, none of them prevails, and most are not inter-operable [21].

Machine learning (ML), natural language processing (NLP), and computer vision problems related to *Document Understanding* and *Document IE* have been studied for decades. Despite the major potential of IE from semi-structured business documents, published research on *Document IE* often focuses on other

---

[1] The term *semi-structured documents* is commonly used in different meanings: Some use it for text files containing semi-structured data [94], such as XML files. We use the term to refer to visually rich documents without a fixed layout [66].

A. Barrón-Cedeño et al. (Eds.): CLEF 2022, LNCS 13390, pp. 105–117, 2022.
https://doi.org/10.1007/978-3-031-13643-6_8

domains [13,14,38,39,73,100], and many of the defined tasks and benchmarks do not reflect domain-specific ML aspects and pitfalls of IE from semi-structured business documents. Publications dealing with business documents typically use private datasets [19,34,40,47,59,60,69,97], hindering reproducibility and cross-evaluation. This is caused by the absence of a large public dataset of semi-structured business documents, noted by several authors [20,41,60,77].

The contributions of this position paper are threefold: first, we provide a review of IE problems, datasets and benchmarks relevant to semi-structured business documents. Second, we identify unaddressed aspects of the tasks and formulate new definitions for *Key Information Localization and Extraction* and *Line Item Recognition*. Third, we stress the lack of a large-scale dataset of semi-structured business documents and we discuss potential sources of documents for such dataset.

2   Document Information Extraction Problems

2.1   Key Information Extraction (KIE) and Localization (KILE)

Most formulations of KIE come from NLP, where it is usually defined as the extraction of values of a fixed set of entities/classes from an unstructured text source into a structured form [35,43,56,96,96]. Based on the document representation, Garncarek et al. [25] categorize KIE into three groups: (i) sequence-based (working with serialized text [37]), (ii) graph-based (modeling each doc./page as a graph with nodes corresponding to textual segments and their features [17,32,41]), and (iii) grid-based (treating documents as a 2D grid of token embeddings [19,40]). Sequence-based KIE is closely related to *Named Entity Recognition* (NER) [43]—a sub-task of KIE [48,96] dealing with sequence tagging problems. Borchmann et al. [6] say that (end-to-end) KIE, unlike NER, does not assume that token-level annotations are available. The task is also referred to as Slot Filling [60], meaning that a pre-defined slot is filled with the extracted text.

The common definitions of KIE, as well as some of the datasets [6,73], do not require the location of the extracted information within the document. While the localization is typically not crucial w.r.t. to the downstream task, it plays a vital role in applications that require human validation. We extend the definition by explicitly including the localization:

Definition 1 (KILE). Given a document, the goal of *Key Information Localization and Extraction* (KILE) is to localize (e.g., by a bounding box) fields of each pre-defined category (*key*), to read out their values, and to aggregate the values to extract the key information of each category.

Compared to *Semantic Entity Recognition*, as defined by Xu et al. [93], bounding boxes in KILE are not limited to individual words (tokens).

2.2   Table Extraction and Line Items

Table Understanding [33] and *Table Extraction* (TE) [26,98] are problems where the tabular structure is crucial for IE. Unlike KIE, which outputs individual fields

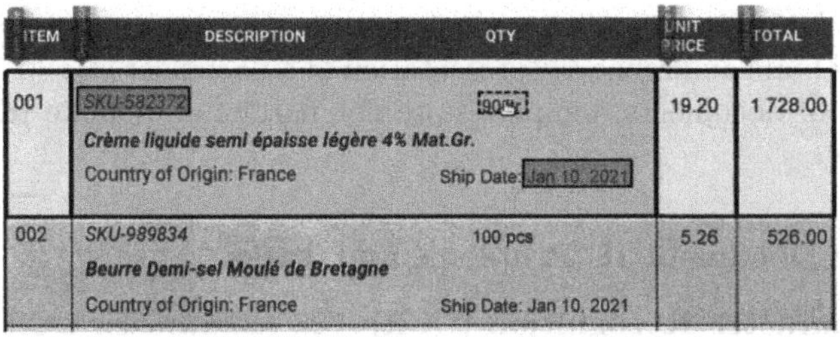

Fig. 1. Example of a table structure where field type is not uniquely determined by its column. Source: https://rossum.ai/help/article/extracting-nested-values-line-items.

independently, TE typically deals with a list of (line) items [4,19,33,48,60], each consisting of a tuple of fields (e.g., *goods* and *price*).

In simple tables, columns determine the field type and rows determine which item the value belongs to. The table can therefore be represented as a grid [68,78]. A bottom-up approach [62,98] can handle more complex tables as in Fig. 1, without relying on a row or column detection. Detected cells or fields can be converted to table structure (determining the line items and columns) in a post-processing step, e.g., spatial clustering [98]. Other works [46,99] tackle the table extraction by directly solving an image-to-markup (e.g., XML or TEX) problem.

We argue that the problem definition should not rely on the structure but rather reflect the information to be extracted and stored. This is close to the problem of detecting the area belonging to a single line item [19]. We define *Line Item* (LI) and the task of *Line Item Recognition* (LIR) as follows:

Definition 2 (LI). A *Line Item* is a tuple of fields describing a single object instance to be extracted, e.g., a row in a table.

Definition 3 (LIR). Given an image of a document page or of a table, the goal of *Line Item Recognition* is to detect all LI present in the section, classify them into a fixed set of classes (e.g., *ordered item*, *discount*, . . .) and for every detected LI, localize and extract key information (as in Definition 1) related to it.

Note that this definition of LIR allows: (i) detection of several tables with different item types, as well as different item types within a single table; (ii) a single field (e.g., a date) to belong to several line items.

2.3   One-Shot Learning for Information Extraction

Layouts of business documents vary greatly, even within a single document type. There are thousands of invoice templates available, and vendors often further adjust them to their needs. Systems without the ability of fast re-training are at risk of degraded performance when faced with a shift in the incoming data distribution [29], such as when presented with previously unseen layouts.

Improving IE with each processed document is known as a one-shot [20]/ online [74] template matching, case-based reasoning [29], or configuration-free IE

[69]. This includes systems that reuse annotations of similar documents [20,69] or iteratively build and refine a representation of a document class [17,29,53,67]. Annotations of documents' templates are not part of any public IE dataset of sufficient size.

2.4 Other Document IE Problems and Tasks

Optical Character Recognition (OCR) [72], handwritten OCR [31], scene text recognition [2,95], including (sub)word or text-line level predictions, are standard problems with reviews and comparisons available [28,36,54,58]. While highly relevant to the document IE, this paper aims at the "higher-level" document IE problems, often assuming text extracted from PDF or OCR is available.

Document Layout Analysis (DLA) is typically posed as an object detection problem: given a document page, find the minimum bounding boxes (or other area representation [1,15]) of layout elements such as *Paragraph, Heading, Table, Figure,* or *Caption*. Most DLA datasets [1,15,23,100] contain such layout annotations for scientific/technical publications and magazines.

Extraction of Key-Value Pairs (KVP) refers to recognizing pairs of linked data items where the key is used as a unique identifier for the value. This task usually consists of semantic labeling and semantic linking [92,93]. Contrary to KIE, KVP extraction does not require the set of keys to be fixed. It also assumes that both key and value are present in the document. This may be useful, e.g., to extract data from unknown forms. However, in semi-structured business documents, it is pretty standard that the keys of interests (known in advance) are not explicitly present in the document.

Question Answering (QA), also known as *Machine Reading Comprehension*, is a common problem in information retrieval and NLP. The goal is to automatically answer questions formulated in natural language. Many NLP tasks can be reformulated as QA [42,51]. Similar to KIE, QA can be extended to incorporate visual information to Visual Question Answering (VQA) [50]. VQA system may also interpret and extract content from the figures, diagrams, and other non-textual elements.

KIE can be formulated as an instance of VQA. However, we typically know which classes of key information should be extracted, rendering the natural language interface unnecessary.

3   Semi-structured Business Document Datasets

Publications on business document IE are often based on private datasets [19,34,40,47,59,60,69,97]. Due to the documents' sensitive content, authors are typically not allowed to share the experimental data. Large third-party sources like common crawl are publicly available; however, re-publishing such data may pose legal issues. For example, a large common crawl dataset of PDF documents by Xu et al. [93] was not published, while pre-training on it was crucial for the proposed method, and the C4 dataset [64] is shared in the form of code that extracts it directly from Common Crawl.

Table 1. Datasets related to KI(L)E from semi-structured business documents.

name	document type	docs	fieldtypes	source	multipage	lang.	type
WildReceipt [76]	receipts	1740	25	photo	no	en	KILE
Ghega [53]	patents/datasheets	246	11/8	scan	yes	en	KILE
EPHOIE [79]	chinese forms	1494	10	scan	no	zh	KILE
CORD [61]	receipts	11000	42	photo	no	ind	KILE
DeepForm [75]	invoices, orders	1000	6	scan	yes	en	KILE
Kleister Charity [73]	financial reports	2788	8	scan	yes	en	KIE
Kleister NDA [73]	NDA documents	540	4	scan	yes	en	KIE
SROIE [35]	receipts	973	4	scan	no	en	KIE

Publicly available datasets for KI(L)E from business documents are summarized in Table 1. However, most of them are relatively small and only contain a few annotated field types. The two largest datasets consist entirely of receipts. Table 1 does not include datasets without KIE annotations—RVL-CDIP [30] (classification), FUNSD [27] and XFUND [93] (no fieldtypes), NIST [84] (forms identification) and DocVQA [50] (QA)—and datasets we were not able to download[2] [3,8,65,101].

Borchmann et al. [6] recently joined and re-formulated several existing document IE datasets to build the DUE benchmark for several document understanding tasks on different document domains. DeepForm [75] and Kleister Charity [73] are the only subsets of DUE with business document KIE annotations.

While there are many datasets for Table Detection and LIR [15,18,22,24,26, 57,63,70,71,87,90,98–100], some of them are not accessible anymore [18,24,26, 70]. We find only FinTabNet [98] and SynthTabNet [57] to be relevant to us by covering complex financial tables.

4 Where to Get More Documents

Publicly Available Documents. Business documents are typically not shared publicly due to their private content, often including confidential and personal information. There are exceptions to this rule—e.g., institutions such as governments or charities have to make certain financial documents publicly available for transparency reasons. Databases of such documents have already been used to create public datasets for document IE: Several datasets[3]—IIT-CDIP [44], RVL-CDIP [30], FUNSD [27], and DocVQA [50]—were built from documents from the UCSF Industry Documents Library[4] [83]. Annual Reports of the S&P 500 companies [88] were used to create FinTabNet [98]. Non-disclosure agreements from the EDGAR[5] database [82], collected for the U.S. Securities and Exchange

---

[2] For some only the annotations are available, without the original PDFs/images.

[3] Some of the datasets are subsets: FUNSD [27] ⊂ RVL-CDIP [30] ⊂ IIT-CDIP [44].

[4] A large proportion of the UCSF Industry Document Library are old documents, often written on a typewriter, which presents a domain shift w.r.t. today's documents.

[5] Automated crawling of the site not allowed: https://www.sec.gov/os/accessing-edgar-data.

Commission, were used for the Kleister-NDA [73] dataset. The DeepForm dataset [75] consists of documents related to broadcast stations from the FCC Public Inspection Files [86]. Financial records from the Charity Commission [81] were used to create the Kleister-Charity dataset [73]. Several QA datasets [9,55,101] were also collected from open data sources [80,85,91]. Other datasets were build via web search [10,49,76], from Common Crawl[6] [64,93], Wikipedia [11–13], or platforms for sharing scientific papers [38,39,100].

Synthetic Documents. Manual annotation is expensive, and the collection of data from public sources may be limited by the presence of personal data or intellectual property. This reasoning calls for leveraging synthetic datasets. Xu et al. [93] manually replaced the content of publicly available documents with synthetic data. Bensch et al. [4] generate synthetic invoice documents automatically. However, we observe that the generated invoices have a plain style and do not resemble the distribution of visual layouts of real business documents. Nassar et al. [57] synthesized a table dataset of four appearance styles based on existing datasets [46,98,99].

We consider three ways to define layouts to be filled with synthetic data: (i) manual design of layouts which can be used to generate a high number of documents with different semantically-matching values, but costly at scale with increasing num. of layouts, (ii) extraction from public documents followed by sensitive anonymization like in [93], and (iii) using a generative model, e.g., to generate realistic layouts dissimilar to those already present in the dataset. We consider such a problem statement an interesting open research problem.

5   Discussion and Future Work

We argue that the problems of KILE and LIR, as defined in Sect. 2, are crucial for automating B2B document communication where key information must be extracted from localized fields and line items. The review of public datasets in Sect. 3 shows that—except for receipts [35,61,76]—semi-structured business documents like invoices, orders, and delivery notes are underrepresented in document IE. Based on manual inspection of selected documents from publicly available sources in Sect. 4, we noticed the distribution of documents differs significantly among different sources. An ideal dataset should cover a large variety of visual styles and layouts and provide diagnostic subsets [6] to differentiate errors in various special cases. Due to high annotation costs and possibly legally protected content of business documents, synthetic data are a potentially affordable alternative for building a large-scale dataset. While synthetic data have been proven successful for OCR [45], the potential of data synthesis for business document IE has not yet been fulfilled: existing attempts either target other tasks and document types [93] or do not reflect the rich visual distribution of semi-structured business documents [4]. An advantage of generating synthetic

---

[6] CC-MAIN-2022-05 contains almost 3 billion documents out of which 0.84% are PDFs [89] – however, most of them are not semi-structured business documents.

documents of a given layout is the known layout annotation for benchmarking one-shot information extraction.

To enable benchmarking of information extraction on data and tasks highly relevant to real-world application scenarios, in our future work, we are preparing a large-scale public dataset of semi-structured business documents, following the observations and points made in this paper.

References

1. Antonacopoulos, A., Bridson, D., Papadopoulos, C., Pletschacher, S.: A realistic dataset for performance evaluation of document layout analysis. In: Proceedings of ICDAR, pp. 296–300. IEEE (2009)
2. Baek, Y., Lee, B., Han, D., Yun, S., Lee, H.: Character region awareness for text detection. In: Proceedings of the IEEE/CVF CVPR, pp. 9365–9374 (2019)
3. Baviskar, D., Ahirrao, S., Kotecha, K.: Multi-layout invoice document dataset (MIDD): a dataset for named entity recognition. Data (2021). https://doi.org/10.3390/data6070078
4. Bensch, O., Popa, M., Spille, C.: Key information extraction from documents: evaluation and generator. In: Abbès, S.B., et al. (eds.) Proceedings of DeepOntoNLP and X-SENTIMENT. CEUR Workshop Proceedings, vol. 2918, pp. 47–53. CEUR-WS.org (2021)
5. Berge, J.: The EDIFACT Standards. Blackwell Publishers, Inc. (1994)
6. Borchmann, L., et al.: DUE: End-to-end document understanding benchmark. In: Proceedings of NeurIPS (2021)
7. Bosak, J., McGrath, T., Holman, G.K.: Universal business language v2. 0. Organization for the Advancement of Structured Information Standards (OASIS), Standard (2006)
8. Cesarini, F., Francesconi, E., Gori, M., Soda, G.: Analysis and understanding of multi-class invoices. Doc. Anal. Recogn. **6**(2), 102–114 (2003)
9. Chaudhry, R., Shekhar, S., Gupta, U., Maneriker, P., Bansal, P., Joshi, A.: LEAF-QA: locate, encode & attend for figure question answering. In: Proceedings of WACV, pp. 3501–3510. IEEE (2020). https://doi.org/10.1109/WACV45572.2020.9093269
10. Chen, L., et al.: WebSRC: a dataset for web-based structural reading comprehension. CoRR (2021)
11. Chen, W., Chang, M., Schlinger, E., Wang, W.Y., Cohen, W.W.: Open question answering over tables and text. In: Proceedings of ICLR (2021)
12. Chen, W., et al.: TabFact: a large-scale dataset for table-based fact verification. In: Proceedings of ICLR (2020)
13. Chen, W., Zha, H., Chen, Z., Xiong, W., Wang, H., Wang, W.Y.: HybridQA: a dataset of multi-hop question answering over tabular and textual data. In: Cohn, T., He, Y., Liu, Y. (eds.) Findings of the Association for Computational Linguistics: EMNLP. Findings of ACL, vol. EMNLP 2020, pp. 1026–1036. Association for Computational Linguistics (2020). https://doi.org/10.18653/v1/2020.findings-emnlp.91
14. Cho, M., Amplayo, R.K., Hwang, S., Park, J.: Adversarial TableQA: attention supervision for question answering on tables. In: Zhu, J., Takeuchi, I. (eds.) Proceedings of ACML. Proceedings of Machine Learning Research, vol. 95, pp. 391–406 (2018)

15. Clausner, C., Antonacopoulos, A., Pletschacher, S.: ICDAR 2019 competition on recognition of documents with complex layouts-RDCL2019. In: Proceedings of ICDAR, pp. 1521–1526. IEEE (2019)
16. Cristani, M., Bertolaso, A., Scannapieco, S., Tomazzoli, C.: Future paradigms of automated processing of business documents. Int. J. Inf. Manag. **40**, 67–75 (2018)
17. d'Andecy, V.P., Hartmann, E., Rusinol, M.: Field extraction by hybrid incremental and a-priori structural templates. In: 2018 13th IAPR International Workshop on Document Analysis Systems (DAS), pp. 251–256. IEEE (2018)
18. Deng, Y., Rosenberg, D.S., Mann, G.: Challenges in end-to-end neural scientific table recognition. In: Proceedings of ICDAR, pp. 894–901. IEEE (2019). https://doi.org/10.1109/ICDAR.2019.00148
19. Denk, T.I., Reisswig, C.: BERTgrid: contextualized embedding for 2D document representation and understanding. arXiv preprint arXiv:1909.04948 (2019)
20. Dhakal, P., Munikar, M., Dahal, B.: One-shot template matching for automatic document data capture. In: Proceedings of Artificial Intelligence for Transforming Business and Society (AITB), vol. 1, pp. 1–6. IEEE (2019)
21. Directive 2014/55/EU of the European parliament and of the council on electronic invoicing in public procurement, April 2014. https://eur-lex.europa.eu/eli/dir/2014/55/oj
22. Fang, J., Tao, X., Tang, Z., Qiu, R., Liu, Y.: Dataset, ground-truth and performance metrics for table detection evaluation. In: Blumenstein, M., Pal, U., Uchida, S. (eds.) Proceedings of IAPR International Workshop on Document Analysis Systems, DAS, pp. 445–449. IEEE (2012). https://doi.org/10.1109/DAS.2012.29
23. Ford, G., Thoma, G.R.: Ground truth data for document image analysis. In: Symposium on Document Image Understanding and Technology, pp. 199–205. Citeseer (2003)
24. Gao, L., Yi, X., Jiang, Z., Hao, L., Tang, Z.: ICDAR2017 competition on page object detection. In: Proceedings of ICDAR, pp. 1417–1422 (2017). https://doi.org/10.1109/ICDAR.2017.231
25. Garncarek, Ł., et al.: LAMBERT: layout-aware language modeling for information extraction. In: Lladós, J., Lopresti, D., Uchida, S. (eds.) ICDAR 2021. LNCS, vol. 12821, pp. 532–547. Springer, Cham (2021). https://doi.org/10.1007/978-3-030-86549-8_34
26. Göbel, M.C., Hassan, T., Oro, E., Orsi, G.: ICDAR 2013 table competition. In: Proceedings of ICDAR, pp. 1449–1453. IEEE Computer Society (2013). https://doi.org/10.1109/ICDAR.2013.292
27. Jaume, G., Ekenel, H.K., Thiran, J.P.: FUNSD: a dataset for form understanding in noisy scanned documents. In: ICDAR-OST (2019, accepted)
28. Hamad, K.A., Mehmet, K.: A detailed analysis of optical character recognition technology. Int. J. Appl. Math. Electron. Comput. **1**(Special Issue-1), 244–249 (2016)
29. Hamza, H., Belaïd, Y., Belaïd, A.: Case-based reasoning for invoice analysis and recognition. In: Weber, R.O., Richter, M.M. (eds.) ICCBR 2007. LNCS (LNAI), vol. 4626, pp. 404–418. Springer, Heidelberg (2007). https://doi.org/10.1007/978-3-540-74141-1_28
30. Harley, A.W., Ufkes, A., Derpanis, K.G.: Evaluation of deep convolutional nets for document image classification and retrieval. In: International Conference on Document Analysis and Recognition (ICDAR) (2015)
31. He, S., Schomaker, L.: Beyond OCR: multi-faceted understanding of handwritten document characteristics. Pattern Recogn. **63**, 321–333 (2017)

32. Holeček, M.: Learning from similarity and information extraction from structured documents. Int. J. Doc. Anal. Recogn. (IJDAR) 1–17 (2021)

33. Holeček, M., Hoskovec, A., Baudiš, P., Klinger, P.: Table understanding in structured documents. In: 2019 International Conference on Document Analysis and Recognition Workshops (ICDARW), vol. 5, pp. 158–164. IEEE (2019)

34. Holt, X., Chisholm, A.: Extracting structured data from invoices. In: Proceedings of the Australasian Language Technology Association Workshop 2018, pp. 53–59 (2018)

35. Huang, Z., et al.: ICDAR2019 competition on scanned receipt OCR and information extraction. In: Proceedings of ICDAR, pp. 1516–1520. IEEE (2019). https://doi.org/10.1109/ICDAR.2019.00244

36. Islam, N., Islam, Z., Noor, N.: A survey on optical character recognition system. arXiv preprint arXiv:1710.05703 (2017)

37. Jiang, J.: Information extraction from text. In: Aggarwal, C., Zhai, C. (eds.) Mining Text Data, pp. 11–41. Springer, Cham (2012). https://doi.org/10.1007/978-1-4614-3223-4_2

38. Jobin, K.V., Mondal, A., Jawahar, C.V.: DocFigure: a dataset for scientific document figure classification. In: 13th IAPR International Workshop on Graphics Recognition, GREC@ICDAR, pp. 74–79. IEEE (2019). https://doi.org/10.1109/ICDARW.2019.00018

39. Kardas, M., et al.: AxCell: automatic extraction of results from machine learning papers. arXiv preprint arXiv:2004.14356 (2020)

40. Katti, A.R., et al.: Chargrid: towards understanding 2D documents. arXiv preprint arXiv:1809.08799 (2018)

41. Krieger, F., Drews, P., Funk, B., Wobbe, T.: Information extraction from invoices: a graph neural network approach for datasets with high layout variety. In: Ahlemann, F., Schütte, R., Stieglitz, S. (eds.) WI 2021. LNISO, vol. 47, pp. 5–20. Springer, Cham (2021). https://doi.org/10.1007/978-3-030-86797-3_1

42. Kumar, A., et al.: Ask me anything: dynamic memory networks for natural language processing. In: Balcan, M., Weinberger, K.Q. (eds.) Proceedings of ICML, vol. 48, pp. 1378–1387. JMLR.org (2016)

43. Lample, G., Ballesteros, M., Subramanian, S., Kawakami, K., Dyer, C.: Neural architectures for named entity recognition. In: Knight, K., Nenkova, A., Rambow, O. (eds.) Proceedings of NAACL HLT, pp. 260–270 (2016). https://doi.org/10.18653/v1/n16-1030

44. Lewis, D., Agam, G., Argamon, S., Frieder, O., Grossman, D., Heard, J.: Building a test collection for complex document information processing. In: Proceedings of the 29th Annual International ACM SIGIR Conference on Research and Development in Information Retrieval, pp. 665–666 (2006)

45. Li, J., Wang, S., Wang, Y., Tang, Z.: Synthesizing data for text recognition with style transfer. Multimed. Tools Appl. **78**(20), 29183–29196 (2019)

46. Li, M., Cui, L., Huang, S., Wei, F., Zhou, M., Li, Z.: TableBank: table benchmark for image-based table detection and recognition. In: Calzolari, N., et al. (eds.) Proceedings of The 12th Language Resources and Evaluation Conference, LREC. pp. 1918–1925 (2020)

47. Liu, W., Zhang, Y., Wan, B.: Unstructured document recognition on business invoice. Machine Learning, Stanford iTunes University, Stanford, CA, USA, Technical report (2016)

48. Majumder, B.P., Potti, N., Tata, S., Wendt, J.B., Zhao, Q., Najork, M.: Representation learning for information extraction from form-like documents. In: Jurafsky,

D., Chai, J., Schluter, N., Tetreault, J.R. (eds.) Proceedings of the 58th Annual Meeting of the Association for Computational Linguistics, ACL, pp. 6495–6504 (2020). https://doi.org/10.18653/v1/2020.acl-main.580

49. Mathew, M., Bagal, V., Tito, R., Karatzas, D., Valveny, E., Jawahar, C.: InfographicVQA. In: Proceedings of the IEEE/CVF Winter Conference on Applications of Computer Vision, pp. 1697–1706 (2022)

50. Mathew, M., Karatzas, D., Jawahar, C.V.: DocVQA: a dataset for VQA on document images. In: Proceedings of WACV, pp. 2199–2208. IEEE (2021). https://doi.org/10.1109/WACV48630.2021.00225

51. McCann, B., Keskar, N.S., Xiong, C., Socher, R.: The natural language decathlon: multitask learning as question answering. CoRR (2018)

52. Meadows, B., Seaburg, L.: Universal business language 1.0. Organization for the Advancement of Structured Information Standards (OASIS) (2004)

53. Medvet, E., Bartoli, A., Davanzo, G.: A probabilistic approach to printed document understanding. Int. J. Doc. Anal. Recogn. **14**(4), 335–347 (2011). https://doi.org/10.1007/s10032-010-0137-1

54. Memon, J., Sami, M., Khan, R.A., Uddin, M.: Handwritten optical character recognition (OCR): a comprehensive systematic literature review (SLR). IEEE Access **8**, 142642–142668 (2020)

55. Methani, N., Ganguly, P., Khapra, M.M., Kumar, P.: PlotQA: reasoning over scientific plots. In: Proceedings of WACV, pp. 1516–1525 (2020). https://doi.org/10.1109/WACV45572.2020.9093523

56. Nadeau, D., Sekine, S.: A survey of named entity recognition and classification. Lingvisticæ Investigationes, pp. 3–26 (2007). https://doi.org/10.1075/li.30.1.03nad

57. Nassar, A., Livathinos, N., Lysak, M., Staar, P.W.J.: TableFormer: table structure understanding with transformers. CoRR abs/2203.01017 (2022). https://doi.org/10.48550/arXiv.2203.01017

58. Nayef, N., et al.: ICDAR 2019 robust reading challenge on multi-lingual scene text detection and recognition-RRC-MLT-2019. In: Proceedings of ICDAR, pp. 1582–1587. IEEE (2019)

59. Palm, R.B., Laws, F., Winther, O.: Attend, copy, parse end-to-end information extraction from documents. In: 2019 International Conference on Document Analysis and Recognition (ICDAR), pp. 329–336. IEEE (2019)

60. Palm, R.B., Winther, O., Laws, F.: CloudScan - a configuration-free invoice analysis system using recurrent neural networks. In: Proceedings of ICDAR, pp. 406–413. IEEE (2017). https://doi.org/10.1109/ICDAR.2017.74

61. Park, S., el al.: Cord: a consolidated receipt dataset for post-OCR parsing. In: Workshop on Document Intelligence at NeurIPS 2019 (2019)

62. Prasad, D., Gadpal, A., Kapadni, K., Visave, M., Sultanpure, K.: CascadeTabNet: an approach for end to end table detection and structure recognition from image-based documents. In: Proceedings of CVPRw, pp. 2439–2447 (2020). https://doi.org/10.1109/CVPRW50498.2020.00294

63. Qasim, S.R., Mahmood, H., Shafait, F.: Rethinking table recognition using graph neural networks. In: Proceedings of ICDAR, pp. 142–147. IEEE (2019). https://doi.org/10.1109/ICDAR.2019.00031

64. Raffel, C., et al.: Exploring the limits of transfer learning with a unified text-to-text transformer. arXiv preprint arXiv:1910.10683 (2019)

65. Rastogi, M., et al.: Information extraction from document images via FCA based template detection and knowledge graph rule induction. In: Proceedings

of CVPRw, pp. 2377–2385 (2020). https://doi.org/10.1109/CVPRW50498.2020.
00287

66. Riba, P., Dutta, A., Goldmann, L., Fornés, A., Ramos, O., Lladós, J.: Table
detection in invoice documents by graph neural networks. In: 2019 International
Conference on Document Analysis and Recognition (ICDAR), pp. 122–127. IEEE
(2019)

67. Rusinol, M., Benkhelfallah, T., Poulain dAndecy, V.: Field extraction from administrative documents by incremental structural templates. In: 2013 12th International Conference on Document Analysis and Recognition, pp. 1100–1104. IEEE
(2013)

68. Schreiber, S., Agne, S., Wolf, I., Dengel, A., Ahmed, S.: DeepDeSRT: deep learning for detection and structure recognition of tables in document images. In:
Proceedings of ICDAR, pp. 1162–1167 (2017). https://doi.org/10.1109/ICDAR.
2017.192

69. Schuster, D., et al.: Intellix-end-user trained information extraction for document
archiving. In: 2013 12th International Conference on Document Analysis and
Recognition, pp. 101–105. IEEE (2013)

70. Shahab, A., Shafait, F., Kieninger, T., Dengel, A.: An open approach towards the
benchmarking of table structure recognition systems. In: Doermann, D.S., Govin-
daraju, V., Lopresti, D.P., Natarajan, P. (eds.) The Ninth IAPR International
Workshop on Document Analysis Systems, DAS, pp. 113–120 (2010). https://
doi.org/10.1145/1815330.1815345

71. Siegel, N., Lourie, N., Power, R., Ammar, W.: Extracting scientific figures with
distantly supervised neural networks. In: Chen, J., Gonçalves, M.A., Allen, J.M.,
Fox, E.A., Kan, M., Petras, V. (eds.) Proceedings of the 18th ACM/IEEE on
Joint Conference on Digital Libraries, JCDL, pp. 223–232 (2018). https://doi.
org/10.1145/3197026.3197040

72. Smith, R.: An overview of the tesseract OCR engine. In: Ninth International
Conference on Document Analysis and Recognition (ICDAR 2007), vol. 2, pp.
629–633. IEEE (2007)

73. Stanisławek, T., et al.: Kleister: key information extraction datasets involving long
documents with complex layouts. In: Lladós, J., Lopresti, D., Uchida, S. (eds.)
ICDAR 2021. LNCS, vol. 12821, pp. 564–579. Springer, Cham (2021). https://
doi.org/10.1007/978-3-030-86549-8_36

74. Stockerl, M., Ringlstetter, C., Schubert, M., Ntoutsi, E., Kriegel, H.P.: Online
template matching over a stream of digitized documents. In: Proceedings of the
27th International Conference on Scientific and Statistical Database Management,
pp. 1–12 (2015)

75. Stray, J., Svetlichnaya, S.: DeepForm: extract information from documents (2020).
https://wandb.ai/deepform/political-ad-extraction, benchmark

76. Sun, H., Kuang, Z., Yue, X., Lin, C., Zhang, W.: Spatial dual-modality graph
reasoning for key information extraction. arXiv preprint arXiv:2103.14470 (2021)

77. Sunder, V., Srinivasan, A., Vig, L., Shroff, G., Rahul, R.: One-shot information
extraction from document images using neuro-deductive program synthesis. arXiv
preprint arXiv:1906.02427 (2019)

78. Tensmeyer, C., Morariu, V.I., Price, B., Cohen, S., Martinez, T.: Deep splitting
and merging for table structure decomposition. In: 2019 International Conference
on Document Analysis and Recognition (ICDAR), pp. 114–121. IEEE (2019)

79. Wang, J., et al.: Towards robust visual information extraction in real world: new
dataset and novel solution. In: Proceedings of the AAAI Conference on Artificial
Intelligence (2021)

80. Web: Annual reports. https://www.annualreports.com/. Accessed 28 Apr 2022
81. Web: Charity Commission for England and Wales. https://apps.charitycommission.gov.uk/showcharity/registerofcharities/RegisterHomePage.aspx. Accessed 22 Apr 2022
82. Web: EDGAR. https://www.sec.gov/edgar.shtml. Accessed 22 Apr 2022
83. Web: Industry Documents Library. https://www.industrydocuments.ucsf.edu/. Accessed 22 Apr 2022
84. Web: NIST Special Database 2. https://www.nist.gov/srd/nist-special-database-2. Accessed 25 Apr 2022
85. Web: Open Government Data (OGD) Platform India. https://visualize.data.gov.in/. Accessed 22 Apr 2022
86. Web: Public Inspection Files. https://publicfiles.fcc.gov/. Accessed 22 Apr 2022
87. Web: Scitsr. https://github.com/Academic-Hammer/SciTSR. Accessed 26 Apr 2022
88. Web: S&P 500 Companies with Financial Information. https://www.spglobal.com/spdji/en/indices/equity/sp-500/#data. Accessed 25 Apr 2022
89. Web: Statistics of Common Crawl Monthly Archives – MIME Types. https://commoncrawl.github.io/cc-crawl-statistics/plots/mimetypes. Accessed 22 Apr 2022
90. Web: Tablebank. https://github.com/doc-analysis/TableBank. Accessed 26 Apr 2022
91. Web: World Bank Open Data. https://data.worldbank.org/. Accessed 22 Apr 2022
92. Xu, Y., Li, M., Cui, L., Huang, S., Wei, F., Zhou, M.: LayoutLM: pre-training of text and layout for document image understanding. In: Gupta, R., Liu, Y., Tang, J., Prakash, B.A. (eds.) Proceedings on KDD, pp. 1192–1200 (2020). https://doi.org/10.1145/3394486.3403172
93. Xu, Y., et al.: LayoutXLM: multimodal pre-training for multilingual visually-rich document understanding. CoRR (2021)
94. Yi, J., Sundaresan, N.: A classifier for semi-structured documents. In: Proceedings of the Sixth ACM SIGKDD International Conference on Knowledge Discovery and Data Mining, pp. 340–344 (2000)
95. Yu, D., et al.: Towards accurate scene text recognition with semantic reasoning networks. In: Proceedings of the IEEE/CVF Conference on Computer Vision and Pattern Recognition, pp. 12113–12122 (2020)
96. Yu, W., Lu, N., Qi, X., Gong, P., Xiao, R.: PICK: processing key information extraction from documents using improved graph learning-convolutional networks. In: Proceedings of ICPR, pp. 4363–4370. IEEE (2020). https://doi.org/10.1109/ICPR48806.2021.9412927
97. Zhao, X., Wu, Z., Wang, X.: CUTIE: learning to understand documents with convolutional universal text information extractor. CoRR abs/1903.12363 (2019). http://arxiv.org/abs/1903.12363
98. Zheng, X., Burdick, D., Popa, L., Zhong, X., Wang, N.X.R.: Global table extractor (GTE): a framework for joint table identification and cell structure recognition using visual context. In: Proceedings of WACV, pp. 697–706. IEEE (2021). https://doi.org/10.1109/WACV48630.2021.00074
99. Zhong, X., ShafieiBavani, E., Jimeno Yepes, A.: Image-based table recognition: data, model, and evaluation. In: Vedaldi, A., Bischof, H., Brox, T., Frahm, J.-M. (eds.) ECCV 2020. LNCS, vol. 12366, pp. 564–580. Springer, Cham (2020). https://doi.org/10.1007/978-3-030-58589-1_34

100. Zhong, X., Tang, J., Jimeno-Yepes, A.: PubLayNet: largest dataset ever for document layout analysis. In: Proceedings of ICDAR, pp. 1015–1022. IEEE, September 2019. https://doi.org/10.1109/ICDAR.2019.00166

101. Zhu, F., et al.: TAT-QA: a question answering benchmark on a hybrid of tabular and textual content in finance. In: Zong, C., Xia, F., Li, W., Navigli, R. (eds.) Proceedings International Joint Conference on Natural Language Processing, pp. 3277–3287 (2021). https://doi.org/10.18653/v1/2021.acl-long.254

An Analysis of Logic Rule Dissemination in Sentiment Classifiers

Shashank Gupta[✉][iD], Mohamed Reda Bouadjenek[iD],
and Antonio Robles-Kelly[iD]

School of Information Technology, Deakin University, Waurn Ponds Campus,
Geelong, VIC 3216, Australia
guptashas@deakin.edu.au

Abstract. Disseminating and incorporating logic rules in deep neural networks has been extensively explored for sentiment classification. Methods that are proposed for that goal rely on a component that aims to capture and model logic rules, followed by a sequence model to process the input sequence. While these methods claim to effectively capture syntactic structures that affect sentiment, they only show improvement in terms of accuracy to support their claims with no further analysis. Focusing on the *A-but-B* rule, we use the PERCY metric (a recently developed Post-hoc Explanation-based score for logic Rule dissemination ConsistencY assessment) to analyze and study the ability of these methods to identify the *A-but-B* structure, and to make their classification decision based on the *B* conjunct. PERCY proceeds by estimating feature attribution scores using LIME, a model-agnostic framework that aims to explain the predictions of any classifier in an interpretable and faithful manner. Our experiments show that (a) accuracy is misleading in assessing these methods, (b) not all these methods are effectively capturing the *A-but-B* structure, (c) often, the underlying sequence model is what captures the syntactic structure, and (d) the best method classifies less than 25% of test examples based on the *B* conjunct.

Keywords: Sentiment Classification · Logic Rules · Explainable AI

1 Introduction

Methods of disseminating and incorporating logic rules in Deep Neural Networks have been extensively explored for sentiment classification. The two main methods developed for that purpose are: (i) Iterative Knowledge Distillation method [1] and (ii) the Contextualized Word Embeddings approach [2]. Briefly, these methods rely on a component aimed at capturing and modeling logic rules (e.g., the teacher network in the Iterative Distillation method and the ELMo model [3] in the Contextualized Word Embeddings approach), followed by a sequence model to process the input sequence, (e.g., a RNN).

The authors of these two methods claim that they effectively capture syntactic structures in the input sentence that affect its sentiment, but they have

A. Barrón-Cedeño et al. (Eds.): CLEF 2022, LNCS 13390, pp. 118–124, 2022.
https://doi.org/10.1007/978-3-031-13643-6_9

only used the improvement in terms of accuracy to support their claim with no further analysis. However, achieving a high classification accuracy does not necessarily mean that a method has effectively captured and encoded rules and other textual syntactic structures. For example, let's consider the sentence *"the casting was not bad but the movie was awful"* that has an *A-but-B* structure – a component *A* followed by *but* which is then followed by a component *B*. In this example, the conjunction is interpreted as an argument for the second conjunct, with the first functioning concessively [4–6]. While a sentiment classifier can correctly identify that this sentence has a negative sentiment, it may fail to infer it's decision based *exclusively* on the *B* part of the sentence (i.e., *"the movie was awful"*), but instead, it may based it's decision on individual negative words also present in Part *A* (i.e., *"bad"*).

While focusing on the *A-but-B* syntactic structure and sentiment classification, we propose in this paper to study the ability of the aforementioned methods to: (i) effectively identifying the *A-but-B* structure in an input sentence, and to (ii) make their classification decision based on the *B* conjunct of a sentence. Specifically, we rely on the PERCY metric [7], a recently developed Post-hoc Explanation-based score for logic Rule dissemination ConsistencY assessment. PERCY estimates feature attribution scores using LIME [8], a model-agnostic framework that aims to explain predictions of any classifier in an interpretable and faithful manner. We validate our findings with an exhaustive experimental evaluation using the SST2 dataset [6] by testing various sentiment classifiers designed for logic rules dissemination. Among numerous findings, we show that: (a) accuracy is misleading in assessing methods for capturing logic rules, (b) not all methods are effectively capturing the *A-but-B* structure, (c) their sequence model is often what captures the syntactic structure, and (d) the best method bases its decision on the *B* conjunct in less than 25% of test examples.

2   Logic Rule Dissemination Methods

In this section, we first describe the neural network architecture we use for sequence modeling, before discussing the main methods we analyse for logic rules dissemination in that architecture.

2.1   Network Architecture

The backbone neural network [9,10] we use throughout this paper is depicted in Fig. 1. Three 1D CNN sequence layers (kernel size

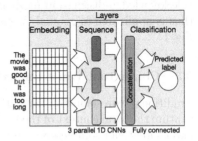

Fig. 1. Neural network.

of 3, 4, and 5) process the word embeddings of an input sequence in parallel in order to extract diverse features and pass the concatenated features into a feed-forward binary classification layer with a sigmoid activation to extract the sentiment of the input sentence – 0 for a negative sentiment and 1 for a positive sentiment. In the next subsections, we will discuss the methods we analyze in this article that aim to incorporate and disseminate logic rules in the neural network architecture depicted in Fig. 1.

2.2 Iterative Rule Knowledge Distillation

The Iterative rule knowledge distillation method proposed by Hu et al. [1] aims to transfer the domain knowledge encoded in first order logic rules into a neural network defined by a conditional probability $p_\theta(y|x)$ where θ is a parameter to learn. Specifically, during training, a posterior $q(y|x)$ is constructed by projecting $p_\theta(y|x)$ into a subspace constrained by the rules to encode the desirable properties, by using the following loss:

$$\min_{q,\xi \geq 0} \quad KL(q(y|x)||p_\theta(y|x)) + C \sum_{x \in X} \xi_x$$

$$s.t. \quad (1 - \mathbb{E}_{y \leftarrow q(\bullet|x)}[r_\theta(x,y)]) \leq \xi_x$$

where $q(y|x)$ denotes the distribution of (x,y) when x is drawn uniformly from the train set X and y is drawn according to $q(\bullet|x)$, and $r_\theta(x,y) \in [0,1]$ is a variable that indicates how well labeling x with y satisfies the rule. The closed form solution for $q(y|x)$ is used as soft targets to imitate the outputs of a rule-regularized projection of $p_\theta(y|x)$, which explicitly includes rule knowledge as regularization terms.

Next, the rule knowledge is transferred to the posterior $p_\theta(y|x)$ through knowledge distillation optimization objective:

$$(1 - \pi) \times \mathcal{L}(p_\theta, P_{true}) + \pi \times \mathcal{L}(p_\theta, q)$$

where P_{true} denotes the distribution implied by the ground truth, $\mathcal{L}(\bullet, \bullet)$ denotes the cross-entropy function, and π is a hyperparameter that needs to be tuned to calibrate the relative importance of the two objectives. Overall, the Iterative rule knowledge distillation method is agnostic to the network architecture, and thus is applicable to general types of neural models such as the one depicted in Fig. 1.

2.3 Contextual Word Embeddings

Traditional word embeddings methods like Word2Vec [11] and Glove [12] do not capture the local context of the word in a sentence. However, language is complex and context can completely change the meaning of a word in a sentence. Hence, contextual word embeddings methods have emerged as a way to capture the different nuances of the meaning of words given the surrounding text. Krishna et al. [2] have advocated that contextualized word embeddings might capture logic rules and thus disseminate that latent information in the 1D CNN sequence models of the neural network in Fig. 1. In the following, we briefly review two of the main contextual word embedding methods we use in our experiments.

ELMo: stands for Embeddings from Language Models is a pre-trained model developed by Peters et al. [3]. Instead of using a fixed embedding for each word, ELMo looks at the entire sentence before assigning each word in it an embedding. It uses a bi-directional LSTM trained on a specific task to be able to create those embeddings. Krishna et al. [2] proposed to use ELMo in their method.

BERT: stands for Bidirectional Encoder Representations from transformers. This is also a pre-trained model developed by Devlin et al. [13]. Briefly, the BERT is a model based on Encoder Transformer blocks [14], which processes each element of the input sequence by incorporating and estimating the influence of other elements in the sequence to create embeddings.

To further test the hypothesis proposed by Krishna et al. [2], we conduct experiments with two different context-free word embeddings namely Word2vec developed by Mikolov et al. [11] and Glove developed by Pennington et al. [12] in which each token is mapped to a unique vector independent of its context. These word embeddings are used as an ablation study to analyze the effectiveness of the rule knowledge distillation method discussed in the previous section.

3 Methodology

As mentioned earlier, our main goal in this paper is to assess each sentiment classifier for it's ability to correctly classify a test example with an A-*but*-B structure only on the basis of the B conjunct. For this purpose, we use a metric called **PERCY** [7], which stands for *Post-hoc Explanation-based Rule ConsistencY assessment Score*. Specifically, given a sentence S which is an ordered sequence of terms $[t_1 t_2 \cdots t_n]$, PERCY relies on LIME to assign a weight w_n to each term t_n in S where a positive weight indicates that t_n contributes and supports the positive class, and a negative weight indicates how much t_n supports the negative class. In order to estimate how much a term t_n contributes to the final decision of the classifier, PERCY normalizes its weight as follows:

$$
\tilde{w}_n = \begin{cases} w_n \times P(y = 1\,S), & \text{if } w_n \geq 0 \\ |w_n| \times P(y = 0\,S), & \text{otherwise} \end{cases}
\tag{1}
$$

where $P(y = c|S)$ is the probability to predict class c given sentence S. Hence, every sentence in our test set is mapped to a vector $[\tilde{w}_1 \tilde{w}_2 \cdots \tilde{w}_n]$ with $\tilde{w}_n$ indicating how much the word t_n contributed to the final decision of the classifier. Next, given a sentence that contains an A-*but*-B structure, PERCY defines the normalized weights $\tilde{W}(A) = [\tilde{w}_0 \cdots \tilde{w}_{i-1}]$ and $\tilde{W}(B) = [\tilde{w}_{i+1} \cdots \tilde{w}_n]$ as respectively the left and right sub-sequences w.r.t the word *"but"* indexed by i. Finally, PERCY computes an expectation over weights as follows: $\mathbb{E}_A(W) = \sum_{\tilde{w}_k \in \tilde{W}(A)} \tilde{w}_k$ and $\mathbb{E}_B(W) = \sum_{\tilde{w}_k \in \tilde{W}(B)} \tilde{w}_k$, and concludes that a classifier has based its classification prediction by relying on the B conjunct if: $\mathbb{E}_B(W) > \mathbb{E}_A(W)$ ***and*** p-value ≤ 0.05 – this condition aims to make sure that the observed difference is statistically significant.

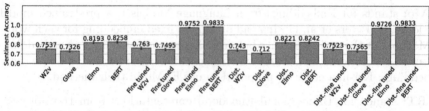

(a) Sentiment Accuracy of Classifiers with 95% confidence interval.

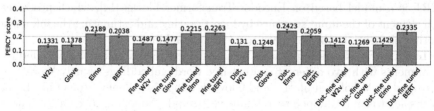

(b) Proportion of test examples that have been *correctly* classified based on the *B* conjunct according to the *PERCY* score with 95% confidence interval.

Fig. 2. Results obtained on SST2 dataset.

4   Experimental Evaluation

In this section, we first describe the dataset we have used in our evaluation before discussing the obtained results.

4.1   Dataset

Our experiments (as well as those presented in Hu et al. [1] and Krishna et al. [2]) are based on the Stanford Sentiment Treebank (SST2) dataset [6], which is a binary sentiment classification dataset. The dataset consists of 9,613 single sentences extracted from movie reviews, where sentences are labelled as either positive or negative each accounting for about 51.6% and 48.3%. A total of 1,078 sentences contain the *A-but-B* syntactic structure which accounts for about 11.2% of the dataset. We report our results only on test examples that contain an *A-but-B* syntactic structure to demonstrate the ability of a classifier to capture *A-but-B* pattern. Hence, all classifier are trained, tuned, and tested using stratified nested *k*-fold cross-validation and evaluated primarily according to accuracy. These sentences are identified simply by searching for the word "but" as proposed in [1,2,6].

4.2   Performance Evaluation

In this section, we discuss the results of our analysis of logic rules dissemination methods in sentiment classifiers. The configuration options that were considered are the following: {Word2vec, Glove, ELMo, BERT} × {Static, Fine-tuning} × {no distillation, distillation}, which gives a total of 16 classifier analysed on sentences with an *A-but-B* structure. To summarize all the results obtained over all the above configurations, Figs. 2a and 2b show the accuracy and the ability of the methods to base their classification decisions on the *B* conjunct. From these results, we make the following observations:

Accuracy Analysis: In Fig. 2a, we observe that the distillation model described in Hu et al. [1] is ineffective as it gives almost no improvement in terms of accuracy as also noted in [2]. Second, we note that fine-tuning all embeddings provides a statistically significant improvement of accuracy for almost all methods. Finally, it is clear that the best method is BERT, followed by ELMo, followed by either Glove or Word2vec.

Rule Dissemination Analysis: In Fig. 2b we show the proportion of test examples that have been *correctly* classified based on the *B* conjunct using PERCY score described in Sect. 3. Briefly, we first observe that for all methods, less than 25% of the test examples are effectively classified based on the *B* conjunct, which shows that the intent of these methods as described by their authors in [1,2] is far from being achieved. This suggests that there is still a lot of research to be done on this NLP topic. Second, we again note that there is almost no improvement between for instance Word2vec with and without distillation (Figs. 2a and 2b), which simply suggests that in [1] it is the 1D CNN sequence model that is capturing to some extent the *A-but-B* structure. Finally, we note that some models although have higher sentiment accuracy perform poorly on rule dissemination performance and vice-versa. For example, Dist. Elmo and Dist. BERT have similar sentiment accuracy in Fig. 2a but Dist. Elmo outperforms Dist. BERT by a statistically significant margin on rule dissemination performance in Fig. 2b. Similar phenomenon can be observed for Dist. fine-tuned Elmo and BERT models where later outperforms former even though having similar sentiment accuracy. This indicates that accuracy is misleading and there is no correlation between sentiment accuracy and actual rule dissemination performance.

5   Conclusion

This paper gives an analysis and a study of logic rules dissemination methods on their ability to identify *A-but-B* structures while making their classification decision based on the *B* conjunct. We use a rule consistency assessment metric called PERCY for that goal. Our experimental evaluation shows that (a) accuracy is misleading to assess whether the classifier based its decision as per B conjunct (b) not all methods are effectively capturing *A-but-B* structure, (c) that their underlying sequence model is often the one that captures to some extent the syntactic structure, and (d) that for the best method, less than 25% of test examples are effectively classified based on the *B* conjunct, indicating that a lot of research needs to be done in this topic.

References

1. Hu, Z., Ma, X., Liu, Z., Hovy, E., Xing, E.: Harnessing deep neural networks with logic rules. In: Proceedings of the 54th Annual Meeting of the Association for Computational Linguistics, Berlin, Germany, August 2016, vol. 1: Long Papers, pp. 2410–2420. Association for Computational Linguistics (2016)
2. Krishna, K., Jyothi, P., Iyyer, M.: Revisiting the importance of encoding logic rules in sentiment classification. In: Proceedings of the 2018 Conference on Empirical Methods in Natural Language Processing, Brussels, Belgium, October–November 2018, pp. 4743–4751. Association for Computational Linguistics (2018)
3. Peters, M.E., Neumann, M., Iyyer, M., Gardner, M., Clark, C., Lee, K., Zettlemoyer, L.: Deep contextualized word representations. In Proceedings of NAACL (2018)
4. Lakoff, R.: If's, and's and but's about conjunction. In: Fillmore, C.J., Langndoen, D.T. (eds.) Studies in Linguistic Semantics, Irvington, pp. 3–114 (1971)
5. Blakemore, D.: Denial and contrast: a relevance theoretic analysis of "but". Linguist. Phil. **12**(1), 15–37 (1989)
6. Socher, R., et al.: Recursive deep models for semantic compositionality over a sentiment treebank. In: Proceedings of the 2013 Conference on Empirical Methods in Natural Language Processing, Seattle, Washington, USA, October 2013, pp. 1631–1642. Association for Computational Linguistics (2013)
7. Gupta, S., Bouadjenek, M.R., Robles-Kelly, A.: PERCY: a post-hoc explanation-based score for logic rule dissemination consistency assessment in sentiment classification. Technical report, Deakin University, School of Information Technology (2022)
8. Ribeiro, M.T., Singh, S., Guestrin, C.: "why should i trust you?": explaining the predictions of any classifier. In: Proceedings of the 22nd ACM SIGKDD International Conference on Knowledge Discovery and Data Mining, KDD 2016, New York, NY, USA, pp. 1135–1144. Association for Computing Machinery (2016)
9. Kim, Y.: Convolutional neural networks for sentence classification. In: Proceedings of the 2014 Conference on Empirical Methods in Natural Language Processing (EMNLP), Doha, Qatar, October 2014, pp. 1746–1751. Association for Computational Linguistics (2014)
10. Zhang, Y., Wallace, B.: A sensitivity analysis of (and practitioners' guide to) convolutional neural networks for sentence classification. In: Proceedings of the Eighth International Joint Conference on Natural Language Processing, Taipei, Taiwan, November 2017, vol. 1: Long Papers, pp. 253–263. Asian Federation of Natural Language Processing (2017)
11. Mikolov, T., Sutskever, I., Chen, K., Corrado, G.S., Dean, J.: Distributed representations of words and phrases and their compositionality. In: Burges, C.J.C., Bottou, L., Welling, M., Ghahramani, Z., Weinberger, K.Q. (eds.) Advances in Neural Information Processing Systems, vol. 26. Curran Associates Inc. (2013)
12. Pennington, J., Socher, R., Manning, C.D.: Glove: global vectors for word representation. In: Empirical Methods in Natural Language Processing (EMNLP), pp. 1532–1543 (2014)
13. Devlin, J., Chang, M.W., Lee, K., Toutanova, K.: BERT: Pre-training of deep bidirectional transformers for language understanding. In: Proceedings of the 2019 Conference of the North American Chapter of the Association for Computational Linguistics: Human Language Technologies, vol. 1 (Long and Short Papers), Minneapolis, Minnesota, June 2019, pp. 4171–4186. Association for Computational Linguistics (2019)
14. Vaswani, A., et al.: Attention is all you need. In: Guyon, I., et al. (eds.) Advances in Neural Information Processing Systems, vol. 30. Curran Associates Inc. (2017)

Using Entities in Knowledge Graph Hierarchies to Classify Sensitive Information

Erlend Frayling$^{(\boxtimes)}$, Craig Macdonald$^{(\boxtimes)}$, Graham McDonald, and Iadh Ounis

Univerity of Glasgow, Glasgow G12 8QQ, UK
{erlend.frayling,craig.macdonald,graham.mcdonald,
iadh.ounis}@glasgow.ac.uk

Abstract. Text classification has been shown to be effective for assisting human reviewers to identify sensitive information when reviewing documents to release to the public. However, automatically classifying sensitive information is difficult, since sensitivity is often due to contextual knowledge that must be inferred from the text. For example, the mention of a specific named entity is unlikely to provide enough context to automatically know if the information is sensitive. However, knowing the conceptual role of the entity, e.g. if the entity is a politician or a terrorist, can provide useful additional contextual information. Human sensitivity reviewers use their prior knowledge of such contextual information when making sensitivity judgements. However, statistical or contextualized classifiers cannot easily resolve these cases from the text alone. In this paper, we propose a feature extraction method that models entities in a hierarchical structure, based on the underlying structure of Wikipedia, to generate a more informative representation of entities and their roles. Our experiments, on a test collection containing real-world sensitivities, show that our proposed approach results in a significant improvement in sensitivity classification performance (2.2% BAC, McNemar's Test, p < 0.05) compared to a text based sensitivity classifier.

1 Introduction

Technology Assisted Review (TAR) [2] has been shown to improve the efficiency of government sensitivity reviewing processes through use of text classifiers to recognise sensitivities, as the classifiers can assist reviewers with predictions as to whether documents contain sensitivity or not [10]. However, training a classifier to predict sensitivities is a complex task. Sensitivity identification is not a topic-oriented task [1], and sensitivity itself can arise from factors that are implicit to the text and are not exposed in an individual textual term. Indeed, sensitivity, like the background knowledge of the concepts and entities mentioned in documents, can be latent to the text. An expert human reviewer's prior knowledge enables them to deduce latent sensitivities using their knowledge of the subject matter. On the other hand, text classifiers that are trained using the

© The Author(s), under exclusive license to Springer Nature Switzerland AG 2022
A. Barrón-Cedeño et al. (Eds.): CLEF 2022, LNCS 13390, pp. 125–132, 2022.
https://doi.org/10.1007/978-3-031-13643-6_10

distributions of terms in the text [11], or even those trained with contextualised embeddings [4], are limited to learning from the distributions of textual features and, as such, may fail to identify latent sensitivities (even contextualised language models such as BERT [4] do not experience sensitive data).

Entities such as people, places or organisations are a rich source of latent contextual information. In this work, we propose a sensitivity classification approach that aims to integrate information that is representative of what a human reviewer might possess through their prior knowledge. For example, a reviewer might know that two entities are both political leaders, and that they represent opposing political parties - a subtlety that a contextualised classifier model may not so easily pick up. Sensitivity can often be nuanced in this way. For example, in a 'who said what about who' situation, the specifics of 'who' can matter more than the 'what' [9] - hence, recognising that the 'who' are both political entities might be informative for (sensitivity) classification. To this end, we propose a novel approach to build a hierarchical relationship model of entities present in a collection of government documents, using the underlying hierarchical structure of Wikipedia. We use this structure to infer latent information about entities in documents for classification. Specifically, we attempt to identify how certain entities in documents are related by underlying hierarchical concepts; For example, that two identified politicians, though different entities by name, are both leaders of communist regimes. Experiments conducted on a collection of 1000 real government documents with actual sensitivities demonstrate that we can attain significant improvements in accuracy of sensitivity classification.

2   Related Work

Several techniques have been proposed by McDonald et al. to improve sensitivity classification performance, including using Part of Speech (PoS) tagging and semantic word embedding features [8,9]. To our knowledge, there has been little work concerning the central importance of entities for classifying sensitivity. In the closest work to our own, [11], the authors feature engineered an opinionated numerical score representing diplomatic risk associated with some countries mentioned in the text, from the perspective of the UK. There have been several attempts to improve models in the more general category of text classification machine learning by enhancing entity representations. E-BERT [13] is a good example, which modified the original BERT model [4] to handle entities as unique tokens and unique vector representations showed improved performance over the original model. However, it is not yet clear if the entity representation within models such as E-BERT can learn to reflect well the genericism/specialism structure that can be encapsulated in knowledge bases.

Indeed, the use of knowledge bases in classification is most prevalent in domains where substantial and specialised knowledge bases already exist, e.g. in biomedicine. One work [7] utilised a pre-existing hierarchical knowledge graph of symptoms and diseases to learn a graph convolution neural network, which improved the effectiveness of medical diagnosis. BLUEBERT [12], which follows the BERT [4] architecture, was pre-trained on abstracts from the PubMed

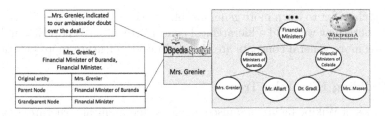

Fig. 1. Process of identifying entities in text & enriching with hierarchical tree entities.

knowledge base. This model was designed to perform the Biomedical Language Understanding Evaluation (BLUE) benchmark [12] and showed improved performance in BLUE tasks over a model pre-trained on more general datasets.

Training on specific knowledge bases for specific tasks has shown significant performance benefits versus training on general knowledge bases [12]. However, in sensitivity review, we lack a publicly available knowledge base structure to use for training models. Therefore, we focus on using a general knowledge base in this work. Notably, our focus is on enhancing representations of entities in sensitive documents using knowledge base information for two reasons. Firstly, as discussed in Sect. 1, we hypothesise that entities are a rich source of latent information and, in some cases, sensitivity in documents. Secondly, because knowledge bases provide information about entities, this kind of information is the most easily accessible. Flisar et al. [5] applied the DBpedia semantic knowledge base, by using the DBpedia Spotlight [3] to identify DBpedia entities, and then modelling them as key concepts in short texts for classification.

DBpedia, as a semantic knowledge base that has been derived from Wikipedia encodes a plethora of semantic relationships between concepts and links to the wider semantic web. On the other hand, Kapanipathi et al. [3] used the simpler structure of Wikipedia's Category Graph (WCG), which aims to group similar Wikipedia pages in a hierarchical relationship, and hence is a self-contained knowledge structure. Instead, we propose a simpler Wikipedia-based knowledge graph, instantiated from the Wikipedia pages themselves. In the next section, we introduce our model for hierarchical modelling of entities for sensitivity review.

3   Hierarchy Modelling and Features

We aim to enrich the representation of documents with additional entities that can assist classifiers in identifying sensitive text, by allowing the inference of more general sensitivity rules - for instance, rather than a person's name said something, the classifier can learn that a minister in a foreign government said something, which may be more significant. To this end, we derive a knowledge base that allows to generalise from linked entities.

Figure 1 provides an overview of our approach - a sentence about "Mrs Grenier" is indicating something is being told to an ambassador in confidence (and hence may be sensitive, due to a need to preserve international relations), but a classifier that is aware that Mrs Grenier is a finance minister in "Buranda" may help that classifier to learn more generalisable classifier rules. In the following, we describe both how we build a hierarchical concept tree from Wikipedia, and also how these more general concepts are encoded into the feature representations.

3.1 Building Hierarchy Tree

The articles of Wikipedia are organised in a loose hierarchical structure, separate from that exhibited by WCG. The central principle of this alternate structure is that, for any Wikipedia article, clicking on the first linked article in the text, recursively, will, in most cases, eventually bring the user to the article for Philosophy. This forms a tree structure over the nodes (or articles), where more abstract Wikipedia articles like Science and Rational are intermediate nodes close to the root node Philosophy. More specific entities like countries and people are farther from the root.

On the other hand, while WCG has hierarchical properties, it is not fundamentally a tree structure, as each Wikipedia page can have multiple categories. From our experience in this work, the "first-link" observation creates a usable tree with the desired properties, which we call Philosophy Hierarchy Structure (PHS).

Following [5,6], we use a Named Entity Linking tool to identify entities in documents, before generating classification features to avoid building the entire tree structure available in PHS. Indeed, in our task, we are not concerned with knowing all entities in the hierarchy, just those presently identifiable in the documents being reviewed, and the entities in their path to Philosophy. Moreover, as there are more than 6 million articles on Wikipedia, building the entire tree structure would be unnecessarily cumbersome. Therefore, we build only a local tree.

More specifically, we use the DBpedia Spotlight [3] NEL to identify all unique people, places and organisation entities. Spotlight provides a disambiguated link to the Wikipedia page for each detected entity in the document collection, which we use to retrieve the article's content. We retrieve the first link to the next (parent) article from that content. We consider this initial set of detected entities as the set of leaf nodes in our tree structure. We iteratively retrieve parent nodes for all Wikipedia articles in the initial leaf set, then for the intermediate nodes. We stop when all branches reach the Philosophy Node. In reality, the tree structure has imperfections – when creating a branch three outcomes are possible: (i) A generated branch reaches the node for Philosophy correctly, and the recursive parsing cycle is stopped; (ii) A branch of nodes forms a loop where one node in the branch points to a node further down; (iii) The branch breaks when the upper-most node cannot be parsed to obtain the next node. However, imperfect branches still contain the hierarchical information we need about entities present in the document collection and can still be used.

3.2 Feature Development

Having described the production of a tree structure object, we now describe our approach to extracting features from this tree. Key to our hypothesis discussed

Table 1. Results from experiment on 1000 record collection. Significant improvements over the text-only baseline classifier are denoted with * (McNemar's test, $p < 0.05$).

Features	P	R	F1	BAC
$\downarrow N \setminus$ baseline $\rightarrow$	0.363	0.657	0.468	0.636
0	0.369	0.661	0.474	0.641
1	0.371	0.661	0.476	0.643
2	0.370	0.657	0.473	0.641
3	0.369	0.665	0.475	0.642
4	0.373	**0.669**	0.479	0.646
5 *	**0.378**	**0.669**	**0.483**	**0.650**
6	0.374	0.665	0.479	0.646
7	0.374	**0.669**	0.480	0.647
8	0.373	**0.669**	0.479	0.646
9	0.372	0.665	0.477	0.644

in Sect. 1, we argue that certain entities in documents sharing parent nodes in the tree represents a hierarchical relationship that could be useful for classification. We identify and model these relationships for entities in a collection of documents as text features in our approach.

To generate features for a given document we find the associated set of DBpedia entities present in the text and their corresponding nodes in the tree produced in Sect. 3.1. For each node in the tree we identify the next N parent nodes, where N is some integer number of nodes to climb into the tree. We combine the original set of DBpedia entities for each document with the additional parent nodes to form a new extended set of entities. We expect that across a corpus of documents, parent nodes will appear in documents for which the detected DBpedia entities are different, revealing that the different entities have underlying connections represented by the parent nodes in the extended collection of entities. This extended set of entities for each document can be used as additional features in a classification task. For example, referring to Fig. 1, if Mrs. Grenier retires from her position as financial minister, and a new individual (Mr Allart) takes over, the surface form name of the individual will change in newer documents. However, using the extended features would still provide the common connection of 'Financial Ministers of Buranda'. In this sense, generalisation is achieved, and a classifier may make the connection that both Mr. Allart and Mrs. Grenier share equal importance across old and new documents.

4   Experiments

We perform experiments to address two research questions, namely:

RQ1: Can a text classifier use our hierarchically enriched entity features to predict sensitivity in government documents more accurately?

RQ2: Does changing the number of added parent nodes N of the hierarchically enriched features, detailed in Sect. 3.2, affect classification effectiveness, and which number N is most effective in this task?

4.1   Experimental Setup

We use a collection of 1000 government documents that have been reviewed for sensitivity by experienced government reviewers. The data collection was assessed for sensitivities relating to international relations and personal information, which are common types of sensitivities defined in freedom of information settings. The collection contains 251 (25%) sensitive documents in total, across both categories of sensitivity assessed. We use a 10-fold cross-validation setup, averaging Precision (P), Recall (R), Balanced Accuracy Score (BAC) and F1 measure across folds.

We generate a hierarchical relationship tree using the process described in Sect. 3.1. DBpedia Spotlight detects 2226 entities in the collection, and the total number of nodes in the final tree structure is 5129. We extract several feature sets for each document. Firstly, the text of each document alone. Secondly, we extract a set of entities directly detected by DBpedia's Spotlight tool for each document (denoted $N = 0$). Further, we use this entity set for each document to feature engineer hierarchically-enhanced representations for tree depth values of $1 \leq N \leq 9$, as described in Sect. 3.2. We use the original entity set and all nine hierarchically enhanced sets as ten separate feature sets. Finally, we combine each of the ten sets of entities with the original text of each document to produce combined text and entity features. For classification, we apply a Multinomial Naive Bayes model. Words are represented using term frequency only, removing stopwords that occur in the Sci-Kit Learn's English stopword list.

4.2   Results

Table 1 presents the effectiveness of the Multinomial Naive Bayes classifier using different combinations of features. The table presents effectiveness in terms of Precision, Recall, F1, and Balance Accuracy for each configuration. We also test each configuration for statistical significance ($p < 0.05$) compared to the baseline that classifies documents on only their text features (denoted 'text'). Firstly, from Table 1, we note that all sets of entity features improve classifier performance when combined with the text features. The best performance increase w.r.t. the baseline occurs when classifying document text with our entity features when considering a hierarchy depth (N) of 5. This result is a 2.2% improvement in BAC score over the baseline and a 3.2% improvement in F1, which is statistically significant according to a McNemar's test ($p < 0.05$). Moreover, all experiments combining text features with entity sets outperform the baseline of BAC 0.636. Among precision and recall, we note that precision is enhanced by 4% ($0.353 \rightarrow 0.378$), while recall is enhanced by 2% ($0.657 \rightarrow 0.669$). Indeed, in an assistive classification task such as sensitivity review, precision is important, as false positive may cause reviewers to loose confidence in the predictions.

Thus, we answer our research questions as follows: for **RQ1**, we find that entity features making use of the PHS hierarchy can be used to identify sensitivities more accurately when used in addition to the textual features of the documents. For **RQ2**, we find that adding five levels of parent nodes to the enriched set of entities for each original entity occurring in text achieves the best performance, but all $N > 0$ outperform adding only the original entities.

5 Conclusions

In this work, we proposed a novel approach to provide a sensitivity classifier with a hierarchical representation of entities that allows a classifier to infer new generalised rules about entities and sensitivity. Moreover, we evaluated the effectiveness of our features for sensitivity classification and showed that our enhanced entity features allow a classifier to make more successful predictions about sensitivities. We showed that significant improvements can be obtained compared to a baseline text classification approach (McNemar's test, $p < 0.05$), particularly improving precision. In future work, we will apply Graph Neural Networks in conjunction with the hierarchical graph structures, which we expect to result in further classification improvements.

Acknowledgements. E. Frayling, C. Macdonald and I. Ounis acknowledge the support of Innovate UK through a Knowledge Transfer Partnership (# 12040). All authors thank SVGC Ltd. for their support.

References

1. Berardi, G., Esuli, A., Macdonald, C., Ounis, I., Sebastiani, F.: Semi-automated text classification for sensitivity identification. In Proceedings of CIKM (2015)
2. Cormack, G.V., Grossman, M.R.: Evaluation of machine-learning protocols for technology-assisted review in electronic discovery. In Proceedings of SIGIR (2014)
3. Daiber, J., Jakob, M., Hokamp, C., Mendes, P.N.: Improving efficiency and accuracy in multilingual entity extraction. In Proceedings of I-SEMANTICS (2013)
4. Devlin, J., Chang, M.W., Lee, K., Toutanova, K.: BERT: pre-training of deep bidirectional transformers for language understanding. arXiv preprint arXiv:1810.04805 (2018)
5. Flisar, J., Podgorelec, V.: Improving short text classification using information from DBpedia ontology. Fundamenta Informaticae **172**(3), 261–297 (2020)
6. Kapanipathi, P., Jain, P., Venkataramani, C., Sheth, A.: User interests identification on Twitter using a hierarchical knowledge base. In: Proceedings of ESWC (2014)
7. Liu, B., Zuccon, G., Hua, W., Chen, W.: Diagnosis ranking with knowledge graph convolutional networks. In: Hiemstra, D., Moens, M.-F., Mothe, J., Perego, R., Potthast, M., Sebastiani, F. (eds.) ECIR 2021. LNCS, vol. 12656, pp. 359–374. Springer, Cham (2021). https://doi.org/10.1007/978-3-030-72113-8_24
8. McDonald, G., Macdonald, C., Ounis, I.: Using part-of-speech n-grams for sensitive-text classification. In: Proceedings of ICTIR (2015)

9. McDonald, G., Macdonald, C., Ounis, I.: Enhancing sensitivity classification with semantic features using word embeddings. In: Proceedings of ECIR (2017)
10. McDonald, G., Macdonald, C., Ounis, I.: Towards maximising openness in digital sensitivity review using reviewing time predictions. In: Proceedings of ECIR (2018)
11. McDonald, G., Macdonald, C., Ounis, I., Gollins, T.: Towards a classifier for digital sensitivity review. In: Proceedings of ECIR (2014)
12. Peng, Y., Yan, S., Lu, Z.: Transfer learning in biomedical natural language processing: an evaluation of BERT and ELMo on ten benchmarking datasets. In: Proceedings of BioNLP Workshop and Shared Task (2019)
13. Poerner, N., Waltinger, U., Schütze, H.: E-BERT: efficient-yet-effective entity embeddings for BERT. arXiv preprint arXiv:1911.03681 (2019)

Best of 2021 Labs

Evaluating Research Dataset Recommendations in a Living Lab

Jüri Keller$^{(\boxtimes)}$ and Leon Paul Mondrian Munz

Technische Hochschule Köln, Ubierring 48, 50678 Cologne, Germany
{jueri.keller,leon_paul_mondrian.munz}@smail.th-koeln.de

Abstract. The search for research datasets is as important as laborious. Due to the importance of the choice of research data in further research, this decision must be made carefully. Additionally, because of the growing amounts of data in almost all areas, research data is already a central artifact in empirical sciences. Consequentially, research dataset recommendations can beneficially supplement scientific publication searches. We formulated the recommendation task as a retrieval problem by focussing on broad similarities between research datasets and scientific publications. In a multistage approach, initial recommendations were retrieved by the BM25 ranking function and dynamic queries. Subsequently, the initial ranking was re-ranked utilizing click feedback and document embeddings. The proposed system was evaluated live on real user interaction data using the STELLA infrastructure in the LiLAS Lab at CLEF 2021. Our experimental system could efficiently be fine-tuned before the live evaluation by pre-testing the system with a pseudo test collection based on prior user interaction data from the live system. The results indicate that the experimental system outperforms the other participating systems.

Keywords: Living Labs · (Online) Evaluation in IR · Recommender System · Research Dataset Retrieval

1 Introduction

Due to the continuing flood of information and the steadily growing number of scientific publications and research datasets, the ability to find them is an ongoing challenge. Since the search for datasets, even using designated search engines, can be tedious, a possible solution may be to recommend relevant research datasets directly to corresponding publications.

The proposed system makes use of the broad similarities between scientific publications and research datasets and is based on the probabilistic BM25 ranking function to determine the similarity between index and query [18]. Results from the TREC-COVID Challenge[1] described by Roberts et al. show that almost all top-performing systems used BM25 as first stage ranker to produce already

---

[1] https://ir.nist.gov/covidSubmit/index.html.

© The Author(s), under exclusive license to Springer Nature Switzerland AG 2022
A. Barrón-Cedeño et al. (Eds.): CLEF 2022, LNCS 13390, pp. 135–148, 2022.
https://doi.org/10.1007/978-3-031-13643-6_11

good baselines [17]. By treating publications and datasets both as documents, these established retrieval techniques can be used to create dataset recommendations. The publications are used to generate queries dynamically that are subsequently used to query datasets. This initial baseline is advanced by re-ranking techniques utilizing cross-data type document embeddings and user interaction data as relevance indicators.

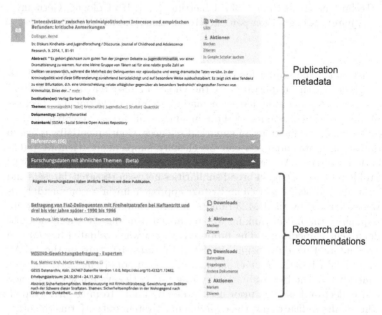

Fig. 1. This figure shows an example publication detail website. Below the details of the publication, a ranking of recommended datasets is presented.

The system was evaluated by the Living Labs for Academic Search (LiLAS) CLEF Challenge as a type B submission to Task 2 [20]. While task one focused on ad-hoc retrieval of scientific publications, Task 2 demanded for dataset recommendations related to scientific publications. Both tasks could be submitted as a pre-computed ranking (Type A) or as a live system (Type B) that retrieves the ranking ad-hoc. The STELLA infrastructure emulates pre-computed runs as live systems for queries available in the run and only utilizes the system for those queries. By that, STELLA enables the comparison of live systems with pre-computed ones [8]. Task 2 anticipated recommender systems suggesting research datasets as a supplement to the scientific publication pages in the social science database GESIS-Search[2]. Since all publications are known, dataset recommendations for all possible queries could be pre-computed. As a beta service solely created for the Living Lab, the website presented a ranking of a maximum of six datasets and additional metadata below the details of a publication on its

---

[2] https://search.gesis.org/.

overview site. The layout of the website is shown in Fig. 1. More detailed task explanations and the general evaluation can be found in the lab overview [20]. Living Labs differ strongly from retrieval experiments following the Cranfield paradigm and bring several unique challenges. Therefore, more authentic results can be gained. To account for the lack of initial relevance assessments and efficiently utilize the valuable user interaction feedback, we created pseudo test collections to pre-test the proposed recommender system.

The main contribution of this work is the proposal of a BM25 based dataset recommender which is pretested with a pseudo test collection and evaluated in an online evaluation. The remainder of this paper is structured as follows. After this introduction, the related work in the adjacent research fields is outlined in Sect. 2. The system itself, including the pre-processing pipeline, initial ranking and rerankings, are described in more detail in Sect. 3. Subsequently, the pre-testing process and process of the pseudo test collections are characterized in Sects. 4. After a description of results in Sect. 5 this paper closes with a Conclusion in Sect. 6.

2   Related Work

Compared to conventional Cranfield paradigm information retrieval (IR) experiments, Living Lab IR experiments aim to evaluate search systems in a real-world fashion through actual user interactions. Therefore, experimental systems extend existing search systems and are evaluated based on collected user feedback [20]. By that, the Living Labs for Academic Search (LiLAS) lab followed a series of labs dedicated to the living lab approach like NewsREEL [16], LL4IR [23] and TREC OpenSearch [4] did before. Through the STELLA infrastructure the experimental systems of both types could be integrated into the live systems. Further, STELLA creates an interleaved ranking by systematically combining the results from two systems [8,22]. More lifelike results and insights are expected by utilizing real user interactions to assess search systems. Azzopardi and Balog describe different stages of IR experiment environments, from the traditional test collection to the living lab [3].

Based on the metadata available, the dataset recommendation task was formulated as a dataset search task, recommending the retrieved datasets given a query constructed from a seed document. Chapman et al. provide a good overview of the field of dataset retrieval [9]. They differentiate between two types of dataset search systems, the first, most similar to document retrieval, returns existing datasets given a user query. In contrast, the second method composes a dataset based on the user query from existing data. Chapman et al. also described commonalities of datasets and documents initial retrieval systems can focus on, which may serve as a starting point in this new field [9]. These findings are the foundation for the proposed recommender system relying on the BM25 ranking function and a cross-datatype collection. Kren and Mathiak analyzed dataset retrieval in the Social Sciences. They concluded that the choice of dataset is a more important and, therefore, more time-consuming decision than the choice of literature [15]. Further, even though research datasets are increasingly accessible, the connection between publications and research datasets remains often unclear [13]. Kacprzaka et al. hypothesized based on large log analyses of four open data search portals that dataset search is mainly explorative motivated [14]. This strengthens the use-case of supplementing document search results with related datasets to create a more complete overview or serve as a starting point for more exhaustive searches. The main objective should be to recommend datasets mentioned in or closely related to the seed publication.

Recommender systems are a well-discussed topic because of their broad application and the necessity to keep up with increasing information. While Bobadilla et al. [7] provides an overview of the general field, Beel et al. [5] focus specifically on research-paper. Recommender systems use data analysis techniques to help users find the content of individual relevance. These recommender systems are often categorized into three overarching approaches based on the information source utilized to generate recommendations of, namely: Content-based recommendation, collaborative recommendation, and hybrid approaches. Content-based recommender systems primarily source the available metadata of items for recommendation. In contrast, the collaborative approaches recommend items based on user interactions with items. In conjunction of both worlds, the hybrid approaches combine collaborative, and content-based methods [1].

As initially mentioned, intermediate evaluation in a living lab setting is especially challenging because of the lacking test collection. To provide a starting point for the LiLAS lab, Schaer et al. provided head queries and candidate documents from the two real-world academic search systems, allowing the construction of pseudo test collections [21]. Pseudo test collections are a long-established method to create synthetic queries, and relevance judgments [6]. Motivated by reducing the cost of test collection creation, features are computed offline from global document information [2]. The provided head queries and candidates partially resemble the live system and also contain its relevance scores. Used as a pseudo test collection, they are suited to compare the experimental with the live system.

3   Research Dataset Recommendations

For the proposed system, the content-based approach appears to be most suitable for the prevalent use-case of recommending scientific datasets as a supplementary service during literature search. Compared to other recommendation tasks where extensive user interaction data is produced, saved in user profiles and available to fuel recommendation algorithms, ad-hoc searches, often performed without connected user profiles, provide only limited user interaction data. However, the well-established publishing practices in the scientific sector provide rich metadata for both publication and datasets. Mainly focusing on the available metadata additionally brings two advantages: First, even for niche items, in this use-case, both publications and datasets, sufficient metadata is available as a recommendation foundation. Second, while systems solely rely on

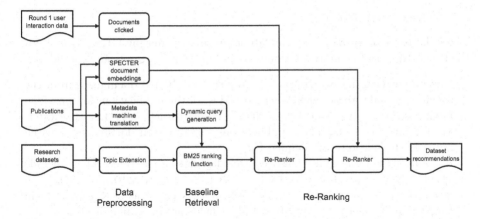

Fig. 2. Visualization of the full system used to pre-compute the recommendations, from data input on the left to the final output on the right. Curvy boxes represent data inputs, rectangular boxes processing steps.

user interaction data, they naturally suffer from the cold start problem where no information is available for new items; here again, the metadata of the item provides enough information.

Since recommendations are provided in addition and related to a publication, the recommendations can be pre-computed for all available publications and need only to be updated if a new item is added. If a publication is added, new recommendations need to be calculated specifically for this item. However, all recommendations need to be updated if a dataset is added. Nevertheless, since the, admittedly by now outdated, recommendations still preserve certain relevance, they do not need to be updated right away. Therefore this approach is well suited to be pre-computed.

The presented approach relies on the broad similarity between the provided publications and datasets, which will be described in more detail in the following subsection. By focusing on these similarities, the research dataset recommendation task is formulated as the well known and explored retrieval task. The publications research datasets need to be recommended for are used as query information, like in TREC style evaluations. By sourcing various metadata fields of a publication, a query is composed and used to retrieve the datasets which will be recommended. Initially, a baseline is retrieved using the BM25 ranking function and re-ranked by incorporating the few but available user interaction data from the first evaluation round. Since the recommendations are pre-computed but evaluated in a live environment, run time can be neglected, allowing to apply a second, more resource-intensive, neural re-ranker.

In the following, the dataset and the recommender system are described. Figure 2 gives a schematic overview of the whole process.

3.1   The GESIS Corpus

Three datasets are provided by the lab organizers originating GESIS Search for the Research Data Recommendation task: publications, datasets and candidates.

- The publication dataset contains metadata for 110,420 documents from the social science database GESIS-Search[3]. Most of the publications are provided in English or German and have textual metadata such as TITLE, ABSTRACT, TOPICS and PERSONS. These publications serve as seed documents research datasets should be recommended for.
- In addition, metadata for 99,541 research datasets are provided. These include TITLE, TOPICS, ABSTRACT, DATA TYPE, COLLECTION METHOD, temporal and geographical coverage, PRIMARY INVESTIGATORS as well as CONTRIBUTORS in English and or German. Not all metadata fields are available for all datasets.
- The GESIS corpus also contains collections of candidates. The top 1000 most used seed documents and their dataset recommendations are listed there. They were retrieved from the live recommender system and contain, besides the dataset identifier also, TF-IDF relevance scores.

Not all metadata fields were set for all datasets. To improve metadata completeness, we added machine translations of the missing title and abstract fields and systematically add missing topics.

3.2   Data Pre-processing

Further investigations showed that the publication metadata fields for title and abstract are inconsistent in multiple ways. Not all fields are available in all languages, and not always is the actual language of a field corresponding with its label. While a German publication may have an English or partially English title and abstract, the actual language of the metadata field is of interest to correctly apply text processing to it. The publications dataset showed similar but less strong divergences. To guarantee at least one match between the related fields of a publication and a dataset, all titles and abstracts of the publications dataset are machine translated into both languages using Deep_translator[4].

Not all metadata records have topics assigned. The assigned topics originate from a controlled vocabulary managed by and named after the Consortium of European Social Science Data Archives (CESSDA)[5]. To assign appropriate topics automatically, only existing topics are considered for assignment. A collection of all assigned topics in the corpus was created and then translated into German and English depending on their source language. To preserve the original metadata two additional fields were added: TOPIC_EXT_GER and TOPIC_EXT_EN. To maintain a high topic relevance for the newly assigned topics, a topic was assigned only if it appeared in the title of the research dataset. Through this procedure, 556 German topics and 2359 English topics could be assigned.

---

[3] https://search.gesis.org/.

[4] https://pypi.org/project/deep-translator/.

[5] https://www.cessda.eu/.

3.3   Baseline Retrieval

By separating fields with multiple languages into separate fields for each language, language depending text processing could be applied to one index. The publication is used as a query to search the created index of research datasets to generate recommendations for a publication. As baseline search, Apache Solr BM25 ranking function with the default parameters $k1 = 1.2$ and $b = 0.75$ and various field combinations and boosting factors are used. Since not all fields are available for all seed publications, queries are generated dynamically considering all available fields and therefore differ in length and complexity for each publication. Each field of a seed publication is used to query the corresponding field of the research datasets. Only the topic field is queried by a concatenation of the TITLE, ABSTRACT and TOPIC fields of the seed publication. With these queries the fields TITLE, ABSTRACT and TOPIC as well as their language variations TITLE_EN, TITLE_DE, ABSTRACT_EN, ABSTRACT_DE, TOPIC_EN and TOPIC_DE and the extended topic fields EXT_TOPIC_DE and EXT_TOPIC_EN of the research datasets are queried if available. Each of these fields is boosted individually, considering its ability to describe the searched dataset. In general, title fields are boosted higher than abstract fields, for example. Static factors between zero and one are used as boosts to weight the scores of the fields individually.

3.4   Re-ranking

The baseline results are re-ranked in two ways to improve the recommendation quality. First, a re-ranker based on the results from round one is applied. The lab was structured in two rounds with intermediate evaluation to allow further system adjustments during the experiment. Since the described approach was only active in the second round, the results from the first round were available as additional information to re-rank the results. On top of the results re-ranked by the first re-ranker, a second re-ranker is applied, considering similarity based on document embeddings.

As a signal of relevance, the click feedback from the first evaluation round is used to boost certain datasets. Given a ranking from the baseline, system datasets that were clicked in round one are boosted, considering the same query publications. Due to click sparsity and importance, a strong, static boost is added to rank the affected datasets to the top of the ranking. Incorporating user interaction data into the recommender system transforms the content-based approach into a hybrid one. This means that the recommendations need to be updated more regularly to account for ongoing variance in user interactions.

Since publications and datasets have broad similarities in structure and nature, the overall document similarity is considered another factor of relevance. To measure similarity across documents and datasets, document embeddings and the k-nearest neighbors (k-NN) [12] algorithm are used. The document embeddings are calculated using SPECTER [10] a transformer-based SciBERT language model through its available web API[6]. From the title and abstract of a document, the language model calculates a vector that represents the document. With vectors for all documents, the documents can be mapped in a multidimensional space, and the distances between them can be measured. The closer the documents are, the bigger the similarity between them. The k-NN algorithm uses the euclidean distance to measure the distance between the documents. The closest dataset to each seed publication is calculated. Given a baseline ranking, the most similar datasets are calculated for that query publications, and all matches gain a strong static boost.

4   System Pre-testing

Multiple experiments were conducted to test different system configurations and determine the optimal metadata combinations and parameter settings for the field booster and re-ranker. While the predominant IR experiment type following the Cranfield paradigm relies on expensive annotated test collections, more real-world use-cases lack these amenities. These use-cases often cannot provide the required resources in terms of time and money to create a comprehensive test collection but have access to real user interaction data from live systems to evaluate experimental systems on. To maximize the efficiency of experiments and simultaneously minimize the risk of exposing potential customers to unpleasant results, it is most important to pretest the experimental systems as good as possible. Therefore, pseudo test collections may help to pretest IR experiments offline before an online evaluation with real users.

To pretest the general system and analyze the behavior of the system to specific adjustments, we created pseudo test collections from the provided head queries and candidate datasets. For each head query, the datasets recommended by the live system are provided as candidates. The pseudo test collection is constructed from all provided head queries and candidates. As most useful appeared to use the TF-IDF scores from the live system directly as relevance scores in the pseudo test collection. This pseudo test collection allowed to compare our experimental system with the TF-IDF based live system as baseline offline [20].

This data holds no ground truth but can help put the results in context. Following the premise of the Living Lab evaluating experimental IR systems based on real user interactions, it is most likely that an existing system should be improved. However, since the flaws of the live system might not be known, the experimental system can only be evaluated with the live data. Therefore, differences between the live and experimental systems will be minimized at first to create a neutral starting point and then systematically deviate from that. In conclusion, the overall goal of pre-testing is to determine system settings, returning results not too far off from the baseline system but still providing enough variation for different results. All runs are evaluated using pytrec_eval[7].

---

[6] https://github.com/allenai/paper-embedding-public-apis.
[7] https://github.com/cvangysel/pytrec_eval.

Table 1. Evaluation results for different system settings achieved during pre-testing based on the second pseudo test collection. The system producing the run number six, highlighted as italic, was submitted as final system. Except from the two results marked with the dagger (†) the not re-ranked runs perform better then the re-ranked versions.

Run	re-ranked	topic boost	abstract boost	map	nDCG	P@5	P@10	R@10	rel_ret
1	False	0.5	1	0.077	0.281	0.273	0.241	0.024	0.434
2	True	0.5	1	0.077	0.280	0.269	0.239	0.024	0.434
3	False	0.7	1	0.070	0.266	0.256	0.224	0.023	0.411
4	True	0.7	1	0.070	0.266	0.255	†0.225	†0.023	0.411
5	False	0.3	1	0.082	0.292	**0.278**	0.249	**0.025**	0.452
6	*True*	*0.3*	*1*	*0.082*	*0.291*	*0.272*	*0.246*	*0.025*	*0.452*
7	False	0.3	0.3	0.074	0.274	0.263	0.230	0.023	0.425
8	True	0.3	0.3	0.073	0.273	0.253	0.229	0.023	0.425
9	False	0.3	0.5	**0.083**	**0.293**	0.277	0.246	**0.025**	**0.453**
10	True	0.3	0.5	0.082	0.292	0.266	0.243	**0.025**	**0.453**

In early experiments the construction of the query was tested. Fields were added gradually to improve overall datasets retrieved and relevant datasets retrieved. Finally, the fields TITLE, TITLE_EN and TOPIC from seed publications where used to generate the query. With this dynamically constructed query the recommendation datasets were retrieved using the original fields TITLE, ABSTRACT, TOPIC and the newly created field EXT_TOPIC_DE additionally the English fields TITLE_EN, ABSTRACT_EN, EXT_TOPIC_EN were used. A second set of experiments were conducted to pretest the field boosting and re-ranker. Selected results from these experiments are shown in Table 1. All experiments were evaluated with and without re-ranking to evaluate both system components individually. Even runs, also indicated by the re-ranked field, are based on the same base system as their predecessor, but the initial results are re-ranked additionally. The first six runs compare the three different boosting for the topic fields of the dataset metadata. The boosts 0.5, 0.7 and 0.3 are tested. Surprisingly, boosting the topics down to 0.3, tested in run 6, showed the best results. In the remaining four runs, seven to ten negative boosts of different strengths are applied to the abstract field to account for the higher amount of words. In runs seven and eight, abstract fields are boosted down to 0.3, and in runs nine and ten, slightly less harsh, down to 0.5. Results, in general, are close to each other, as shown by runs five and nine, which are almost the same. Even though run nine performed slightly better in overall metrics like *nDCG*, the *P@5* for run five was slightly better. Since just a few recommendations can be provided on the document search result page, these metrics were prioritized, and the highlighted run six configuration was used for the final system tekma_n. Both runs are based on the same initial ranking, but for run six the re-ranker was applied.

For the experiments shown in Table 1 almost all runs without re-ranking performed slightly better than their re-ranked versions. Only the metrics *P@10* and *R@10* from run four except this observation and are therefore marked with an dagger. Regardless of this, run number six was submitted as the final run to test the re-ranking approach in the live system and on the full dataset. This decision was strengthened by the observation that re-ranking could be applied to few datasets only during pre-testing.

5   Experimental Evaluation

Schaer et al. provide a comprehensive evaluation of the different systems, including also weighted results accounting for the interleaved experiment setup where two systems merge their results into one result page ranking [20]. Additionally, this analysis focuses more specifically on the recommender system itself. STELLA, the infrastructure through which the Living Lab experiments are realized, provides detailed result feedback [8]. To quantify the individual performance of the systems in an interleaved experimental setting, Schuth et al. proposed a set of interleaving metrics [23]. Depending on the sum of results from one system clicked for one interleaved ranking in one session, the system wins, loses or results in a tie. In conjunction with the experimental system and the pre-testing runs described in Sect. 4, the effectiveness of all ranking stages can be evaluated. Over the course of six weeks, from 12. April to 24. May 2021 the system tekma_n received 3097 Impressions. Compared to the other systems the described experimental system tekma_n wins 42 times, the other experimental system gesis_rec_pyterrier wins 26 times and the baseline system gesis_rec_pyserini wins 51 times. However, the rankings are always composed of one experimental system and the baseline system, the baseline is utilized for twice as many sessions as an experimental system. Both experimental systems resulted in one tie and the system tekma_n loses one more ranking compared to the other experimental system gesis_rec_pyterrier. All results are summarized in Table 2.

Table 2. Final results of Round 2, reproduced from Schaer et al. [19]. The dagger symbol (†) indicates the baseline system.

System	Win	Loss	Tie	Outcome	Session	Impression	Clicks	CTR
gesis_rec_pyserini†	51	68	2	0.43	3288	6034	53	0.0088
gesis_rec_pyterrier	26	25	1	0.51	1529	2937	27	0.0092
tekma_n	42	26	1	0.62	1759	3097	45	0.0145

The datasets clicked from the interleaved result page ranking are unevenly distributed over the ranking favoring the first positions. Since the recommendations are presented as a ranking, as illustrated in Fig. 1, this exemplarily shows the position-bias in rankings [11]. While datasets in the first position were clicked 21 times, datasets ranked lower were clicked less often. The full distribution of ranking positions documents that were clicked is shown in Fig. 3. Considering just clicked recommendation lists, both systems, the baseline and the experimental, were utilized almost equally for the first ranking. The baseline system could rank 11 times first and the experimental system 10 times. Comparing all recommendation rankings, this finding amplifies slightly, resulting in 1021 by 958 in favor of the baseline system.

Distribution of datasets clicked per ranking posiotion

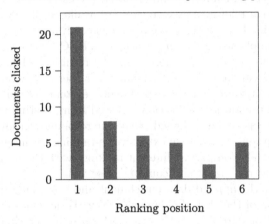

Fig. 3. Distribution of recommended datasets clicked per ranking position, reproduced from Schaer et al. [19]. The distribution shows a position bias where high ranked datasets are clicked more often.

To further analyze the individual recommendations the experimental system tekma_n performed worse than the comparative system, the submitted recommendations are compared with the actually clicked recommendations. The experimental system tekma_n does not rank nine clicked datasets at all but ranked four datasets at the exact same position they were ranked by the baseline system and were clicked.

One main aspect of the proposed system is the data pre-processing endeavors to account for the multi-lingual data and queries described in the previous Sect. 3.2. To measure any effects of these approaches, namely the machine translation of the title, abstract and the systematical topic expansions, the evaluated rankings are compared to rankings created during pre-testing for the same query publication. If a dataset is ranked lower without data pre-processing applied, this directly impacts being clicked for that query. Surprisingly no applied data pre-processing method, neither the translations nor the new assigned, formerly missing, topics resulted in changed positions for the clicked documents. Remembering the small basis of data, data pre-processing did not affect the results.

Following the same evaluation method, the re-ranking techniques are analyzed. Both re-rankers were assessed individually and in conjunction, but the results stayed the same. No clicked documents were re-ranked. Given these observations, the system performance observed solely relies on the query construction and initial BM25 ranking function. To achieve more comprehensive or even significant results, more user interaction is needed.

6   Conclusion

We applied well-established IR techniques and concepts to the fairly new field of scientific dataset recommendation and explored the early stages of IR experiments before extensive relevance assessments are available. By evaluating our endeavor in a live Living Lab experiment environment, advantages and challenges could be explored, emphasizing the differences to TREC style evaluations. By relying on real user interaction data, more authentic results can be achieved and real-world constraints can be faced. Additionally, the user interaction data provide an additional data source for the experimental system. Through extensive pre-testing based on pseudo test collections created from existing systems, the experimental system was initially aligned. Since the recommendation task could completely be pre-computed, resource extensive re-ranking techniques could be tested. By retrospectively comparing different rankings with the user click data, the impact of different ranking stages could be observed.

Results show that the applied data enrichment and re-ranking methods did not affect the position of the clicked documents. Nevertheless, our experimental system with a CTR of 0.0145 and 42 wins performed better than the other experimental system with 26 wins and a CTR of 0.0092. The results showing that the baseline only achieved 51 wins but was used in twice as many sessions indicate that our approach might outperform the baseline as well. These results must be attributed to the BM25 function and dynamic query generation. Similarities in metadata of research datasets and scientific publications allow applying these retrieval methods to create a sufficient recommender baseline. However, based on the little available user interaction data, no statistically significant results could be achieved. Nevertheless, extensive pre-testing proved to be an effective tool for achieving good results in online evaluations. Through a pseudo test collection, the recommender could be initially fine-tuned even before the online evaluation was started.

These findings can be used as reference points for future experiments at the intersection of live evaluated but pre-computed systems. Additionally, they can function as a gateway for initial systems recommending research datasets. In future works, multiple ranking stages could be extended. The data pre-processing could be improved to support more languages or add more topics. The user interaction data could be incorporated more distinctively by differentiating between types of interaction. This would also avoid the popularity bias, which would harm the results after a while if only clicked items are boosted. Especially interesting would be to test the system on a larger scale or longer online period to attract more user interactions as more data is required for reliable results.

References

1. Adomavicius, G., Tuzhilin, A.: Toward the next generation of recommender systems: a survey of the state-of-the-art and possible extensions. IEEE Trans. Knowl. Data Eng. **17**, 734–749 (2005). https://doi.org/10.1109/TKDE.2005.99
2. Asadi, N., Metzler, D., Elsayed, T., Lin, J.: Pseudo test collections for learning web search ranking functions. In: Ma, W., Nie, J., Baeza-Yates, R., Chua, T., Croft, W.B. (eds.) Proceeding of the 34th International ACM SIGIR Conference on Research and Development in Information Retrieval, SIGIR 2011, Beijing, China, 25–29 July 2011, pp. 1073–1082. ACM (2011). https://doi.org/10.1145/2009916.2010058
3. Azzopardi, L., Balog, K.: Towards a living lab for information retrieval research and development. In: Forner, P., Gonzalo, J., Kekäläinen, J., Lalmas, M., de Rijke, M. (eds.) CLEF 2011. LNCS, vol. 6941, pp. 26–37. Springer, Heidelberg (2011). https://doi.org/10.1007/978-3-642-23708-9_5
4. Balog, K., Schuth, A., Dekker, P., Schaer, P., Chuang, P.Y., Tavakolpoursaleh, N.: Overview of the trec 2016 open search track. In: Voorhees, E.M., Ellis, A. (eds.) TREC, vol. Special Publication 500–321. National Institute of Standards and Technology (NIST) (2016)
5. Beel, J., Gipp, B., Langer, S., Breitinger, C.: Research-paper recommender systems: a literature survey. Int. J. Digit. Libr. **17**(4), 305–338 (2015). https://doi.org/10.1007/s00799-015-0156-0
6. Berendsen, R., Tsagkias, M., Weerkamp, W., de Rijke, M.: Pseudo test collections for training and tuning microblog rankers. In: Jones, G.J.F., Sheridan, P., Kelly, D., de Rijke, M., Sakai, T. (eds.) The 36th International ACM SIGIR Conference on Research and Development in Information Retrieval, SIGIR 2013, Dublin, Ireland - July 28 - August 01 2013, pp. 53–62. ACM (2013). https://doi.org/10.1145/2484028.2484063
7. Bobadilla, J., Ortega, F., Hernando, A., Gutiérrez, A.: Recommender systems survey. Knowl. Based Syst. **46**, 109–132 (2013). https://doi.org/10.1016/j.knosys.2013.03.012
8. Breuer, T., Schaer, P., Tavakolpoursaleh, N., Schaible, J., Wolff, B., Müller, B.: STELLA: towards a framework for the reproducibility of online search experiments. In: Clancy, R., Ferro, N., Hauff, C., Lin, J., Sakai, T., Wu, Z.Z. (eds.) Proceedings of the Open-Source IR Replicability Challenge co-located with 42nd International ACM SIGIR Conference on Research and Development in Information Retrieval, OSIRRC@SIGIR 2019, Paris, France, 25 July 2019. CEUR Workshop Proceedings, vol. 2409, pp. 8–11. CEUR-WS.org (2019), http://ceur-ws.org/Vol-2409/position01.pdf
9. Chapman, A., et al.: Dataset search: a survey. VLDB J. **29**(1), 251–272 (2019). https://doi.org/10.1007/s00778-019-00564-x
10. Cohan, A., Feldman, S., Beltagy, I., Downey, D., Weld, D.S.: SPECTER: document-level representation learning using citation-informed transformers. In: Jurafsky, D., Chai, J., Schluter, N., Tetreault, J.R. (eds.) Proceedings of the 58th Annual Meeting of the Association for Computational Linguistics, ACL 2020, 5–10 July 2020, pp. 2270–2282. Association for Computational Linguistics (2020). https://doi.org/10.18653/v1/2020.acl-main.207
11. Craswell, N., Zoeter, O., Taylor, M.J., Ramsey, B.: An experimental comparison of click position-bias models. In: Najork, M., Broder, A.Z., Chakrabarti, S. (eds.) Proceedings of the International Conference on Web Search and Web Data Mining,

WSDM 2008, Palo Alto, California, USA, 11–12 February 2008, pp. 87–94. ACM (2008). https://doi.org/10.1145/1341531.1341545

12. Fix, E., Hodges, J.L.: Discriminatory analysis. Nonparametric discrimination: consistency properties. Int. Stat. Rev. Rev. Int. de Stat. **57**(3), 238–247 (1989). http://www.jstor.org/stable/1403797

13. Hienert, D., Kern, D., Boland, K., Zapilko, B., Mutschke, P.: A digital library for research data and related information in the social sciences. In: Bonn, M., Wu, D., Downie, J.S., Martaus, A. (eds.) 19th ACM/IEEE Joint Conference on Digital Libraries, JCDL 2019, Champaign, IL, USA, 2–6 June 2019, pp. 148–157. IEEE (2019). https://doi.org/10.1109/JCDL.2019.00030

14. Kacprzak, E., Koesten, L., Ibáñez, L., Blount, T., Tennison, J., Simperl, E.: Characterising dataset search - an analysis of search logs and data requests. J. Web Semant. **55**, 37–55 (2019). https://doi.org/10.1016/j.websem.2018.11.003

15. Kern, D., Mathiak, B.: Are there any differences in data set retrieval compared to well-known literature retrieval? In: Kapidakis, S., Mazurek, C., Werla, M. (eds.) TPDL 2015. LNCS, vol. 9316, pp. 197–208. Springer, Cham (2015). https://doi.org/10.1007/978-3-319-24592-8_15

16. Lommatzsch, A., Kille, B., Hopfgartner, F., Ramming, L.: Newsreel multimedia at mediaeval 2018: news recommendation with image and text content. In: Larson, M.A. (eds.) Working Notes Proceedings of the MediaEval 2018 Workshop, Sophia Antipolis, France, 29–31 October 2018. CEUR Workshop Proceedings, vol. 2283. CEUR-WS.org (2018). http://ceur-ws.org/Vol-2283/MediaEval_18_paper_5.pdf

17. Roberts, K., et al.: Searching for scientific evidence in a pandemic: an overview of TREC-COVID. CoRR abs/2104.09632 (2021). https://arxiv.org/abs/2104.09632

18. Robertson, S.E., Walker, S., Jones, S., Hancock-Beaulieu, M., Gatford, M.: Okapi at TREC-3. In: Harman, D.K. (ed.) Proceedings of The Third Text Retrieval Conference, TREC 1994, Gaithersburg, Maryland, USA, 2–4 November 1994. NIST Special Publication, vol. 500–225, pp. 109–126. National Institute of Standards and Technology (NIST) (1994). http://trec.nist.gov/pubs/trec3/papers/city.ps.gz

19. Schaer, P., Breuer, T., Castro, L.J., Wolff, B., Schaible, J., Tavakolpoursaleh, N.: Overview of lilas 2021 – living labs for academic search. In: Candan, K.S. (ed.) CLEF 2021. LNCS, vol. 12880, pp. 394–418. Springer, Cham (2021). https://doi.org/10.1007/978-3-030-85251-1_25

20. Schaer, P., Breuer, T., Castro, L.J., Wolff, B., Schaible, J., Tavakolpoursaleh, N.: Overview of lilas 2021 - living labs for academic search (extended overview). In: Faggioli, G., Ferro, N., Joly, A., Maistro, M., Piroi, F. (eds.) Proceedings of the Working Notes of CLEF 2021 - Conference and Labs of the Evaluation Forum, Bucharest, Romania, September 21st - to - 24th, 2021. CEUR Workshop Proceedings, vol. 2936, pp. 1668–1699. CEUR-WS.org (2021). http://ceur-ws.org/Vol-2936/paper-143.pdf

21. Schaer, P., Schaible, J., Müller, B.: Living labs for academic search at CLEF 2020. In: ECIR 2020. LNCS, vol. 12036, pp. 580–586. Springer, Cham (2020). https://doi.org/10.1007/978-3-030-45442-5_75

22. Schaible, J., Breuer, T., Tavakolpoursaleh, N., Müller, B., Wolff, B., Schaer, P.: Evaluation Infrastructures for Academic Shared Tasks. Datenbank-Spektrum **20**(1), 29–36 (2020). https://doi.org/10.1007/s13222-020-00335-x

23. Schuth, A., Balog, K., Kelly, L.: Overview of the living labs for information retrieval evaluation (LL4IR) CLEF lab 2015. In: Mothe, J. (ed.) CLEF 2015. LNCS, vol. 9283, pp. 484–496. Springer, Cham (2015). https://doi.org/10.1007/978-3-319-24027-5_47

Analysing Moral Beliefs for Detecting Hate Speech Spreaders on Twitter

Mirko Lai(✉)(iD), Marco Antonio Stranisci(iD), Cristina Bosco(iD), Rossana Damiano(iD), and Viviana Patti(iD)

Dipartimento di Informatica, Università degli Studi di Torino, Turin, Italy
{mirko.lai,marco.stranisci,cristina.bosco,rossana.damiano, viviana.patti}@unito.it

Abstract. The Hate and Morality (HAMOR) submission for the *Profiling Hate Speech Spreaders on Twitter* task at PAN 2021 ranked as the 19th position - over 67 participating teams - according to the averaged accuracy value of 73% over the two languages - English (62%) and Spanish (84%). The method proposed four types of features for inferring users attitudes just from the text in their messages: HS detection, users morality, named entities, and communicative behaviour. In this paper, since the test set is now available, we were able to analyse false negative and false positive prediction with the aim of shed more light on the hate speech spreading phenomena. Furthermore, we fine-tuned the features based on users morality and named entities showing that semantic resources could help in facing Hate Speech Spreaders detection on Twitter.

Keywords: Hate Speech · Moral Foundation Theory · Twitter

1 Introduction

The Profiling Hate Speech (HS) Spreaders on Twitter is an Author Profiling task [17] organised at PAN [4]. Teams are invited to develop a model that, given a Twitter feed of 200 messages, determines whether its author spreads hatred contents. The task is multilingual, and covers Spanish and English languages. The training set is composed of 200 users per language, 100 of them annotated as haters by having posted at least one HS in their feeds; the annotation of single tweets is not available, though. All the information about users, mentions, hashtags, and urls are anonymised, making not replicable in this context approaches based on demographic features [22], or community detection [3,13].

Our team participated to the task with a system called *The Hate and Morality* (HAMOR). The name of the model refers to the combined use of HS and moral values detection [7] for analysing a feed of tweets, in order to infer a general attitude of a user towards people vulnerable to discrimination. Our approach relies on the moral

M. Lai and M. A. Stranisci—Contributed equally to this work.

pluralistic hypothesis (Cfr [6, 18, 19]), according to which moral foundations are many and people more prioritise some values than other ones. This can lead to divergent and often conflicting points of view on debated facts, and might also be a factor in HS spreading [8]. More specifically, we considered a group-bound moral judgement as the signal of a potential negative stance against minorities, and used it as a feature to classify HS spreaders together with a HS detection model. The paper is structured as follow: Sect. 2 brings again the attention on the description of the features used in the task, and Sect. 3 is devoted to an error analysis focusing on a better understanding of false positive cases. Section 4 proposes a qualitative analysis of the proposed features. Then, Sect. 4.3 describes the improvements made to our system for better predicting HS Spreaders on Twitter. In Conclusions (Sect. 5) the contribution of our approach on this phenomena are discussed.

2 Feature Selection

Four types of features for inferring users attitudes just from the text in their messages have been selected to train our model: HS detection (Sect. 2.1), users morality (Sect. 2.2), Named Entities (Sect. 2.3), Communicative behaviour (Sect. 2.4). We employed a manual ensemble-based feature selection method combining multiple feature subsets for selecting the optimal subset of features that improves classification accuracy for each language.

2.1 Hate Speech Detection

HS detection is the automated task of detecting whether a piece of text contains HS. Several shared tasks on HD detection have taken place and large annotated corpora are available in different languages. For example, the *HatEval* dataset for *hate speech detection against immigrants and women in Spanish and English tweets* has been released to be used at the Task 5 of the SemEval-2019 workshop [1]. We decided to use the entire *HatEval* dataset for training three models and we proposed the following features:

- SemEvalSVM (*SESVM*): 1-dimensional feature that counts - for each user - the number of hateful tweets predicted by a linear SVM trained using a text 1–3 g bag-of-words representation.
- Atalaya (*ATA*) [16]: 1-dimensional feature that counts - for each user - the number of hateful tweets predicted by a linear-kernel SVM trained on a text representation composed of bag-of-words, bag-of-characters and tweet embeddings, computed from fastText word vectors. We were inspired from the Atalaya team's system that achieved the best scores in the *HatEval* Spanish sub-task.
- Fermi (*FER*) [10]: a 1-dimensional feature that counts - for each user - the number of hateful tweets predicted by SVM with the RBF kernel trained on tweet embeddings from Universal Sentence Encoder. We were inspired by the Fermi team's system that obtained the best result at the *HatEval* English sub-task.

Furthermore, the growing interest on this topic leads the research community (and not only) to develop some lexica of hateful words such as HurtLex [2],

NoSwearing[1], and The Racial Slur Database[2]. HurtLex is a lexicon of offensive, aggressive, and hateful words in over 50 languages (including English and Spanish). The words are divided into 17 different categories. Then, NoSwearing is a list of English swear words, bad words, and curse words. The Spanish translation was made by Pamungkas et al. [15]. Finally, the Racial Slur Database is a list of words that could be used against someone - of a specific race, sex, gender etc. - divided into more then 150 categories. The list is only available in English, we thus computed the Spanish translation using Babelnet's API [14]. We also take advantage of spaCy[3] models *en_core_web_lg*, and *es_core_news_lg* for expanding the three lexica. Indeed, we used the tok2vec embedding representation for including in the three lists the 10 most similar tokens of each word. We can thus propose the following features:

- HurtLex (*HL*): a 18-dimensional feature that evaluates the number of hateful words used by each user, the mean of hateful words in each tweet, and the standard deviation. We exploited the following 6 categories: negative stereotypes ethnic slurs, moral and behavioural defects, words related to prostitution, words related to homosexuality, words related to the seven deadly sins of the Christian tradition, felonies and words related to crime and immoral behaviour (we exclusively considered the conservative level).
- No Swearing (*NoS*): a 3-dimensional feature that evaluates the number of swearing words used by each user, the mean of swearing words in each tweet, and the standard deviation.
- The Racial Slur Database (*RSdb*): a 27-dimensional feature that evaluates the number of swearing words used by each user, the mean of swearing words in each tweet, and the standard deviation for each of the following 9 categories: Asians, Arabs, Black people, Chinese, Hispanics, Jews, Mexicans, mixed races, Muslims.

2.2 Moral Values Detection

According to many scholars, moral beliefs are not universal, but reside on a plurality of "irreducible basic elements" [21]. Several configuration of values are possible, and some of them are in conflict, such as autonomy *versus* community [19], or conservation *versus* openness to change [18]. The Moral Foundation Theory (MFT) [6] shares this approach since it distinguishes five dyads leading to people morality: care/harm, fairness/cheating, which relies on individualisation, and loyalty/betrayal, authority/subversion and purity/degradation, which are binding foundations. Some of these combinations may correlate with specific political positions, as emerges from experimental results [5]: liberals seem to agree on individualisation values, whereas conservatives could be more likely to follow binding dyads.

In building our model, we considered binding moral dyads as a potential feature characterising a HS spreader. More specifically, we claimed that users who rely on loyalty/betrayal and authority/subversion might be inclined to post hatred contents online.

---

[1] https://www.noswearing.com/.

[2] http://www.rsdb.org/full.

[3] https://spacy.io/.

Hence, we referred to two existing resources: the extended Moral Foundations Dictionary (eMFD) [9], and the Moral Foundations Twitter Corpus (MFTC) [7].

The eMFD is a dictionary of 2,965 terms categorised by a specific moral foundation. We chose all those related to loyalty/betrayal and authority/subversion moral concerns, and translated them in Spanish scripting babbel.com and wordreferences.com (the translated dictionary amounts to 4,622 words). Finally, we expanded the words list using the same methodology explained in Sect. 2.1. The result is the following feature: for each user we computed the mean, the standard deviation, and the total amount of terms occurring in her/his tweet.

- extended Moral Foundations Dictionary (*eMFD*): a 12-dimensional feature that includes the mean, the standard deviation, and the total amount of terms occurring in her/his tweets for the four categories loyalty/betrayal and authority/subversion.

The MFTC is a collection of 35, 000 tweets annotated for their moral domains, and organised in 7 subcorpora, each focusing on a specific discourse domain (e.g.: the Black Lives Matters, and #metoo movements, and the US 2016 presidential elections). Using transfer learning as a label assignment method, we converted the original multi-label annotation schema in a binary-label one: 9, 000 texts annotated as loyalty, betrayal, authority or subversion were considered as potentially correlated with HS (*true*), while the other not (*false*). Using the resulting corpora as training set, we thus proposed the following feature.

- Moral Foundations Twitter Corpus (MFTC): a 1-dimensional feature that counts - for each user - the number of hateful tweets predicted by a linear SVM trained using a text 1–3 g bag-of-words representation.

2.3   Named Entity Recognition of HS Target

In a message, the mention of a person belonging to a group vulnerable to discrimination might be seen as a signal of hatred contents, since the clear presence of a target in this kind of expressions allows discriminating between what is HS and what is not. Thereby, we implemented a feature aimed at detecting the presence of a potential HS target within a tweet.

We first collected all the entities of type PERSON in the whole training set detected by the transition-based named entity recognition component of spaCy. Then, we searched the retrieved entities on Wikipedia through the Opensearch API[4]. The example below shows the Wikipedia pages returned by the Opensearch API when the entity *Kamala* is requested.

```
['Kamala','Kamala Harris','Kamal Haasan',
'Kamala (wrestler)','Kamala Khan','Kamala Surayya',
'Kamala Harris 2020 presidential campaign',
'Kamaladevi Chattopadhyay','Kamala Mills fire',
'Kamalani Dung']
```

---

[4] https://www.mediawiki.org/wiki/API:Opensearch.

However, this operation is revealed to be not accurate. In fact, it does not return a unique result for each entity detected by spaCy, but a set of 10 potential candidates. Therefore, we decided to create two lists - one for each language - of HS targets including only persons that belong to categories that could be subject to discrimination.

With the aim of detecting the relevant categories, we scraped the *category box* from the Wikipedia pages of all entities of type PEOPLE detected by spaCy (3, 996 English, and 5, 089 Spanish). The result is a list of Wikipedia's categories per language, which needed to be filtered to avoid not relevant results.

The Fig. 1 shows a partial selection of *Kamala Harris* category box, which contains several references to unnecessary information, such as '1964 births', or 'Writers from Oakland, California', but also usefully ones, such us '*African-American* candidates for President of the United States' or '*Women* vice presidents'.

Categories: Kamala Harris | 1964 births | 21st-century American memoirists
| 21st-century American politicians | 21st-century American women politicians
| 21st-century American women writers | African-American candidates for President of the United States
| African-American candidates for Vice President of the United States
| African-American members of the Cabinet of the United States | African-American memoirists
| African-American people in California politics | African-American United States senators
| African-American women in politics | African-American women lawyers
| American people of Indian Tamil descent | American politicians of Indian descent
| American politicians of Jamaican descent | American prosecutors | American women lawyers
| American women memoirists | Asian-American members of the Cabinet of the United States
| Asian-American United States senators | Baptists from California | Women vice presidents
| Writers from Oakland, California

Fig. 1. A selection of categories for Kamala Harris on Wikipedia's category box

After a manual analysis of the two lists, we thus narrowed them by a regex filtering, in order to obtain only a set of relevant categories: 279 for English, and 415 for Spanish. Finally, we collected all the individuals who are their members. As final result, we obtained two gazetteers of potential HS targets (7, 5890 entities for English, and 31, 235 for Spanish) in the following format.

```
{Margaret Skirving Gibb : Scottish feminists,
 Melih Abdulhayoğlu : Turkish emigrants to the USA,
 James Adomian : LGBT people from Nebraska [...]}
```

We thus proposed a feature that counts the mentions towards persons belonging to a group vulnerable to discrimination.

– Named Entity Recognition of HS target (NER): a 5-dimensional feature expressing the total number of potential HS targets mentioned in her/his tweets, the mean, the standard deviation, and the ratio between the number of HS target, and all the HS targets mentioned by the user.

2.4 Communicative Behaviour

Under the label 'Communicative behaviour' a set of features related to the structure of the tweet and to the user's style has been grouped. The total number, the mean, and the standard deviation have been computed for each feature over all users feeds.

- Uppercase Words (UpW): this feature refers to the amount of words starting with a capital letter and the number of words containing at least two uppercase characters.
- Punctuation Marks (PM): a 6-dimensional feature that includes the frequency of exclamation marks, question marks, periods, commas, semicolons, and finally the sum of all the punctuation marks mentioned before.
- Length (Len): 3 different features were considered to build a vector: number of words, number of characters, and the average of the length of the words in each tweet.
- Communicative Styles (CoSty): a 3-dimensional feature that computes the fraction of retweets, of replies, and of original tweets over all user's feed.
- Emoji Profile (EPro): this feature tries to distinguish some user's traits from the emoji her/his used. We implemented a one-hot encoding representation of the modifiers used in the emoji ZWJ sequences (e.g. *man*: *medium skin tone*, beard) that includes the 5 different skin tone modifiers and the gender modifiers, in addition to the religious emojis (e.g. Christian Cross) and the national flags.

We finally employed bag-of-words models as feature:

- Bag of Words (BoW): binary 1–3 g of all user's tweets.
- Bag of Emojis (BoE): binary 1–2 g of all user's tweets only including emojis.

3 Error Analysis

The organisers provided a dataset for training participant systems including 400 Twitter's feeds - 200 in English and 200 in Spanish - binary labelled with HS Spreader. The distribution is perfectly balanced among the true and false labels. In order to assess the performance of the participating systems, a test set of 200 unlabelled Twitter's feeds - 100 for each language - was also provided.

The current availability of the correct labels for the test set allows us to perform an error analysis that we focus on better understanding the false positive cases. The test set is balanced for both languages (50% of the users are hate speech spreaders).

Table 1 shows confusion matrix of our submission for both languages. For each of the languages, the entry in row 0 and column 1 indicates the amount of *false positives*, i.e. samples that our system erroneously predicted as HS spreader (1) while they weren't. The entry in row 1 and column 0 indicates the amount of *false negatives*.

For both languages, the number of false positives is similar to the amount of false negatives, while in Spanish fewer errors in the prediction of HS spreaders can be observed with respect to English.

We aim to perform a manual error analysis mostly evaluating the tweets of the users that are not HS spreaders, but that have been predicted as such by our model. Unfortunately, also observing the correct labels provided by the organizers, we cannot

Table 1. Confusion Matrix

	EN			ES		
	Predicted			*Predicted*		
	0	*1*		*0*	*1*	
Actual 0	33	17	50	41	9	50
1	21	29	50	7	43	50
	54	46		48	52	

check whether a single tweet is HS or not, hence labels only indicate whether the user that generated the feed (where the tweet is included) is a hater or not. Since then a user feed is composed by several tweets, we decided to filter them by automatically predicting whether each single tweet is hateful or not using one of the models proposed in Sect. 2.1: SESVM, ATA, and FER.

Figure 2 shows the number of users y having at least x tweets that have been predicted as hateful in their feeds by our models.

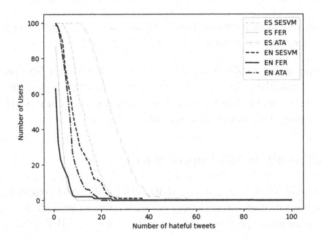

Fig. 2. The number of users y having at least x hateful tweets

FER is the model that shows a more conservative trend: it predicts not more than one tweet as hateful in 62 English and 84 Spanish users' feeds. Furthermore, it does not predict more than 10 hateful tweets in any Spanish users' feeds and it follows a similar trend on both languages. On the contrary, ATA and SESVM are more inclusive predicting at least 1 hateful tweets in all users' feeds. Furthermore, ATA seems to be more conservative on English than on Spanish tweets. For such reasons, we decided to use FER for automatically predicting HS in individual tweets.

The results provided by FER allows us to better understand the motivation of the erroneous classification of some user as HS spreader, that wasn't according to the correct labels provided by the organisers for the test set. For what concerns English, we find several tweets in hate-less speech spreaders feeds containing profanities and slurs. As an illustrative example, a feed of a black woman that wrote:

My nigga just came home with a Lush 2[5]. Goodnight bitches 😊

Although the author of the tweet uses emojis that include skin tones and the female sign in her feed, these signs do not help the model to understand that she is a black woman that uses the words *nigga* (racism) and *bitches* (misogyny) in a funny way for communicating with her followers.

Also in some Spanish case, although the false positive entries are very few, we found several profanities and bad words in false positive hate-less speech spreaders feeds. Following an illustrative example of three tweets extracted from the same false positive user's feed:

"#USER# Para toda la mierda femiorca[6], que os jodan hdps" (*For all the femiorca shit, fuck you son of a beach*)

Pinta negro para cualquier persona a día de hoy. Esto es vivir en un imposible. #URL# (*Paint black for anyone today. This is living in an impossible. #URL#*)

Y los que no son junden son masones. Que asco de UE. #URL# (*And those who are not Jew are Masons. What a mess of the EU. #URL#*)

Although the author of the tweet has not been considered a HS spreader by the organisers of the task, these three tweets express very strong and questionable opinions against feminist movements, black people rights, and Jews. For our model it is therefore difficult to not predict this user as a HS spreader.

4   Features Analysis and Improvement

Experimental results showed a significant delta between the two languages, despite both relied on a similar set of features. Hence, in this section we provide a deeper analysis of features adopted in our proposal, with a specific focus on MFT Values, and Named Entities.

4.1   MFT Values

In our experimental setting, we selected only two moral dyads from the MFT. This choice relied on psychological studies claiming for a correlation between the political stance of a person and certain moral configurations. However, such assumption is derived from psychological surveys rather than from NLP experiments. Thereby, we analysed how MFT dimensions correlate with HS spreading. We used the eMFT dictionary [9], to count all the occurrences of words expressing MFT values for each user. Then, we computed the Spearman's correlation between each value and HS spreaders

---

[5] Lush 2 is a Sex Toys.
[6] Femiorca is a feminist community.

labels in order to observe which were more significant for the task. As it can be observed in Table 2, the role of MFT values may be more relevant for Spanish, since the average ρ for this language is 0.26 while the average ρ for English is 0.09. In both languages there is always one element in each dyad that better correlates with HS. For instance, Harm obtains a higher Spearman's ρ score than Care. This may suggest the existence of a set of different moral frames adopted by users (Cfr [11]) that shape their communicative behaviour. A closer look into the dyads shows some interesting trends about the correlation between moral values and HS. Harm and Subversion are predominant in their respective dyads for English and Spanish, suggesting a moral configuration in which binding and individualisation values interact in determining users stance. On the opposite, the Purity/Degradation dyad behave differently between languages. English HS spreaders seem to focus on the violation of the dyad (Degradation), while Spanish users do the opposite. Finally, none of the Fairness/Cheating and Loyalty/Betrayal values significantly correlates with HS in English. Such distribution seems to confirm that moral stance is topic-sensitive, as demonstrated by [7]. Further investigation in existing corpora may shed more light on this phenomenon.

Table 2. The Spearman's correlation of each MFT values with HS Spreader in the dataset.

Moral Value	Spearman's ρ (en)	Spearman's ρ (es)
Loyalty	0.003	0.276
Betrayal	0.069	0.134
Purity	0.027	0.406
Degradation	0.181	0.329
Care	−0.035	0.144
Harm	0.137	0.404
Fairness	0.075	0.337
Cheating	0.015	0.038
Authority	0.012	0.174
Subversion	0.143	0.359

We then proposed a feature that includes the full spectrum of moral values: eMFD+.

4.2 Named Entities

In our original submission, the creation of gazetteers with named entities who are potentially target of HS was based on a manual selection of Wikipedia categories containing some target words related to vulnerable groups (e.g.: American women non-fiction writers). This led to sparse representations of this feature, since we obtained 11, 480 categories of people for Spanish, and 36, 366 for English and most of them were not mentioned by users in their tweets. We decided to remove all categories of named entities that were mentioned by less than 20 users in the data set, dramatically reducing the number of categories to 204 for Spanish and 225 for English. Finally, we computed the Spearman's correlation between the occurrences of each category and HS spreaders labels. Table 3 shows the 5 categories which best correlate with HS. As it can be observed, women are a shared target across languages, while religious minorities are a significant target for English and LGBT for Spanish. As for MFT values, it seems that the distribution and relevance of vulnerable categories for HS detection is strongly influenced by current events. For instance, several mentions of Kamala Harris appear to be correlated with the 2020 US elections.

Table 3. The Spearman's correlation of each category of people vulnerable to HS and HS Spreader in the dataset.

Category of people (en)	Spearman's ρ	Category of people (es)	Spearman's ρ
American_women_podcasters	0.200	Feministas_de_Madrid	0.440
American_women_rock_singers	0.189	Mujeres_guerreras_ficticias	0.267
American_women_non-fiction_writers	0, 175	Mujeres	0.220
Kenyan_Muslims	0.171	Artistas_LGBT_de_España	0.214
American_women_memoirists	0.165	Mujeres_del_siglo_XX	0, 206

We propose an enhanced version of the NER feature (NER+) that exclusively takes in consideration the entities belonging to this filtered set of categories.

4.3   Fine-Tuning

Our official submission obtained 84% and 62% in terms of accuracy on HS Spreader identification respectively for Spanish and English. The final score, used in determining the final ranking, is the averaged accuracy values per language which corresponds to 73% [17]. Here, we verify the contribution of the fine-tuned featured described in Sect. 4.

English. Our submission for the English subtask employed the features NER, eMFD, RSdb, HatEval, and FER. The dimensional space representation of each user's feed was relatively simple and the obtained results was 12% points below the highest one (the UO-UPV [12] team obtained 74%).

Thereby, we tried to increase the complexity representation adding the Communicative Behaviour feature BoW to this configuration. The model achieves 65% in term of accuracy, still very much below the state of the art. We then employed the features NER+, eMFD+, and replaced ATA with FER which has been shown to be more skewed on precision in detecting HS (see Sect. 3). The obtained accuracy increased of other 2%points. Table 4 shows the contribution of each fine-tuned features. Replacing FER with ATA does not affect the result, as well as the enhanced NER feature seems to not improving the prediction. However, the effectiveness of feature based on the full spectrum of moral values (eMFD+) is showed.

Table 4. Evaluation of the contribute of enhanced on English subtask

Feature Set	Accuracy
NER, eMFD, RSdb, HatEval, and FER	0.67
replacing NER with NER+	*0.67*
replacing eMFD+ with eMFD	0.63
replacing FER with ATA	0.67

Spanish. For Spanish submission, we employed two Communicative Behaviour features (BoW and BoE), NER, eMFD, HL, NoS, ATA. We then applied the enhanced version of eMFD and NER and we also tried to replace ATA with FER in order to test a more conservative feature. The obtained accuracy increased of 1%point achieving the highest result obtained by the team SIINODINUOVO [20]. Table 5 shows the contribution of each fine-tuned features:

Table 5. Evaluation of the contribute of enhanced on Spanish subtask

Feature Set	Accuracy
BoW, BoE, NER, eMFD, HL, NoS, ATA	0.85
replacing NER with NER+	*0.85*
replacing eMFD+ with eMFD	0.78
replacing FER with ATA	0.83

Also in this case, the effectiveness of feature based on the full spectrum of moral values (eMFD+) is showed. Then, a conservative feature based on HS detection such as FER better affected the result. Finally, we could employed NER+ without making any significant changes.

Cross-Language. We finally have given some thought to how the decision to propose different features set for each language had been a good choice. We therefore trained the English model with the features set used for Spanish employing the enhanced version of NER and eMFD. The performance increased further to 71% accuracy for the English subtask. It would have meant the achievement of 78% average accuracy (85% and 70% respectively for Spanish and English) over the two languages (in other words, 2th position in the official ranking with a detachment of only 1% points from the 1st position).

Therefore, the choice to use different set of features for the two languages was inauspicious. However, the effectiveness of features based on lexica (HL, NoS), morality values (eMFD), and Named Entity Recognition (NER) in a multilingual perspective is therefore confirmed and leaves opportunities for further future exploration open.

5 Conclusions

In this paper we presented a detailed analysis of the HAMOR submission for the *Profiling Hate Speech Spreaders on Twitter* task at PAN-2021. Our approach, chiefly based on external resources such as other annotated corpora, lexica, and semi-structured content, proved to be highly successful concerning the task of HS Spreader identification in both languages, as our system ranked as the 19th position among 67 participating teams. The results show that the use of external resources preserves stable values of accuracy between the experimental setting and the prevision of the test set on Spanish sub-task. The proposed lexica gave a considerable contribution for obtaining these results and the use of named entity recognition for detection potential target of HS looks promising. In the future, we plan to employ the features discarded from the submitted run for a prediction on the test set. We also deeper explored the features base on named entity recognition and proposed a finer grained approach for employing MFT features, considering different combination of moral values, and analyzing how moral attitudes may vary across different countries. Finally, we propose a cross lingual set of features that improve the result obtained by our model in term of accuracy. All the code used for on this work is available on GitHub for further exploration and for allowing reproducibility of our experiments[7].

References

1. Basile, V., et al.: SemEval-2019 Task 5: multilingual detection of hate speech against immigrants and women in twitter. In: Proceedings of the 13th International Workshop on Semantic Evaluation, pp. 54–63. Association for Computational Linguistics (2019)
2. Bassignana, E., Basile, V., Patti, V.: Hurtlex: a multilingual lexicon of words to hurt. In: 5th Italian Conference on Computational Linguistics, CLiC-it 2018, vol. 2253, pp. 1–6. CEUR-WS (2018)
3. Cignarella, A.T., Lai, M., Bosco, C., Patti, V., Rosso, P.: Sardistance@evalita2020: overview of the task on stance detection in Italian tweets. In: Proceedings of the 7th Evaluation Campaign of Natural Language Processing and Speech Tools for Italian (EVALITA 2020), CEUR Workshop Proceedings, vol. 2765. CEUR-WS.org, Aachen (2020)
4. Faggioli, G., Ferro, N., Joly, A., Maistro, M., Piroi, F. (eds.): Proceedings of the Working Notes of CLEF 2021 - Conference and Labs of the Evaluation Forum, Bucharest, Romania, 21–24 September 2021, CEUR Workshop Proceedings, vol. 2936. CEUR-WS.org (2021)
5. Graham, J., Haidt, J., Nosek, B.A.: Liberals and conservatives rely on different sets of moral foundations. J. Pers. Soc. Psychol. **96**(5), 1029 (2009)
6. Haidt, J., Joseph, C., et al.: The moral mind: how five sets of innate intuitions guide the development of many culture-specific virtues, and perhaps even modules. Innate Mind **3**, 367–391 (2007)
7. Hoover, J., et al.: Moral foundations twitter corpus: a collection of 35k tweets annotated for moral sentiment. Social Psychol. Pers. Sci. **11**(8), 1057–1071 (2020)
8. Hoover, J., et al.: Bound in hatred: the role of group-based morality in acts of hate. PsyArXiv (2019)

---

[7] https://github.com/mirkolai/PAN2021_HaMor.

9. Hopp, F.R., Fisher, J.T., Cornell, D., Huskey, R., Weber, R.: The extended Moral Foundations Dictionary (eMFD): development and applications of a crowd-sourced approach to extracting moral intuitions from text. Behav. Res. Methods **53**(1), 232–246 (2021)

10. Indurthi, V., Syed, B., Shrivastava, M., Chakravartula, N., Gupta, M., Varma, V.: FERMI at SemEval-2019 task 5: using sentence embeddings to identify hate speech against immigrants and women in Twitter. In: Proceedings of the 13th International Workshop on Semantic Evaluation, pp. 70–74. Association for Computational Linguistics (2019)

11. Kwak, H., An, J., Jing, E., Ahn, Y.Y.: Frameaxis: characterizing microframe bias and intensity with word embedding. PeerJ Comput. Sci. **7**, e644 (2021)

12. Labadie, R., Castro-Castro, D., Ortega Bueno, R.: Deep modeling of latent representations for twitter profiles on hate speech spreaders identification: notebook for PAN at CLEF 2021. In: Faggioli, G., Ferro, N., Joly, A., Maistro, M., Piroi, F. (eds.) Proceedings of the Working Notes of CLEF 2021 - Conference and Labs of the Evaluation Forum, Bucharest, Romania, 21–24 September 2021, CEUR Workshop Proceedings, vol. 2936, pp. 2035–2046. CEUR-WS.org (2021)

13. Mishra, P., Del Tredici, M., Yannakoudakis, H., Shutova, E.: Author profiling for abuse detection. In: Proceedings of the 27th International Conference on Computational Linguistics, pp. 1088–1098 (2018)

14. Navigli, R., Ponzetto, S.P.: Babelnet: the automatic construction, evaluation and application of a wide-coverage multilingual semantic network. Artif. Intell. **193**, 217–250 (2012)

15. Pamungkas, E.W., Cignarella, A.T., Basile, V., Patti, V., et al.: 14-ExLab@ UniTo for AMI at IberEval2018: exploiting lexical knowledge for detecting misogyny in English and Spanish tweets. In: 3rd Workshop on Evaluation of Human Language Technologies for Iberian Languages, IberEval 2018, vol. 2150, pp. 234–241. CEUR-WS (2018)

16. Pérez, J.M., Luque, F.M.: Atalaya at SemEval 2019 task 5: robust embeddings for tweet classification. In: Proceedings of the 13th International Workshop on Semantic Evaluation, pp. 64–69. Association for Computational Linguistics (2019)

17. Rangel, F., De la Peña Sarracén, G.L., Chulvi, B., Fersini, E., Rosso, P.: Profiling hate speech spreaders on twitter task at PAN 2021. In: Faggioli, G., Ferro, N., Joly, A., Maistro, M., Piroi, F. (eds.) Proceedings of the Working Notes of CLEF 2021 - Conference and Labs of the Evaluation Forum, Bucharest, Romania, 21–24 September 2021, CEUR Workshop Proceedings, vol. 2936, pp. 1772–1789. CEUR-WS.org (2021)

18. Schwartz, S.H.: An overview of the schwartz theory of basic values. Online Read. Psychol. Cult. **2**(1), 2307–0919 (2012)

19. Shweder, R.A., Much, N.C., Mahapatra, M., Park, L.: The "big three" of morality (autonomy, community, divinity) and the "big three" explanations of suffering. Morality Health, 119–169 (1997)

20. Siino, M., Di Nuovo, E., Tinnirello, I., La Cascia, M.: Detection of hate speech spreaders using convolutional neural networks. In: Faggioli, G., Ferro, N., Joly, A., Maistro, M., Piroi, F. (eds.) Proceedings of the Working Notes of CLEF 2021 - Conference and Labs of the Evaluation Forum, Bucharest, Romania, 21–24 September 2021, CEUR Workshop Proceedings, vol. 2936, pp. 2126–2136. CEUR-WS.org (2021)

21. Stranisci, M., De Leonardis, M., Bosco, C., Viviana, P.: The expression of moral values in the twitter debate: a corpus of conversations. IJCoL - Special Issue: Comput. Dial. Model. Role Pragmatics Common Ground Interact. **7**(1,2), 113–132 (2021)

22. Waseem, Z., Hovy, D.: Hateful symbols or hateful people? predictive features for hate speech detection on Twitter. In: Proceedings of the NAACL Student Research Workshop. Association for Computational Linguistics (2016)

Query Expansion, Argument Mining
and Document Scoring for an Efficient
Question Answering System

Alaa Alhamzeh[1,2(✉)], Mohamed Bouhaouel[2], Előd Egyed-Zsigmond[1],
Jelena Mitrović[2,3], Lionel Brunie[1], and Harald Kosch[2]

[1] INSA de Lyon, 20 Avenue Albert Einstein, 69100 Villeurbanne, France
{Alaa.Alhamzeh,Elod.Egyed-zsigmond,Lionel.Brunie}@insa-lyon.fr
[2] Universität Passau, Innstraße 41, 94032 Passau, Germany
{Mohamed.Bouhaouel,Jelena.Mitrovic,Harald.Kosch}@uni-passau.de
[3] Institute for AI Research and Development, Fruškogorska 1, 21000 Novi Sad, Serbia

Abstract. In the current world, individuals are faced with decision making problems and opinion formation processes on a daily basis. Nevertheless, answering a comparative question by retrieving documents based only on traditional measures (such as TF-IDF and BM25) does not always satisfy the need. In this paper, we propose a multi-layer architecture to answer comparative questions based on arguments. Our approach consists of a pipeline of query expansion, argument mining model, and sorting of the documents by a combination of different ranking criteria. Given the crucial role of the argument mining step, we examined two models: DistilBERT and an ensemble learning approach using stacking of SVM and DistilBERT. We compare the results of both models using two argumentation corpora on the level of argument identification task, and further using the dataset of CLEF 2021 Touché Lab shared task 2 on the level of answering comparative questions.

Keywords: Comparative Question Answering · Computational Argumentation · Argument Search

1 Introduction

Argumentation is a fundamental aspect of human communication and decision making. According to van Eemeren and Grootendoorst [1], *"Argumentation is a verbal, social, and rational activity aimed at convincing a reasonable critic of the acceptability of a standpoint by putting forward a constellation of propositions justifying or refuting the proposition expressed in the standpoint"*.

Computational argumentation considers mainly three aspects: argument mining, argument quality assessment and argument generation. Recently, those topics have been studied by scholars from different interdisciplinary fields (e.g., fact checking and automated decision making). Furthermore, the arguments can serve as the keystone of an intelligent web search engine. In this

© The Author(s), under exclusive license to Springer Nature Switzerland AG 2022
A. Barrón-Cedeño et al. (Eds.): CLEF 2022, LNCS 13390, pp. 162–174, 2022.
https://doi.org/10.1007/978-3-031-13643-6_13

regard, the Webis Group has organized an argumentation retrieval event "Touché Lab at CLEF" (2020-present)[1] that consists of two independent shared tasks: 1) Argument Retrieval for Controversial Questions and 2) Argument Retrieval for Comparative Questions. We present in this paper our adopted approach for participation as **"Rayla Team"** in the second shared task 2021. The main objective of this task is to help users facing some choice problem in their daily life: Given a comparative question (for instance, "Which is better, a Mac or a PC?"), the objective is to retrieve documents from the ClueWeb12 corpus[2], and rank them based on different criteria, mainly, the arguments they provide.

In order to have more granularity control, our architecture incorporates several units, each one is dedicated to perform a specific sub-task, namely: query expansion, argument mining, scoring, and sorting.

The Touché organizers offer the participants to use their own TARGER [2] tool, for the argument retrieval sub-task. Since this is the main engine for an argument-based search engine, we decided to develop our own module based on the latest developments in the field of computational argument mining. Therefore, we implemented, for our participation, a new transfer learning model for argument identification based on DistilBERT [3]. However, the extraction of the relevant documents depends on different factors and criteria like query expansion and the ranking algorithm. Subsequently, we aim at testing the real impact of improving the argument mining model itself. Hence, we injected our ensemble learning model [4] instead of DistilBERT in the same submitted global architecture and we additionally test its outcome on the shared task data. Moreover, to foster the research in this area, we provide our source code publicly through the GitHub repository[3].

The remainder of the paper is organized as follows: In Sect. 2, we go through a conceptual background of argument mining and argument-based question answering. In Sect. 3, we present our retrieval system architecture as well as the function of each unit. We present and discuss the results of our approach with respect to the two proposed argument identification models in Sects. 4 and 5, respectively. Finally, we summarize the main outcomes and conclusions in Sect. 6.

2   Related Work

2.1   Argument Mining

Argumentative text identification is generally the first step of a complete argument mining system. The literature reports mainly classical machine learning models and few attempts to use deep learning models (e.g., [5,6]). In a classical machine learning model, the training and testing data are used for the same

---

[1] https://webis.de/events/touche-21/index.html.
[2] https://lemurproject.org/clueweb12/.
[3] https://github.com/bouhao01/arg-search-engine.

task and follow the same distribution. However, transfer learning aims to transfer previous learned knowledge from one source task (or domain) to a different target one, considering that source and target tasks and domains may be the same or may be different but related [7]. This premise can be useful in the argument mining domain from different perspectives. First of all, common knowledge about the language is obviously appreciated. Second, transfer learning can solve or at least help to solve one of the biggest challenges in the argument mining field, the lack of labeled datasets. Third, even available datasets are often of small size and very domain and task dependent. They may follow different annotations, argument schemes, and various feature spaces. This means that for each potential application of argument mining, we need argument experts to label a significant amount of data for the task at hand, which is definitely an expensive work in terms of time and human-effort. Hence, transfer learning will fine-tune pre-trained knowledge on big data to serve another problem, and that is why we built on it for the argument identification stage.

Similarly, Wambsganss et al. [5], proposed an approach for argument identification using BERT [8] and they achieved good performance in different known argumentation corpora. However, our transfer learning model is based on DistilBERT [3], a distilled version of BERT, which retains 97% of the language understanding capabilities of the base BERT model, being 40% smaller in size, and 60% faster. Nevertheless, we observe that in a small training set regime (which is usually the case in available AM datasets), a simple classical machine learning model may outperform a more complex deep learning one. In addition, training a classical model and interpreting it is much faster and easier than a neural network based one. On the other hand, traditional machine learning algorithms fall short as soon as the testing data distribution or the target task are not the same as the training data distribution or the learned task. In contrast, in transfer learning, the learning of new tasks relies on previously learned ones. Hence, we aimed at achieving a trade-off between those two learning methods by assembling the DistilBERT model with a traditional machine learning one in our new experiments based on our previous work regarding cross-domain argument identification [4].

2.2   Comparative Question Answering

Question answering is a sophisticated form of information retrieval that started to develop as a field decades ago. However, with the growth of world wide web data, as well as of private databases, the need for more precise, well-expressed and shortly formulated answers is growing, too. Hence, several studies are devoted to the representation of natural language stated in the query and in the documents. Extracting the arguments stated in the document is one way to clearly capture the grounded statements (premises) and the final conclusion (claim) presented in the text. Therefore, many recent works focus on the arguments as a potential tool for improving comparative question answering [9–11], and more generally, on building an argument-based search engine as in the work of Daxenberger et al. [12] with respect to their *summetix* project (formely known as ArgumenText)[4]. Improving a retrieval model for argument-based comparison systems is the target of the Touché Task 2, which has been also addressed in 2020 [13]. The best result of the 2020 shared task was introduced by the team "Bilbo Baggins" of Abye et al. [14]. They used a two-step approach: 1) query expansion by antonyms and synonyms, 2) document ranking based on three measures: relevance, credibility, and support features. An overall score is deduced by summing up those three scores after weights multiplication. Eventually this approach can be improved by keeping the same pipeline and enhancing the inner sub-tasks. For instance, instead of using a static classifier 'XGBoost' for the comparative sentence classification, a BERT-based model can ameliorate the robustness of the classifier and extend its operating range. With respect to 2021 edition, the "Katana" team [15] scored similar results to ours, according to the official metrics of the competition (for measure nDCG@5 score). They keep the original questions as queries and re-rank the top-100 ChatNoir results. Their best results were obtained by gradient boosting methods, training on ranking cost function: XGBoost and LightGBM.

3   Approach and Implementation

In this section, we present our proposed approach and adopted methods to build a search engine for answering comparative questions based on argumentation identification. The overall architecture of our approach is presented in Fig. 1. It consists of a sequence of seven stages. We extend on them individually in the upcoming sections. We used the same architecture to submit four runs with different configurations via TIRA platform [16].

3.1   Query Expansion

Query Expansion (QE) is the process of reformulating a given query to improve the retrieval performance and increase the recall of the candidate retrieval [17]. Our query expansion module involves a variety of techniques in order to generate three different queries to be passed to the next step, as the following:

- *Query 1:* is the original query itself.
- *Query 2:* is generated from the original query by: (1) removing English stop words, punctuation marks, and comparison adjectives also called comparison operators. (2) Stemming of the remaining words to their base forms, and aggregating them together with conjunctive AND operator.
- *Query 3:* will be generated from the original query only if the latter contains a comparison operator, as follows:
  - Search for synonyms and/or antonyms of the comparison operator of the query to get what is called the *context of the comparison operator*, whose size is 5 synonyms/antonyms in our case.
  - Remove English stop words and punctuation marks.

---
[4] https://www.summetix.com/.

166   A. Alhamzeh et al.

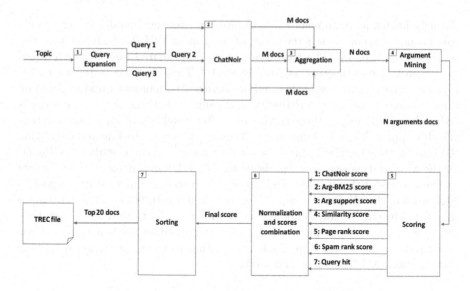

Fig. 1. Global architecture of the submitted approach

- Eliminate the comparison operator from the original query and stemming the remaining words/terms to their base form.
- We create 5 queries out of the original query, and each time adding one of the 5 synonyms/antonyms from the *context of the comparison operator*. Those 5 output queries are sent to ChatNoir as one disjunctive OR-query: Query 3.

For more details, you may find a complete example of the query expansion output in our original notebook paper of this work presented in [18].

3.2 Document Retrieval by ChatNoir API

The ClueWeb12 document dataset for this task is easily accessible through the ChatNoir API[5] [19,20] that is based on the BM25F ranking algorithm.

3.3 Document Aggregation

As Query 3 may be empty, we have a minimum of $2 \times M$ and a maximum of $3 \times M$ retrieved documents (M: retrieved docs per query). Every document has a unique id (uuid), which we use to remove the redundant documents returned by more than one query. For instance, if a document is retrieved by Query 1 and Query 2, with different ChatNoir scores for each (*score*, *page_rank* and *spam_rank*), the document aggregation component will output one document with scores deduced by the sum of scores from Q1 and Q2 (*score* = $score^1 + score^2$, ...).

---

[5] https://www.chatnoir.eu/doc/api/.

3.4   Argument Mining

In our particular task, we seek to detect the comparative sentences in the document, therefore, argument identification can be sufficient. Hence, we take the sentences from the document aggregation step and apply binary classification using the model presented in Fig. 2 to label every sentence as an argument or non-argument. We searched for datasets that contain both argument and non-argument labels. Student Essays [21] and Web discourse [22] are public and very common used corpora which satisfy well this purpose.

The **Student Essays corpus** contains 402 Essays. The annotation covers the following argument components: 'major claim', 'claim' and 'premise'.

The **User-generated Web Discourse corpus** is a smaller dataset that contains 340 documents about 6 controversial topics. The document may refer to an article, blog post, comment, or forum posts. The annotation has been done by [22] according to Toulmin's model [23].

In order to unify data for both corpora, we label any argument component (premise, claim or major claim) as an 'argument', and the rest of the text sentences as 'non-argument'.

The choice of BERT-based model is justified by the fact that among different existent transformers, BERT [8] has achieved the state of the art results in several NLP tasks [5,6,24]. For our particular task, we performed different experiments using several BERT-like models (BERT base, RoBERTa-base [25], DistilRoBERTa , DistilBERT)[6] and we achieved very similar results. Hence, we decided to use the DistilBERT model given that it is 40% less than BERT in size with a relevant in-line performance and a faster training/testing time [3].

Fig. 2. Transfer learning model architecture

Figure 2 shows our transfer learning model architecture to perform the argument identification task using DistilBERT. The first block is the Tokenizer that takes care of all the BERT input requirements. The second block is the DistilBERT fine-tuned model that outputs mainly a vector of length of 768 (default length). In order to adapt the output of this pre-trained model to our specific task, the linear layer, which is a fully connected layer that helps with dimension changes, is applied on top of the DistilBERT transformer and outputs a vector of size 2. In this vector, the index of the maximum value reflects the predicted class id.

In addition to the single DistilBERT model, we have tested our stacked model presented in [4] for argument identification. The model is based on combining

---

[6] We used Transformers from huggingface.com for our experiments.

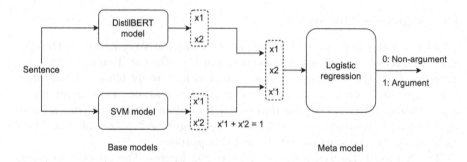

Fig. 3. Stacked model architecture [4]

two approaches using the **stacking** ensemble method: 1) A traditional machine learning approach which uses textual features. 2) A transfer learning approach that consists of the DistilBERT model (cf. Fig. 2).

As shown in Fig. 3, the two models are trained in parallel to produce outputs for the meta-model to learn from, in order to determine the final prediction. In order to feed the meta-model with independent input-array, we take only $x'1$ from the output of SVM ($x'1$ and $x'2$, are two probabilities where $x'1 + x'2 = 1$). On the other hand, DistilBERT outputs two independent logits, x1 and x2, consequently, they are both taken into consideration. For the task of argument identification, we found in our previous work [4] that when training on the same set of corpora, the stacked model outperforms the two individual learners in terms of accuracy, recall, and F1-score.

3.5 Scoring

The scoring or ranking step is essential for any search engine system because many users tend to check out the top results without spending time to carefully review the later ones. Subsequently, our objective now is to estimate the best matching between the query and the candidate answers, in order to sort them at the final stage. To this end, we investigate different scores based on different aspects. First of all, the document relevance which can be checked simply by ChatNoir BM25 score and Query hit count.

However, even if a document content is relevant to the query, it may be fake or biased, thus we inspect the credibility of the document itself by considering: Page Rank score as well as Spam rank score.

Moreover, as we built our retrieval system based on arguments, we take into consideration the argument quality level by three different scores: argument support, query-argument similarity, and argument BM25 score.

We refer to each of our ranking scores by a score-id from (1) to (7) to be further used in Table 2. The complete details of those scores are addressed in the following:

- (1) ChatNoir score: returned form ChatNoir API indicating BM25 measure.
- (2) Arg-BM25 score: calculated on argumentative sentences of each document with respect to the original query. This is done through re-indexing

the retrieved documents by creating new ones that contain only argumentative sentences. Then the arg-BM25 score of each document is calculated by querying the new argumentative documents with the original topic.

- (3) Argument support score: represents the ratio of argument sentences among all existent sentences in the document.
- (4) Similarity score: evaluates the similarity of two sentences based on the context and English language understanding using the *Sentence Transformer*[7] library [26]. We calculate the similarity between the original query and every argumentative sentence in the document, and consider the average.
- (5) Page Rank score: returned form ChatNoir API measuring the importance of the source website pages.
- (6) Spam rank score: returned form ChatNoir API indicating the probability of the website to be a spam.
- (7) Query hit count: indicates how many times the document is retrieved by the three queries.

3.6   Normalization and Scores Combination

For the final score, we normalize all previously calculated scores, so that all values are between 0 and 1. These scores are aggregated using particular weights which we set up manually based on the announced relevance judgments of the 2020-Touche task 2. Thereby, we have done several experiments on that edition topics, while changing each time manually the values of weights. Then, we took the best values of weights and applied them in different runs for our submission.

3.7   Sorting

At this stage, the documents are sorted based on the final score to get the top 20 documents that are highly relevant to answer the comparative query. The final output is inserted into a text file while respecting the standard TREC format proposed by the Touché organizers.

4   Evaluation

4.1   On the Level of Argument Identification

To evaluate the argument mining component, we trained and tested DistilBERT and the stacking model on the two datasets presented in Sect. 3.4: Student Essays and the User-generated Web Discourse as well as the combination of those two corpora. Table 1 shows the results of (1) DistilBERT and (2) stacking model to accomplish the binary classification task (argument/non-argument) at the sentence level on the different corpora.

Our DistilBERT model results are in-line with the findings of Wambsganss et al. [5] using BERT. However, we can obviously see that the stacking model is outperforming the two base models and increasing the overall classification performance [4].

---

[7] https://github.com/UKPLab/sentence-transformers.

Table 1. Argument identification results on different datasets using (1) DistilBERT and (2) stacking model. The best results are marked in boldface.

	Accuracy		Precision		Recall		F1-score	
Dataset	(1)	(2)	(1)	(2)	(1)	(2)	(1)	(2)
Student Essays	0.8727	**0.9162**	0.8016	**0.8890**	0.7477	**0.8195**	0.7697	**0.8483**
Web Discourse	0.7799	**0.7855**	**0.7718**	0.7449	0.6484	**0.6958**	0.6655	**0.7113**
Merged Corpora	0.8587	**0.8780**	0.7887	**0.8326**	0.7529	**0.7659**	0.7683	**0.7921**

4.2   On the Level of Touché Shared Task 2

The shown architecture in Fig. 1 represents our base approach, from which we derive four submissions to the task-2 of Touché 2021 by experimentally modifying score weights and the number of retrieved documents. In Table 2, we present the used score weights for each run in addition to the number of submitted documents per topic.

Table 2. Configurations of each run: scores are defined in Sect. 3.5 with respect to the score-ids (1) to (7)

	Score Weights							
Run Tag	(1)	(2)	(3)	(4)	(5)	(6)	(7)	Docs
DistilBERT_argumentation_bm25	0	1	0	0	0	0	0	30
DistilBERT_argumentation_advanced_ranking_r1	15	25	25	15	20	0	0	20
DistilBERT_argumentation_advanced_ranking_r2	10	10	50	20	10	5	5	40
DistilBERT_argumentation_advanced_ranking_r3	10	15	10	50	10	0	0	40
Stack_argumentation_bm25	0	1	0	0	0	0	0	30

Table 3. Results of each run

	Relevance		Quality	
Run Tag	nDCG@5	Rank/20	nDCG@5	Rank/20
DistilBERT_argumentation_bm25	0.466	6	**0.688**	1
DistilBERT_argumentation_advanced_ranking_r1	**0.473**	3	0.670	5
DistilBERT_argumentation_advanced_ranking_r2	0.458	8	0.630	11
DistilBERT_argumentation_advanced_ranking_r3	0.471	4	0.625	13
Touché baseline	0.422	6	0.636	6
Stack_argumentation_bm25	0.444	N	0.640	N

Table 3 shows the outcome of each run. Those results conducted by the Touché committee through the manual labeling of the documents with the help of human assessors. For each run, two evaluation scores are calculated that define the rhetorical quality and the relevance of the submitted documents [27]. By comparing our results with the Touché baseline, which is a TF-IDF-like algorithm, most of our runs outperform the baseline in terms of relevance and quality. This indeed confirms that the problem of answering comparative questions should not be addressed as a traditional documents retrieval problem.

Consequently, our participation system using DistilBERT substantially outperform the proposed baseline and scored first place in quality and third in relevance by achieving a score of $nDCG@5 = 0.688$ and $nDCG@5 = 0.473$ respectively [27].

Moreover, based on the official announced relevance and quality judgments, we were able to further produce the results of our system using the stacking model instead of DistilBERT: *Stack_argumentation_bm25* as shown in Table 3.

5   Discussion

We have presented an advanced web search engine for answering comparative questions. Every component in the architecture (cf. Figure 1) plays an important role to achieve the best retrieval results. For instance, the query expansion component makes a first selection and build a set of topic-related documents. The three different queries generated from the original topic increase the coverage of the related documents and this works very well with the ChatNoir API since it is a basic BM25 retrieval system.

When plugging the stacked model instead of the DistilBERT in the overall architecture (*Stack_argumentation_bm25*), we observe that it achieves good results in terms of quality, but it did not improve scores as it was expected. One reason could be that the type of text retrieved from the ClueWeb12 is widely varied from the text that SVM was trained on. In fact, the training of the SVM is based on textual features, such as 1–3 gram Bag of Words (BoW) and Named Entity Recognition (NER), which are limited to the context and the type of the text during the training, unlike the DistilBERT model which generalize over text. Hence, the results in this run are held back by the SVM model against the heterogeneous text and the new themes of the retrieved documents. As a conclusion, we note a score degradation compared to the the DistilBERT-based submissions.

On the other hand, the fact of having seven components chained in series could generate easily an error propagation and amplification. For instance, by only increasing the initial number of retrieved documents with a very high value, the final score is negatively degraded. This is due to the inclusion of too many unrelated documents that influence the rest of components. This explains the poor quality scores of runs *DistilBERT_argumentation_advanced_ranking_r2* and *DistilBERT_argumentation_advanced_ranking_r3*.

6   Conclusion

In this paper, we presented our submitted system to the shared task of argument retrieval for answering comparative questions. Our approach consists of a pipeline of several components including query expansion, document retrieval and aggregation, argument mining, and scoring. A main contribution is the transfer learning model we developed based on the DistilBERT transformer using two different datasets for adapting it to the argument identification task. Additionally, the presented stacked model represents a step towards the interpretability of models. For sorting the documents, we consider different criteria to re-rank the output documents with respect to their relevance, the argument they contain, their trustworthiness, and credibility. In our future work, we plan to improve the ranking stage by using an argument quality assessment model. We also plan to use more advanced techniques for query expansion, in addition to using a machine learning model that learns the best weights of the scores.

Acknowledgements. The project on which this report is based was partly funded by the German Federal Ministry of Education and Research (BMBF) under the funding code 01—S20049. The author is responsible for the content of this publication.

References

1. Van Eemeren, F., Grootendorst, R.: The Pragma-Dialectical Approach. Cambridge University Press, A systematic theory of argumentation (2004)
2. Chernodub, A., et al.: Targer: neural argument mining at your fingertips. In: Proceedings of the 57th Annual Meeting of the Association for Computational Linguistics: System Demonstrations, pp. 195–200 (2019)
3. Sanh, V., Debut, L., Chaumond, J., Wolf, T.: DistilBERT, a distilled version of BERT: smaller, faster, cheaper and lighter. arXiv preprint arXiv:1910.01108 (2019)
4. Alhamzeh, A., Bouhaouel, M., Egyed-Zsigmond, E., Mitrović, J., Brunie, L., Kosch, H.: A stacking approach for cross-domain argument identification. In: Strauss, C., Kotsis, G., Tjoa, A.M., Khalil, I. (eds.) DEXA 2021. LNCS, vol. 12923, pp. 361–373. Springer, Cham (2021). https://doi.org/10.1007/978-3-030-86472-9_33
5. Wambsganss, T., Molyndris, N., Söllner, M.: Unlocking transfer learning in argumentation mining: a domain-independent modelling approach. In: 15th International Conference on Wirtschaftsinformatik (2020)
6. Reimers, N., Schiller, B., Beck, T., Daxenberger, J., Stab, C., Gurevych, I. Classification and clustering of arguments with contextualized word embeddings. arXiv preprint arXiv:1906.09821 (2019)
7. Pan, S.J., Yang, Q.: A survey on transfer learning. IEEE Trans. Knowl. Data Eng. **22**(10), 1345–1359 (2009)
8. Devlin, J., Chang, M. W., Lee, K., Toutanova, K. BERT: pre-training of deep bidirectional transformers for language understanding. arXiv preprint arXiv:1810.04805 (2018)

9. Panchenko, A., Bondarenko, A., Franzek, M., Hagen, M., Biemann, C.: Categorizing comparative sentences. arXiv preprint arXiv:1809.06152 (2018)
10. Schilwächter, M., Bondarenko, A., Zenker, J., Hagen, M., Biemann, C., Panchenko, A.: Answering comparative questions: better than ten-blue-links? In: Proceedings of the 2019 Conference on Human Information Interaction and Retrieval, pp. 361–365 (2019)
11. Bondarenko, A., Panchenko, A., Beloucif, M., Biemann, C., Hagen, M.: Answering comparative questions with arguments. Datenbank-Spektrum **20**(2), 155–160 (2020)
12. Daxenberger, J., Schiller, B., Stahlhut, C., Kaiser, E., Gurevych, I.: Argumentext: argument classification and clustering in a generalized search scenario. Datenbank-Spektrum **20**(2), 115–121 (2020)
13. Bondarenko, A., et al.: Overview of touché 2021: argument retrieval. In: Hiemstra, D., Moens, M.-F., Mothe, J., Perego, R., Potthast, M., Sebastiani, F. (eds.) ECIR 2021. LNCS, vol. 12657, pp. 574–582. Springer, Cham (2021). https://doi.org/10.1007/978-3-030-72240-1_67
14. Abye, T., Sager, T., Triebel, A.J.: An open-domain web search engine for answering comparative questions. In: Cappellato, L., Eickhoff, C., Ferro, N., Néveol, A. (eds.), Working Notes of CLEF 2020-Conference and Labs of the Evaluation Forum, Thessaloniki, Greece, 22–25 September 2020, volume 2696 of CEUR Workshop Proceedings. CEUR-WS.org (2020)
15. Chekalina, V., Panchenko, A.: Retrieving comparative arguments using ensemble methods and neural information retrieval. Working Notes of CLEF (2021)
16. Potthast, M., Gollub, T., Wiegmann, M., Stein, B.: TIRA integrated research architecture. In: Information Retrieval Evaluation in a Changing World. TIRS, vol. 41, pp. 123–160. Springer, Cham (2019). https://doi.org/10.1007/978-3-030-22948-1_5
17. Azad, H.K., Deepak, A.: Query expansion techniques for information retrieval: a survey. Inf. Process. Manag. **56**(5), 1698–1735 (2019)
18. Alhamzeh, A., Bouhaouel, M., Egyed-Zsigmond, E., Mitrović, J.: DistilBERT-based argumentation retrieval for answering comparative questions. Working Notes of CLEF (2021)
19. Pasi, G., Piwowarski, B., Azzopardi, L., Hanbury, A. (eds.): ECIR 2018. LNCS, vol. 10772. Springer, Cham (2018). https://doi.org/10.1007/978-3-319-76941-7
20. Potthast, M., et al.: A search engine for the ClueWeb09 corpus. In: Hersh, B., Callan, J., Maarek, Y., Sanderson, M. (eds.), 35th International ACM Conference on Research and Development in Information Retrieval (SIGIR 2012), p. 1004. ACM, August 2012
21. Stab, C., Gurevych, I.: Annotating argument components and relations in persuasive essays. In: Proceedings of COLING 2014, the 25th International Conference on Computational Linguistics: Technical Papers, pp. 1501–1510 (2014)
22. Habernal, I., Gurevych, I.: Argumentation mining in user-generated web discourse. Comput. Linguist. **43**(1), 125–179 (2017)
23. Toulmin, S.E.: The Uses of Argument, 2nd edn. Cambridge University Press, Cambridge (2003)
24. Niven, T., Kao, H.Y.: Probing neural network comprehension of natural language arguments. arXiv preprint arXiv:1907.07355 (2019)
25. Liu, Y., et al.: Roberta: a robustly optimized BERT pretraining approach. arXiv preprint arXiv:1907.11692 (2019)

174 A. Alhamzeh et al.

26. Reimers, N., Gurevych, I.: Sentence-BERT: sentence embeddings using siamese BERT-networks. In: Proceedings of the 2019 Conference on Empirical Methods in Natural Language Processing. Association for Computational Linguistics, p. 11 (2019)
27. Bondarenko, A., et al.: Overview of touché 2021: argument retrieval. In: Candan, K.S., et al. (eds.) CLEF 2021. LNCS, vol. 12880, pp. 450–467. Springer, Cham (2021). https://doi.org/10.1007/978-3-030-85251-1_28

Transformer-Encoder-Based Mathematical Information Retrieval

Anja Reusch[1]([ID]), Maik Thiele[2][ID], and Wolfgang Lehner[1][ID]

[1] Database Systems Group, Technische Universität Dresden, Dresden, Germany
{anja.reusch,wolfgang.lehner}@htw-dresden.de
[2] Hochschule für Technik und Wirtschaft Dresden, Dresden, Germany
maik.thiele@htw-dresden.de

Abstract. Mathematical Information Retrieval (MIR) deals with the task of finding relevant documents that contain text and mathematical formulas. Therefore, retrieval systems should not only be able to process natural language, but also mathematical and scientific notation to retrieve documents.

In this work, we evaluate two transformer-encoder-based approaches on a Question Answer retrieval task. Our pre-trained ALBERT-model demonstrated competitive performance as it ranked in the first place for p'@10. Furthermore, we found that separating the pre-training data into chunks of text and formulas improved the overall performance on formula data.

Keywords: Mathematical Language Processing · Information Retrieval · BERT-based Models · ARQMath Lab

1 Introduction

With the rising number of scientific publications and mathematics-aware online communities available Mathematical Information Retrieval (MIR) has become crucial for the exploration of documents mixing natural language and mathematical notation. However, only interpreting natural language alone is not sufficient for retrieval in such documents anymore since the usage of mathematical notation is crucial to understanding the information conveyed by the author. Hence, to search or retrieve information from these platforms, a retrieval system needs to understand, represent and interpret the notation of mathematical expressions.

For Natural Language-based Information Retrieval systems applying large pre-trained language models such as BERT [3] have been found to be effective and out-performing previous, traditional IR systems that were based on string matching methods [17]. In previous work, we showed that our approach of using an ALBERT-based classifier as a similarity measure is beneficial for Mathematical Question Answering when answers depend on the written text [20]. However, when answering the question relies on a proper interpretation of the formulas, traditional methods are still more suitable.

The second disadvantage is that the average query latency of BERT-based systems is a few orders of magnitude higher compared to non-neural methods [8,13] due to the fact that for each query-document pair an entire forward pass through the deep network needs to be performed. A recent advance in terms of speed without neglecting performance is ColBERT [8], where the authors applied a late interaction mechanism to assess the relevance of a document given a query. This approach made offline indexing of the collection and a faster evaluation possible since only one forward pass of the query is necessary.

In this work we would like to address the following two challenges: (1) We will perform an analysis of pre-training adjustments for ALBERT to increase the models' performance on formula understanding. Here, we evaluate further pre-training on in-domain data as well as pre-training from scratch. Furthermore, we analyze the granularity on which the data is split during pre-training. (2) Three models based on ColBERT are applied to MIR for a more efficient evaluation. Here, we make use of readily available BERT-based models as well as a model further pre-trained on in-domain data.

We will evaluate our approaches for MIR on the question-answer retrieval task of ARQMath Labs 2020 [14] and 2021 [15]. This task deals with the retrieval of relevant answers given a question from the Mathematics StackExchange Community and involves understanding the problem of the question poster in terms of natural language in combination with mathematical notation in form of LaTeX.

This work is structured as follows: We will first introduce the ARQMath 2021 Lab and then review relevant literature for Information Retrieval and BERT-based systems for natural language and multi-modal tasks. In Sect. 4 the overall architecture of our approaches will be explained along with the data set used to pre-train and fine-tune the models. Section 5 describes the experiments and discusses their results. Finally, the last section summarizes our work.

2   ARQMath 2021 Lab

The aim of ARQMath Lab 2021 [15] is to accelerate the research in mathematical Information Retrieval. Task 1 of the Lab involves the retrieval of relevant answer posts for a question asked on the Mathematics StackExchange[1], which is a platform where users post questions to be answered by the community. The questions should be related to mathematics topics at any level.

ARQMath 2021 provides data from the Mathematics StackExchange including question and answer posts from 2010 to 2018. In total, the collection contains 1 M questions and 1.4 M answers. Furthermore, users may use mathematical formulas to clarify their posts. These formulas written in LaTeX notation were extracted and parsed into Symbol Layout Trees and Operator Trees. Each formula got assigned a formula id and a visual ids. Formulas sharing the same visual appearance received the same visual id. Apart from this corpus of posts and formulas that are available for training and evaluating models, also a test set of queries is released by the organizers of ARQMath.

---

[1] https://math.stackexchange.com.

The goal of Task 1 is the retrieval of an answer post from 2010–2018 to questions that were posted in 2019. The query topics of 2020 and 2021 contain 99 and 100 topics, respectively, which are question posts including title, text, and tags. In the 2020 test set 77 queries were evaluated for Task 1, while its evaluation in 2021 included 71 queries. The optimal answers retrieved by the participants are expected to answer the complete question on their own. The relevance of each question was assessed by reviewers during the evaluation process.

The participating teams submitted for each topic a ranked list of 1,000 documents retrieved by their systems, which were scored by Normalized Discounted Cumulative Gain (nDCG), but with unjudged documents removed before assessment (nDCG'). The graded relevance scale used for scoring ranged from 0 (not relevant) to 3 (highly relevant). Two additional measures, mAP' and p'@10, were also reported using binarized relevance judgments (0 and 1: not relevant, 2 and 3: relevant) after removing unjudged documents. The relevance assessment was performed by pooling after the teams submitted their results.

3   Related Work

Bidirectional Encoder Representations from Transformers (BERT) is an architecture based on the encoder of a Transformer model which was designed for language modeling [3]. Due to the success of this and other pre-trained, Transformer-based language models, BERT has been a basis in many systems for Natural Language Understanding (NLU) tasks and applications in Information Retrieval. Also, there exist several advanced versions such as RoBERTa [12] or ALBERT [10] with the goal to optimize BERT's performance.

The influence of in-domain pre-training has been analyzed by Gururangan et al. [5] who found that this is especially valuable when the domain vocabulary has a low overlap with the originally applied pre-training data. As a consequence, various models for different domains have been developed, such as BioBERT [11], ClinicalBERT [1,6] or SciBERT [2] for scientific domains or CuBERT [7] and CodeBERT [4], which could all demonstrate their improvements compared to the original models without domain-specific pre-training.

BERT-based models for mathematical domains have also been studied with the most recent example being MathBERT [19]. In addition, during the last ARQMath Lab 2020, two teams submitted systems based on BERT and RoBERTa [18,21]. Both teams used the models to generate post embeddings for a given question and all answers. Their similarity is calculated by comparing the vectors using cosine similarity.

Shortly after BERT outperformed previous approaches in various NLU tasks, it was also successfully applied to Information Retrieval. The model by Nogueira et al. classified its input consisting of a query and one document for their relevance resulting in a score that can be used to rank multiple documents [17]. This approach achieved state-of-the-art performance, but was much slower and computationally expensive than previous systems, because one forward pass through the entire deep neural network was necessary to score one query-document pair.

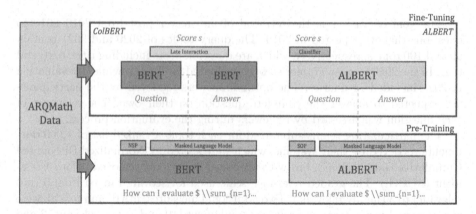

Fig. 1. Overview of our pre-training (details in Fig. 2) and fine-tuning procedure (details in Fig. 3).

Nevertheless, this approach has also been proven to be effective for the multi-modal retrieval of source code [4] and was also applied to Mathematical Question Answering using an ALBERT model trained and evaluated on the ARQMath Lab 2020 test set [20]. The evaluation results were also broken down to the categories determining which part of the question influenced answering it the most. The model showed the best performance when answering the question depended on the written text. But for questions relying on formulas the results were worse than systems based on non-neural methods. Therefore, the modeling capability of formulas needs to be improved to also be able to capture their semantics in a better way.

Due to the fact that the ranking model by Nogueira et al. came with a steep increase in computational cost, recent research focused on improving the evaluation time without neglecting its performance gains. Despite there being more than one model dealing with this challenge, we will focus in this work on the approach by Khattab et al. called ColBERT [8]. ColBERT uses a BERT model to separately encode a query and a document and then apply a novel late interaction mechanism to calculate the similarity. This way they achieved competitive results when re-ranking on the popular MSMARCO data set [16] with a latency of 61 ms compared to 107,000 ms using the BERT-based approach by Nogueira et al.

4   Transformer-Encoder-Based MIR

Models based on Transformer-Encoders like BERT have shown to be effective in Natural Language Understanding and Information Retrieval tasks. Their strength was also shown in scenarios where not only natural language plays an important role, such as Code Retrieval or Mathematical Language Processing as in this lab [4,19,20]. Building on top of these achievements, we apply two deep neural models based on the popular BERT in this work: ALBERT and Col-BERT. ALBERT is a recent model, which is optimized by factorization of the embeddings and parameter sharing between layers. These optimizations result in fewer training parameters and therefore a lower memory consumption and accelerated training speed compared to BERT which is why we chose ALBERT over BERT. The general idea of our first approach is to employ the ALBERT model to determine the similarity score between two snippets, a question and an answer. This is achieved by fine-tuning the pre-trained ALBERT model with a classifier on top which predicts how well the two snippets match. The second method that we apply for Task 1 uses a BERT model as a basis of ColBERT. The query and each document are passed through BERT separately in order to encode their respective content. This way an offline computation of the representations of each document is possible beforehand. The late interaction mechanism in form of the L2 distance is applied to aggregate and compare the contextualized embeddings. Finally, the documents are ranked by this computed L2 distance.

The success of BERT and BERT-based models is attributed to their transformer architecture [23] and also to the self-supervised pre-training on large amounts of data. In this work, we will focus on the latter aspect and pre-train models on different data highlighting its influence. The overall process of our approach is depicted in Fig. 1. We will present details about the pre-training and fine-tuning in the next sections.

4.1 Pre-training

As mentioned previously, BERT and also ALBERT rely on pre-training on rather simple tasks. BERT is pre-trained using two objectives to obtain a general understanding of language: the masked language model (MLM) and the next sentence prediction (NSP).

Pre-training is performed on a sentence-level granularity. Each sentence S is split into tokens: $S = w_1 w_2 \cdots w_N$. Before inputting the sentence into the model, each token w_i is embedded using a sum of three different embeddings, the word embedding t_i encoding the semantic of the token, the position embedding p_i denoting its position within the sentence, and the segment embedding s_i in order to discern between the first and the second segment when the model is presented e.g., two sentences as for the NSP task. The segment embeddings will also help our model to differentiate between the query and document as the two segments later. All three embeddings are added up to form the input embedding E_i for each token: $E_i = t_i + p_i + s_i$. In order to obtain a representation of the entire input, we prepend the sentence S with a classification token $w_S = \langle CLS \rangle$. It is embedded in the same way as the other tokens and will be used for the NSP task and also for fine-tuning tasks that rely on a representation of the input such as classification.

The first pre-training task is the masked language model meaning tokens from the input sentence are randomly replaced by a $\langle MASK \rangle$ token, a different token, or is not changed at all. After embedding the input, it is fed into the

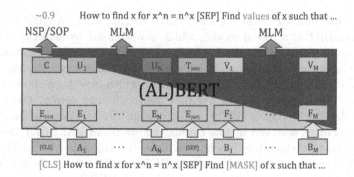

Fig. 2. BERT's and ALBERT's pre-training process, 0.9 symbolizes the NSP or SOP score for the two sentences, the red word 'values' is the predicted word for the masked token. (Color figure online)

BERT model, consisting of 12 layers of transformer encoder blocks, resulting in a contextualized output embedding U_i for each input token:

$CU_1U_2 \cdots U_N = \text{BERT}(E_{CLS}E_1E_2 \cdots E_N)$, where E_{CLS} and C are the input and output embeddings of the $\langle CLS \rangle$ token. Afterward, a simple classifier is applied in order to predict the original word from the input:

$P(w_j|S) = \text{softmax}(U_i \cdot W_{MLM} + b_{MLM})_j$, where w_j is the j-th word from the vocabulary. This determines the probability that the i-th input word was w_j given the input sentence S. The weight matrix W_{MLM} and its bias b_{MLM} are only used for this pre-training task and are not reused afterward.

The next sentence prediction objective predicts whether the sentence given to the model as the first segment S_A appears in a text before the sentences given to the model as the second segment S_B (label 1) or whether the second sentence is a random sentence from another document (label 0). This task is performed as a binary classification using the output embedding C as its input:

$p(label = i|S) = \text{softmax}(C \cdot W_{NSP} + b_{NSP})_i$, where the matrix W_{NSP} and the bias b_{NSP} are only used for the NSP and are not used otherwise later.

ALBERT also makes use of the MLM objective, but it has been found that NSP, predicting whether the second sentence in the input is swapped with a sentence from another document from the corpus, is a relatively challenging task and was changed to the sentence order prediction (SOP) [10]. Here, the model is asked to determine what the correct order of two presented sentences is. Hence, the model is presented with two sentences and performs their classification in the same way as BERT's NSP. Therefore, the formulas for NSP as introduced above apply as well.

The pre-training process of BERT and ALBERT is depicted together in Fig. 2. Note, that BERT applies a classification on the output embedding C for the NSP objective, while ALBERT does the same for the SOP objective. Both models use the MLM objective.

Pre-training Data. Before pre-training we applied the official tool provided by ARQMath to read the posts, wrapped formulas in \$, and removed other HTML markup, yielding a list of paragraphs for each post. BERT and ALBERT models rely on sentence-separated data during pre-processing for the NSP and SOP tasks. Two different strategies were tested: (1) split the text into sentences, (2) split it into chunks of text and formulas. The SOP task is designed to work with sentences. Hence, (1) is usually used in various NLP tasks. On the other hand, our goal was to increase the model's understanding of formulas. Therefore, strategy (2) splits a paragraph first into sentences, but also when a sentence contains a formula (with more than three LaTeX tokens to avoid splitting at e.g., definitions of symbols). In case the remaining text is too short (less than ten characters), it is concatenated to the formula before, separated by a \$ sign. Before inputting the data into the models, tokenizing, creating the pre-training data for each task, i.e., masking tokens and assembling pairs of sentences, and further pre-processing was performed by the pre-processing scripts provided in the official BERT and ALBERT repositories[2]. For the models that started from official checkpoints, we used the released sentencepiece vocabulary [9]. For the models that started from scratch, we trained our own sentencepiece model using the parameters recommended in the ALBERT repository which had a vocabulary overlap of 32.1% compared to the released sentencepiece vocabulary for ALBERT. Sentencepiece tokenizes the input into subwords using byte-pair-encoding [22]. Details can be found in Appendix A.

4.2 ALBERT Model

In order to predict whether an answer $A = A_1 A_2 \cdots A_M$ is relevant to a question $Q = Q_1 Q_2 \cdots Q_N$ the pre-trained ALBERT model with a classifier on top is trained as depicted in Fig. 3. The input string $\langle CLS \rangle Q_1 Q_2 \cdots Q_N \langle SEP \rangle A_1 A_2 \cdots A_M$, with $\langle CLS \rangle$ being the classification token and $\langle SEP \rangle$ the separation token, is presented to the model:

$CU_1 U_2 \cdots U_N = \text{ALBERT}(E_{CLS} E_1 E_2 \cdots E_{N+M})$, where E_i and E_{CLS} are the input embeddings for each input token and the $\langle CLS \rangle$ token, respectively, calculated as explained in Sect. 4.1. After the forward pass through the model, the output vector of the $\langle CLS \rangle$ token C is given into a classification layer:

$p(label = i|Q, A) = \text{softmax}(C \cdot W_{MIR} + b_{MIR})_i$, where the label 1 stands for a matching or correct answer for the query and label 0 otherwise. During evaluation, the resulting probability of the classification layer for label 1, is the assigned similarity score s for the answer A to a question Q and is then used to rank the answers in the corpus: $s(Q, A) = p(label = 1|Q, A)$.

Fine-Tuning Data. In order to fine-tune the ALBERT models, we paired each question with one correct answer and one incorrect answer. The correct answer was randomly chosen from the answers of the question. Each question

---

[2] https://github.com/google-research/bert, https://github.com/google-research/AL
BERT.

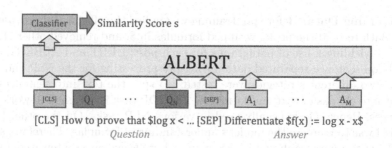

Fig. 3. Architecture of ALBERT's Fine-Tuning.

in the corpus comes along with tags, i.e., categories indicating the topic of a question such as *sequences-and-series* or *limits*. As an incorrect answer for each question, we picked a random answer from one question sharing at least one tag with the original question by chance. Following this procedure, we yielded 1.9M examples, of which 90% were used as training data for the fine-tuning task. We presented the model the entire text of the questions and answers using the structure introduced in the previous section.

4.3   ColBERT Model

Our second approach was to train ColBERT on top of a pre-trained BERT model. In each training step, the model is presented the query Q and two answers: one being a relevant answer A, the second being an answer B that should be regarded as non-relevant by the model. All three strings, Q, A and B are prepended with a token denoting the string as either question (query), $\langle Q \rangle$ or answer (document) $\langle D \rangle$, and are passed through the BERT model individually to create contextualized embeddings for each post:
$C_Q Q U_1 U_2 \cdots U_N = \mathrm{BERT}(E_{CLS} E_Q E_1 E_2 \cdots E_N),$
$C_D D V_1 V_2 \cdots V_M = \mathrm{BERT}(E_{CLS} E_D F_1 F_2 \cdots F_M),$ where E_i, F_i, E_{CLS}, E_Q, and E_D are the input embeddings for each input token, the $\langle CLS \rangle$ token, $\langle Q \rangle$ token and the $\langle D \rangle$ token, respectively, calculated as explained in Sect. 4.1. Using the late interaction mechanism as specified by Khattab et al. [8] a relevance or similarity score is calculated for each question-answer pair and optimized by applying softmax cross-entropy loss over the scores: $s(Q, A) = \sum_{i=1}^{N} \max_{j \in \{1,\ldots,M\}} U_i V_j^{T}.$ More implementation-specific detail can be found in the work by Khattab et al. [8].

Fine-Tuning Data. We use the same procedure to generate training data for the ColBERT-based models, but with the difference that we used up to $N_{answers} = 10$ correct and incorrect answers in case a question had that many submitted answers. If fewer answers were present, the minimum of correct and incorrect answers was used such that the number of correct and incorrect answers matched. We paired each correct answer with all incorrect answers, generating at

Table 1. Overview of Pre-Training Configurations of ALBERT models

Model	Initialization	Pre-Training Data	Steps
BASE 750K	ALBERT base	(1) sentence split	750k
BASE 250K	ALBERT base	(1) sentence split	250k
BASE COMBINED	ALBERT base	(1)+(2) combined	135k
SCRATCH 1M	Random	(1) sentence split	1M
SCRATCH 2M	Random	(1) sentence split	2M
SCRATCH SEP	Random	(2) separated	1M

most $10 \times 10 = 100$ samples for each question. We experimented with $N_{answers} = 1$ and $N_{answers} = 5$, but we achieved best results with $N_{answers} = 10$.

4.4 Evaluation Data

During evaluation we exploited the tag information from the queries in order to rank only the answers that shared at least one tag with the query question. In this way, we saved large amounts of computation time for the ALBERT-based models. Each question and the answers were pre-processed and paired in the same way as during fine-tuning.

For ColBERT, we generated an index based on all answers whose question had at least one tag that was associated with at least one query question.

For each query the organizers of the Lab annotated whether answering the question mostly depends on its text, its formulas or both. We used these categories for the interpretation of our results.

5 Experiments

We tested various scenarios for training ALBERT of which we report six in this work: The models BASE 750K, BASE COMBINED and BASE 250K are initialized from the official weights of the ALBERT base model released by the ALBERT authors while the weights of SCRATCH 1M, SCRATCH SEP and SCRATCH 2M were initialized randomly. Four models were further pre-trained on ARQMath data using strategy (1), i.e., sentence split text (see Sect. 4.1). The data preprocessed with strategy (2), i.e., data split into text and LaTeX, was mixed with the aforementioned data to pre-train BASE COMBINED. SCRATCH SEP was only pre-trained on the separated data of strategy (2). The models were trained for different numbers of steps. A summary of the different combination of initialization model, pre-training data and number of steps for each model can be found in Table 1. We submitted results of four ALBERT-based models to the ARQ-Math 2021 Lab and evaluated BASE 250K and SCRATCH 2M using the official evaluation tools.

ColBERT can be seen as an extension of BERT whose performance depends on its pre-training [8]. Therefore, we apply three differently pre-trained models

Table 2. Results of MIR on ARQMath Lab 2020 and 2021

Model	Official Identifier	2020			2021		
		nDCG'	mAP'	p'@10	nDCG'	mAP'	p'@10
BASE 750K	TU_DBS_P	0.380	0.198	0.316	0.377	0.158	**0.227**
SCRATCH 1M	TU_DBS_A1	0.362	0.178	0.304	0.353	0.132	0.180
BASE COMBINED	TU_DBS_A3	0.359	0.173	0.299	0.357	0.141	0.194
SCRATCH SEP	TU_DBS_A2	0.356	0.173	0.291	0.367	0.147	0.217
COLSCIBERT	TU_DBS_A4	0.045	0.016	0.071	0.028	0.004	0.009
Additional Experiments							
BASE 250K	-	0.375	0.193	0.311			
SCRATCH 2M	-	0.359	0.177	0.297			
COLARQBERT	-	0.225	0.073	0.131			
COLBERT	-	0.183	0.053	0.110			
ARQMath Competitors							
Best '20&'21	MathDowsers-primary	**0.433**	0.191	0.249	**0.434**	**0.169**	0.211
Best '20	DPRL-RRF	0.422	**0.247**	**0.386**	0.347	0.101	0.132
Best Baseline	linked_results	0.279	0.194	**0.386**	0.203	0.120	**0.282**

as the basis for ColBERT: COLBERT uses the weights of the original BERT-base, COLSCIBERT uses SciBERT [2], which was trained on a large corpus of scientific publications from multiple domains and finally, we pre-trained our own BERT model for COLARQBERT. The last model was initialized using the original BERT weights and was then further pre-trained on the sentence split data (1) described earlier. The pre-training of COLARQBERT was performed using the code published by the BERT authors, while the ColBERT repository was slightly adapted to support different checkpoints than BERT base in order to train the other models. Finally, COLSCIBERT model was submitted to the competition, while COLBERT and COLARQBERT were evaluated later using the official evaluation guide. More details about our hyperparameters can be found in Appendix B.

5.1 Evaluation

The results of our ALBERT and ColBERT-based models are shown in Table 2 together with additional experiments that were not submitted and results of other models from the ARQMath 2021 Lab for comparison. We report the scores of the 2020 and 2021 test sets. In addition, we break down the nDCG' score results of 2020 by the categories on which part answering the question depends. These categories are either text, formula, or both in combination and were annotated by the organizers of the lab. The scores for each category are reported in Table 3.

Pre-training Adjustments. In general, our results can be seen as competitive. Regarding nDCG', all ALBERT-based models could outperform the baseline

Table 3. nDCG' Scores by Category on the 2020 Test Set

	Base 750k	Scratch 1M	Base Combined	Scratch Sep	Base 250k
Both	0.365	0.365	0.364	0.321	**0.370**
Formula	**0.382**	0.354	0.338	0.367	0.366
Text	0.411	0.375	0.399	**0.421**	0.408

systems in both years. On the 2020's test set, our ALBERT-based models are all in the range of the top four teams for mAP'. In 2021, our best model ranks second among all teams regarding mAP'. Our results for p'@10 are not as high as the best baseline, but there was not a single system from any of the teams that could beat the baseline results for p'@10. Comparing to the other participants, our system receives the highest score for p'@10 in 2021.

The reason why our Precision is relatively high, but the nDCG' is lower compared to the other teams that received higher scores could be that our systems do not rank all answers for each topic due to the too time consuming evaluation. Possibly, our results would have been better if all answers would have been scored for their relevance.

We will now take a deeper look at the differences between the models we trained. When comparing Base 750k and Base 250k, the overall score is slightly increased by the longer pre-training. In Table 3 we see that with longer pre-training the model learned a better understanding of text and formulas on their own, but for category 'both' the results decreased. On the other hand, pre-training for too many steps shows effects of over-fitting as the scores start to decrease again as we see in the difference between Scratch 1M and Scratch 2M.

The comparison of Scratch 1M and Scratch Sep shows that the separation of text and mathematical formulas leads to better nDCG' scores for queries dependent on formulas and text separately, but the performance degrades on question-answers pairs that depend on both, which is expected since the model was not pre-trained on data that involved both in one example. Base Combined has a much lower nDCG' value for the formula category in comparison to the other models. This can be explained by the fact that it was pre-trained for a much lower number of steps. The same effect is visible when viewing Base 750k and Base 250k. Therefore, we hypothesize that a pre-training of 750k or even more steps could even outperform Base 750k and Scratch Sep in all three categories.

BERT-base models generally benefit from a long pre-training on a large corpus. In our experiments, we could not observe this behavior. We experimented with models trained for 2M steps on data from 41 StackExchange communities supporting LaTeX, but the results are worse than the ones presented in Table 2.

ColBERT. ColSciBERT is the fifth model we submitted for the 2020 ARQ-Math Task 1 and it was trained using ColBERT. As can be seen from the results table, its performance is not optimal hinting at a substantial problem during training or evaluation. This could be caused by using SciBERT as the basis for ColBERT. Two other models that were not officially submitted to the Lab received higher scores, but are still not on par with our other ALBERT-based approaches regarding all three metrics. This confirms the hypothesis that SciB-ERT is not suitable in this scenario.

Nevertheless, with ColBERT the time required to evaluate all 100 topics of 2020 took around six minutes using two NVIDIA GTX 1080 while evaluating one query using our ALBERT-based classification approach took between ten minutes and one hour on one NVIDIA V100. Therefore, further research in this direction is worthwhile for speeding up the evaluation while receiving competitive scores at the same time. Future work here should further analyze the performance and determine the best training scenario for a ColBERT-based system.

6   Conclusion

Mathematical Information Retrieval (MIR) deals with the retrieval of documents from a corpus, which are relevant to a query, where documents and queries may include both, natural language and mathematical formulas. One instance of such an objective is Task 1 of the ARQMath Lab, whose goal is to retrieve answers given a mathematical question. Since this challenge includes not only text written in English, but also formulas, approaches from Natural Language Processing and Information Retrieval have to be adapted in order to be able to interpret the semantics of mathematical formulas as well. In this work, we performed an analysis of different pre-training variations for an ALBERT-based approach for MIR. Our best model was built up on the initialization from ALBERT base and was further trained on the data provided by ARQMath. Furthermore, we showed that separating large chunks of natural language text and LaTeX notation in one sentence increased the model's performance on formula-only and text-only dependent questions, respectively. The second contribution of this work was to explore the application of ColBERT to accelerate the evaluation of queries because our classification-based approach is too time-consuming. Thereby, we trained and evaluated a ColBERT model and showed that further improvements are necessary before this approach can reach state-of-the-art performance. To improve the modeling capabilities of mathematical formulas, we recommended strategies involving several pre-training methods that include syntactical features of formulas which we have not yet taken into account. To facilitate research based on our work, we release the code for data pre-processing and the training of the models in the project's repository[3]. The source code for training the ColBERT-based models was forked from the official ColBERT repository and slightly adjusted[4].

Acknowledgments. This work was supported by the DFG under Germany's Excellence Strategy, Grant No. EXC-2068–390729961, Cluster of Excellence "Physics of Life" of TU Dresden. Furthermore, the authors are grateful for the GWK support for funding

---

[3] https://github.com/AnReu/ALBERT-for-Math-AR.
[4] https://github.com/AnReu/ColBERT-for-Formulas.

this project by providing computing time through the Center for Information Services and HPC (ZIH) at TU Dresden.

A Tokenization

We applied the sentencepiece tokenizer which splits the input into subwords using byte-pair-encoding, e.g., the sentence 'how can i evaluate $ \ sum_{n=1}^\ infty \ frac{2n}{3^{n+1}} $?' would be tokenized into 'how can i evaluate $ \ sum _ { n = 1 } ^\ in ##ft ##y \ fra ##c { 2 ##n } { 3 ^ { n + 1 } } $?' by the BERT tokenizer, where single tokens are separated by spaces. Input sequences whose length after tokenization exceeded the maximum number of input tokens were truncated to the maximum length. In case two segments together exceeding the maximum length during e.g., NSP or fine-tuning, token by token was deleted from the longest sequence until the sum of the number of both segments equaled the maximum length.

B Hyperparameters

B.1 ALBERT Models

All six models followed the recommendations for hyperparameter configuration during pre-training, with 12M parameters, using the LAMP optimizer [24], 3,125 warm up steps, maximum sequence length of 512 and a vocabulary size of 30,000. Furthermore, we used a batch size of 32 and a learning rate of 0.0005. After pre-training, each classification model was fine-tuned for 125k steps using a batch size of 32, a learning rate of 2e−5 and 200 warm-up steps. Both pre-training and fine-tuning was performed using the code published in the official ALBERT repository.

B.2 ColBERT Models

The hyperparameters recommended by the BERT authors in their repository were used to pre-train this model: The learning rate was set to 2e-05, one batch contained 16 samples and the models were trained for 500k steps. In contrast to the recommendations we set the maximum length of the input to 512, because we did not start to train the model from scratch, where the initial sequence length was set to 128, but rather further trained the already fully pre-trained model on additional data. The training of all three ColBERT models made use of the same hyperparameter configuration. We optimized the L2 similarity between 128-dimensional vectors with a batch size of 128 for 75k steps. Other parameters kept their default values. Punctuation tokens were masked, but we also experimented with models that did not mask them, but we could not see a significant difference in the results. We also started to incorporate ALBERT as a base model for ColBERT, but did not yet find a configuration for a successful training.

References

1. Alsentzer, E., et al.: Publicly available clinical BERT embeddings. In: NAACL HLT 2019, p. 72 (2019)
2. Beltagy, I., Lo, K., Cohan, A.: SciBERT: a pretrained language model for scientific text. In: Proceedings of the 2019 Conference on Empirical Methods in Natural Language Processing and the 9th International Joint Conference on Natural Language Processing (EMNLP-IJCNLP), pp. 3606–3611 (2019)
3. Devlin, J., Chang, M.W., Lee, K., Toutanova, K.: BERT: pre-training of deep bidirectional transformers for language understanding. arXiv preprint arXiv:1810.04805 (2018)
4. Feng, Z., et al.: CodeBERT: a pre-trained model for programming and natural languages. arXiv preprint arXiv:2002.08155 (2020)
5. Gururangan, S., et al.: Don't stop pretraining: adapt language models to domains and tasks. In: Proceedings of the 58th Annual Meeting of the Association for Computational Linguistics, pp. 8342–8360 (2020)
6. Huang, K., Altosaar, J., Ranganath, R.: ClinicalBERT: modeling clinical notes and predicting hospital readmission. arXiv preprint arXiv:1904.05342 (2019)
7. Kanade, A., Maniatis, P., Balakrishnan, G., Shi, K.: Learning and evaluating contextual embedding of source code. In: International Conference on Machine Learning, pp. 5110–5121. PMLR (2020)
8. Khattab, O., Zaharia, M.: ColBERT: Efficient and effective passage search via contextualized late interaction over BERT. In: Proceedings of the 43rd International ACM SIGIR Conference on Research and Development in Information Retrieval, pp. 39–48 (2020)
9. Kudo, T., Richardson, J.: Sentencepiece: a simple and language independent subword tokenizer and detokenizer for neural text processing. arXiv preprint arXiv:1808.06226 (2018)
10. Lan, Z., Chen, M., Goodman, S., Gimpel, K., Sharma, P., Soricut, R.: ALBERT: a lite BERT for self-supervised learning of language representations. arXiv preprint arXiv:1909.11942 (2019)
11. Lee, J., et al.: BioBERT: a pre-trained biomedical language representation model for biomedical text mining. Bioinformatics **36**(4), 1234–1240 (2020)
12. Liu, Y., et al.: RoBERTa: a robustly optimized BERT pretraining approach. arXiv preprint arXiv:1907.11692 (2019)
13. MacAvaney, S., Yates, A., Cohan, A., Goharian, N.: CEDR: contextualized embeddings for document ranking. In: Proceedings of the 42nd International ACM SIGIR Conference on Research and Development in Information Retrieval, pp. 1101–1104 (2019)
14. Mansouri, B., Agarwal, A., Oard, D., Zanibbi, R.: Finding old answers to new math questions: the ARQMath lab at CLEF 2020. In: Jose, M., et al. (eds.) ECIR 2020. LNCS, vol. 12036, pp. 564–571. Springer, Cham (2020). https://doi.org/10.1007/978-3-030-45442-5_73
15. Mansouri, B., Agarwal, A., Oard, D., Zanibbi, R.: Advancing math-aware search: the ARQMath-2 lab at clef 2021, pp. 631–638 (2021)
16. Nguyen, T., et al.: MS MARCO: a human generated machine reading comprehension dataset. In: CoCo@ NIPS (2016)
17. Nogueira, R., Cho, K.: Passage re-ranking with BERT. arXiv preprint arXiv:1901.04085 (2019)

18. Novotný, V., Sojka, P., Štefánik, M., Lupták, D.: Three is better than one. In: CEUR Workshop Proceedings, Thessaloniki, Greece (2020)
19. Peng, S., Yuan, K., Gao, L., Tang, Z.: MathBERT: a pre-trained model for mathematical formula understanding. arXiv preprint arXiv:2105.00377 (2021)
20. Reusch, A., Thiele, M., Lehner, W.: An ALBERT-based similarity measure for mathematical answer retrieval. In: Proceedings of the 44rd International ACM SIGIR Conference on Research and Development in Information Retrieval (2021)
21. Rohatgi, S., Wu, J., Giles, C.L.: PSU at CLEF-2020 ARQMath track: unsupervised re-ranking using pretraining. In: CEUR Workshop Proceedings, Thessaloniki, Greece (2020)
22. Sennrich, R., Haddow, B., Birch, A.: Neural machine translation of rare words with subword units. In: Proceedings of the 54th Annual Meeting of the Association for Computational Linguistics (Volume 1: Long Papers), pp. 1715–1725. Association for Computational Linguistics, Berlin, August 2016. https://doi.org/10.18653/v1/P16-1162, https://www.aclweb.org/anthology/P16-1162
23. Vaswani, A., et al.: Attention is all you need. Adv. Neural. Inf. Process. Syst. **30**, 5998–6008 (2017)
24. You, Y., et al.: Large batch optimization for deep learning: training BERT in 76 minutes. In: International Conference on Learning Representations (2019)

ImageCLEF 2021 Best of Labs:
The Curious Case of Caption Generation
for Medical Images

Aaron Nicolson$^{(\boxtimes)}$ ⓘ, Jason Dowling ⓘ, and Bevan Koopman ⓘ

Australian e-Health Research Centre, Commonwealth Scientific and Industrial
Research Organisation, Herston 4006, QLD, Australia
{aaron.nicolson,jason.dowling,bevan.koopman}@csiro.au

Abstract. As part of Best of Labs, we have been invited to conduct
further investigation on the ImageCLEFmed Caption task of 2021. The
task required participants to automatically compose coherent captions
for a set of medical images. The most popular means of doing this is
with an encoder-to-decoder model. In this work, we investigate a set
of choices with regards to aspects of an encoder-to-decoder model. Such
choices include what pre-training data should be used, what architecture
should be used for the encoder, whether a natural language understanding (e.g., BERT) or generation (e.g., GPT2) checkpoint should be used
to initialise the parameters of the decoder, and what formatting should
be applied to the ground truth captions during training. For each of these
choices, we first made assumptions about what should be used for each
choice and why. Our empirical evaluation then either proved or disproved
these assumptions—with the aim to inform others in the field. Our most
important finding was that the formatting applied to the ground truth
captions of the training set had the greatest impact on the scores of the
task's official metric. In addition, we discuss a number of inconsistencies in the results that others may experience when developing a medical
image captioning system.

Keywords: Medical image captioning · Encoder-to-decoder ·
Multi-modal · Warm-starting

1 Introduction

ImageCLEFmed Caption 2021 is an international challenge where teams develop
a system that automatically generates a coherent caption for a given medical image (for example, X-ray, computed tomography, magnetic resonance, or
ultrasonography) [8,19]. To succeed, the system must not only identify medical
concepts but also their interplay. As with most medical image analysis tasks, a
deep learning model was the key component of the participants' systems. The
model was trained using the provided dataset, containing medical images and
their associated ground truth captions. Its training set was relatively small (2.8K

A. Barrón-Cedeño et al. (Eds.): CLEF 2022, LNCS 13390, pp. 190–203, 2022.
https://doi.org/10.1007/978-3-031-13643-6_15

examples), adding complexity to the task. A training example from the task is shown in Fig. 1. The most popular model for medical image captioning is the encoder-to-decoder model: the encoder produces features from a given image which are then used to condition the decoder when generating the caption [18].

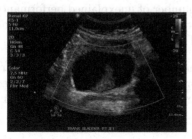

(a) Medical image

"This image is a transverse evaluation of the bladder and right ureteral jet. Renal ultrasound studies also include evaluation of the ureterovesical junction through Color Flow Doppler study of fluid movement of the ureteral jet."

(b) Ground truth caption

Fig. 1. The task was to develop an automated system that, given a medical image, could predict the ground truth caption. Training example *synpic100306* from the Image-CLEFmed Caption 2021 dataset is shown, where **(a)** is the medical image and **(b)** is its ground truth caption.

Our approach to ImageCLEFMed Caption 2021 was to use a Vision Transformer (ViT) [4] as the encoder and PubMedBERT [6] as the decoder (both are detailed in Sect. 2) [14]. Neither a ViT nor a domain-specific natural language checkpoint such as PubMedBERT had previously been explored for medical image captioning. As such, we have been invited to conduct a further investigation on the previously mentioned task as part of Best of Labs.

For this work, we aim to investigate a set of important choices for an encoder-to-decoder model:

Choice 1: Pre-training data—The choice in question is what pre-training data to use for warm-starting. Warm-starting refers to the initialisation of a models parameters with those of a pre-trained checkpoint. A checkpoint includes the values of all the learned parameters of a trained model. The pre-training data could be from the general domain (e.g., Wikipedia articles used for BERT [3]); or domain specific (e.g., biomedical corpora used for PubMed-BERT [6]). Moreover, does warm-starting with a checkpoint from a related task (e.g., Chest X-Ray (CXR) report generation) improve performance?

Choice 2: Encoder—The architecture of the encoder. Specifically, whether to add convolutional layers to the ViT or not.

Choice 3: Decoder—The type of pre-training task of the decoder checkpoint. The pre-training task could be a Natural Language Understanding (NLU) task (e.g., the self-supervised learning tasks used to form BERT), or a Natural Language Generation (NLG) task (e.g., the language modelling task used to form GPT2 [21]). NLU is the comprehension of natural language through grammar and context while NLG is the construction of natural language based on a given input.

Choice 4: Formatting—How should the captions be formatted for training? The official metric (described in Subsect. 3.2) employs a series of natural language formatting steps, such as removing punctuation and stopwords. These steps may seem innocuous and are rarely reported in other studies, but as part of our submissions we had a number of unexplained performance differences that we posit were a result of the differences between the caption formatting during training and that used for the official metric [14].

Anyone setting out to develop medical image caption generation systems are faced with the above choices, as we were before participating in ImageCLEFmed Caption 2021. From these choices and one's intuition, the following assumptions may be held:

Assumption 1: Pre-train data—Warm-starting with a domain-specific checkpoint, such as PubMedBERT, would outperform warm-starting with a general-domain checkpoint, such as BERT. Moreover, one would assume that an encoder-to-decoder model warm-started with a checkpoint from a related task (e.g., CXR report generation) would outperform a model with its encoder and decoder warm-started with general-domain checkpoints. This is based on our expectations that transferring knowledge learned on a related task to the final task typically results in an improvement in performance—especially when the training set of the final task is relatively small.

Assumption 2: Encoder—That a ViT with convolutional layers would outperform one without. This is based on the fact that convolutional layers (with small kernel sizes) have an inductive bias towards local spatial regions—an advantage for modelling the fine details present in medical images.

Assumption 3: Decoder—That an NLG checkpoint, such as GPT2, would outperform an NLU checkpoint, such as PubMedBERT. This based on the intuition that the pre-training task of an NLG checkpoint would be more transferable to the task of caption generation.

Assumption 4: Formatting—That formatting the ground truth captions of the training set does not have an impact on the performance of a model with regards to the official metric. This is based on the fact that the metric used for the challenge applies a series of formatting steps to both the predicted and ground truth captions—potentially rendering any formatting applied during training redundant.

Curiously, our experience was that many of these intuitive assumptions were not supported by our empirical evaluation.

The remainder of this paper is around an empirical evaluation of a set of models on the ImageCLEFmed Caption 2021 task, where the models were selected to prove or disprove the above assumptions. The results will help to inform others working on similar tasks who may share the same assumption. We test each assumption individually and discuss why they do or do not hold.

2   Background and Related Work

Prior to ImageCLEFmed Caption 2021, a Convolutional Neural Network (CNN) [7] and a decoder-only Transformer [27] were typically employed as the encoder and decoder, respectively. Convolutional layers have an inductive bias towards local spatial regions owing to their small kernel sizes, making them ideal for modelling the fine details present in medical images. Transformers, which leverage the attention mechanism, have the ability to model the relationship between all of its inputs simultaneously, lending themselves to modelling the free text of medical captions [27].

General-domain ImageNet checkpoints were also frequently employed to warm-start the encoder (where ImageNet is a large general-domain image classification task) [10, 25]. The transfer of knowledge from the pre-training task to the final task can provide a significant performance boost, particularly when the pre-training dataset is significantly larger than that of the final task. Furthermore, warm-starting is particularly effective when the domain of the pre-training task is similar to that of the final task. However, there is a lack of medical image checkpoints outside of CXR tasks [22]. Furthermore, warm-starting the decoder was not common practice prior to the competition.

Prior to the challenge, ViTs were investigated for computer vision tasks and demonstrated the ability to model the relationship between patches of an image. However, ViTs lack the inductive biases that enable CNNs to perform well on such tasks. Surprisingly, ViTs are able to overcome this deficiency at larger dataset sizes (14M-300M images) [4]. Subsequently, it was demonstrated that a ViT encoder warm-started with an ImageNet checkpoint outperformed its CNN counterpart on general-domain image captioning [12]. Given this, and the aptitude of ImageNet checkpoints at warm-starting medical image tasks, we selected the ViT and its ImageNet checkpoint for the encoder of our original system.

While medical image checkpoints were scarce in the literature, many medical text checkpoints were available. Several large pre-trained NLU encoder-only Transformer checkpoints were formed via the self-supervised learning strategies of BERT [3] and large biomedical corpora. One instance is PubMedBERT [6]—an encoder-only Transformer pre-trained on biomedical articles from the PubMed corpus.[1] However, a decoder is typically warm-started with an NLG decoder-only Transformer checkpoint, such as GPT2 [21]. Despite this, Rothe *et al.* demonstrated that the decoder warm-started with an NLU checkpoint could outperform its NLG counterpart on several sequence-to-sequence NLG tasks [24]. This suggested that PubMedBERT would be ideal to warm-start the decoder.

With our encoder-to-decoder model, ViT2PubMedBERT, we had nine submissions. Amongst the submissions we attempted additional pre-training of the encoder on medical images (from four X-ray datasets), pre-training of the encoder and decoder on a larger medical image captioning datset (ROCO [20]), and additional fine-tuning using reinforcement learning [23]. However, there was

---

[1] https://pubmed.ncbi.nlm.nih.gov/.

a discrepancy between the validation scores we were attaining on our metrics versus the test scores attained using the official metric: a validation score improvement did not correlate with a test score improvement [14]. One possible reason for this is that the formatting used on the ground truth captions of the training set was different to that used for the official metric.

Our subsequent work following ImageCLEFmed Caption 2021 investigated a range of architectures and checkpoints for warm-starting the encoder and decoder for a related task: CXR report generation. Other than the fact that this is a more specific task (i.e., only one modality is considered), the biggest difference is the size of the datset, with the MIMIC-CXR dataset including 270K examples in the training set [9]. We also found that CNNs such as ResNets outperformed ViT [7]. We also investigated improvements to the ViT and found that a Convolutional vision Transformer (CvT) encoder produced the highest performance. We also found that GPT2 and DistilGPT2 [26] outperformed domain-specific NLU checkpoints such as PubMedBERT—possibly due to the fact that GPT2 is an NLG checkpoint. This was different to the finding of Rothe *et al.*, likely due to one key difference: the task of the encoder for CXR report generation is to produce features from images rather than natural language. Another finding was that PubMedBERT—a domain-specific NLU checkpoint—was able to outperform BERT—a general-domain NLU checkpoint [16]. These findings have influenced some of the aforementioned assumptions—as our ancillary aim of this study is to determine if the findings on MIMIC-CXR are upheld on the ImageCLEFmed Caption 2021 task.

3   Methodology

In this section, we describe the dataset, metrics, models, fine-tuning strategy, image pre-processing, and text formatting.

3.1   Task Description and Dataset

For ImageCLEFmed Caption 2021, participants were tasked with developing a system that could generate a caption for a given medical image. The motivation behind this task is to help develop tools that can aid medical experts with interpreting and summarising medical images, a task that is often time-consuming and a bottleneck in clinical diagnosis pipelines. Each example from the dataset consisted of a medical image and its associated ground truth caption, as shown in Fig. 1. The data was divided into training ($n = 2\,756$), validation ($n = 500$), and test ($n = 444$) sets. Evaluation was performed by comparing the predicted captions to the annotations provided by medical doctors (i.e., the ground truth captions).

3.2 Metrics

We adopted the official metric of ImageCLEFmed Caption 2021 for the validation and test sets: CLEF-BLEU. It was computed as follows for each predicted and ground truth caption:

1. **Lowercased:** The caption was first converted to lower-case.
2. **Remove punctuation:** All punctuation was then removed and the caption was tokenized into its individual words.
3. **Remove stopwords:** Stopwords were then removed using NLTK's English stopword list (NLTK v3.2.2).
4. **Stemming:** Stemming was next applied using NLTK's Snowball stemmer (NLTK v3.2.2).
5. The score was then calculated as the average score of BLEU-1, BLEU-2, BLEU-3, and BLEU-4 between the predicted and ground truth captions [17].

Note that each caption was always considered as a single sentence, even if it contained several sentences.

Furthermore, the following metrics were used for evaluation on the validation set: BLEU-1, BLEU-2, BLEU-3, and BLEU-4 [17], ROUGE-L [11], and CIDEr [28]. This was to aid with understanding how formatting the ground truth captions impacted the performance each model. The formatting is detailed in Subsect. 3.5.

3.3 Models

The encoder-to-decoder models investigated in this work are listed below. An example of one is shown in Fig. 2. The input to the encoder is a medical image. The output of the encoder (CvT-21) is fed to the cross-attention module of the decoder (DistilGPT2), which then generates a caption in an autoregressive fashion—conditioned on the encoders output. It should be noted that each model employs a linear layer that projects the last hidden state of the encoder to the hidden size of the decoder.

ViT2BERT—ViT (86M parameters) is the encoder [4]. It was warm-started with a checkpoint pre-trained on ImageNet-22K (14M images, 21 843 classes) at a resolution of 224×224 and then additionally trained on ImageNet-1K (1M images, 1 000 classes) at resolution of 384×384. BERT (110M parameters) is the decoder, which is pre-trained on BookCorpus [31] and Wikipedia articles in an uncased manner using self-supervised learning [3]. Both ViT and BERT are 12 layers with a hidden size of 768.

ViT2PubMedBERT—Identical to ViT2BERT, except that PubMedBERT (110M parameters) is the decoder. Its main difference to BERT is the pre-training data: abstracts from PubMed (4.5B words) and articles from PubMed Central (13.5B words).

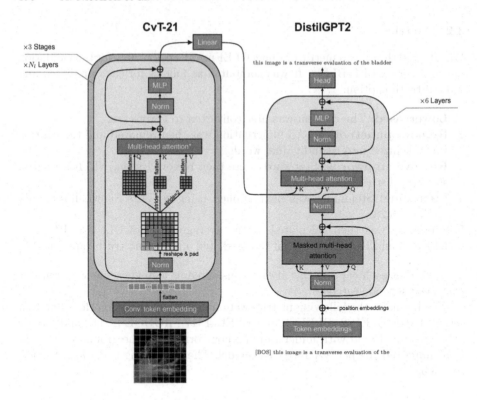

Fig. 2. CvT2DistilGPT2. Q, K, and V are the queries, keys, and values, respectively, for multi-head attention [27]. $*$ indicates that the linear layers for Q, K, and V are replaced with the convolutional layers depicted below the multi-head attention module. [BOS] is the beginning-of-sentence special token. N_l is the number of layers for each stage, where $N_l = 1$, $N_l = 4$, and $N_l = 16$ for the first, second, and third stage, respectively. The head for DistilGPT2 is the same used for language modelling.

ViT2DistilGPT2—Identical to ViT2BERT, except that DistilGPT2 (82M parameters) is the decoder. It is pre-trained using knowledge distillation where DistilGPT2 was the student and GPT2 was the teacher. OpenWeb-Text, a reproduction of OpenAI's WebText corpus, was used as the pre-training data [5]. DistilGPT2 includes 6 layers with a hidden size of 768.

CvT2DistilGPT2—Identical to ViT2DistilGPT2, except that CvT-21 (32M parameters) is the encoder. CvT-21 was warm-started with an ImageNet-22K checkpoint with a resolution of 384×384 [30]. It has three stages, with a combined 21 layers.

CvT2DistilGPT2·MIMIC-CXR—This is CvT2DistilGPT2 warm-started with a MIMIC-CXR checkpoint [15,16]. The checkpoint was not additionally fine-tuned with reinforcement learning on MIMIC-CXR.

3.4   Medical Image Pre-processing

Each medical image $X \in \mathbb{R}^{C \times W \times H}$ (where C, W, and H denote the number of channels, the width, and height, respectively) had an 8-bit pixel depth and three channels ($C = 3$). The image was first resized using bilinear interpolation to a size of $\mathbb{R}^{3 \times 384 \times 384}$. During training, the image was also rotated at an angle sampled from $\mathcal{U}[-5°, 5°]$. Finally, the image was standardised using the mean and standard deviation of each channel provided with the encoder checkpoint.

3.5   Caption Formatting and Generation

We investigated five different formatting strategies for the ground truth captions of the training and validation sets, to determine their impact on CLEF-BLEU:

1. No formatting.
2. Lowercased.
3. Lowercased + no punctuation.
4. Lowercased + no punctuation + no stopwords.
5. Lowercased + no punctuation + no stopwords + stemming.

These formatting steps were described in Subsect. 3.2. When generating the captions during validation and testing, beam search with a beam size of four and a maximum number of 128 subwords was used.

3.6   Fine-Tuning

Teacher forcing was used for fine-tuning [29]. Each model was implemented in PyTorch version 1.10.1 and trained with 4×NVIDIA P100 16 GB GPUs. To reduce memory consumption, we employed PyTorch's automatic mixed precision (a combination of 16-bit and 32-bit floating point variables). For fine-tuning, the following configuration was used: categorical cross-entropy as the loss function; a mini-batch size of 32; early stopping with a patience of 20 epochs and a minimum delta of $1e-4$; $AdamW$ optimiser for gradient descent optimisation [13]; an initial learning rate of $1e-5$ and $1e-4$ for the encoder and all other parameters, respectively, following [2]. All other hyperparameters for $AdamW$ were set to their defaults. To select the best epoch for a model, the highest validation BLEU-4 score was used.

Table 1. Results on the validation and test sets of ImageCLEFmed Caption 2021. A higher colour saturation indicates a higher score. For CLEF-BLEU, the full formatting described in Subsect. 3.2 was applied to both the predicted and ground truth captions for every row. For the other metrics and for training, the indicated formatting was applied to the ground truth captions. ↪ MIMIC-CXR indicates CvT2DistilGPT2·MIMIC-CXR.

Model	Validation Set							Test Set
	BLEU-1	BLEU-2	BLEU-3	BLEU-4	ROUGE-L	CIDEr	CLEF BLEU	CLEF BLEU
Strategy 1: No formatting								
ViT2BERT	0.315	0.258	0.227	0.206	0.275	1.462	0.405	0.406
ViT2PubMedBERT	0.348	0.291	0.258	0.236	0.306	1.703	0.432	0.406
ViT2DistilGPT2	0.363	0.319	0.299	0.288	0.328	2.243	0.428	0.384
CvT2DistilGPT2	**0.370**	**0.326**	**0.305**	**0.293**	**0.332**	**2.297**	0.433	0.400
↪MIMIC-CXR	0.348	0.304	0.283	0.272	0.320	2.212	0.427	0.407
Strategy 2: Lowercased								
ViT2BERT	0.358	0.304	0.277	0.261	0.317	1.932	0.405	0.406
ViT2PubMedBERT	**0.395**	**0.342**	**0.314**	**0.297**	0.353	2.243	0.432	0.406
ViT2DistilGPT2	0.370	0.322	0.299	0.287	0.336	2.340	0.437	0.408
CvT2DistilGPT2	0.380	0.332	0.308	0.295	**0.356**	**2.466**	0.448	0.405
↪MIMIC-CXR	0.354	0.304	0.281	0.269	0.318	2.094	0.402	0.397
Strategy 3: Lowercased + no punctuation								
ViT2BERT	0.378	0.325	0.299	0.286	0.347	2.328	0.444	0.404
ViT2PubMedBERT	**0.417**	**0.364**	**0.337**	**0.323**	**0.379**	**2.591**	0.453	0.426
ViT2DistilGPT2	0.387	0.338	0.314	0.301	0.351	2.400	0.441	0.394
CvT2DistilGPT2	0.388	0.338	0.315	0.302	0.356	2.521	0.446	0.401
↪MIMIC-CXR	0.373	0.320	0.296	0.283	0.333	2.267	0.414	0.400
Strategy 4: Lowercased + no punctuation + no stopwords								
ViT2BERT	0.355	0.319	0.302	0.292	0.327	2.430	0.451	**0.430**
ViT2PubMedBERT	**0.374**	**0.337**	**0.319**	**0.308**	**0.347**	**2.601**	0.458	0.423
ViT2DistilGPT2	0.342	0.308	0.292	0.283	0.311	2.356	0.421	0.410
CvT2DistilGPT2	0.332	0.301	0.286	0.277	0.315	2.409	0.430	0.400
↪MIMIC-CXR	0.322	0.290	0.274	0.264	0.308	2.342	0.422	0.398
Strategy 5: Lowercased + no punctuation + no stopwords + stemming								
ViT2BERT	0.364	0.321	0.301	0.290	0.328	2.355	0.419	0.396
ViT2PubMedBERT	**0.393**	**0.346**	**0.323**	**0.310**	**0.366**	**2.593**	0.416	0.410
ViT2DistilGPT2	0.355	0.317	0.299	0.289	0.326	2.444	0.399	0.383
CvT2DistilGPT2	0.355	0.316	0.298	0.288	0.330	2.462	0.416	0.391
↪MIMIC-CXR	0.353	0.313	0.295	0.285	0.328	2.457	0.425	0.394

4   Results and Discussion

Table 1 presents results from our empirical evaluation. We shall discuss the results as they relate to the four assumptions detailed in the introduction.

4.1 Assumption 1: Pre-training Data

The first assumption was that PubMedBERT as the decoder would outperform BERT—as it is a domain-specific checkpoint. In terms of the validation scores, this assumption stood, as ViT2PubMedBERT outperformed ViT2BERT (except for validation CLEF-BLEU on Strategy 5). However, this finding was not consistent with the test scores, with ViT2BERT producing the highest score of any model (0.430). This contradiction indicates that the results on the validation set do not generalise to the test set.

The next assumption was that warm-starting the encoder-to-decoder model with a CXR report generation checkpoint would improve performance, especially given the small size of the training set. The performance of CvT2DistilGPT2·MIMIC-CXR was not significantly different from CvT2DistilGPT2 in terms of the test scores. However, the validation scores refute the assumption, as CvT2DistilGPT2 consistently produced higher validation scores. One explanation is that X-rays are not the dominant modality in the ImageCLEFmed Caption 2021 training set, where computed tomography and magnetic resonance are more represented [1, Table 1].

4.2 Assumption 2: Encoder

Here, we determine if including convolutional layers in the encoder, i.e., choosing CvT over ViT, improves performance. When no formatting is used during training, CvT2DistilGPT2 attains higher validation and test scores than ViT2DistilGPT2. However, when formatting is used during training, the picture becomes unclear. For example, CvT2DistilGPT2 attains higher validation and test scores for Strategies 3 and 5, while ViT2DistilGPT2 attains higher validation and test scores for Strategies 2 and 4. Therefore, it is unclear if adding convolutional layers to ViT (i.e., using CvT instead) is advantageous for this task, refuting the findings in [16]. However, it should be noted that CvT consumes drastically fewer parameters than ViT—demonstrating parameter efficiency.

4.3 Assumption 3: Decoder

The assumption made for the decoder was that an NLG checkpoint would outperform an NLU checkpoint. Comparing ViT2BERT to ViT2DistilGPT2 on the validation scores for no formatting, DistilGPT2 as the decoder outperforms BERT by a large margin. However, this margin decreases as the number of formatting steps increases—BERT as the decoder even outperforms DistilGPT2 on the validation set in certain cases. This indicates that BERT is less sensitive to the formatting steps applied to the ground truth captions of the training set. However, their scores on the test set tell a different story. BERT as the decoder attained a higher test score than DistilGPT2 for each strategy, except Strategy 2. Again, the results on the validation set are misleading, as they do not generalise to the test set [16].

4.4   Assumption 4: Formatting

Originally, we assumed that formatting the ground truth captions of the training set would have no impact on performance. However, the results indicate that, in fact, it does have an impact. ViT2BERT experienced an absolute test CLEF-BLEU improvement of 2.4% when Strategy 3 was used instead of no formatting. This is opposite to the original assumption made—applying formatting to the predicted and ground truth captions before evaluation does not mean that there is no benefit to using formatted ground truth captions as the training target.

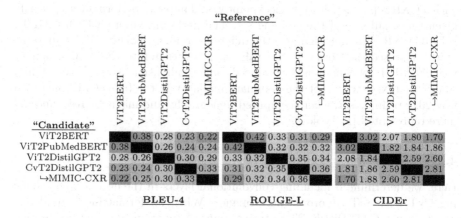

Fig. 3. The *similarity* between the predicted captions of the models on the validation set. Each metric requires *reference* and *candidate* captions. Here, the predicted captions of one model are used as the reference captions (instead of the ground truth captions) and the predicted captions of the other model as the candidate captions. No formatting was applied to the ground truth or predicted captions during training or evaluation. The presented matrices are not symmetric as each metric treats the candidate and reference captions differently. ↪ MIMIC-CXR indicates CvT2DistilGPT2·MIMIC-CXR.

On another note, BERT and PubMedBERT appear to be either less sensitive to formatting, or benefit from formatting, especially with the third and fourth formatting strategies. This could be caused by multiple factors; an NLU checkpoint may be more robust than an NLG checkpoint to formatting. Moreover, DistilGPT2 may be disadvantaged by the fact that it is cased, rather than uncased like BERT and PubMedBERT.

4.5   Similarity Between Predicted Captions

The results in Table 1 are solely focus on model differences according to their effectiveness on the ImageCLEFmed Caption 2021 task. While this provides some insight, we also want to understand how similar the captions generated by the models (i.e., the predicted captions) are to one another. Specifically, two models may have a similar effectiveness on the ImageCLEFmed Caption task, but they may generate significantly different captions. To compute the similarity between a pair of models, we give their generated captions to a metric. Each metric consumes *candidate* and *reference* captions and treats each differently. Hence, we conduct a pair-wise comparison between the generated captions of a pair of models. The results of this are shown in Fig. 3.

It can be seen that the generated captions of ViT2PubMedBERT and ViT2BERT were the most similar to one another; CvT2DistilGPT2 and CvT2DistilGPT2·MIMIC-CXR also attained a high similarity. This is somewhat surprising given that the pre-training data of the checkpoints in each comparison are different. However, the high similarity makes sense from an architectural point of view as the models in each comparison employ the same (or very similar) encoder and decoder architectures. The most dissimilar models are ViT2BERT and CvT2DistilGPT2·MIMIC-CXR. This makes sense as they are the most dissimilar in terms of their pre-training data, encoder, and decoder. Finally, ViT2DistilGPT2 versus CvTDistilGPT2 had a higher similarity than ViT2DistilGPT2 and ViT2BERT, indicating that the decoder has a larger impact on dissimilarity than the encoder.

5   Conclusion

For our Best of Labs contribution, we posed a set of assumptions regarding choices pertaining to an encoder-to-decoder model for medical image captioning, and then set out to prove or disprove them through an empirical evaluation. Our key finding was that the type of formatting applied to the ground truth captions of the training set had the greatest impact on the scores obtained on the official metric of the task. The results also indicate that BERT and PubMedBERT as the decoder are less sensitive to additional formatting steps than DistilGPT2. Unfortunately, assumptions made about the pre-training data, encoder, and decoder could not be proved or disproved, as the results were inconclusive. A key problem was that the hierarchy of performance amongst the models on the validation set did not generalise to the test set. This could be due to the limited size of the dataset or significant differences between the validation and test sets.

Acknowledgement. This work was partially funded by CSIRO's Machine Learning and Artificial Intelligence Future Science Platform (MLAI FSP).

References

1. Charalampakos, F., Karatzas, V., Kougia, V., Pavlopoulos, J., Androutsopoulos, I.: AUEB NLP group at ImageCLEFmed caption tasks 2021. In: Proceedings of the 12th International Conference of the CLEF Association, Bucharest, Romania, pp. 1–17, September 2021
2. Chen, Z., Song, Y., Chang, T., Wan, X.: Generating radiology reports via memory-driven transformer. In: Proceedings of the 2020 Conference on Empirical Methods in Natural Language Processing (EMNLP), pp. 1439–1449. Association for Computational Linguistics (2020). https://doi.org/10.18653/v1/2020.emnlp-main.112

3. Devlin, J., Chang, M., Lee, K., Toutanova, K.: BERT: pre-training of deep bidirectional transformers for language understanding. In: Proceedings of the 2019 Conference of the North American Chapter of the Association for Computational Linguistics: Human Language Technologies (Long and Short Papers), Minneapolis, Minnesota, vol. 1, pp. 4171–4186. Association for Computational Linguistics, June 2019. https://doi.org/10.18653/v1/N19-1423. https://www.aclweb.org/anthology/N19-1423

4. Dosovitskiy, A., et al.: An image is worth 16×16 words: transformers for image recognition at scale. arXiv:2010.11929 [cs.CV], October 2020

5. Gokaslan, A., Cohen, V.: OpenWebText Corpus (2019). http://Skylion007.github.io/OpenWebTextCorpus

6. Gu, Y., et al.: Domain-specific language model pretraining for biomedical natural language processing. arXiv:2007.15779 [cs.CL], July 2020

7. He, K., Zhang, X., Ren, S., Sun, J.: Deep residual learning for image recognition. In: 2016 IEEE Conference on Computer Vision and Pattern Recognition (CVPR). IEEE, June 2016. https://doi.org/10.1109/cvpr.2016.90

8. Ionescu, B., et al.: Overview of the ImageCLEF 2021: multimedia retrieval in medical, nature, internet and social media applications. In: Candan, K.S., et al. (eds.) CLEF 2021. LNCS, vol. 12880, pp. 345–370. Springer, Cham (2021). https://doi.org/10.1007/978-3-030-85251-1_23

9. Johnson, A.E.W., et al.: MIMIC-CXR-JPG, a large publicly available database of labeled chest radiographs. arXiv:1901.07042 [cs.CV], January 2019

10. Ke, A., Ellsworth, W., Banerjee, O., Ng, A.Y., Rajpurkar, P.: CheXtransfer: performance and parameter efficiency of ImageNet models for chest X-Ray interpretation. In: Proceedings of the Conference on Health, Inference, and Learning, pp. 116–124. ACM, April 2021. https://doi.org/10.1145/3450439.3451867

11. Lin, C., Och, F.J.: Automatic evaluation of machine translation quality using longest common subsequence and skip-bigram statistics. In: Proceedings of the 42nd Annual Meeting of the Association for Computational Linguistics (ACL-2004), Barcelona, Spain, pp. 605–612, July 2004. https://doi.org/10.3115/1218955.1219032. https://aclanthology.org/P04-1077

12. Liu, W., Chen, S., Guo, L., Zhu, X., Liu, J.: CPTR: full transformer network for image captioning. arXiv:2101.10804 [cs.CV], January 2021

13. Loshchilov, I., Hutter, F.: Decoupled weight decay regularization. In: International Conference on Learning Representations (2019). https://openreview.net/forum?id=Bkg6RiCqY7

14. Nicolson, A., Dowling, J., Koopman, B.: AEHRC CSIRO at ImageCLEFmed caption 2021. In: Proceedings of the 12th International Conference of the CLEF Association, Bucharest, Romania, pp. 1–12, September 2021

15. Nicolson, A., Dowling, J., Koopman, B.: Chest X-Ray report generation checkpoints for CvT2DistilGPT2 (2022). https://doi.org/10.25919/64WX-0950

16. Nicolson, A., Dowling, J., Koopman, B.: Improving chest X-Ray report generation by leveraging warm-starting, January 2022

17. Papineni, K., Roukos, S., Ward, T., Zhu, W.: BLEU: a method for automatic evaluation of machine translation. In: Proceedings of the 40th Annual Meeting of the Association for Computational Linguistics, Philadelphia, Pennsylvania, USA, pp. 311–318. Association for Computational Linguistics, July 2002. https://doi.org/10.3115/1073083.1073135. https://www.aclweb.org/anthology/P02-1040

18. Pavlopoulos, J., Kougia, V., Androutsopoulos, I., Papamichail, D.: Diagnostic captioning: a survey, January 2021. arXiv:2101.07299 [cs.CV]

19. Pelka, O., Ben Abacha, A., García Seco de Herrera, A., Jacutprakart, J., Friedrich, C.M., Müller, H.: Overview of the ImageCLEFmed 2021 concept & caption prediction task. In: CLEF2021 Working Notes. CEUR Workshop Proceedings, CEUR-WS.org, Bucharest, Romania, 21–24 September 2021
20. Pelka, O., Koitka, S., Rückert, J., Nensa, F., Friedrich, C.M.: Radiology Objects in COntext (ROCO): a multimodal image dataset. In: Stoyanov, D., et al. (eds.) LABELS/CVII/STENT -2018. LNCS, vol. 11043, pp. 180–189. Springer, Cham (2018). https://doi.org/10.1007/978-3-030-01364-6_20
21. Radford, A., Wu, J., Child, R., Luan, D., Amodei, D., Sutskever, I.: Language models are unsupervised multitask learners. OpenAI Blog 1(8), 9 (2019)
22. Rajpurkar, P., et al.: CheXNet: radiologist-level pneumonia detection on chest X-Rays with deep learning. arXiv:1711.05225 [cs.CV], November 2017
23. Rennie, S.J., Marcheret, E., Mroueh, Y., Ross, J., Goel, V.: Self-critical sequence training for image captioning. In: 2017 IEEE Conference on Computer Vision and Pattern Recognition (CVPR). IEEE, July 2017. https://doi.org/10.1109/cvpr.2017.131
24. Rothe, S., Narayan, S., Severyn, A.: Leveraging pre-trained checkpoints for sequence generation tasks. Trans. Assoc. Comput. Linguist. 8, 264–280 (2020). https://doi.org/10.1162/tacl_a_00313
25. Russakovsky, O., et al.: ImageNet large scale visual recognition challenge. Int. J. Comput. Vision 115(3), 211–252 (2015). https://doi.org/10.1007/s11263-015-0816-y
26. Sanh, V., Debut, L., Chaumond, J., Wolf, T.: DistilBERT, a distilled version of BERT: smaller, faster, cheaper and lighter. arXiv:1910.01108 [cs.CL], October 2019
27. Vaswani, A., et al.: Attention is all you need. In: Proceedings of the 31st International Conference on Neural Information Processing Systems, NIPS 2017, pp. 6000–6010. Curran Associates Inc., Red Hook (2017)
28. Vedantam, R., Lawrence Zitnick, C., Parikh, D.: CIDEr: consensus-based image description evaluation. In: Proceedings of the IEEE Conference on Computer Vision and Pattern Recognition (CVPR), June 2015
29. Williams, R.J., Zipser, D.: A learning algorithm for continually running fully recurrent neural networks. Neural Comput. 1(2), 270–280 (1989). https://doi.org/10.1162/neco.1989.1.2.270
30. Wu, H., et al.: CvT: introducing convolutions to vision transformers. arXiv:2103.15808 [cs.CV], March 2021
31. Zhu, Y., et al.: Aligning books and movies: towards story-like visual explanations by watching movies and reading books. In: Proceedings of the IEEE International Conference on Computer Vision (ICCV), December 2015

Data-Centric and Model-Centric Approaches for Biomedical Question Answering

Wonjin Yoon[1] ⓘ, Jaehyo Yoo[1] ⓘ, Sumin Seo[1] ⓘ, Mujeen Sung[1] ⓘ,
Minbyul Jeong[1] ⓘ, Gangwoo Kim[1] ⓘ, and Jaewoo Kang[1,2(✉)] ⓘ

[1] Korea University, Seoul 02841, South Korea
{wjyoon,jaehyoyoo,suminseo,mujeensung,
minbyuljeong,gangwoo_kim,kangj}@korea.ac.kr
[2] AIGEN Sciences Inc., Seoul 04778, South Korea

Abstract. Biomedical question answering (BioQA) is the process of automated information extraction from the biomedical literature, and as the number of accessible biomedical papers is increasing rapidly, BioQA is attracting more attention. In order to improve the performance of BioQA systems, we designed strategies for the sub-tasks of BioQA and assessed their effectiveness using the BioASQ dataset. We designed data-centric and model-centric strategies based on the potential for improvement for each sub-task. For example, model design for the factoid-type questions has been explored intensely but the potential of increased label consistency has not been investigated (data-centric approach). On the other hand, for list-type questions, we apply the sequence tagging model as it is more natural for the multi-answer (i.e. multi-label) task (model-centric approach).

Our experimental results suggest two main points: scarce resources like BioQA datasets can be benefited from data-centric approaches with relatively little effort; and a model design reflecting data characteristics can improve the performance of the system.

The scope of this paper is majorly focused on applications of our strategies in the BioASQ 8b dataset and our participating systems in the 9th BioASQ challenges. Our submissions achieve competitive results with top or near top performance in the 9th challenge (Task b - Phase B).

Keywords: BioNLP · Biomedical Natural Language Processing · BioASQ · Biomedical Question Answering

1 Introduction

Recently, the number of accessible biomedical literature has been increasing rapidly with more than a million articles newly indexed to MEDLINE in 2021 [1]. A plethora of biomedical papers is making it impossible for researchers to manually read and extract valuable information from the entire literature library. Biomedical question answering (BioQA) task is a specific category of Question Answering (QA) tasks where questions and/or the related documents are in the context of the biomedical domain, and has emerged as a means to assist the automated extraction of information.

The BioASQ challenge [21] is a series of annual competitions for biomedical literature, which encompasses document classification, document retrieval, and QA tasks, at scale. The challenge and datasets, which are the outputs of the challenge, are one of the richest sources for BioQA research. Questions of the challenge are categorized in 4 groups: Factoid-type, List-type, Yes/No-type, and Summary-type questions. Answers of QA tasks are in two formats: exact answer and ideal answer. For an exact answer, the expected output is a word or a short phrase, whereas for an ideal answer, the output is composed of one or multiple complete sentences.

In this paper, we present approaches for the four different categories of the questions and our participation in QA tasks of the 9th BioASQ challenge [11]. Our strategies were decided based on the potential for improvement of existing approaches for each question type.

For exact answers, we utilize both data-centric approach (factoid-type questions) and model-centric approach (list-type questions) to improve our previous systems [7] for BioASQ 8b. For factoid questions, model designs for this type of question have been intensely explored. However, the effect of increased label consistency in the dataset has not been investigated, despite that Jeong et al. [7] suggested that some questions are not answerable within the setting of extractive QA setting, which is a predominant setting in answering factoid questions [7,13,22,25]. For list questions, our previous work [24] suggested that the questions with multiple answers, which are abundant in the biomedical setting, can be benefited from adopting a sequence tagging structure. We have applied the sequence tagging setting for solving the list-type question.

For ideal answers, we apply abstractive summarization method using large-scale language model, BART [14], and a pipeline approach to alleviate the problem of factual inconsistency problem, which is one of the weak points of abstractive summarization methods [6,12,28]. For the challenge (Task 9b - Phase B), our submissions are in both exact and ideal answer formats.

Our experimental results suggest two main points: scarce resources like BioQA datasets can be benefited from data-centric approaches with relatively little effort; and a model design reflecting data characteristics can improve the performance of the system.

2   Methods

In this section, we describe details of our approaches. In the Sect. 2.1, 2.2, and 2.3, we describe our approaches for the exact answers of factoid-type, list-type, and yes/no-type questions. In the Sect. 3.2, we describe our unified model for ideal answers, which can answer all four types of questions: 3 aforementioned types and summary-type questions. For all types, we use fine-tuned BioBERT [13], which has been proven to be effective on various NLP tasks in the biomedical or clinical domain [2,3,8,9,18,25], as a backbone model.

Table 1. Data cleaning operations for the answers.

Operation	Question	Original Answer	Normalized Answer
Sentence to phrase	What family do mDia proteins belong in?	['mDia proteins are members of the formin family']	['formin family']
	What does polyadenylate-binding protein 4 (PABP4) bind to?	['PABP4 binds mRNA poly(A) tails.']	['mRNA poly(A) tails']
	Where are the orexigenic peptides synthesized?	['The orexigenic peptides are sythesized in the hypothalamus.']	['the hypothalamus']
Punctuation marks	What is the effect of CRD-BP on the stability of c-myc mRNA?	['To protect c-myc CRD from endonucleolytic attack.']	['To protect c-myc CRD from endonucleolytic attack']
	What is gingipain?	['A keystone periodontal pathogen. ']	['A keystone periodontal pathogen']
	What is the role of the UBC9 enzyme in the protein sumoylation pathway?	['SUMO-conjugating enzyme']	['SUMO-conjugating enzyme', 'SUMO conjugating enzyme']

Table 2. Data cleaning operation for the questions. Misspelled words are corrected.

Original Question	Cleaned Question
Which **trancription** factor activates the betalain pathway?	Which **transcription** factor activates the betalain pathway?
What happens to H2AX upon DNA **bouble** strand breaks?	What happens to H2AX upon DNA **double** strand breaks?
What is a popular **mesaure** of gene expression in RNA-seq experiments?	What is a popular **measure** of gene expression in RNA-seq experiments?

2.1 Data-Centric Approach; Factoid Questions

Recently, pre-trained models [4,19] have achieved dramatic improvements of downstream tasks in both general and biomedical domain by harnessing large-scale models with transfer-learning methods [4,10,18,20]. BioQA models have also benefited from transfer-learning [7,22,25]. However, utilizing the maximum of scarce resources is susceptible to the rare error of the training samples, as opposed to the case with relatively rich datasets where a few erroneous samples can be ignored by the model. Moreover, Jeong et al. [7] measured the proportion of questions, which is unanswerable if transformed to the extractive QA setting, in the test dataset of BioASQ 8b. Hence, the models trying to solve the task under the extractive QA setting are suffering from unexpected noise.

In this section, we introduce our data-centric approach for factoid-type questions of the BioASQ 9b challenge. Data-centric approach is a concept of improving the quality of training data to make it better fit into the model. The term *data-centric* forms a binary opposition term pair with *model-centric* approach which focuses more into improving model to achieve better performance. The concept is introduced by Ng [16] at a seminar, where he argued the benefit of data-centric approaches and showed that, for some datasets, larger improvements in performance can be made with data-centric approach than model-centric approach.

Our main aim of data-centric approach is to increase labeling consistency and exclude or clean noisy data points. Table 1 and 2 shows the data cleaning operations and the examples of them. The answers from the BioASQ 9b training samples are mostly in the format of noun phrases. However, some data points have answers in sentences format. We manually modify such sentence answers to a noun phrase format. Additional normalization processes are made to correct misspelled word and to remove punctuation marks. Word correction and minor normalization processes are made to the questions. We do not modify grammatical structure of the questions and we count both British and American spelling as correct.

152 changes are made to the 9b training dataset including 22 question corrections and 24 dropped data points. For the evaluation of our models, we do not apply normalization steps.[1]

2.2 Model-Centric Approach; List Questions

Extractive question answering is a task of finding answer spans of a question in the given passage (i.e. the related documents). List-type questions are questions with multiple answers whereas factoid questions are questions that can be answered with one phrase. For list-type question, the number of answer for a given question is uninformed (i.e. not available as a metadata). Hence, deciding it remains a challenging and key operation to participating systems.

Previous works utilize factoid models with an additional steps to decide the number of answers for the questions. Factoid models are designed to predict a single answer span, and thus, they can not be trained on multi-label setup directly. In other words, for a training data point of list-type question, one answer span is trained for a training step and the other answers are acting as a noise since they are considered as non-answer tokens. We call this setting as "start-end span prediction", which is commonly used in biomedical extractive QA [7,22,25].

Following the approach of Yoon et al. [24], our systems for list-type questions are based on the sequence tagging approach. Specifically, a question and its corresponding passage are concatenated to construct a sequence, which is a training data point. Our systems adopt either BIO or IO scheme to annotate answer spans. For each tokens in the passage is tagged as B, I, or O tag which stands for Beginning, Inside, Outside, respectively.

Sequence tagging approach has two significant benefits over the previous models. First, the model can be trained on multiple answers simultaneously

---

[1] Resources for our data cleaning operations (our annotations) are available at https:// github.com/dmis-lab/bioasq9b-dmis.

which is more natural model design for the multi-answer task. As all available training labels (i.e. answers) are used for the training, rather than acting as training noise, the model learn the maximum of the dataset. Second, the approach is an end-to-end model that finds all the answers in the passage, whereas the previous models require complex post-processing steps to decide the number of answers. Models for the previous BioASQ challenges [7,25] decided the number of answers by threshold-based answer decision process, where the threshold value is a hyperparameter that needs to be tuned. Additionally, rule-based answer number detection are adopted under the assumption that if the numeric value exists in the question, the value is highly likely to be the required number of answers (ex. Question: List *3* apoE mimetics). In contrast, our approach does not require additional hyperparameter searching nor have to rely on the weak assumption.

2.3 Yes/No Questions

Following the systems of our participation for BioASQ 8b [7], we use BioBERT [13] with an additional pre-training step on MNLI dataset [23]. For BioASQ 9b task, our systems are based on BioBERT$_{LARGE}$, which has more model parameters than BioBERT$_{BASE}$ model from the previous participation. For yes/no-type questions, we adjust the ratio of *yes* questions and *no* questions in the training set to 1:1. The original training dataset is heavily skewed towards question with *yes* answers. As shown in Table 3, original 8b training set consists of 10,284 *yes* and 1,691 *no* question and passage pair samples. After our pre-processing steps, approximately 85.8% of answers for the yes/no-type questions are *yes*. Our systems are trained on the down-sampled training dataset to alleviate the class imbalance problem.

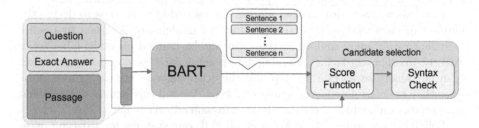

Fig. 1. Overview of our systems for ideal answers. Question, passage and the exact answer form an input sequence. *Candidate selection* module scores the candidate sentences and select the best candidate as an ideal answer for a given question.

2.4 Ideal Answers

Abstractive summarization models have been suffered from factual inconsistency, occasional distortion or fabrication of facts [6,12,28]. To mitigate the problem,

Table 3. Proportion of question samples having yes or no as its answers. Samples are pre-processed from original questions and consist of question and passage pairs.

	Yes	No
Original Training 8b	10284 (85.8%)	1691 (14.1%)
Down-sampled Training 8b	1691 (50.0%)	1691 (50.0%)

we utilized the exact answers as one of the inputs and the candidate selection stage.

Figure 1 shows an overview of our approach for the ideal answers. Our model utilize the predicted exact answer as a input for generating a ideal answer. In detail, we generate all the combinations of triples of an exact answer (A) and all the passages ($P_1, \ldots, P_n$) available for a question (Q). Then using the triples, we generate a candidate sentences ($C_1, \ldots, C_n$) for a question. We then select one ideal answer from the candidates using candidate selection process.

Candidate Selection. Our model is designed to generate one answer candidate for a input sequences. In the original dataset, multiple articles/snippets are provided as corresponding passages for a question. Hence multiple answer candidates are generated for a question. We need a candidate selection process to submit one answer, and the quality of this process can largely impact the performance of the model. Our candidate selection process is composed of a scoring function and a syntax checker.

Score Function. We have tried to score the generated ideal answer candidates ($C_1, \ldots, C_n$) by checking the presence of (candidate) exact answer(s). In order to check the presence, we employ BERN [9], a BioBERT based named entity recognition (NER) and linking system, to detect all the entities in the ideal answer candidates. For each candidate sentences for factoid and list questions, we calculate the F1-score using the tentative exact answer(s) and recognize entities from the candidate sentences. If one of the tentative answer(s) exists in the generated sentences, we add it to the set of recognized entities even if it is not detected by BERN.

Syntax Checker. We use **language-check** python package[2], an automated syntax checker, to correct or filter out grammatically wrong candidate sentences. From the list of ideal answer candidates, we select one with the highest score and check it with the syntax checker. If grammar errors are detected, we try to correct it with the checker. However, if the checker finds impossible to fix it, we then skip the candidate and move to the second highest candidate.

---

[2] https://github.com/myint/language-check.

3   Results

In this section, we present our experimental results on proposed strategies. In the first part of this section (Sect. 3.1), we report performance for exact answers. In the second part of this section (Sect. 3.2), we provide official results of our proposed ideal answer system in Table 8, which are evaluated by human experts.

3.1   Exact Answers

In this subsection, we describe comparative results between existing approach and our proposed approach on Factoid-type, List-type, and Yes/no-type questions in Table 4, Table 5, and Table 6, respectively.

Table 4 presents our experimental results on the factoid dataset. We report the statistics of 10 independent experiments with different random seeds, to minimise the performance variation caused by the random initialization of the model parameters and the ordering of the training data points as suggested by Dror et al. [5], Owing to our data-centric approach (noisy data points exclusion and increasing label consistency), our model performance shows a 0.79% score improvement on the mean reciprocal rank (MRR) score with statistical significance (p-value < 0.05; equal-variance t-test), which is an official evaluation metric for the factoid-type question.

Table 4. Performance of our factoid-type model on original training dataset and cleaned dataset of BioASQ 8B. The results are statistics on 10 independent experiments with different random seeds. s (SD) denotes standard deviation. The performance is based on the macro average of 5 batches. We only describe results on full-abstract setting.

	Original 8b		Cleaned 8b		p-value
	Mean	s (SD)	**Mean**	s (SD)	
Strict Accuracy (SAcc)	**0.3980**	0.0096	**0.4026**	0.0106	*0.3430*
Lenient Accuracy (LAcc)	**0.5781**	0.0060	**0.5934**	0.0079	*0.0002*
Mean Reciprocal Rank (MRR)	**0.4686**	0.0058	**0.4765**	0.0064	*0.0133*

Table 5 shows the experimental results on the list dataset for sequence tagging model and the baseline model, namely Start-End model [7] (reported results using snippet datasets). Performance of Start-End model is evaluated without hyperparameter tuning (include threshold searching) and used values from GitHub repository[3]. Our model outperforms the baseline model, which we used for the last year, with large gap.

In the challenge, we also submitted ensemble system, which consists of sequence tagging model and start-end model.

---

[3] https://github.com/dmis-lab/bioasq8b.

Table 5. Performance of list-type question models on different settings. The results are statistics on 10 independent experiments with different random seeds. The scores in the table are reported based on the macro average of 5 batches in BioASQ 8B test sets. F1 score is the official evaluation metric for list-type questions. Statistics with asterisks (*) are from our previous work [24]. Performance of Start-End prediction is evaluated without hyperparameter tuning (include threshold searching) and used values from GitHub repository [7].

	Precison		Recall		F1-score	
	Mean	*s* (SD)	**Mean**	*s* (SD)	**Mean**	*s* (SD)
Start-End prediction*	**0.4581**	(0.0071)	**0.3335**	(0.0049)	**0.3428**	(0.0054)
Sequence Tagging (IO)	**0.3790**	(0.0096)	**0.5900**	(0.0086)	**0.4258**	(0.0052)
Sequence Tagging (BIO)*	**0.3888**	(0.0105)	**0.5936**	(0.0126)	**0.4355**	(0.0083)

Table 6. Performance of Yes/No-type question model on different sampling strategy. The performance is based on the macro average of 5 batches in BioASQ 8B test sets. We only describe results on snippet setting. Macro F1 score is the official evaluation metric for Yes/No type questions.

	Acc	Macro F1	Yes F1	No F1
Original Training 8b	0.82	0.80	0.87	0.73
Down-sampled Training 8b	**0.94**	**0.93**	**0.95**	**0.92**

For yes/no-type questions, we introduce the down-sampling method which we balance the number of yes and no questions from the training set by sampling 1,691 yes questions out of 10284 yes questions (i.e. sampled 16% of yes questions). We have empirically shown that a down-sampled dataset alleviates the underlying class imbalance issue of yes/no-type questions. The results of yes/no model trained on down-sampled and original training data are described in Table 6. For models trained on down-sampled data, macro F1 score is improved from 0.80 to 0.93 (0.13 improvement). We have found that the down sampling method achieves better model performance for imbalanced yes/no-type questions. We conduct the experiments following the same hyperparameter settings, including learning rate and the number of learning steps.

Finally, Table 7 shows the result of our participation and the best performing system in the challenge. If our system scored the highest performance for a given batch, we marked it using bold font. The scores are obtained from the leaderboard[4].

3.2 Ideal Answers

The official evaluation metric for the ideal models is *human evaluation score*, which is scored by the human experts on the following four criteria: recall,

---

[4] Last checked on 2022 May.

Table 7. BioASQ 9B results for our systems. The scores are obtained from the leaderboard. Score in **bold font** denotes that our system achieved the highest score from the leaderboard.

Test Batch Number	Factoid (MRR)		List (F1)		Yes/No (MaF1)	
	Ours	Top-1	Ours	Top-1	Ours	Top-1
9B Batch 1	38.79	46.32	**53.39**		**92.58**	
9B Batch 2	52.94	55.39	46.44	48.92	88.54	94.54
9B Batch 3	42.34	61.49	54.21	58.87	90.23	95.32
9B Batch 4	57.26	69.29	**70.61**		**94.80**	
9B Batch 5	49.07	60.19	46.37	53.06	75.64	80.81

precision, repetition, and readability. We provide the official evaluation result (Table 8)[5] and the qualitative analysis of the model (Table 9) in the following paragraphs.

Table 8 shows the official evaluation scores of top-performing systems and our submissions. The Numbers in the table are human evaluation scores and a higher score means a better answer quality (Maximum score of 5). We compared the systems of University of California at San Diego (UCSD) [17], Macquarie University (MQ) [15], National Central University (NCU) [27], and our system.

UCSD and our systems generates summaries using abstractive summarization methods while MQ and NCU systems utilized extractive summarization methods. Abstractive summarization systems including our system and UCSD shows better quality in terms of Readability and Repetition criteria as their training objectives are more weighted in reproducing natural sentences. On the contrary, abstractive summarization models shows relatively sub-optimal performances in recall due to aforementioned factual inconsistency problem.

Table 9 shows examples of system outputs and the utility of the tentative exact answers in generating ideal answers. As we denoted in Sect. 3.2, tentative exact answers are used as one of the inputs to generate ideal answer candidates and to select best candidates for the questions. By utilizing the exact answer as one of the input to the ideal answer generation model, the model is imposed to consider the given exact answer in the sentence generation step and potentially include the exact answer in the generated sentence. The examples in the Table 9 show that the generated ideal answer sentences successfully include the given exact answers (potential exact answers). We did not exclude the snippets that do not contain the exact answer since our model is expected to include the given exact answers in the generated sentences even if they dose not exist in the passage.

---

[5] The official result (human evaluation) is on: http://participants-area.bioasq.org/results/9b/phaseB/.

Table 8. The official results for ideal answers on BioASQ 9B challenge. The best performing systems (based on the leaderboard) of each team are reported in this table. Numbers are human evaluation scores (maximum score of 5) and a higher score means a better quality answer.

		Batch 1	Batch 2	Batch 3	Batch 4	Batch 5	Average
Readability	**Ours**	**4.41**	**4.52**	**4.48**	4.41	4.49	**4.462**
	UCSD [17]	4.39	4.41	4.26	4.39	4.37	4.364
	MQ [15]	4.13	4.09	4.02	4.10	4.02	4.072
	NCU [27]	-	-	-	**4.53**	4.53	-
Recall	**Ours**	3.58	4.05	3.85	3.99	3.94	3.882
	UCSD [17]	4.12	4.43	4.33	4.18	4.18	4.248
	MQ [15]	**4.52**	**4.54**	**4.53**	**4.47**	**4.68**	**4.548**
	NCU [27]	-	-	-	4.18	4.38	-
Precision	**Ours**	3.65	3.95	3.96	3.92	3.87	3.870
	UCSD [17]	**4.28**	**4.41**	**4.31**	4.30	4.28	**4.316**
	MQ [15]	4.07	4.15	4.08	4.06	4.05	4.082
	NCU [27]	-	-	-	**4.38**	**4.41**	-
Repetition	**Ours**	4.71	**4.79**	4.75	4.75	4.69	4.738
	UCSD [17]	**4.79**	4.78	**4.84**	4.78	4.79	**4.796**
	MQ [15]	4.15	4.25	4.15	4.17	4.12	4.168
	NCU [27]	-	-	-	**4.89**	**4.88**	-

Table 9. Qualitative analysis of ideal answers on BioASQ 9B test sets. Tentative exact answers, which are the outputs of our exact models, are denoted as *Exact Answer* in the table.

Type	Question - Exact Answer - Ideal Answer
Factoid	Question: Which disease is monitored in the BIOCURA cohort? Exact Answer: **rheumatoid arthritis** Ideal Answer: Baseline disease activity in patients with **rheumatoid arthritis** is monitored in the BIOCURA cohort.
List	Question: What are the years of the initiation and completion of the Human Genome project? Exact Answer: **1990, 2003** Ideal Answer: The Human Genome Project (HGP) was initiated in **1990**, and the completion of the genome project was in **2003**.
Yes/No	Question: Is Tocilizumab (Actemra) used to block/antagonize the IL-6 receptor? Exact Answer: **yes** Ideal Answer: **Yes**, Tocilizumab (Actemra) is a humanized monoclonal antibody that binds to and inhibits the IL-6 receptor. It is used for treatment of type 1 diabetes.

4   Conclusion

In this paper, we propose data-centric and model-centric approaches for sub-types of biomedical question answering tasks, We showed the applications of our strategies in the BioASQ 8b dataset and our participating systems in the 9th BioASQ challenges. Our data-centric approach for factoid-type questions is to increased label consistency by fixing misplaced labels. We found out that the strategy is effective for scarce resources like BioQA datasets since it improves the performance of our model with minimal human effort. For list-type questions, we apply the sequence tagging model and achieve better performance with minimal intervention of human (i.e. heuristics and complex pre- and post-processing). Furthermore, for yes/no-type questions in the BioASQ 9b, our models have shown the best performance by utilizing down-sampling. We conclude that it is beneficial for BioQA models to use data-centric and/or model-centric approaches in consideration of the features of questions and answers.

Acknowledgements. We express gratitude towards Dr. Jihye Kim and Dr. Sungjoon Park from Korea University for their invaluable insight into our systems' output. This research is supported by National Research Foundation of Korea (NRF-2020R1A2C3010638) and a grant of the Korea Health Technology R&D Project through the Korea Health Industry Development Institute (KHIDI), funded by the Ministry of Health & Welfare, Republic of Korea (grant number: HR20C0021).

Author Note. This work is submitted to the 2022 CLEF - *Best of 2021 Labs* track. Our work originates from our challenge participation in the 9th BioASQ (2021 CLEF Labs), presented under the title *KU-DMIS at BioASQ 9: Data-centric and model-centric approaches for biomedical question answering (Yoon et al. 2021 [26])*.

References

1. Medline PubMed Production Statistics. https://www.nlm.nih.gov/bsd/medline_pubmed_production_stats.html. Accessed 19 June 2022
2. Alsentzer, E., et al.: Publicly available clinical BERT embeddings. In: Proceedings of the 2nd Clinical Natural Language Processing Workshop, pp. 72–78. Association for Computational Linguistics, Minneapolis, June 2019. https://doi.org/10.18653/v1/W19-1909, https://www.aclweb.org/anthology/W19-1909
3. Beltagy, I., Lo, K., Cohan, A.: SciBERT: a pretrained language model for scientific text. In: Proceedings of the 2019 Conference on Empirical Methods in Natural Language Processing and the 9th International Joint Conference on Natural Language Processing (EMNLP-IJCNLP), pp. 3615–3620 (2019)
4. Devlin, J., Chang, M.W., Lee, K., Toutanova, K.: BERT: pre-training of deep bidirectional transformers for language understanding. In: Proceedings of the 2019 Conference of the North American Chapter of the Association for Computational Linguistics: Human Language Technologies, Volume 1 (Long and Short Papers), pp. 4171–4186. Association for Computational Linguistics, Minneapolis, June 2019. https://doi.org/10.18653/v1/N19-1423, https://aclanthology.org/N19-1423

5. Dror, R., Peled-Cohen, L., Shlomov, S., Reichart, R.: Statistical significance testing for natural language processing. Synthesis Lect. Hum. Lang. Technol. **13**(2), 1–116 (2020)
6. Falke, T., Ribeiro, L.F., Utama, P.A., Dagan, I., Gurevych, I.: Ranking generated summaries by correctness: an interesting but challenging application for natural language inference. In: Proceedings of the 57th Annual Meeting of the Association for Computational Linguistics, pp. 2214–2220 (2019)
7. Jeong, M., et al.: Transferability of natural language inference to biomedical question answering. arXiv preprint arXiv:2007.00217 (2020)
8. Jin, Q., Dhingra, B., Cohen, W.W., Lu, X.: Probing biomedical embeddings from language models. arXiv preprint (2019)
9. Kim, D., et al.: A neural named entity recognition and multi-type normalization tool for biomedical text mining. IEEE Access **7**, 73729–73740 (2019). https://doi.org/10.1109/ACCESS.2019.2920708
10. Kim, N., et al.: Probing what different NLP tasks teach machines about function word comprehension. In: Proceedings of the Eighth Joint Conference on Lexical and Computational Semantics (*SEM 2019), pp. 235–249. Association for Computational Linguistics, Minneapolis, June 2019. https://doi.org/10.18653/v1/S19-1026, https://www.aclweb.org/anthology/S19-1026
11. Krithara, A., Nentidis, A., Paliouras, G., Krallinger, M., Miranda, A.: BioASQ at CLEF2021: large-scale biomedical semantic indexing and question answering. In: Hiemstra, D., Moens, M.-F., Mothe, J., Perego, R., Potthast, M., Sebastiani, F. (eds.) ECIR 2021. LNCS, vol. 12657, pp. 624–630. Springer, Cham (2021). https://doi.org/10.1007/978-3-030-72240-1_73
12. Kryściński, W., McCann, B., Xiong, C., Socher, R.: Evaluating the factual consistency of abstractive text summarization. arXiv preprint arXiv:1910.12840 (2019)
13. Lee, J., Yoon, W., Kim, S., Kim, D., Kim, S., So, C.H., Kang, J.: BioBERT: a pre-trained biomedical language representation model for biomedical text mining. Bioinformatics **36**(4), 1234–1240 (2020)
14. Lewis, M., et al.: BART: denoising sequence-to-sequence pre-training for natural language generation, translation, and comprehension (2019)
15. Mollá, D., Khanna, U., Galat, D., Nguyen, V., Rybinski, M.: Query-focused extractive summarisation for finding ideal answers to biomedical and COVID-19 questions. arXiv preprint arXiv:2108.12189 (2021)
16. Ng, A.Y.: A Chat with Andrew on MLOps: from model-centric to data-centric AI (2021). https://www.youtube.com/06-AZXmwHjo
17. Ozyurt, I.B.: End-to-end biomedical question answering via bio-answerfinder and discriminative language representation models. In: CLEF (Working Notes) (2021)
18. Peng, Y., Yan, S., Lu, Z.: Transfer learning in biomedical natural language processing: an evaluation of BERT and ELMo on ten benchmarking datasets. arXiv preprint (2019)
19. Peters, M.E., et al.: Deep contesxtualized word representations (2018)
20. Phang, J., Févry, T., Bowman, S.R.: Sentence encoders on STILTs: supplementary training on intermediate labeled-data tasks (2019)
21. Tsatsaronis, G., et al.: An overview of the BIOASQ large-scale biomedical semantic indexing and question answering competition. BMC Bioinform. **16**(1), 1–28 (2015)
22. Wiese, G., Weissenborn, D., Neves, M.: Neural domain adaptation for biomedical question answering. In: Proceedings of the 21st Conference on Computational Natural Language Learning (CoNLL 2017), pp. 281–289. Association for Computational Linguistics, Vancouver, August 2017. https://doi.org/10.18653/v1/K17-1029, https://www.aclweb.org/anthology/K17-1029

23. Williams, A., Nangia, N., Bowman, S.: A broad-coverage challenge corpus for sentence understanding through inference. In: Proceedings of the 2018 Conference of the North American Chapter of the Association for Computational Linguistics: Human Language Technologies, Volume 1 (Long Papers), pp. 1112–1122. Association for Computational Linguistics, New Orleans, June 2018. https://doi.org/10.18653/v1/N18-1101, https://www.aclweb.org/anthology/N18-1101

24. Yoon, W., Jackson, R., Lagerberg, A., Kang, J.: Sequence tagging for biomedical extractive question answering. Bioinformatics (2022). https://doi.org/10.1093/bioinformatics/btac397

25. Yoon, W., Lee, J., Kim, D., Jeong, M., Kang, J.: Pre-trained language model for biomedical question answering. In: Cellier, P., Driessens, K. (eds.) ECML PKDD 2019. CCIS, vol. 1168, pp. 727–740. Springer, Cham (2020). https://doi.org/10.1007/978-3-030-43887-6_64

26. Yoon, W., et al.: KU-DMIS at BioASQ 9: data-centric and model-centric approaches for biomedical question answering. In: CLEF (Working Notes), pp. 351–359 (2021)

27. Zhang, Y., Han, J.C., Tsai, R.T.H.: NCU-IISR/AS-GIS: results of various pre-trained biomedical language models and linear regression model in BioASQ task 9b phase B. In: CEUR Workshop Proceedings (2021)

28. Zhu, C., et al.: Enhancing factual consistency of abstractive summarization. arXiv preprint arXiv:2003.08612 (2020)

Did I Miss Anything? A Study on Ranking Fusion and Manual Query Rewriting in Consumer Health Search

Giorgio Maria Di Nunzio[1,2](✉) [ID] and Federica Vezzani[3] [ID]

[1] Department of Information Engineering, University of Padua, Padua, Italy
`giorgiomaria.dinunzio@unipd.it`
[2] Department of Mathematics, University of Padua, Padua, Italy
[3] Department of Linguistic and Literary Studies, University of Padua, Padua, Italy
`federica.vezzani@unipd.it`

Abstract. In this paper, we describe the methodology and experimental analysis of a twofold strategy for the retrieval of medical relevant information: a ranking fusion and a query reformulation approach. In particular, the query reformulation approach is based on the idea that a query is composed of two parts: the primary term and the secondary term of the query, and that these two parts can be substituted with alternative terms to create a reformulation of the original query. The goal of our work is to evaluate the performances of a search engine over 1) manual query variants; 2) different retrieval functions; 3) w/out pseudo-relevance feedback; 4) reciprocal ranking fusion. We describe the experiments based on the CLEF eHealth 2021 Consumer Health Search Task dataset. The results show that 1) a ranking fusion approach of the baseline models improves MAP significantly; 2) manual query variants open new questions about possible an unintentional bias in the pool of documents that were selected for relevance assessment.

1 Introduction

According to a number of studies and surveys [8], searching for medical information on the Web has grown exponentially in the last ten years, and the vast majority of people seeks health information using only the Internet alone [14]. Given the current situation, understanding and evaluating how consumers search for health information is an important research activity with different facets. Query formation, for example, is a major aspect of consumer health search; in fact, if queries do not reflect users' specific information needs, they will lead to results that do not address the required information [17]. A key problem when searching the Web for health information is that this can be too technical, unreliable, generally misleading. In this sense, people's health literacy has been proven to play a crucial role in the context of health information seeking [5,11].

A. Barrón-Cedeño et al. (Eds.): CLEF 2022, LNCS 13390, pp. 217–229, 2022.
https://doi.org/10.1007/978-3-031-13643-6_17

CLEF eHealth[1] is an evaluation challenge in the medical and biomedical domain in the context of the Conference and Labs of the Evaluation Forum (CLEF) initiative[2]. Since 2012, the goal of the CLEF eHealth evaluation challenge has been to provide researchers with datasets, evaluation frameworks, and events in order to evaluate the performance of search engines across different medical IR tasks. In the CLEF eHealth 2021 edition [16], the "Consumer Health Search" [7] task focused on the analysis of the performance of search engines that support the needs of health consumers who are confronted with a health issue. The three proposed subtasks were: Ad-hoc IR, Weakly Supervised IR, and Document credibility prediction. In this, paper we will mainly focus on the first subtask. Ad-hoc IR. The purpose of this subtask is to evaluate IR systems abilities to provide users with relevant, understandable and credible documents. Similarly to previous years, this subtask revolves around realistic use cases.

1.1 Research Proposal

In this paper, we describe the methodology and experimental analysis of a twofold strategy for the retrieval of medical relevant information:

- An evaluation of a ranking fusion approach [2,6] on different document retrieval strategies, with or without pseudo-relevance feedback [13].
- A study of a manual query variation approach [4,9];

The reason for a manual reformulation approach is that we want to test our research hypothesis before implementing the automatic term extraction approach. In particular, the idea is to isolate and reformulate two parts of a query: the primary term of the query (mainly the pathology), and the secondary term of the query (the desired information about the pathology). In addition, by means of this query reformulation and a ranking fusion of different retrieval strategies we expect to get most of the relevant information that can be subsequently filtered according to the health literacy of the user and/or the quality of the retrieved document. The remainder of the paper will first introduce the methodology for the query rewriting approach and then a summary of the experimental settings that we used in order to fuse the different ranking.

The paper is organized as follows: Sect. 2 defines the approach we propose to create query variants consistently based on a set of simple rules; in Sect. 3, we describe the experimental settings for the analysis of the results that are described in Sect. 4. We conclude with Sect. 5 and we give our future directions.

2 Methodology

The main idea of the proposed manual query rewriting approach is based on the following hypothesis: when a user writes a health-related query, there are

---

[1] https://clefehealth.imag.fr.

[2] http://www.clef-initiative.eu.

Table 1. Examples of query variation. The primary term is highlighted in bold while the secondary term in italics. The alternative terms are underlined.

type	query
original	*Reading problems* in **MS**
variant 1	Reading problems in multiple sclerosis
variant 2	Dyslexia in MS
variant 3	Dyslexia in multiple sclerosis

two parts: 1) the main object (often a pathology, such as diabetes or sclerosis), and 2) the specific aspect we are searching for (for example, a particular diet for diabetic people or the diagnosis of multiple sclerosis). We refer to the first part of the query, the main object, as the *primary term* while the second part, the specific aspect, as the *secondary term.*

Once the primary and secondary terms are identified, we proceed with the rewriting of three query variants in the following way:

1. we find an alternative term for the primary term and substitute it to the original query;
2. we find a alternative term for the secondary term and substitute it to the original query;
3. we substitute both alternative terms to the original query.

In Table 1, we show an example of query rewriting for the original query "Reading problems in MS" (one of the official queries in the CLEF 2021 CHS Task, see Sect. 3). The identified primary term is **MS** which refers to the pathology (the acronym of multiple sclerosis); the secondary term, *reading problems,* is the specific information related we want to search with this query. The table shows the terms, underlined, that were substituted in the three variants.

Given these premises, we need to choose a procedure to look for alternative terms. Since this task reflects a consumer health search task, we want to mimic a situation where laypeople try to find possible replacements of terms that they may not know or understand. For this reason, instead of suggesting the use of medical glossaries, we propose to use Wikipedia to find alternatives to the primary term. In particular, we used the Infobox available for a specific term to see alternative names. For example, for diabetes[3], we can use the 'diabetes mellitus' alternative; or for 'type 2 diabetes'[4], the alternatives 'Diabetes mellitus type 2', 'adult-onset diabetes', or 'noninsulin-dependent diabetes mellitus (NIDDM)'. When the primary term is an acronym (for example, MS), we expand it (see the example in Table 1. See Fig. 1 for an example of these two infoboxes. For the secondary term, we relied on the ability of the user (in our case the authors of the experiment themselves) to find a term that they know without looking at specialized dictionaries. The choice of adopting this double

---

[3] https://en.wikipedia.org/wiki/Diabetes.
[4] https://en.wikipedia.org/wiki/Type_2_diabetes.

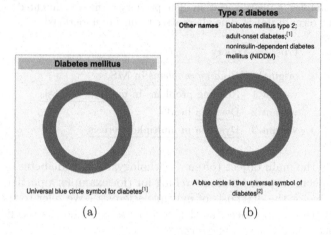

Fig. 1. Wikipedia Infobox for the terms diabetes (a) and type 2 diabetes (b).

reformulation methodology (wikipedia-based and personal knowledge-based) is purposely made in order to include two possible research cases by non-expert users in the medical field.

3   Experimental Setting

In this section, we describe the setting and the choices we made in this paper, especially the differences with the initial experiments presented at CLEF 2021 [10].

3.1   Dataset

The 2021 CLEF eHealth Consumer Health Search document collection consists of two separate crawls of documents: web documents acquired from the CommonCrawl and social media documents composed by Reddit and Twitter submissions [16]. The number of documents in this dataset is 4,896,997 for a total of 5,126,738 terms. For our experiments, we used the indexed version of the collection in Terrier[5] which was provided by the organizers.

3.2   Topics

The 55 topics prepared by the CLEF 2021 eHealth organizers aimed at being representative of medical information needs of laypeople in various scenarios. The set of topics was collected from two sources: 30 topics were based on a consultation between the organizers and laypeople who had experienced multiple sclerosis (MS) or diabetes; 25 topics were based on use cases from Reddit health

---

[5] http://terrier.org.

forums. The organizers also provided, for each topic, the so-called 'narrative', a text that gives some context to understand the information needs behind the short query.

In this paper, we only used the 50 test topics (we never used at any point in the experiments the additional 5 training topics available), and we never used the narratives at any point.

3.3   Baselines

The organizers of the task provided the results of six baselines built with a default Terrier configuration for the following models [1, Chapter 3]: BM25, Dirichlet Language Model (DLM), and TFIDF with or without pseudo relevance feedback using the divergence from randomness (DFR Bo1 model [25] on three documents, selecting ten terms). This resulted in the $3 \times 2 = 6$ baseline systems. The systems are implemented using Terrier version 5.4 [26].

3.4   Relevance Judgements

For each query, a pool of 250 documents was created based on the documents retrieved by the participants and the baselines provided by the organizers[6]. A total of 12,500 assessments were made on 11,357 documents: 7,400 Web documents, and 3,957 social media documents [16] (at this point in time, it is not clear the difference between the assessments and the number of documents provided in the overview of the Lab).

As we will discuss later in the paper, the question of how the pool of documents was created is something to take into consideration. Since both the participants to the Lab and the baselines used the same retrieval models (essentially BM25) with the original queries only, the set of documents pooled for the relevance assessments might be biased. We will discuss this matter in Sect. 4.

3.5   Experiments

For all the experiments, we used the Terrier indexes provided by the organizers of the task together with the PyTerrier[7] software. We used the default parameter settings for the same models: BM25, DLM, and TFIDF. In order to be consistent with the baselines, we used the same pseudo relevance feedback model to perform automatic query expansion, the DFR Bo1 with default parameters.

For the rank fusion approach, we kept the same setting of the preliminary experiments: the reciprocal ranking fusion (RRF) [3] approach with k = 60 to merge the produced rankings. In particular, we will analyze the experiment created with the fusion of the ranking lists of the three models with or without the query expansion.

All the data and source code, runs and query variants, will be made available on GitHub for reproducibility purposes[8].

---

[6] For a survey on test collection based evaluation see [15].
[7] https://pyterrier.readthedocs.io/en/latest/.
[8] https://github.com/gmdn.

3.6 Evaluation Measures

We use the three main measures presented by the organizers in the Ad-hoc retrieval subtask for the comparison of the runs: Mean Average Precision (MAP), preference-based BPref metric, normalized Discounted Cumulative Gain for the top 10 documents (nDCG@10) [1, Chapter 4].

3.7 Additional Notes

In order to perform a better comparison with the baselines provided by the organizers, as well as the pseudo-relevance feedback approach, we changed one of the models we used in the original experiments [10]: we substituted the Poisson model with Laplace after effect normalization (PL2) with DLM, and changed the PRF query expansion approach from RM3 to DFR Bo1.

4   Results

Before diving into the analysis of the results, we want to make some preliminary considerations about the values of the evaluation measures that we will present.

In the overview paper, the organizers provide some information about how the six baselines were created. In particular, they used Terrier, version 5.4, with the following command line:

```
terrier batchretrieve -t topics.txt -w [TF_IDF|DirichletLM|BM25] [-q|]
```

Since we are using the Pyterrier implementation, version 0.8.1, with Terrier, version 5.6, it is important to build the same baselines to have some reference points about the values that we will analyze. In Table 2, we compare the values obtained by the original baselines with Terrier and the same baselines running PyTerrier. The difference between the two values is shown in the last column. In almost all cases, the original baselines appear to be consistently greater than what we can get with PyTerrier with the same parameters. The only exception is the BM25 with query expansion that shows an almost identical performance between the two settings. There may be many reasons why this difference exists, but this is not the main objective of the paper.

4.1   Reciprocal Ranking Fusion

The first set of results we want to compare is that concerning the ranking fusion approach. In our previous experiments, we obtain the best results in the task in terms of MAP, one of the best (considering the issue of the offset in the values of the metrics) in terms of Bpref, and the best among the participants for nDCG@10. In this paper, in order to compare exactly the same models proposed by the organizers, we run the ranking fusion approach substituting the PL2 model with the TFIDF weighting schema and substitute the RM3 pseudo-relevance feedback formula for query expansion with DFR Bo1.

Table 2. Compare average performance baselines. The model with or without query expansion (qe) is compared across the three evaluation measures. In the last column, the difference with baselines obtained by the organizers with Terrier (T) and those obtained in this paper with PyTerrier (P) is shown.

model	measure	Terrier [7]	PyTerrier	$\Delta(T - P)$
BM25	MAP	.364	.360	.004
TFIDF	MAP	.366	.357	.009
DLM	MAP	.369	.348	.021
BM25 qe	MAP	.390	.390	.000
TFIDF qe	MAP	.397	.390	.007
DLM qe	MAP	.242	.229	.013
BM25	Bpref	.471	.465	.006
TFIDF	Bpref	.474	.464	.010
DLM	Bpref	.472	.451	.021
BM25 qe	Bpref	.499	.500	-.001
TFIDF qe	Bpref	.511	.506	.005
DLM qe	Bpref	.369	.359	.010
BM25	nDCG@10	.654	.632	.022
TFIDF	nDCG@10	.646	.630	.016
DLM	nDCG@10	.595	.579	.016
BM25 qe	nDCG@10	.635	.633	.002
TFIDF qe	nDCG@10	.654	.642	.012
DLM qe	nDCG@10	.536	.518	.018

In Table 3, we show the best baseline of the task (TFIDF with query expansion, TFIDF qe) which was also one of the best runs of the task overall across different measures. We compare the performance of this run with our two runs: one obtained with the reciprocal ranking fusion that combines the three baseline models without query expansion (original rrf), and the other one obtained by combining all the models with query expansion (original qe rrf).

The ranking fusion approach confirms to be an effective strategy to achieve a better MAP maintaining a sufficiently good Bpref and nDCG. Very similar results were obtained in our original experiments with a different PRF approach for query expansion and a different retrieval model out of three [10].

We performed a non-parametric statistical test to understand whether these differences in the averaged performances are significant or not. The results of the test are shown in Fig. 2 where the value of the measures of the 50 topics are plotted with boxplots and violin plots realized by means of the ggstatsplot package [12][9]. The dotted lines between the two distributions connect the values

---

[9] The violin plot is the shape around the boxplot that shows the distribution of points by means of a density function.

Table 3. Comparison of the average performance of the best run of CLEF 2021 eHealth adHoc Retrieval [16] (TFIDF qe) with the two runs that use reciprocal ranking fusion without query expansion (original rrf) or with query expansion (original qe rrf).

run	measure	PyTerrier
TFIDF qe	MAP	.390
original rrf	MAP	.410
original qe rrf	MAP	.444
TFIDF qe	Bpref	.506
original rrf	Bpref	.474
original qe rrf	Bpref	.504
TFIDF qe	nDCG@10	.642
original rrf	nDCG@10	.612
original qe rrf	nDCG@10	.615

Table 4. Results of the manual query rewriting (without query expansion). Variant 1, 2, or 3, are indicated with v1, v2, v3 respectively.

model	MAP	Bpref	nDCG@10
BM25 v1	.093	.269	.245
BM25 v2	.083	.282	.157
BM25 v3	.032	.153	.097
TFIDF v1	.097	.277	.251
TFIDF v2	.088	.285	.169
TFIDF v3	.033	.159	.099
DLM v1	.128	.286	.276
DLM v2	.152	.353	.334
DLM v3	.053	.195	.164

of the performance measure that refer to the same topic. The only difference statistically significant is between MAP performances (Wilcoxon signed rank paired test, p-value = $3.17e^{-5}$), the other differences are not significant from a statistical point of view (Bpref p-value = .51, nDCG@10 p-value 0.23).

For pure speculation, we tried to fuse the results of all the runs (with and without query expansion) obtaining an even better overall run (MAP = .466, Bpref = .509, nDCG@10 = .629). However, this cannot be considered the best performance as we are currently 'peeking' the test data and the comparison with the other runs would not be fair.

4.2   Manual Query Rewriting

The preliminary results of the three manual query variants are shown in Table 4. As it can be immediately seen, the performances are much worse than the original

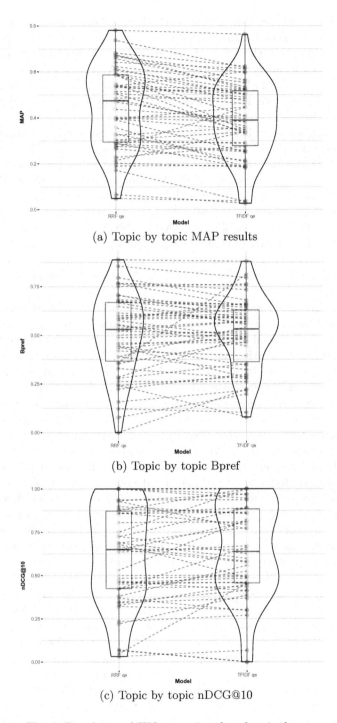

(a) Topic by topic MAP results

(b) Topic by topic Bpref

(c) Topic by topic nDCG@10

Fig. 2. Boxplots and Wilcoxon signed rank paired tests

query. At that point, we immediately carried out a failure analysis before doing any query expansion or ranking fusion that would not get any good result.

After checking that the low performances were not due to a bug in the software, we started to analyze the performances on Average Precision (AP) query by query between the TFIDF baseline (without query expansion) and our simplest variation, the second one, where we keep the primary term and substitute the secondary term.

The results, not included here for space reasons, show that the baseline consistently outperforms the variant on all the queries but one. This query, with identifier 79 in the dataset, is originally written as "Can I pass multiple sclerosis to other family members" and the baseline achieves an AP of .275. When we modify it with "Can I transmit multiple sclerosis to other family members", maintaining the primary term 'multiple sclerosis' and changing the secondary term 'pass' to 'transmit', we obtain an AP of .308. On the other hand, the first variant, "Can I pass MS to other family members", where we substitute the full form with the acronym, achieves an AP for of only .033. This last situation may sound intuitive, since the acronym may obfuscate the main objective of the query.

By following this reasoning, we browsed for original queries containing an acronym and compare the performance with the variant containing the full form. For example, the original query with identifier 92 asks for "causes of fatigue in MS" which achieves an AP equal to .425. We would expect the variant with the full form "causes of fatigue in multiple sclerosis" to perform at least as good as the original version. However, the AP for variant v1 achieves only .110. This odd pattern of significant differences in performances recurs across different queries, such as "Can diabetes be cured", original query with AP .228, compared to "Can diabetes mellitus be cured" that obtains an AP of only .014.

This significant drop in the AP measure requires a deeper investigation. Since we are building the query manually by slightly changing one or two terms with carefully selected alternative terms, such a decrease in the number of retrieved relevant documents should not happen. We might expect some differences in the top ranked documents (maybe MS is truly a better term to interpret the information need related to multiple sclerosis, only a user study would probably give some answers), but not the kind of difference we are observing in the experiments. Consequently, we tried to analyze the distribution of relevance assessments for the different runs. In Table 5, we show for each query variation the number of documents that are in the pool (n) and the number of relevant documents (rel). In many cases, we have a very small number of documents—a few units or even zero in some cases—that are in the pool. On average (last row of the table), the original query obtains around on average 154 documents in the pool with 93 relevant documents. Query variant 1 and 2 obtain about half of the documents in the pool and about half relevant documents, variant 3 around one fourth for pooled documents and relevant documents. Our hypothesis is the use of the same retrieval by both the participants and the organizers without query variations introduced unintentionally some bias in the selection of documents that were pooled and judged.

Table 5. Number of documents that are in the pool (n) and number of relevant documents (rel) for the first ten queries of the CLEF 2021 eHealth Task 2 dataset. In bold, the runs that obtain a greater number of relevance assessments.

query	original		variant 1		variant 2		variant 3	
	n	rel	n	rel	n	rel	n	rel
1	**153**	110	76	57	119	99	66	54
8	**115**	43	72	25	70	27	32	5
22	**184**	127	7	7	37	24	4	4
35	147	35	**150**	38	4	2	5	2
45	**114**	63	17	13	11	9	4	4
51	**142**	65	97	88	135	52	96	86
52	**149**	56	75	27	30	18	15	9
53	**150**	32	81	19	45	13	13	5
54	162	77	89	39	**170**	78	41	28
55	187	83	132	43	**193**	86	0	0
57	**177**	160	47	47	105	100	29	29
58	**163**	159	98	96	127	123	21	21
59	81	47	38	12	**89**	50	39	14
62	**162**	97	60	27	87	41	33	19
63	**188**	185	137	134	20	20	21	21
68	**208**	46	167	30	189	33	4	2
72	65	1	**67**	13	42	1	42	9
77	**183**	3	110	1	53	3	0	0
79	**180**	57	45	24	172	54	41	28
81	66	11	31	7	**74**	11	7	1
83	159	75	**158**	81	77	35	41	26
85	**189**	113	157	98	74	53	21	14
86	**139**	108	121	96	56	35	1	1
92	**157**	66	125	48	17	12	75	33
93	**152**	107	24	12	132	103	14	9
94	**186**	167	59	55	159	144	6	5
95	**144**	26	136	29	1	0	60	10
96	**108**	19	89	6	2	0	0	0
97	**130**	111	92	77	8	8	8	8
98	**139**	137	101	101	45	45	97	97
101	**174**	54	70	12	128	45	27	16
102	**180**	172	78	74	34	33	61	59
105	**182**	116	121	50	217	120	101	41
107	**135**	20	104	23	34	7	99	22
108	**123**	103	41	38	41	38	41	38
109	**180**	180	145	145	174	174	140	140
112	**147**	124	14	11	6	5	3	3
113	**99**	82	27	10	27	10	27	10
114	**184**	119	116	92	32	27	40	33
116	**167**	145	115	87	45	42	71	52
117	**173**	78	116	43	154	72	108	36
118	171	127	81	50	**181**	133	82	53
119	186	87	**196**	77	146	61	94	20
121	**180**	150	43	41	**180**	150	43	41
123	154	145	124	113	**156**	144	118	107
126	**172**	72	134	62	153	66	99	47
127	**170**	101	37	9	56	27	0	0
130	**195**	195	55	55	163	162	61	61
131	**134**	91	46	40	94	76	0	0
132	**123**	81	102	55	119	50	51	3
mean	154	93	88	49	90	54	42	27
	n	rel	n	rel	n	rel	n	rel
query	original		variant 1		variant 2		variant 3	

5   Final Remarks and Future Work

In this paper, we have discussed in detail the experimental results of two approaches for the retrieval of health information in the context of the CLEF 2021 Consumer Health Search. Our main intent was to confirm the advantages of a ranking fusion approach based on strong baselines and to analyze a manual query reformulation approach that identifies two parts of a query. While we were able to confirm the positive effect of a rank fusion approach, the analysis of the results of the manual query rewriting approach showed that there may be some kind of bias in the pool of documents selected for relevance judgements. This bias may have been introduced inadvertently by the use of the same standard retrieval model by both the organizers of the track and the participants. In particular, we showed that a simple type of reformulation (for example, transforming an acronym in its full form) makes the evaluation very unstable and unrealistically low in terms of retrieval performances. Despite the impossibility to make any conclusion about the query reformulation approach, we are convinced that this work is significant to shed the light, once again, on a correct evaluation process and a critical approach to the comparison of the retrieval models.

There are still some open questions that we want to tackle in the future: our hypothesis about the bias in the pool can only be confirmed by an additional round of relevance assessments that includes the top k documents retrieved by our runs. We intentionally did not want to assess documents by ourselves, as this might have added additional bias in the analysis; however, we could not find in time some reviewers (at least three to have a reasonable inter-annotator agreement).

Acknowledgments. We thank the anonymous reviewers for their suggestions and their comments to improve the paper and give the directions for additional investigations.

References

1. Baeza-Yates, R., Ribeiro-Neto, B.A.: Modern Information Retrieval - the Concepts And Technology Behind Search, 2nd edn. Pearson Education Ltd., Harlow, England (2011)
2. Clipa, T., Nunzio, G.M.D.: A study on ranking fusion approaches for the retrieval of medical publications. Information **11**(2), 103 (2020)
3. Cormack, G. V., Clarke, C. L., Buettcher, S.: Reciprocal rank fusion outperforms condorcet and individual rank learning methods. In: Proceedings of the 32nd International ACM SIGIR Conference on Research and Development in Information Retrieval, SIGIR 2009, pp. 758–759, New York, NY, USA, 2009. Association for Computing Machinery
4. Di Nunzio, G. M., Marchesin, S., Vezzani, F.: A study on reciprocal ranking fusion in consumer health search. IMS unipd ad CLEF ehealth 2020 task 2. In: Cappellato, L., Eickhoff, C., Ferro, N., Névéol, A. (eds.), Working Notes of CLEF 2020-Conference and Labs of the Evaluation Forum, Thessaloniki, Greece, September 22–25, 2020, volume 2696 of CEUR Workshop Proceedings. CEUR-WS.org (2020)

5. Diviani, N., van den Putte, B., Giani, S., van Weert, J.C.: Low health literacy and evaluation of online health information: a systematic review of the literature. J. Med. Internet Res. **17**(5), e112 (2015)

6. Frank Hsu, D., Taksa, I.: Comparing rank and score combination methods for data fusion in information retrieval. Inf. Retrieval **8**(3), 449–480 (2005). https://doi.org/10.1007/s10791-005-6994-4

7. Goeuriot, L., et al.: Consumer health search at clef eHealth 2021. In: CLEF 2021 Evaluation Labs and Workshop: Online Working Notes, CEUR Workshop Proceedings, September 2021

8. Hochberg, I., Allon, R., Yom-Tov, E.: Assessment of the frequency of online searches for symptoms before diagnosis: analysis of archival data. J. Med. Internet Res. **22**(3), e15065–e15065 (2020)

9. Di Nunzio, G.M., Vezzani, F.: Using R markdown for replicable experiments in evidence based medicine. In: Bellot, P., et al. (eds.) CLEF 2018. LNCS, vol. 11018, pp. 28–39. Springer, Cham (2018). https://doi.org/10.1007/978-3-319-98932-7_3

10. Di Nunzio, G.M., Vezzani, F.: IMS-UNIPD @ CLEF ehealth task 2: reciprocal ranking fusion in CHS. In: Faggioli, G., Ferro, N., Joly, A., Maistro, M., Piroi, F. (eds.), Proceedings of the Working Notes of CLEF 2021-Conference and Labs of the Evaluation Forum, Bucharest, Romania, 21st - to - 24th September 2021, volume 2936 of CEUR Workshop Proceedings, pp. 775–779. CEUR-WS.org (2021)

11. Palotti, J., Zuccon, G., Hanbury, A.: Consumer health search on the web: study of web page understandability and its integration in ranking algorithms. J. Med. Internet Res. **21**(1), e10986 (2019)

12. Patil, I.: Visualizations with statistical details: the 'ggstatsplot' approach. J. Open Source Softw. **6**(61), 3167 (2021)

13. Ruthven, I., Lalmas, M.: A survey on the use of relevance feedback for information access systems. Knowl. Eng. Rev. **18**(2), 95–145 (2003)

14. Finney Rutten, L.J., Blake, K.D., Greenberg-Worisek, A.J., Allen, S.V., Moser, R.P., Hesse, B.W.: Online health information seeking among us adults: Measuring progress toward a healthy people 2020 objective. Public Health Rep. **134**(6), 617–625 (2019)

15. Sanderson, M.: Test collection based evaluation of information retrieval systems. Found. Trends Inf. Retr. **4**(4), 247–375 (2010)

16. Suominen, H., et al.: Overview of the CLEF eHealth evaluation lab 2021. In: Candan, K.S., et al. (eds.) CLEF 2021. LNCS, vol. 12880, pp. 308–323. Springer, Cham (2021). https://doi.org/10.1007/978-3-030-85251-1_21

17. Zeng, Q.T., Crowell, J., Plovnick, R.M., Kim, E., Ngo, L., Dibble, E.: Assisting consumer health information retrieval with query recommendations. J. Am. Med. Inform. Assoc.: JAMIA **13**(1), 80–90 (2006)

Overviews of 2022 Labs

Overview of eRisk 2022: Early Risk Prediction on the Internet

Javier Parapar[1]([✉]) [iD], Patricia Martín-Rodilla[1] [iD], David E. Losada[2] [iD],
and Fabio Crestani[3] [iD]

[1] Information Retrieval Lab, Centro de Investigación en Tecnoloxías da Información
e as Comunicacións (CITIC), Universidade da Coruña, A Coruña, Spain
{javierparapar,patricia.martin.rodilla}@udc.es
[2] Centro Singular de Investigación en Tecnoloxías Intelixentes (CiTIUS),
Universidade de Santiago de Compostela, Losada, Spain
david.losada@usc.es
[3] Faculty of Informatics, Università della Svizzera italiana (USI),
Lugano, Switzerland
fabio.crestani@usi.ch

Abstract. This paper gives an outline of eRisk 2022, the CLEF conference's sixth edition of this lab. Since the first edition, the main goal of our lab is to explore issues of evaluation methodology, effectiveness metrics and other processes related to early risk detection. Early alerting models may be used in a variety of situations, including those involving health and safety. This edition of eRisk had three tasks. The first task focused on early detecting signs of pathological gambling. The second challenge was to spot early signs of depression The third required participants to fill out automatically an eating disorders questionnaire (based on user writings on social media).

Keywords: Early risk · Pathological gambling · Depression · Eating disorder

1 Introduction

eRisk's primary goal is to research topics such as evaluation methodologies, metrics, and other factors relevant to developing research collections and identifying problems for early risk identification. Early detection technologies can be helpful in a wide range of fields, particularly those concerned with safety and health. When a person begins to exhibit symptoms of a mental illness, a sexual abuser begins interacting with an infant, or a suspected criminal begins publishing anti-social threats on the Internet, an automated system may emit early warnings.

While our evaluation methodologies (strategies for developing new research sets, innovative evaluation metrics, etc.) may be applied across multiple domains, eRisk has thus far focused on psychological issues (depression, self-harm, pathological gambling, and eating disorders). In 2017, we conducted an exploratory task on the early detection of depression [5,6]. The evaluation methods and test dataset described in [4] were the focus of this pilot task. In 2018, we continued detecting early signs of depression while also launching a new task of detecting early signs of anorexia [7,8]. In 2019, we ran the continuation of the challenge on early identification of symptoms of anorexia, a challenge on early detection of signs of self-harm, and a third task aimed at estimating a user's responses to a depression questionnaire focused on her social media interactions [9–11]. In 2020, we continued with the early detection of self-harm and the task on severity estimation of depression symptoms [12–14]. Finally, in the last edition in 2021, we presented two tasks on early detection (pathological gambling and self-harm), and one on the severity estimation of depression [17–19].

Over the years, we have had the opportunity to compare a wide range of solutions that employ various technologies and models (e.g. Natural Language Processing, Machine Learning, or Information Retrieval). We discovered that the interaction between psychological diseases and language use is complex and that the effectiveness of most contributing systems is low. For example, most participants had performance levels (e.g., in terms of F1) that were less than 70%. These numbers show that further research into early prediction tasks is required, and the solutions proposed thus far have much room for improvement.

In 2022, the lab had three campaign-style tasks [20]. The first task is the second edition of the pathological gambling domain. This task follows the same organisation as previous early detection challenges. The second task is also a continuation of the early detection of the depression challenge, whose last edition was in 2018. Finally, we provided a new task for the eating disorder severity estimation. Participants were required to analyse the user's posts and then estimate the user's answers to a standard eating disorder questionnaire. We describe these tasks in greater detail in the following sections of this overview article. We had 93 teams registered for the lab. We finally received results from 17 of them: 41 runs for Task 1, 62 runs for Task 2 and 12 for Task 3.

2 Task 1: Early Detection of Pathological Gambling

This is a continuation of Task 1 from 2021. The challenge was to develop new models for the early detection of pathological gambling risk. Pathological gambling (ICD-10-CM code F63.0) is also known as ludomania. It is commonly known as *gambling addiction* (an urge to gamble independently of its negative consequences). Adult gambling addiction had prevalence rates ranging from 0.1% to 6.0% in 2017, according to the World Health Organization [1]. The task entailed sequentially processing evidence and detecting early signs of pathological gambling, also known as compulsive gambling or disordered gambling, as soon as possible. The work is concerned mainly with analyzing Text Mining solutions and focuses on Social Media texts. Participating systems had to read and process the posts on Social Media in the sequence that users wrote them. As a result, systems getting good results from this task might be used to sequentially monitor user interactions in blogs, social networks, and other forms of online media.

Table 1. Task 1 (pathological gambling). Main statistics of the collection

	Train		Test	
	Gamblers	*Control*	*Gamblers*	*Control*
Num. subjects	164	2,184	81	1998
Num. submissions (posts & comments)	54,674	1,073,88	14,627	1,014,122
Avg num. of submissions per subject	333.37	491.70	180.58	507.56
Avg num. of days from first to last submission	≈560	≈662	≈489.7	≈664.9
Avg num. words per submission	30.64	20.08	30.4	22.2

The test collection for this task had the same format as the collection described in [4]. The source of data is also the same used for previous eRisks. It is a collection of writings (posts or comments) from a set of Social Media users. There are two categories of users, pathological gamblers and non-pathological gamblers, and, for each user, the collection contains a sequence of writings (in chronological order). We set up a server that iteratively gave user writings to the participating teams. More information about the server can be found at the lab website[1].

This was a train and a test task. For the training stage, the teams had access to training data where we released the whole history of writings for training users. We indicated which users had explicitly mentioned that they are pathological gamblers. The participants could therefore tune their systems with the training data. In 2022, the training data for Task 1 was composed of all 2021's Task 1 users.

The test stage consisted of participants connecting to our server and iteratively receiving user writings and sending responses. Each participant could stop and issue an alert at any point in the user chronology. After reading each user post, the teams had to choose between: i) alerting about the user (the system predicts the user will develop the risk) or ii) not alerting about the user. Participants had to make this choice for each user in the test split independently. We considered alerts as final (i.e. further decisions about this individual were ignored). In contrast, *no alerts* were considered non-final (i.e. the participants could later submit an alert about this user if they detected the appearance of signs of risk). We used the accuracy of the decisions and the number of user writings required to make the decisions to evaluate the systems (see below). To support the testing stage, we deployed a REST service. The server iteratively distributed user writings to each participant while waiting for their responses (no new user data was distributed to a specific participant until the service received a decision from that team). The service was open for submissions from January 17th, 2022, until April 22nd 2022.

In order to build the ground truth assessments, we followed existing approaches that optimize the use of assessors time [15,16]. These methods allow

---

[1] https://early.irlab.org/server.html.

to build test collections using simulated pooling strategies. Table 1 reports the main statistics of the test collection used for T1. Evaluation measures are discussed in the next sections.

2.1 Decision-Based Evaluation

This form of evaluation revolves around the (binary) decisions taken for each user by the participating systems. Besides standard classification measures (Precision, Recall and F1[2]), we computed $ERDE$, the early risk detection error used in previous editions of the lab. A full description of $ERDE$ can be found in [4]. Essentially, $ERDE$ is an error measure that introduces a penalty for late correct alerts (true positives). The penalty grows with the delay in emitting the alert, and the delay is measured here as the number of user posts that had to be processed before making the alert.

Since 2019, we complemented the evaluation report with additional decision-based metrics that try to capture additional aspects of the problem. These metrics try to overcome some limitations of $ERDE$, namely:

– the penalty associated to true positives goes quickly to 1. This is due to the functional form of the cost function (sigmoid).
– a perfect system, which detects the true positive case right after the first round of messages (first chunk), does not get error equal to 0.
– with a method based on releasing data in a chunk-based way (as it was done in 2017 and 2018) the contribution of each user to the performance evaluation has a large variance (different for users with few writings per chunk vs users with many writings per chunk).
– $ERDE$ is not interpretable.

Some research teams have analysed these issues and proposed alternative ways for evaluation. Trotzek and colleagues [22] proposed $ERDE_o^{\%}$. This is a variant of ERDE that does not depend on the number of user writings seen before the alert but, instead, it depends on the *percentage* of user writings seen before the alert. In this way, user's contributions to the evaluation are normalized (currently, all users weight the same). However, there is an important limitation of $ERDE_o^{\%}$. In real life applications, the overall number of user writings is not known in advance. Social Media users post contents online and screening tools have to make predictions with the evidence seen. In practice, you do not know when (and if) a user's thread of messages is exhausted. Thus, the performance metric should not depend on knowledge about the total number of user writings.

Another proposal of an alternative evaluation metric for early risk prediction was done by Sadeque and colleagues [21]. They proposed $F_{latency}$, which fits better with our purposes. This measure is described next.

Imagine a user $u \in U$ and an early risk detection system that iteratively analyzes u's writings (e.g. in chronological order, as they appear in Social Media)

---

[2] computed with respect to the positive class.

and, after analyzing k_u user writings ($k_u \geq 1$), takes a binary decision $d_u \in \{0, 1\}$, which represents the decision of the system about the user being a risk case. By $g_u \in \{0, 1\}$, we refer to the user's golden truth label. A key component of an early risk evaluation should be the delay on detecting true positives (we do not want systems to detect these cases too late). Therefore, a first and intuitive measure of delay can be defined as follows[3]:

$$\text{latency}_{TP} = \text{median}\{k_u : u \in U, d_u = g_u = 1\} \qquad (1)$$

This measure of latency is calculated over the true positives detected by the system and assesses the system's delay based on the median number of writings that the system had to process to detect such positive cases. This measure can be included in the experimental report together with standard measures such as Precision (P), Recall (R) and the F-measure (F):

$$P = \frac{|u \in U : d_u = g_u = 1|}{|u \in U : d_u = 1|} \qquad (2)$$

$$R = \frac{|u \in U : d_u = g_u = 1|}{|u \in U : g_u = 1|} \qquad (3)$$

$$F = \frac{2 \cdot P \cdot R}{P + R} \qquad (4)$$

Furthermore, Sadeque et al. proposed a measure, $F_{latency}$, which combines the effectiveness of the decision (estimated with the F measure) and the delay[4] in the decision. This is calculated by multiplying F by a penalty factor based on the median delay. More specifically, each individual (true positive) decision, taken after reading k_u writings, is assigned the following penalty:

$$penalty(k_u) = -1 + \frac{2}{1 + \exp^{-p \cdot (k_u - 1)}} \qquad (5)$$

where p is a parameter that determines how quickly the penalty should increase. In [21], p was set such that the penalty equals 0.5 at the median number of posts of a user[5]. Observe that a decision right after the first writing has no penalty (i.e. $penalty(1) = 0$). Figure 1 plots how the latency penalty increases with the number of observed writings.

The system's overall speed factor is computed as:

$$speed = (1 - \text{median}\{penalty(k_u) : u \in U, d_u = g_u = 1\}) \qquad (6)$$

---

[3] Observe that Sadeque et al. (see [21], pg 497) computed the latency for all users such that $g_u = 1$. We argue that latency should be computed only for the true positives. The false negatives ($g_u = 1$, $d_u = 0$) are not detected by the system and, therefore, they would not generate an alert.

[4] Again, we adopt Sadeque et al.'s proposal but we estimate latency only over the true positives.

[5] In the evaluation we set p to 0.0078, a setting obtained from the eRisk 2017 collection.

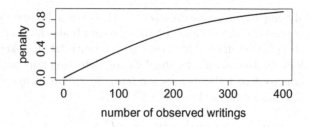

Fig. 1. Latency penalty increases with the number of observed writings (k_u)

where speed equals 1 for a system whose true positives are detected right at the first writing. A slow system, which detects true positives after hundreds of writings, will be assigned a speed score near 0.

Finally, the *latency-weighted* F score is simply:

$$F_{latency} = F \cdot speed \tag{7}$$

Since 2019 user's data were processed by the participants in a post by post basis (i.e. we avoided a chunk-based release of data). Under these conditions, the evaluation approach has the following properties:

- smooth grow of penalties;
- a perfect system gets $F_{latency} = 1$;
- for each user u the system can opt to stop at any point k_u and, therefore, now we do not have the effect of an imbalanced importance of users;
- $F_{latency}$ is more interpretable than $ERDE$.

2.2   Ranking-Based Evaluation

This section discusses an alternative form of evaluation, which was used to complement the evaluation described above. After each release of data (new user writing), the participants had to send back the following information (for each user in the collection): i) a decision for the user (alert/no alert), which was used to compute the decision-based metrics discussed above, and ii) a score that represents the user's level of risk (estimated from the evidence seen so far). We used these scores to build a ranking of users in decreasing estimated risk. For each participating system, we have one ranking at each point (i.e., ranking after one writing, ranking after two writings, etc.). This simulates a continuous re-ranking approach based on the evidence seen so far. In a real-life application, this ranking would be presented to an expert user who could take decisions (e.g. by inspecting the rankings).

Each ranking can be evaluated with standard IR metrics, such as P@10 or NDCG. We, therefore, report the ranking-based performance of the systems after seeing k writings (with varying k).

2.3   Task 1: Results

Table 2. Task 1 (pathological gambling): participating teams, number of runs, number of user writings processed by the team, and lapse of time taken for the entire process.

team	#runs	#user writings processed	lapse of time (from 1st to last response)
UNED-NLP	5	2001	17:58:48
SINAI	3	46	4 days 12:54:03
BioInfo_UAVR	5	1002	22:35:47
RELAI	5	109	7 days 15:27:25
BLUE	3	2001	3 days 13:15:25
BioNLP-UniBuc	5	3	00:37:33
UNSL	5	2001	1 day 21:53:51
NLPGroup-IISERB	5	1020	15 days 21:30:48
stezmo3	5	30	12:30:26

Table 2 shows the participating teams, the number of runs submitted and the approximate lapse of time from the first response to the last response. This time-lapse is indicative of the degree of automation of each team's algorithms. A few of the submitted runs processed the entire thread of messages (2001), but many variants stopped earlier. Some of the teams were still submitting results at the deadline time. Three teams processed the thread of messages reasonably fast (around a day for processing the entire history of user messages). The rest of the teams took several days to run the whole process. Some teams took even more than a week. This extension suggests that they incorporated some form of offline processing.

Table 3 reports the decision-based performance achieved by the participating teams. In terms of Precision, the best performing team was the NLPGroup-IISERB (run 4) but at the expense of a very low recall. In terms of $F1$, $ERDE_{50}$ and latency-weighted $F1$, the best performing run was submitted by the UNED NLD team. Their run (#4) also has a pretty high level of Recall (.938). Many teams achieved perfect Recall at the expense of very low Precision figures. In terms of $ERDE_5$, the best performing runs are SINAI #0 and #1 and BLUE #0. The majority of teams made quick decisions. Overall, these findings indicate that some systems achieved a relatively high level of effectiveness with only a few user submissions. Social and public health systems may use the best predictive algorithms to assist expert humans in detecting signs of pathological gambling as early as possible.

Table 3. Decision-based evaluation for Task 1

Team	Run	P	R	$F1$	$ERDE_5$	$ERDE_{50}$	$latency_{TP}$	$speed$	$latency$-$weighted$ $F1$
UNED-NLP	0	0.285	0.975	0.441	0.019	0.010	2.0	0.996	0.440
UNED-NLP	1	0.555	0.938	0.697	0.019	0.009	2.5	0.994	0.693
UNED-NLP	2	0.296	0.988	0.456	0.019	0.009	2.0	0.996	0.454
UNED-NLP	3	0.536	0.926	0.679	0.019	0.009	3.0	0.992	0.673
UNED-NLP	4	0.809	0.938	**0.869**	0.020	0.008	3.0	0.992	**0.862**
SINAI	0	0.425	0.765	0.546	**0.015**	0.011	1.0	1.000	0.546
SINAI	1	0.575	0.802	0.670	**0.015**	0.009	1.0	1.000	0.670
SINAI	2	0.908	0.728	0.808	0.016	0.011	1.0	1.000	0.808
BioInfo_UAVR	0	0.093	0.988	0.170	0.040	0.017	5.0	0.984	0.167
BioInfo_UAVR	1	0.067	**1.000**	0.126	0.047	0.024	5.0	0.984	0.124
BioInfo_UAVR	2	0.052	**1.000**	0.099	0.051	0.029	5.0	0.984	0.097
BioInfo_UAVR	3	0.050	**1.000**	0.095	0.052	0.030	5.0	0.984	0.094
BioInfo_UAVR	4	0.192	0.988	0.321	0.033	0.011	5.0	0.984	0.316
RELAI	0	0.000	0.000	0.000	0.039	0.039			
RELAI	1	0.000	0.000	0.000	0.039	0.039			
RELAI	2	0.052	0.963	0.099	0.036	0.029	1.0	1.000	0.099
RELAI	3	0.051	0.963	0.098	0.037	0.030	1.0	1.000	0.098
RELAI	4	0.000	0.000	0.000	0.039	0.039			
BLUE	0	0.260	0.975	0.410	**0.015**	0.009	1.0	1.000	0.410
BLUE	1	0.123	0.988	0.219	0.021	0.015	1.0	1.000	0.219
BLUE	2	0.052	**1.000**	0.099	0.037	0.028	1.0	1.000	0.099
BioNLP-UniBuc	0	0.039	**1.000**	0.075	0.038	0.037	1.0	1.000	0.075
BioNLP-UniBuc	1	0.039	**1.000**	0.076	0.038	0.037	1.0	1.000	0.076
BioNLP-UniBuc	2	0.040	**1.000**	0.077	0.037	0.036	1.0	1.000	0.077
BioNLP-UniBuc	3	0.046	**1.000**	0.087	0.033	0.032	1.0	1.000	0.087
BioNLP-UniBuc	4	0.046	**1.000**	0.089	0.032	0.031	1.0	1.000	0.089
UNSL	0	0.401	0.951	0.564	0.041	**0.008**	11.0	0.961	0.542
UNSL	1	0.461	0.938	0.618	0.041	**0.008**	11.0	0.961	0.594
UNSL	2	0.398	0.914	0.554	0.041	**0.008**	12.0	0.957	0.531
UNSL	3	0.365	0.864	0.513	0.017	0.009	3.0	0.992	0.509
UNSL	4	0.052	0.988	0.100	0.051	0.030	5.0	0.984	0.098
NLPGroup-IISERB	0	0.107	0.642	0.183	0.030	0.025	2.0	0.996	0.182
NLPGroup-IISERB	1	0.044	**1.000**	0.084	0.046	0.033	3.0	0.992	0.083
NLPGroup-IISERB	2	0.043	**1.000**	0.083	0.041	0.034	1.0	1.000	0.083
NLPGroup-IISERB	3	0.140	**1.000**	0.246	0.025	0.014	2.0	0.996	0.245
NLPGroup-IISERB	4	**1.000**	0.074	0.138	0.038	0.037	41.5	0.843	0.116
stezmo3	0	0.116	0.864	0.205	0.034	0.015	5.0	0.984	0.202
stezmo3	1	0.116	0.864	0.205	0.049	0.015	12.0	0.957	0.196
stezmo3	2	0.152	0.914	0.261	0.033	0.011	5.0	0.984	0.257
stezmo3	3	0.139	0.864	0.240	0.047	0.013	12.0	0.957	0.229
stezmo3	4	0.160	0.901	0.271	0.043	0.011	7.0	0.977	0.265

Table 4. Ranking-based evaluation for Task 1

Team	Run	1 writing			100 writings			500 writings			1000 writings		
		$P@10$	$NDCG@10$	$NDCG@100$	$P@10$	$NDCG@10$	$NDCG@100$	$P@10$	$NDCG@10$	$NDCG@100$	$P@10$	$NDCG@10$	$NDCG@100$
UNED-NLP	0	0.90	0.88	0.75	0.40	0.29	0.70	0.30	0.20	0.56	0.30	0.19	0.48
UNED-NLP	1	0.90	0.81	0.68	0.80	0.73	0.83	0.50	0.43	0.80	0.50	0.37	0.75
UNED-NLP	2	0.90	0.88	0.76	0.60	0.58	0.79	0.40	0.33	0.55	0.30	0.24	0.46
UNED-NLP	3	0.90	0.81	0.71	0.70	0.66	0.84	0.40	0.35	0.78	0.50	0.42	0.73
UNED-NLP	4	1.00	1.00	0.56	1.00	1.00	0.88	1.00	1.00	0.95	1.00	1.00	0.95
SINAI	0	0.10	0.19	0.56									
SINAI	1	0.70	0.65	0.62									
SINAI	2	1.00	1.00	0.70									
BioInfo_UAVR	0	0.00	0.00	0.03	0.80	0.87	0.33	0.00	0.00	0.00	0.10	0.10	0.03
BioInfo_UAVR	1	0.00	0.00	0.03	0.00	0.00	0.00	0.00	0.00	0.00	0.00	0.00	0.00
BioInfo_UAVR	2	0.00	0.00	0.03	0.40	0.30	0.29	0.00	0.00	0.02	0.10	0.19	0.05
BioInfo_UAVR	3	0.00	0.00	0.03	0.00	0.00	0.10	0.00	0.00	0.00	0.10	0.07	0.02
BioInfo_UAVR	4	0.00	0.00	0.03	0.00	0.00	0.03	0.00	0.00	0.03	0.00	0.00	0.03
RELAI	0	0.30	0.19	0.31	0.20	0.18	0.21						
RELAI	1	0.30	0.19	0.31	0.20	0.13	0.27						
RELAI	2	0.40	0.34	0.41	0.10	0.12	0.36						
RELAI	3	0.40	0.34	0.41	0.50	0.47	0.37						
RELAI	4	0.00	0.00	0.01	0.00	0.00	0.00						
BLUE	0	1.00	1.00	0.76	1.00	1.00	0.81	1.00	1.00	0.89	1.00	1.00	0.89
BLUE	1	1.00	1.00	0.76	1.00	1.00	0.89	1.00	1.00	0.91	1.00	1.00	0.91
BLUE	2	1.00	1.00	0.69	1.00	1.00	0.40	0.00	0.00	0.02	0.00	0.00	0.01
BioNLP-UniBuc	0	0.00	0.00	0.06									
BioNLP-UniBuc	1	0.00	0.00	0.02									
BioNLP-UniBuc	2	0.00	0.00	0.04									
BioNLP-UniBuc	3	0.10	0.19	0.07									
BioNLP-UniBuc	4	0.00	0.00	0.02									
UNSL	0	1.00	1.00	0.68	1.00	1.00	0.90	1.00	1.00	0.93	1.00	1.00	0.95
UNSL	1	1.00	1.00	0.70	1.00	1.00	0.90	1.00	1.00	0.92	1.00	1.00	0.93
UNSL	2	0.90	0.90	0.66	1.00	1.00	0.77	0.90	0.92	0.78	0.90	0.90	0.77
UNSL	3	1.00	1.00	0.69	0.60	0.58	0.72	0.80	0.81	0.77	0.80	0.81	0.78
UNSL	4	0.10	0.07	0.32	0.10	0.07	0.32	0.20	0.13	0.33	0.30	0.22	0.37
NLPGroup-IISERB	0	0.00	0.00	0.02	0.00	0.00	0.03	0.00	0.00	0.03	0.00	0.00	0.03
NLPGroup-IISERB	1	0.00	0.00	0.03	0.00	0.00	0.03	0.00	0.00	0.05	0.00	0.00	0.03
NLPGroup-IISERB	2	0.00	0.00	0.15	0.00	0.00	0.11	0.20	0.13	0.12	0.00	0.00	0.08
NLPGroup-IISERB	3	0.00	0.00	0.01	0.10	0.06	0.10	0.10	0.07	0.12	0.10	0.07	0.12
NLPGroup-IISERB	4	0.20	0.38	0.15	0.00	0.00	0.06	0.00	0.00	0.07	0.00	0.00	0.07
stezmo3	0	0.10	0.06	0.26									
stezmo3	1	0.10	0.06	0.26									
stezmo3	2	0.50	0.58	0.61									
stezmo3	3	0.50	0.58	0.61									
stezmo3	4	0.50	0.58	0.61									

Table 4 presents the ranking-based results. Because some teams only processed a few dozens of user writings, we could only compute their user rankings for the initial number of processsed writings. For those participants providing ties in the scores for the users, we used the traditional *docid* criteria (subject name) for breaking the ties. Some runs (e.g., UNED-NLP #4, BLUE #0 and #1 and UNSL #0, #1 and #2) have very good levels of ranking-based shallow effectiveness over multiple points (after one writing, after 100 writings, and so forth). Regarding the 100 cut-off, the best performing teams after one writing for nDCG are UNED-NLP (#2) and BLUE (#0 and #1). In the other scenarios, both UNED-NLP and UNSL obtain very good results.

3   Task 2: Early Detection of Depression

This is a continuation of the 2017 and 2018 tasks. This task proposes the early risk detection of depression in the very same way as described for pathological gambling in Sect. 2. The test collection for this task also had the same format as the collection described in [4]. The source of data is also the same used for previous eRisks. Here are two categories of users, depressed and non-depressed, and, for each user, the collection contains a sequence of writings (in chronological order). Contrary to the previous editions of the task, this is the first edition where the REST service is used instead of the chuck based release. More information about the server can be found at the lab website[6].

Table 5. Task 2 (Depression). Main statistics of test collection

	Test	
	Depressed	Control
Num. subjects	98	1,302
Num. submissions (posts & comments)	35,332	687,228
Avg num. of submissions per subject	360.53	527,82
Avg num. of days from first to last submission	≈628.2	≈661.7
Avg num. words per submission	27.4	23.5

This was a train and a test task. The test phase followed the same procedure as Task 1 (see Sect. 2). For the training stage, the teams had access to training data where we released the whole history of writings for training users. We indicated what users had explicitly mentioned that they have depression. The participants could therefore tune their systems with the training data. In 2022, the training data for Task 2 was composed of users from the 2017 and 2018 editions.

Again, we followed existing methods to build the assessments using simulated pooling strategies, which optimize the use of assessors time [15,16]. Table 5 reports the main statistics of the test collections used for T2. The same decision and ranking based measures as discussed in Sects. 2.1 and 2.2 were used for this task.

3.1   Task 2: Results

Table 6 shows the participating teams, the number of runs submitted and the approximate lapse of time from the first response to the last response. Most of the submitted runs processed the entire thread of messages (about 2000), but few stopped earlier or were not able to process the users' history in time. Only one team was able to process the entire set of writings in less than a day.

---

[6] https://early.irlab.org/server.html.

Table 6. Task 2 (depression): participating teams, number of runs, number of user writings processed by the team, and lapse of time taken for the whole process.

team	#runs	#user writings processed	lapse of time (from 1st to last response)
CYUT	5	2000	7 days 12:02:44
LauSAn	5	2000	2 days 06:44:17
BLUE	3	2000	2 days 17:16:05
BioInfo_UAVR	5	503	09:38:26
TUA1	5	2000	16:28:49
NLPGroup-IISERB	5	632	11 days 20:35:11
RELAI	5	169	7 days 02:27:10
UNED-MED	5	1318	5 days 13:18:24
Sunday-Rocker2	5	682	4 days 03:54:25
SCIR2	5	2000	1 day 04:52:02
UNSL	5	2000	1 day 09:35:12
E8-IJS	5	2000	3 days 02:36:32
NITK-NLP2	4	6	01:52:57

Table 7. Decision-based evaluation for Task 2

Team	Run	P	R	$F1$	$ERDE_5$	$ERDE_{50}$	$latency_{TP}$	$speed$	$latency\text{-}weighted\ F1$
CYUT	0	0.165	0.918	0.280	0.053	0.032	3.0	0.992	0.277
CYUT	1	0.162	0.898	0.274	0.053	0.032	3.0	0.992	0.272
CYUT	2	0.106	0.867	0.189	0.056	0.047	1.0	1.000	0.189
CYUT	3	0.149	0.878	0.255	0.075	0.040	7.0	0.977	0.249
CYUT	4	0.142	0.918	0.245	0.082	0.041	8.0	0.973	0.239
LauSAn	0	0.137	0.827	0.235	0.041	0.038	1.0	1.000	0.235
LauSAn	1	0.165	0.888	0.279	0.053	0.040	2.0	0.996	0.278
LauSAn	2	0.174	0.867	0.290	0.056	0.031	4.0	0.988	0.287
LauSAn	3	0.420	0.643	0.508	0.059	0.041	6.0	0.981	0.498
LauSAn	4	0.201	0.724	0.315	**0.039**	0.033	1.0	1.000	0.315
BLUE	0	0.395	0.898	0.548	0.047	0.027	5.0	0.984	0.540
BLUE	1	0.213	0.939	0.347	0.054	0.033	4.5	0.986	0.342
BLUE	2	0.106	1.000	0.192	0.074	0.048	4.0	0.988	0.190
BioInfo_UAVR	0	0.222	0.949	0.360	0.071	0.031	9.0	0.969	0.349
BioInfo_UAVR	1	0.091	0.969	0.166	0.101	0.054	8.0	0.973	0.162
BioInfo_UAVR	2	0.171	0.969	0.291	0.083	0.035	11.0	0.961	0.279

(*continued*)

Table 7. (*continued*)

Team	Run	P	R	$F1$	$ERDE_5$	$ERDE_{50}$	$latency_{TP}$	speed	latency-weighted $F1$
BioInfo_UAVR	3	0.090	0.990	0.166	0.101	0.052	6.0	0.981	0.162
BioInfo_UAVR	4	0.378	0.857	0.525	0.069	0.031	16.0	0.942	0.494
TUA1	0	0.155	0.806	0.260	0.055	0.037	3.0	0.992	0.258
TUA1	1	0.129	0.816	0.223	0.053	0.041	3.0	0.992	0.221
TUA1	2	0.155	0.806	0.260	0.055	0.037	3.0	0.992	0.258
TUA1	3	0.129	0.816	0.223	0.053	0.041	3.0	0.992	0.221
TUA1	4	0.159	0.959	0.272	0.052	0.036	3.0	0.992	0.270
NLPGroup-IISERB	0	0.682	0.745	**0.712**	0.055	0.032	9.0	0.969	**0.690**
NLPGroup-IISERB	1	0.385	0.857	0.532	0.062	0.032	18.0	0.934	0.496
NLPGroup-IISERB	2	0.662	0.459	0.542	0.069	0.058	62.0	0.766	0.416
NLPGroup-IISERB	3	0.653	0.500	0.566	0.067	0.046	26.0	0.903	0.511
NLPGroup-IISERB	4	0.000	0.000	0.000	0.070	0.070			
RELAI	0	0.085	0.847	0.155	0.114	0.092	51.0	0.807	0.125
RELAI	1	0.085	0.847	0.155	0.114	0.091	51.0	0.807	0.125
RELAI	2	0.000	0.000	0.000	0.070	0.070			
RELAI	3	0.000	0.000	0.000	0.070	0.070			
RELAI	4	0.000	0.000	0.000	0.070	0.070			
UNED-MED	0	0.119	0.969	0.212	0.091	0.056	18.0	0.934	0.198
UNED-MED	1	0.139	0.980	0.244	0.079	0.046	13.0	0.953	0.233
UNED-MED	2	0.122	0.939	0.215	0.086	0.057	15.0	0.945	0.204
UNED-MED	3	0.131	0.949	0.231	0.084	0.051	15.0	0.945	0.218
UNED-MED	4	0.084	0.163	0.111	0.079	0.078	251.0	0.252	0.028
Sunday-Rocker2	0	0.091	**1.000**	0.167	0.080	0.053	4.0	0.988	0.165
Sunday-Rocker2	1	0.355	0.786	0.489	0.068	0.041	27.0	0.899	0.439
Sunday-Rocker2	2	0.092	0.388	0.149	0.088	0.083	117.5	0.575	0.085
Sunday-Rocker2	3	0.283	0.816	0.420	0.071	0.045	37.5	0.859	0.361
Sunday-Rocker2	4	0.108	**1.000**	0.195	0.082	0.047	6.0	0.981	0.191
SCIR2	0	0.396	0.837	0.538	0.076	0.076	150.0	0.477	0.256
SCIR2	1	0.336	0.878	0.486	0.078	0.078	150.0	0.477	0.232
SCIR2	2	0.235	0.908	0.373	0.051	0.046	3.0	0.992	0.370
SCIR2	3	0.316	0.847	0.460	0.079	**0.026**	44.0	0.834	0.383
SCIR2	4	0.274	0.847	0.414	0.045	0.031	3.0	0.992	0.411
UNSL	0	0.161	0.918	0.274	0.079	0.042	14.5	0.947	0.260
UNSL	1	0.310	0.786	0.445	0.078	0.037	12.0	0.957	0.426
UNSL	2	0.400	0.755	0.523	0.045	**0.026**	3.0	0.992	0.519
UNSL	3	0.144	0.929	0.249	0.055	0.035	3.0	0.992	0.247
UNSL	4	0.080	0.918	0.146	0.099	0.074	5.0	0.984	0.144
E8-IJS	0	**0.684**	0.133	0.222	0.061	0.061	1.0	**1.000**	0.222
E8-IJS	1	0.242	0.959	0.387	0.068	0.036	20.5	0.924	0.357
E8-IJS	2	0.000	0.000	0.000	0.070	0.070			
E8-IJS	3	0.000	0.000	0.000	0.070	0.070			
E8-IJS	4	0.000	0.000	0.000	0.070	0.070			
NITK-NLP2	0	0.138	0.796	0.235	0.047	0.039	2.0	0.996	0.234
NITK-NLP2	1	0.135	0.806	0.231	0.047	0.039	2.0	0.996	0.230
NITK-NLP2	2	0.132	0.786	0.225	0.050	0.040	2.0	0.996	0.225
NITK-NLP2	3	0.149	0.724	0.248	0.049	0.039	2.0	0.996	0.247

Table 7 reports the decision-based performance achieved by the participating teams. In terms of Precision, E8-IJS run #0 obtains the highest values but at the expenses of low Recall. Similarly, Sunday-Rocker systems #0 and #4 obtain and BLUE #2 perfect Recall but with low Precision values. When considering the Precision-Recall trade-off, NLPGroup-IISERB #0 is the best performance being the only run over 0.7 (highest $F1$). Regarding latency-penalized metrics, UNSL #2 and SCIR2 #3 obtained the best $ERDE_{50}$ and LauSAn #4 the best $ERDE_5$ error value. It is again NLPGroup-IISERB #04, the one achieving the best latency-weighted $F1$. This run seems to be quite balanced overall.

Table 8. Ranking-based evaluation for Task 2

Team	Run	1 writing			100 writings			500 writings			1000 writings		
		$P@10$	$NDCG@10$	$NDCG@100$	$P@10$	$NDCG@10$	$NDCG@100$	$P@10$	$NDCG@10$	$NDCG@100$	$P@10$	$NDCG@10$	$NDCG@100$
CYUT	0	0.50	0.49	0.37	0.50	0.52	0.54	0.60	0.59	0.58	0.70	0.72	0.61
CYUT	1	0.70	0.77	0.37	0.60	0.72	0.58	0.60	0.72	0.61	0.70	0.80	0.62
CYUT	2	0.00	0.00	0.16	0.10	0.07	0.25	0.10	0.19	0.31	0.10	0.12	0.29
CYUT	3	0.10	0.07	0.12	0.70	0.70	0.57	0.70	0.72	0.59	**0.80**	0.74	0.60
CYUT	4	0.10	0.06	0.12	0.60	0.68	0.55	0.60	0.69	0.59	**0.80**	0.84	0.61
LauSAn	0	0.60	0.72	0.43	0.30	0.41	0.13	0.20	0.31	0.12	0.10	0.19	0.11
LauSAn	1	0.60	0.66	0.43	0.40	0.33	0.30	0.50	0.50	0.17	0.20	0.15	0.08
LauSAn	2	0.60	0.66	0.43	0.40	0.33	0.29	0.50	0.50	0.18	0.20	0.15	0.13
LauSAn	3	0.60	0.66	0.43	0.40	0.33	0.27	0.50	0.50	0.22	0.20	0.15	0.14
LauSAn	4	0.40	0.38	0.34	0.50	0.49	0.41	0.40	0.27	0.21	0.20	0.22	0.14
BLUE	0	**0.80**	**0.88**	**0.54**	0.60	0.56	0.59	0.80	0.81	0.66	0.80	0.80	0.68
BLUE	1	**0.80**	**0.88**	**0.54**	0.70	0.64	**0.67**	0.80	0.84	**0.74**	0.80	**0.86**	**0.72**
BLUE	2	**0.80**	0.75	0.46	0.40	0.40	0.30	0.30	0.35	0.20	0.30	0.38	0.16
BioInfo_UAVR	0	0.00	0.00	0.04	0.20	0.15	0.15	0.00	0.00	0.09			
BioInfo_UAVR	1	0.00	0.00	0.02	0.20	0.25	0.14	0.20	0.12	0.07			
BioInfo_UAVR	2	0.20	0.13	0.06	0.60	0.60	0.36	0.70	0.78	0.32			
BioInfo_UAVR	3	0.10	0.08	0.08	0.20	0.26	0.14	0.20	0.17	0.08			
BioInfo_UAVR	4	0.10	0.07	0.05	0.00	0.00	0.04	0.00	0.00	0.05			
TUA1	0	**0.80**	**0.88**	0.44	0.60	0.72	0.52	0.60	0.67	0.52	0.70	0.80	0.57
TUA1	1	0.70	0.77	0.44	0.50	0.54	0.39	0.50	0.56	0.42	0.50	0.65	0.43
TUA1	2	**0.80**	**0.88**	0.44	0.60	0.72	0.52	0.60	0.67	0.52	0.70	0.80	0.57
TUA1	3	0.60	0.69	0.43	0.50	0.54	0.39	0.50	0.56	0.42	0.50	0.65	0.43
TUA1	4	0.50	0.37	0.35	0.00	0.00	0.36	0.00	0.00	0.36	0.20	0.12	0.31
NLPGroup-IISERB	0	0.00	0.00	0.02	**0.90**	0.92	0.30	**0.90**	**0.92**	0.33			
NLPGroup-IISERB	1	0.30	0.32	0.13	**0.90**	0.81	0.27	0.80	0.84	0.33			
NLPGroup-IISERB	2	0.70	0.79	0.24	0.00	0.00	0.00	0.00	0.00	0.00			
NLPGroup-IISERB	3	0.00	0.00	0.06	0.10	0.19	0.06	0.00	0.00	0.02			
NLPGroup-IISERB	4	0.00	0.00	0.04	**0.90**	**0.93**	0.66	**0.90**	**0.92**	0.69			
RELAI	0	0.00	0.00	0.07	0.10	0.06	0.20						
RELAI	1	0.00	0.00	0.07	0.20	0.25	0.20						
RELAI	2	0.10	0.12	0.09	0.00	0.00	0.16						
RELAI	3	0.10	0.12	0.09	0.50	0.52	0.31						
RELAI	4	0.10	0.12	0.07	0.00	0.00	0.00						

(continued)

Table 8. (*continued*)

Team	Run	1 writing			100 writings			500 writings			1000 writings		
		P@10	*NDCG@10*	*NDCG@100*	*P@10*	*NDCG@10*	*NDCG@100*	*P@10*	*NDCG@10*	*NDCG@100*	*P@10*	*NDCG@10*	*NDCG@100*
UNED-MED	0	0.70	0.69	0.27	0.80	0.84	0.63	0.60	0.66	0.60	0.50	0.46	0.56
UNED-MED	1	0.50	0.44	0.26	0.70	0.76	0.50	0.60	0.64	0.47	0.80	0.74	0.50
UNED-MED	2	0.70	0.68	0.28	0.50	0.51	0.59	0.80	0.71	0.61	0.50	0.44	0.62
UNED-MED	3	**0.80**	0.82	0.29	0.60	0.44	0.31	0.80	0.73	0.36	0.40	0.51	0.30
UNED-MED	4	0.00	0.00	0.06	0.00	0.00	0.05	0.00	0.00	0.04	0.10	0.19	0.09
Sunday-Rocker2	0	0.40	0.47	0.39	0.40	0.44	0.29	0.50	0.46	0.24			
Sunday-Rocker2	1	0.70	0.81	0.39	**0.90**	**0.93**	0.66	**0.90**	0.88	0.65			
Sunday-Rocker2	2	0.10	0.07	0.23	0.00	0.00	0.11	0.30	0.31	0.17			
Sunday-Rocker2	3	**0.80**	**0.88**	0.41	0.50	0.50	0.23	0.60	0.69	0.34			
Sunday-Rocker2	4	0.30	0.28	0.31	0.30	0.37	0.25	0.40	0.30	0.18			
SCIR2	0	0.10	0.07	0.08	0.00	0.00	0.06	0.00	0.00	0.06	0.10	0.12	0.06
SCIR2	1	0.00	0.00	0.05	0.10	0.07	0.07	0.00	0.00	0.04	0.00	0.00	0.05
SCIR2	2	0.00	0.00	0.06	0.00	0.00	0.05	0.20	0.13	0.07	0.00	0.00	0.06
SCIR2	3	0.10	0.06	0.05	0.00	0.00	0.04	0.00	0.00	0.06	0.00	0.00	0.02
SCIR2	4	0.10	0.19	0.09	0.10	0.07	0.05	0.10	0.10	0.07	0.10	0.06	0.05
UNSL	0	0.60	0.40	0.36	0.20	0.13	0.46	0.30	0.28	0.43	0.60	0.72	0.45
UNSL	1	**0.80**	**0.88**	0.46	0.60	0.73	0.64	0.60	0.73	0.66	0.60	0.71	0.66
UNSL	2	0.70	0.68	0.50	0.50	0.39	0.55	0.70	0.73	0.61	0.70	0.73	0.61
UNSL	3	0.10	0.06	0.15	0.40	0.27	0.43	0.30	0.21	0.42	0.30	0.21	0.42
UNSL	4	0.10	0.12	0.05	0.00	0.00	0.03	0.20	0.19	0.07	0.00	0.00	0.04
E8-IJS	0	0.00	0.00	0.06	0.10	0.07	0.05	0.10	0.12	0.08	0.00	0.00	0.03
E8-IJS	1	0.40	0.58	0.19	0.40	0.41	0.15	0.20	0.15	0.09	0.30	0.38	0.15
E8-IJS	2	0.00	0.00	0.05	0.00	0.00	0.07	0.00	0.00	0.05	0.10	0.19	0.07
E8-IJS	3	0.00	0.00	0.02	0.10	0.10	0.08	0.10	0.06	0.02	0.00	0.00	0.05
E8-IJS	4	0.00	0.00	0.07	0.10	0.10	0.08	0.20	0.31	0.11	0.10	0.06	0.04
NITK-NLP2	0	0.40	0.28	0.15									
NITK-NLP2	1	0.00	0.00	0.01									
NITK-NLP2	2	0.00	0.00	0.02									
NITK-NLP2	3	0.00	0.00	0.02									

Table 8 presents the ranking-based results. Contrary to task 1, no run obtained perfect figures for any of the scenarios. This is worth noting, given that for task 2, there are more positive subjects. Overall, systems #0 and #1 from the BLUE team seem to be the most consistent under the different number of writings among the best-performing ones. Other systems, such as those from NLPGroup-IISERB, show an erratic behaviour going so low as Precision 0 when only one writing was processed but obtaining the best results for the same metrics after 100.

4   Task 3: Measuring the Severity of Eating Disorders

The task consists of estimating the level of different symptoms associated with a diagnosis of eating disorders. To that end, the participants worked from a thread of user submissions. For each user, the participants were given a history of posts and comments on Social Media, and the participants had to estimate the user's responses to a standard eating disorder questionnaire (based on the evidence found in the history of posts/comments).

The questionnaire is defined from the Eating Disorder Examination Questionnaire (EDE-Q)[7], which is a 28-item self-reported questionnaire adapted from the semi-structured interview Eating Disorder Examination (EDE)[8] [3]. This instrument is designed to assess the range and severity of multiple features associated with eating disorders. It employs four subscales (Restraint, Eating Concern, Shape Concern and Weight Concern) and a global score. We only used questions 1–12 and 19–28. Questions are listed in Table 9.

Table 9. Eating Disorder Examination Questionarie

Instructions:

The following questions are concerned with the past four weeks (28 days) only. Please read each question carefully. Please answer all the questions. Thank you..

1. Have you been deliberately trying to limit the amount of food you eat to influence your shape or weight (whether or not you have succeeded) 0. NO DAYS
  1. 1--5 DAYS
  2. 6--12 DAYS
  3. 13--15 DAYS
  4. 16--22 DAYS
  5. 23--27 DAYS
  6. EVERY DAY

2. Have you gone for long periods of time (8 waking hours or more) without eating anything at all in order to influence your shape or weight?
  0. NO DAYS
  1. 1--5 DAYS
  2. 6--12 DAYS
  3. 13--15 DAYS
  4. 16--22 DAYS
  5. 23--27 DAYS
  6. EVERY DAY

3. Have you tried to exclude from your diet any foods that you like in order to influence your shape or weight (whether or not you have succeeded)?
  0. NO DAYS
  1. 1--5 DAYS
  2. 6--12 DAYS
  3. 13--15 DAYS
  4. 16--22 DAYS
  5. 23--27 DAYS
  6. EVERY DAY

(continued)

---

[7] https://www.corc.uk.net/media/1273/ede-q_quesionnaire.pdf.

[8] https://www.corc.uk.net/media/1951/ede_170d.pdf.

Table 9. (*continued*)

4. Have you tried to follow definite rules regarding your eating (for example, a calorie limit) in order to influence your shape or weight (whether or not you have succeeded)?
   0. NO DAYS
   1. 1--5 DAYS
   2. 6--12 DAYS
   3. 13--15 DAYS
   4. 16--22 DAYS
   5. 23--27 DAYS
   6. EVERY DAY

5. Have you had a definite desire to have an empty stomach with the aim of influencing your shape or weight?
   0. NO DAYS
   1. 1--5 DAYS
   2. 6--12 DAYS
   3. 13--15 DAYS
   4. 16--22 DAYS
   5. 23--27 DAYS
   6. EVERY DAY

6. Have you had a definite desire to have a totally flat stomach?
   0. NO DAYS
   1. 1--5 DAYS
   2. 6--12 DAYS
   3. 13--15 DAYS
   4. 16--22 DAYS
   5. 23--27 DAYS
   6. EVERY DAY

7. Has thinking about food, eating or calories made it very difficult to concentrate on things you are interested in (for example, working, following a conversation, or reading)?
   0. NO DAYS
   1. 1--5 DAYS
   2. 6--12 DAYS
   3. 13--15 DAYS
   4. 16--22 DAYS
   5. 23--27 DAYS
   6. EVERY DAY

8. Has thinking about shape or weight made it very difficult to concentrate on things you are interested in (for example, working, following a conversation, or reading)?
   0. NO DAYS
   1. 1--5 DAYS
   2. 6--12 DAYS
   3. 13--15 DAYS
   4. 16--22 DAYS
   5. 23--27 DAYS
   6. EVERY DAY

(*continued*)

Table 9. (*continued*)

9. Have you had a definite fear of losing control over eating
    0. NO DAYS
    1. 1--5 DAYS
    2. 6--12 DAYS
    3. 13--15 DAYS
    4. 16--22 DAYS
    5. 23--27 DAYS
    6. EVERY DAY

10. Have you had a definite fear that you might gain weight?
    0. NO DAYS
    1. 1--5 DAYS
    2. 6--12 DAYS
    3. 13--15 DAYS
    4. 16--22 DAYS
    5. 23--27 DAYS
    6. EVERY DAY

11. Have you felt fat?
    0. NO DAYS
    1. 1--5 DAYS
    2. 6--12 DAYS
    3. 13--15 DAYS
    4. 16--22 DAYS
    5. 23--27 DAYS
    6. EVERY DAY

12. Have you had a strong desire to lose weight?
    0. NO DAYS
    1. 1--5 DAYS
    2. 6--12 DAYS
    3. 13--15 DAYS
    4. 16--22 DAYS
    5. 23--27 DAYS
    6. EVERY DAY

19. Over the past 28 days, on how many days have you eaten in secret (i.e., furtively)? 0... Do not count episodes of binge eating.
    0. NO DAYS
    1. 1--5 DAYS
    2. 6--12 DAYS
    3. 13--15 DAYS
    4. 16--22 DAYS
    5. 23--27 DAYS
    6. EVERY DAY

20. On what proportion of the times that you have eaten have you felt guilty (felt that you've done wrong) because of its effect on your shape or weight? 0... Do not count episodes of binge eating.
    0. NO DAYS
    1. 1--5 DAYS
    2. 6--12 DAYS
    3. 13--15 DAYS
    4. 16--22 DAYS
    5. 23--27 DAYS
    6. EVERY DAY

(*continued*)

Table 9. (*continued*)

21. Over the past 28 days, how concerned have you been about other people seeing you eat?
0... Do not count episodes of binge eating
   0. NO DAYS
   1. 1--5 DAYS
   2. 6--12 DAYS
   3. 13--15 DAYS
   4. 16--22 DAYS
   5. 23--27 DAYS
   6. EVERY DAY

22. Has your weight influenced how you think about (judge) yourself as a person?
   0. NOT AT ALL (0)
   1. SLIGHTY (1)
   2. SLIGHTY (2)
   3. MODERATELY (3)
   4. MODERATELY (4)
   5. MARKEDLY (5)
   6. MARKEDLY (6)

23. Has your shape influenced how you think about (judge) yourself as a person?
   0. NOT AT ALL (0)
   1. SLIGHTY (1)
   2. SLIGHTY (2)
   3. MODERATELY (3)
   4. MODERATELY (4)
   5. MARKEDLY (5)
   6. MARKEDLY (6)

24. How much would it have upset you if you had been asked to weigh yourself once a week (no more, or less, often) for the next four weeks?
   0. NOT AT ALL (0)
   1. SLIGHTY (1)
   2. SLIGHTY (2)
   3. MODERATELY (3)
   4. MODERATELY (4)
   5. MARKEDLY (5)
   6. MARKEDLY (6)

25. How dissatisfied have you been with your weight?
   0. NOT AT ALL (0)
   1. SLIGHTY (1)
   2. SLIGHTY (2)
   3. MODERATELY (3)
   4. MODERATELY (4)
   5. MARKEDLY (5)
   6. MARKEDLY (6)

26. How dissatisfied have you been with your shape?
   0. NOT AT ALL (0)
   1. SLIGHTY (1)
   2. SLIGHTY (2)
   3. MODERATELY (3)
   4. MODERATELY (4)
   5. MARKEDLY (5)
   6. MARKEDLY (6)

(*continued*)

Table 9. (*continued*)

27. How uncomfortable have you felt seeing your body (for example, seeing your shape in the mirror, in a shop window reflection, while undressing or taking a bath or shower)?
    0. NOT AT ALL (0)
    1. SLIGHTY (1)
    2. SLIGHTY (2)
    3. MODERATELY (3)
    4. MODERATELY (4)
    5. MARKEDLY (5)
    6. MARKEDLY (6)

28. How uncomfortable have you felt about others seeing your shape or figure (for example, in communal changing rooms, when swimming, or wearing tight clothes)?
    0. NOT AT ALL (0)
    1. SLIGHTY (1)
    2. SLIGHTY (2)
    3. MODERATELY (3)
    4. MODERATELY (4)
    5. MARKEDLY (5)
    6. MARKEDLY (6)

This task aims to investigate the feasibility of automatically estimating the severity of multiple symptoms associated with eating disorders. The algorithms must estimate the user's response to each individual question based on the user's writing history. We gathered questionnaires completed by Social Media users and their writing history (we extracted each history of writings right after the user provided us with the filled questionnaire). The user-completed questionnaires (ground truth) were used to evaluate the quality of the responses provided by the participating systems.

This was a test only task. No training data was provided to the participants. The participants were given a dataset with 28 users (for each user, his/her history of writings is provided) and they were asked to produce a file with the following structure:

```
username1 answer1 answer2 ... answer22
username2 answer1 answer2 ... answer22
```

Each line has the username and 22 values. These values correspond with the responses to the questions above (the possible values are 0,1,2,3,4,5,6).

4.1   Task 3: Evaluation Metrics

Evaluation is based on the following effectiveness metrics:

– **Mean Zero-One Error** (*MZOE*) between the questionnaire filled by the real user and the questionnaire filled by the system (i.e. fraction of incorrect predictions).

$$MZOE(f,Q) = \frac{|\{q_i \in Q : R(q_i) \neq f(q_i)\}|}{|Q|} \tag{8}$$

where f denotes the classification done by an automatic system, Q is the set of questions of each questionnaire, q_i is the i-th question, $R(q_i)$ is the real user's answer for the i-th question and $f(q_i)$ is the predicted answer of the system for the i-th question. Each user produces a single $MZOE$ score and the reported $MZOE$ is the average over all $MZOE$ values (mean $MZOE$ over all users).

- **Mean Absolute Error** (MAE) between the questionnaire filled by the real user and the questionnaire filled by the system (i.e. average deviation of the predicted response from the true response).

$$MAE(f,Q) = \frac{\sum_{q_i \in Q} |R(q_i) - f(q_i)|}{|Q|} \tag{9}$$

Again, each user produces a single MAE score and the reported MAE is the average over all MAE values (mean MAE over all users).

- **Macroaveraged Mean Absolute Error** (MAE_{macro}) between the questionnaire filled by the real user and the questionnaire filled by the system (see [2]).

$$MAE_{macro}(f,Q) = \frac{1}{7} \sum_{j=0}^{6} \frac{\sum_{q_i \in Q_j} |R(q_i) - f(q_i)|}{|Q_j|} \tag{10}$$

where Q_j represents the set of questions whose true answer is j (note that j goes from 0 to 6 because those are the possible answers to each question). Again, each user produces a single MAE_{macro} score and the reported MAE_{macro} is the average over all MAE_{macro} values (mean MAE_{macro} over all users).

The following measures are based on aggregated scores obtained from the questionnaires. Further details about the EDE-Q instruments can be found elsewhere (e.g. see the scoring section of the questionnaire[9]).

- **Restraint Subscale (RS)**: Given a questionnaire, its restraint score is obtained as the mean response to the first five questions. This measure computes the RMSE between the restraint ED score obtained from the questionnaire filled by the real user and the restraint ED score obtained from the questionnaire filled by the system.

Each user u_i is associated with a real subscale ED score (referred to as $R_{RS}(u_i)$) and an estimated subscale ED score (referred to as $f_{RS}(u_i)$). This metric computes the RMSE between the real and an estimated subscale ED scores as follows:

$$RMSE(f,U) = \sqrt{\frac{\sum_{u_i \in U} (R_{RS}(u_i) - f_{RS}(u_i))^2}{|U|}} \tag{11}$$

where U is the user set.

---

[9] https://www.corc.uk.net/media/1951/ede_170d.pdf.

- **Eating Concern Subscale (ECS)**: Given a questionnaire, its eating concern score is obtained as the mean response to the following questions (7, 9, 19, 21, 20). This metric computes the RMSE (Eq. 12) between the eating concern ED score obtained from the questionnaire filled by the real user and the eating concern ED score obtained from the questionnaire filled by the system.

$$RMSE(f,U) = \sqrt{\frac{\sum_{u_i \in U}(R_{ECS}(u_i) - f_{ECS}(u_i))^2}{|U|}} \qquad (12)$$

- **Shape Concern Subscale (SCS)**: Given a questionnaire, its shape concern score is obtained as the mean response to the following questions (6, 8, 23, 10, 26, 27, 28, 11). This metric computes the RMSE (Eq. 13) between the shape concern ED score obtained from the questionnaire filled by the real user and the shape concern ED score obtained from the questionnaire filled by the system.

$$RMSE(f,U) = \sqrt{\frac{\sum_{u_i \in U}(R_{SCS}(u_i) - f_{SCS}(u_i))^2}{|U|}} \qquad (13)$$

- **Weight Concern Subscale (WCS)**: Given a questionnaire, its weight concern score is obtained as the mean response to the following questions (22, 24, 8, 25, 12). This metric computes the RMSE (Eq. 14) between the weight concern ED score obtained from the questionnaire filled by the real user and the weight concern ED score obtained from the questionnaire filled by the system.

$$RMSE(f,U) = \sqrt{\frac{\sum_{u_i \in U}(R_{WCS}(u_i) - f_{WCS}(u_i))^2}{|U|}} \qquad (14)$$

- **Global ED (GED)**: To obtain an overall or 'global' score, the four subscales scores are summed and the resulting total divided by the number of subscales (i.e. four) [3]. This metric computes the RMSE between the real and an estimated global ED scores as follows:

$$RMSE(f,U) = \sqrt{\frac{\sum_{u_i \in U}(R_{GED}(u_i) - f_{GED}(u_i))^2}{|U|}} \qquad (15)$$

4.2 Task 3: Results

Table 10 presents the results achieved by the participants in this task. To put things in perspective, the table also reports (lower block) the performance achieved by three baseline variants: all 0s and all 6s, which consist of sending the same response (0 or 6) for all the questions, and average, which is the performance achieved by a method that, for each question, sends as a response the answer that is the closest to the mean of the responses sent by all participants (e.g. if the mean response provided by the participants equals 3.7 then this average approach would submit a 4).

Table 10. Task 3 Results. Participating teams and runs with corresponding scores for the metrics.

team	run ID	MZOE	MAE	MAE_{macro}	GED	RS	ECS	SCS	WCS
NLPGroup-IISERB	1	0.92	2.58	2.09	2.04	2.16	1.89	2.74	2.33
NLPGroup-IISERB	2	0.92	**2.18**	**1.76**	**1.74**	**2.00**	**1.73**	**2.03**	**1.92**
NLPGroup-IISERB	3	0.93	2.60	2.10	2.04	2.13	1.90	2.74	2.35
NLPGroup-IISERB	4	0.81	3.36	2.96	3.68	3.69	3.18	4.28	3.82
RELAI	1	0.82	3.31	2.91	3.59	3.65	3.05	4.19	3.74
RELAI	2	0.82	3.30	2.89	3.56	3.65	3.03	4.17	3.71
RELAI	3	0.83	3.15	2.70	3.26	3.04	2.72	4.04	3.61
RELAI	4	0.82	3.32	2.91	3.59	3.66	3.05	4.19	3.74
RELAI	5	0.82	3.19	2.74	3.34	3.15	2.80	4.08	3.64
SINAI	1	0.85	2.65	2.29	2.63	3.29	2.35	2.98	2.40
SINAI	2	0.87	2.60	2.23	2.42	3.01	2.21	2.85	2.31
SINAI	3	0.86	2.62	2.22	2.54	3.15	2.32	2.93	2.36
all 0		0.81	3.36	2.96	3.68	3.69	3.18	4.28	3.82
all 6		**0.67**	2.64	3.04	3.25	3.52	3.72	2.81	3.28
average		0.88	2.72	2.22	2.69	2.76	2.20	3.35	2.85

The results show that the best system was system #2 from NLPGroup-IISERB. It obtained the best results for all metrics but from MZOE where it was not able to surpass the naive *all 6* baseline. Given that this is the first edition of the task and that no training data was provided, the results are not unexpected. We hope that the number of submitted runs and their performance numbers will increase in the following editions of the task.

5   Conclusions

This paper provided an overview of eRisk 2022. The sixth edition of this lab focused on two types of tasks. On the one hand, two tasks were on early detection of pathological gambling and depression (Task 1 and 2, respectively), where participants had sequential access to the user's social media posts and had to send alerts about at-risk individuals. On the other hand, one task was released to measuring the severity of the signs of eating disorders (Task 3), where the participants were given the full user history, and their systems had to automatically estimate the user's responses to a standard depression questionnaire

The proposed tasks received 115 runs from a total of 17 teams. Although the effectiveness of the proposed solutions is still limited, the experimental results show that evidence extracted from social media is valuable, and automatic or semi-automatic screening tools to detect at-risk individuals could be developed. These findings compel us to look into the development of benchmarks for text-based risk indicator screening.

Acknowledgements. This work was supported by projects PLEC2021-007662 (MCIN/AEI/10.13039/ 501100011033, Ministerio de Ciencia e Innovación, Agencia Estatal de Investigación, Plan de Recuperación, Transformación y Resiliencia, Unión Europea-Next Generation EU) and RTI2018-093336-B-C21, RTI2018-093336-B-C22 (Ministerio de Ciencia e Innvovación & ERDF). The first and second authors thank the financial support supplied by the Consellería de Educación, Universidade e Formación Profesional (accreditation 2019–2022 ED431G/01, GPC ED431B 2022/33) and the European Regional Development Fund, which acknowledges the CITIC Research Center in ICT of the University of A Coruña as a Research Center of the Galician University System. The third author also thanks the financial support supplied by the Consellería de Educación, Universidade e Formación Profesional (accreditation 2019–2022 ED431G-2019/04, ED431C 2018/29) and the European Regional Development Fund, which acknowledges the CiTIUS-Research Center in Intelligent Technologies of the University of Santiago de Compostela as a Research Center of the Galician University System.

References

1. Abbott, M.: The epidemiology and impact of gambling disorder and other gambling-related harm. In: WHO Forum on Alcohol, Drugs and Addictive Behaviours, Geneva, Switzerland (2017)
2. Baccianella, S., Esuli, A., Sebastiani, F.: Evaluation measures for ordinal regression, pp. 283–287 (2009). https://doi.org/10.1109/ISDA.2009.230
3. Fairburn, C.G., Cooper, Z., O'Connor, M.: Eating disorder examination, Edition 17.0D, April 2014
4. Losada, D.E., Crestani, F.: A test collection for research on depression and language use. In: Fuhr, N., et al. (eds.) CLEF 2016. LNCS, vol. 9822, pp. 28–39. Springer, Cham (2016). https://doi.org/10.1007/978-3-319-44564-9_3
5. Losada, D.E., Crestani, F., Parapar, J.: eRISK 2017: CLEF lab on early risk prediction on the Internet: experimental foundations. In: Jones, G.J.F., et al. (eds.) CLEF 2017. LNCS, vol. 10456, pp. 346–360. Springer, Cham (2017). https://doi.org/10.1007/978-3-319-65813-1_30
6. Losada, D.E., Crestani, F., Parapar, J.: eRisk 2017: CLEF lab on early risk prediction on the Internet: experimental foundations. In: CEUR Proceedings of the Conference and Labs of the Evaluation Forum, CLEF 2017, Dublin, Ireland (2017)
7. Losada, D.E., Crestani, F., Parapar, J.: Overview of eRisk 2018: early risk prediction on the Internet (extended lab overview). In: CEUR Proceedings of the Conference and Labs of the Evaluation Forum, CLEF 2018, Avignon, France (2018)
8. Losada, D.E., Crestani, F., Parapar, J.: Overview of eRisk: early risk prediction on the Internet. In: Bellot, P., et al. (eds.) CLEF 2018. LNCS, vol. 11018, pp. 343–361. Springer, Cham (2018). https://doi.org/10.1007/978-3-319-98932-7_30
9. Losada, D.E., Crestani, F., Parapar, J.: Early detection of risks on the Internet: an exploratory campaign. In: Azzopardi, L., Stein, B., Fuhr, N., Mayr, P., Hauff, C., Hiemstra, D. (eds.) ECIR 2019. LNCS, vol. 11438, pp. 259–266. Springer, Cham (2019). https://doi.org/10.1007/978-3-030-15719-7_35

10. Losada, D.E., Crestani, F., Parapar, J.: Overview of eRisk 2019 early risk prediction on the Internet. In: Crestani, F., et al. (eds.) CLEF 2019. LNCS, vol. 11696, pp. 340–357. Springer, Cham (2019). https://doi.org/10.1007/978-3-030-28577-7_27

11. Losada, D.E., Crestani, F., Parapar, J.: Overview of eRisk at CLEF 2019: early risk prediction on the Internet (extended overview). In: CEUR Proceedings of the Conference and Labs of the Evaluation Forum, CLEF 2019, Lugano, Switzerland (2019)

12. Losada, D.E., Crestani, F., Parapar, J.: eRisk 2020: self-harm and depression challenges. In: Jose, J.M., et al. (eds.) ECIR 2020. LNCS, vol. 12036, pp. 557–563. Springer, Cham (2020). https://doi.org/10.1007/978-3-030-45442-5_72

13. Losada, D.E., Crestani, F., Parapar, J.: Overview of eRisk 2020: early risk prediction on the Internet. In: Arampatzis, A., et al. (eds.) CLEF 2020. LNCS, vol. 12260, pp. 272–287. Springer, Cham (2020). https://doi.org/10.1007/978-3-030-58219-7_20

14. Losada, D.E., Crestani, F., Parapar, J.: Overview of eRisk at CLEF 2020: early risk prediction on the Internet (extended overview). In: Working Notes of CLEF 2020 - Conference and Labs of the Evaluation Forum, Thessaloniki, Greece, 22–25 September 2020 (2020)

15. Otero, D., Parapar, J., Barreiro, Á.: Beaver: efficiently building test collections for novel tasks. In: Proceedings of the First Joint Conference of the Information Retrieval Communities in Europe (CIRCLE 2020), Samatan, Gers, France, 6–9 July 2020 (2020)

16. Otero, D., Parapar, J., Barreiro, Á.: The wisdom of the rankers: a cost-effective method for building pooled test collections without participant systems. In: The 36th ACM/SIGAPP Symposium on Applied Computing, Virtual Event, SAC 2021, Republic of Korea, 22–26 March 2021, pp. 672–680 (2021)

17. Parapar, J., Martín-Rodilla, P., Losada, D.E., Crestani, F.: eRisk 2021: pathological gambling, self-harm and depression challenges. In: Hiemstra, D., Moens, M.-F., Mothe, J., Perego, R., Potthast, M., Sebastiani, F. (eds.) ECIR 2021. LNCS, vol. 12657, pp. 650–656. Springer, Cham (2021). https://doi.org/10.1007/978-3-030-72240-1_76

18. Parapar, J., Martín-Rodilla, P., Losada, D.E., Crestani, F.: Overview of eRisk 2021: early risk prediction on the Internet. In: Candan, K.S., et al. (eds.) CLEF 2021. LNCS, vol. 12880, pp. 324–344. Springer, Cham (2021). https://doi.org/10.1007/978-3-030-85251-1_22

19. Parapar, J., Martín-Rodilla, P., Losada, D.E., Crestani, F.: Overview of eRisk at CLEF 2021: early risk prediction on the Internet (extended overview). In: Proceedings of the Working Notes of CLEF 2021 - Conference and Labs of the Evaluation Forum, Bucharest, Romania, 21st–24th September 2021, pp. 864–887 (2021)

20. Parapar, J., Martín-Rodilla, P., Losada, D.E., Crestani, F.: eRisk 2022: pathological gambling, depression, and eating disorder challenges. In: Hagen, M. (ed.) ECIR 2022. LNCS, vol. 13186, pp. 436–442. Springer, Cham (2022). https://doi.org/10.1007/978-3-030-99739-7_54

21. Sadeque, F., Xu, D., Bethard, S.: Measuring the latency of depression detection in social media. In: WSDM, pp. 495–503. ACM (2018)

22. Trotzek, M., Koitka, S., Friedrich, C.: Utilizing neural networks and linguistic metadata for early detection of depression indications in text sequences. IEEE Trans. Knowl. Data Eng. **32**, 588–601 (2018)

Overview of LifeCLEF 2022: An Evaluation of Machine-Learning Based Species Identification and Species Distribution Prediction

Alexis Joly[1,8(✉)] , Hervé Goëau[2,8] , Stefan Kahl[6,8] , Lukáš Picek[8,9] ,
Titouan Lorieul[1,8] , Elijah Cole[8,9] , Benjamin Deneu[1,8] ,
Maximilien Servajean[7,8] , Andrew Durso[8,10] , Hervé Glotin[3,8] ,
Robert Planqué[4,8] , Willem-Pier Vellinga[4,8] , Amanda Navine[8,13] ,
Holger Klinck[6,8], Tom Denton[8,11], Ivan Eggel[5,8], Pierre Bonnet[2,8] ,
Milan Šulc[8,12] , and Marek Hrúz[8,9]

[1] Inria, LIRMM, Univ Montpellier, CNRS, Montpellier, France
alexis.joly@inria.fr
[2] CIRAD, UMR AMAP, Montpellier, Occitanie, France
[3] Univ. Toulon, Aix Marseille Univ., CNRS, LIS, DYNI team, Marseille, France
[4] Xeno-canto Foundation, Amsterdam, The Netherlands
[5] HES-SO, Sierre, Switzerland
[6] KLYCCB, Cornell Lab of Ornithology, Cornell University, Ithaca, USA
[7] LIRMM, AMI, Univ Paul Valéry Montpellier, Univ Montpellier, CNRS,
Montpellier, France
[8] Department of Computing and Mathematical Sciences, Caltech, Pasadena, USA
[9] Department of Cybernetics, FAV, University of West Bohemia, Pilsen, Czechia
[10] Department of Biological Sciences, Florida Gulf Coast University,
Fort Myers, USA
[11] Google LLC, San Francisco, USA
[12] Rossum.ai, Prague, Czech Republic
[13] Listening Observatory for Hawaiian Ecosystems, Univ. of Hawai'i at Hilo,
Hilo, USA

Abstract. Building accurate knowledge of the identity, the geographic distribution and the evolution of species is essential for the sustainable development of humanity, as well as for biodiversity conservation. However, the difficulty of identifying plants, animals and fungi is hindering the aggregation of new data and knowledge. Identifying and naming living organisms is almost impossible for the general public and is often difficult even for professionals and naturalists. Bridging this gap is a key step towards enabling effective biodiversity monitoring systems. The LifeCLEF campaign, presented in this paper, has been promoting and evaluating advances in this domain since 2011. The 2022 edition proposes five data-oriented challenges related to the identification and prediction of biodiversity: (i) PlantCLEF: very large-scale plant identification, (ii) BirdCLEF: bird species recognition in audio soundscapes, (iii) GeoLifeCLEF: remote sensing based prediction of species, (iv) SnakeCLEF: snake species identification on a global scale, and

A. Barrón-Cedeño et al. (Eds.): CLEF 2022, LNCS 13390, pp. 257–285, 2022.
https://doi.org/10.1007/978-3-031-13643-6_19

(v) FungiCLEF: fungi recognition as an open set classification problem. This paper overviews the motivation, methodology and main outcomes of that five challenges.

1 LifeCLEF Lab Overview

Accurately identifying organisms observed in the wild is an essential step in ecological studies. Unfortunately, observing and identifying living organisms requires high levels of expertise. For instance, vascular plants alone account for more than 300,000 different species and the distinctions between them can be quite subtle. The world-wide shortage of trained taxonomists and curators capable of identifying organisms has come to be known as the *taxonomic imped-iment*. Since the Rio Conference of 1992, it has been recognized as one of the major obstacles to the global implementation of the Convention on Biological Diversity[1]. In 2004, Gaston and O'Neill [17] discussed the potential of automated approaches for species identification. They suggested that, if the scientific community were able to (i) produce large training datasets, (ii) precisely evaluate error rates, (iii) scale up automated approaches, and (iv) detect novel species, then it would be possible to develop a generic automated species identification system that would open up new vistas for research in biology and related fields.

Since the publication of [17], automated species identification has been studied in many contexts [4,19,20,32,50,76,77,87]. This area continues to expand rapidly, particularly due to advances in deep learning [3,18,51,60,79–82]. In order to measure progress in a sustainable and repeatable way, the LifeCLEF[2] research platform was created in 2014 as a continuation and extension of the plant identification task that had been run within the ImageCLEF lab[3] since 2011 [22–24]. Since 2014, LifeCLEF expanded the challenge by considering animals and fungi in addition to plants, and including audio and video content in addition to images [33–40]. Nearly a thousand researchers and data scientists register yearly to LifeCLEF in order to either download the data, subscribe to the mailing list, benefit from the shared evaluation tools, etc. The number of participants who finally crossed the finish line by submitting runs was respectively: 22 in 2014, 18 in 2015, 17 in 2016, 18 in 2017, 13 in 2018, 16 in 2019, 16 in 2020, 1,022 in 2021 (including the 1,004 participants of the BirdCLEF Kaggle challenge). The 2022 edition proposes five data-oriented challenges: three in the continuity of the 2021 edition (BirdCLEF, GeoLifeCLEF and SnakeCLEF), one new challenge related to fungi recognition with a focus on the combination of visual information with meta-data on an open species set (FungiCLEF), and a considerable expansion of the PlantCLEF challenge towards the identification of the world's flora (about 300K species).

---

[1] https://www.cbd.int/.
[2] http://www.lifeclef.org/.
[3] http://www.imageclef.org/.

The system used to run the challenges (registration, submission, leaderboard, etc.) was the AICrowd platform[4] for the PlantCLEF challenge and the Kaggle platform[5] for the GeoLifeCLEF, BirdCLEF, SnakeCLEF and FungiCLEF challenges. Three of the challenges (GeoLifeCLEF, SnakeCLEF, and FungiCLEF) were organized jointly with FGVC[6], an annual workshop dedicated to Fine-Grained Visual Categorization organized in the context of the CVPR[7] international conference on computer vision and pattern recognition.

In total, 951 people/teams participated to LifeCLEF 2022 edition by submitting runs to at least one of the five challenges (802 only for the BirdCLEF challenge). Only some of them managed to get the results right, and about 30 of them went all the way through the CLEF process by writing and submitting a *working note* describing their approach and results (for publication in CEUR-WS proceedings[8]). In the following sections, we provide a synthesis of the methodology and main outcomes of each of the five challenges. More details can be found in the extended overview reports of each challenge and in the individual working notes of the participants (references provided below).

2   PlantCLEF Challenge: Identify the World's Flora

A detailed description of the challenge and a more complete discussion of the results can be found in the dedicated working note [21].

2.1   Objective

Automated identification of plants has recently improved considerably thanks to the progress of deep learning and the availability of training data with more and more photos in the field. In the context of LifeCLEF 2018, we measured a top-1 classification accuracy over 10K species up to 90% and we showed that automated systems were not so far from human expertise [33]. However, these very high performances are far from being reached at the scale of the world flora. It is estimated that there are about 391,000 vascular plant species currently known to science and new plant species are still discovered and described each year. This plant diversity is a major element in the functioning of ecosystems as well as for the development of human civilization. Unfortunately, the vast majority of these species are very poorly known and the number of training images available is extremely low for the majority of them [67].

The goal of the 2022 edition of PlantCLEF was to take another step towards identifying the world's flora. Therefore, we have built a training set of unprecedented size covering 80K species and containing 4M images. It was shared with the community through a challenge[9] hosted on the AIcrowd platform.

---

[4] https://www.aicrowd.com.
[5] https://www.kaggle.com.
[6] http://www.fgvc.org/.
[7] https://cvpr2022.thecvf.com/.
[8] http://ceur-ws.org/.
[9] https://www.aicrowd.com/challenges/lifeclef-2022-plant.

2.2   Dataset

The training set is composed of two subsets: a trusted training set coming from the GBIF[10] portal (the world's largest biodversity data portal) and a web-based training set containing images collected via web search engines and containing several kinds of noise.

More precisely, the GBIF training dataset is based on a selection of more than 2.9M images covering 80k plant species shared and collected mainly via GBIF (and Encyclopedia Of Life[11] to a lesser extent). These images come mainly from academic sources (museums, universities, national institutions) and collaborative platforms such as inaturalist or Pl@ntNet, implying a fairly high certainty of determination quality (collaborative platforms only share their highest quality data qualified as "research graded"). To limit the size of the training set and limit class imbalance, the number of images was limited to around 100 images per species, favouring types of views adapted to the identification of plants (close-ups of flowers, fruits, leaves, trunks, ...).

The web dataset, on the other side, is based on a collection of web images provided by commercial search engines (Google and Bing). The raw downloaded data has a significant rate of species identification errors and a massive presence of (near)-duplicates and images not adapted for the identification of plant photographs (e.g. herbarium sheets, landscapes, microscopic views, ...). It even contains completely off-topic images such as portrait photos of botanists, maps, graphs, other kingdoms of the living, manufactured objects, etc. Thus, the raw data was cleaned up using a semi-automatic filtering (iterations of CNNs training, inference and human labelling). This filtering process drastically reduced the number of irrelevant pictures and also improved the overall image quality by favoring close-ups of flowers, fruits, leaves, trunks, etc. The web dataset finally contains about 1.1 million images covering about 57k species.

Participants were allowed to use complementary training data (e.g. for pre-training purposes) but at the condition that (i) the experiment is entirely reproducible, i.e. that the used external resource is clearly referenced and accessible to any other research group in the world, (ii) the use of external training data or not is mentioned for each run, and (iii) the additional resource does not contain any of the test observations. External training data was allowed but participants had to provide at least one submission that used only the provided data.

Lastly, the test set was built from multi-image plant observations collected on the Pl@ntNet platform during the year 2021 (observations not yet shared through GBIF, and thus not present in the training set). Only observations that received a very high confidence score in the Pl@ntNet collaborative review process were selected for the challenge to ensure the highest possible quality of determination. This process involves people with a wide range of skills (from beginners to world-leading experts), but these have different weights in the decision algorithms. Finally, the test set contains about 27k plant observations related

---

[10] https://gbif.org/.
[11] https://eol.org/.

to about 55k images (a plant can be associated with several images) covering about 7.3k species.

2.3 Evaluation Protocol

The primary metrics used for the evaluation of the task is be the Mean Reciprocal Rank. The MRR is a statistic measure for evaluating any process that produces a list of possible responses to a sample of queries ordered by probability of correctness. The reciprocal rank of a query response is the multiplicative inverse of the rank of the first correct answer. The MRR is the average of the reciprocal ranks for the whole test set:

$$MRR = \frac{1}{Q} \sum_{q=1}^{Q} \frac{1}{\text{rank}_q} \tag{1}$$

where Q is the total number of query occurrences (plant observations) in the test set. However, the macro-average version of the MRR (average MRR per species in the test set) was used because of the long tail of the data distribution to rebalance the results between under- and over-represented species in the test set.

2.4 Participants and Results

Eight participants registered to the PlantCLEF challenge hosted on AICrowd but only four of them managed to perform well. The four others encountered difficulties mainly related to the very large scale of the challenge (both in terms of the number of images and number of classes) and the need of high ended GPUs for resource-intensive experiments. Details of the methods and systems used are synthesized in the extended overview working note of the challenge [21] and further developed in the individual working notes of participants ([5,8,46, 58,68,86]. We report in Fig. 1 the performance achieved by the different runs of the participants.

The main outcomes we can derive from that results are the following:

– the best results were obtained by the only team which used vision transformers [86] contrary to the others which used convolutional neural networks, i.e. the traditional approach of the state-of-the-art for image-based plant identification. However, this gain in identification quality is paid for by a significant increase of the training time. The winning team reported that they had to stop the training of the model in order to submit their run to the challenge. Thus, better results could have surely been obtained with a few more days of training (as demonstrated through post-challenge evaluations reported in the their working note [86].
– One of the main difficulties of the challenge was the very large number of classes (80K). For most of the models used, the majority of the weights to be trained are those of the last fully connected layer of the classifier. This was an

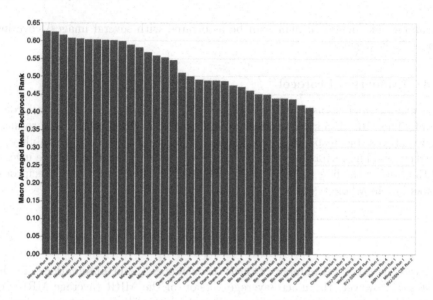

Fig. 1. PlantCLEF 2022 results

important consideration for all participants in their model selection strategy. Some teams have tried to limit this cost through specific approaches. The BioMachina team [5], in particular, used a two-level hierarchical softmax to reduce the number of weights drastically. They reported an considerable training time reduction while maintaining almost the same identification quality.

3 BirdCLEF Challenge: Bird Call Identification in Soundscape Recordings

A detailed description of the challenge and a more complete discussion of the results can be found in the dedicated working note [43].

3.1 Objective

The *LifeCLEF Bird Recognition Challenge* (BirdCLEF) was launched in 2014 and has since become the largest bird sound recognition challenge in terms of dataset size and species diversity, with multiple tens of thousands of recordings covering up to 1,500 species [25,41,42,44]. Birds are ideal indicators to identify early warning signs of habitat changes that are likely to affect many other species. They have been shown to respond to various environmental changes over many spatial scales. Large collections of (avian) audio data are an excellent resource to conduct research that can help to deal with environmental challenges of our time. The community platform Xeno-canto[12] in particular was launched in 2005

---

[12] https://www.xeno-canto.org/.

and hosts bird sounds from all continents. It receives new recordings every day from some of the remotest places on Earth. The Xeno-canto archive currently consists of more than 700,000 focal recordings covering over 10,000 species of birds, making it one of the most comprehensive collections of bird sound recordings worldwide, and certainly the most comprehensive collection shared under Creative Commons licenses. Xeno-canto data was used for BirdCLEF in all past editions to provide researchers with large and diverse datasets for training and testing.

In recent years, research in the domain of bioacoustics shifted towards deep neural networks for sound event recognition [45,73]. In past editions, we have seen many attempts to utilize convolutional neural network (CNN) classifiers to identify bird calls based on visual representations of these sounds (i.e., spectrograms) [26,48,59]. Despite their success for bird sound recognition in focal recordings, the classification performance of CNN on continuous, omnidirectional soundscapes remained low. Passive acoustic monitoring can be a valuable sampling tool for habitat assessments and the observation of environmental niches which often are endangered. However, manual processing of large collections of soundscape data is not desirable and automated attempts can help to advance this process [84]. Yet, the lack of suitable validation and test data prevented the development of reliable techniques to solve this task. Bridging the acoustic gap between high-quality training recordings and soundscapes with high ambient noise levels is one of the most challenging tasks in the domain of audio event recognition. This is especially true when sufficient amounts of training data are lacking. This is the case for many rare and endangered bird species around the globe and despite the vast amounts of data collected on Xeno-canto, audio data for endangered birds is still sparse. However, it is those endangered species that are most relevant for conservation, rendering acoustic monitoring of endangered birds particularly difficult.

The main goal of the 2022 edition of BirdCLEF was to advance automated detection of rare and endangered bird species that lack large amounts of training data. The competition was hosted on Kaggle[13] to attract machine learning experts from around the world to participate and submit. The overall task design was consistent with previous editions, but the focus was shifted towards species with very few training samples.

3.2 Dataset and Evaluation Protocol

As the "extinction capital of the world," Hawai'i has lost 68% of its bird species, the consequences of which can harm entire food chains. Researchers use population monitoring to understand how native birds react to changes in the environment and conservation efforts. But many of the remaining birds across the islands are isolated in difficult-to-access, high-elevation habitats. With physical monitoring difficult, scientists have turned to sound recordings. This approach could provide a passive, low labor, and cost-effective strategy for studying endangered bird populations.

---

[13] https://www.kaggle.com/c/birdclef-2022.

264   A. Joly et al.

Fig. 2. Expert ornithologists provided bounding box labels for all soundscape recordings indicating calling of 21 target species. In this example, all 'I' iwi calls were annotated, while vocalizations of other species were not labeled. This labeling scheme was applied to all test data soundscapes.

Current methods for processing large bioacoustic datasets involve manual annotation of each recording. This requires specialized training and prohibitively large amounts of time. Thankfully, recent advances in machine learning have made it possible to automatically identify bird songs for common species with ample training data. However, it remains challenging to develop such tools for rare and endangered species, such as those in Hawai'i.

Deploying a bird sound recognition system to a new recording and observation site requires classifiers that generalize well across different acoustic domains. Focal recordings of bird species form an excellent base to develop such a detection system. However, the lack of annotated soundscape data for a new deployment site poses a significant challenge. As in previous editions, training data was provided by the Xeno-canto community and consisted of more than 14,800 recordings covering 152 species. Participants were allowed to use metadata to develop their systems. Most notably, we provided detailed location information on recording sites of focal and soundscape recordings, allowing participants to account for migration and spatial distribution of bird species.

In this edition, test data, consisting of 5,356 soundscapes amounting to more than 90 h of recordings, were hidden and only accessible to participants during the inference process. These soundscapes were collected for various research projects by the Listening Observatory for Hawaiian Ecosystems (LOHE) at the University of Hawai'i at Hilo from 7 sites across the islands of Hawai'i, Maui, and Kaua'i. All soundscapes received some level of manual bird vocalization annotation by specially trained members of the LOHE lab using Raven Pro 1.5 software, however some recordings had a select few target species annotated, while others were annotated for every detectable species (see Fig. 2). In light of these uneven annotation strategies, only the subset of species for which every vocalization was annotated were scored for any given file. This resulted in a total of 21 scored bird species in the contest, 15 species endemic to the Hawaiian Islands and 6 introduced species.

The goal of the task was to localize and identify 21 target bird species within the provided soundscape test set. Each soundscape was divided into segments of

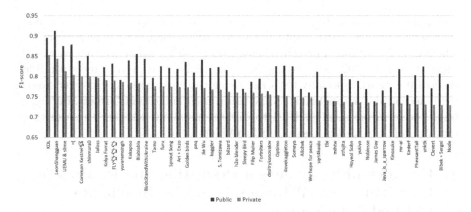

Fig. 3. Scores achieved by the best systems evaluated within the bird identification task of LifeCLEF 2022.

5 s, and a list of audible species had to be returned for each segment. The used evaluation metric was a weighted variant of the macro-averaged F1-score. In previous editions, ranking metrics were used to assess the overall classification performance. However, when applying bird call identification systems to real-world data, confidence thresholds have to be set in order to provide meaningful results. The F1-score as balanced metric between recall and precision appears to better reflect this circumstance. For each 5-second segment, a binary call indication for all 21 scored species had to be returned. Participants had to apply a threshold to determine if a species is vocalizing during a given segment (True) or not (False).

3.3 Participants and Results

1,019 participants from 62 countries on 807 teams entered the BirdCLEF 2022 competition and submitted a total of 23,352 runs. Details of the best methods and systems used are synthesized in the overview working notes paper of the task [43] and further developed in the individual working notes of participants. In Fig. 3 we report the performance achieved by the top 50 collected runs. The private leaderboard score is the primary metric and was revealed to participants after the submission deadline to avoid probing the hidden test data. Public leaderboard scores were visible to participants over the course of the entire challenge.

The baseline F1-score in this year's edition was 0.5112 (public 0.4849) with all scored birds marked as silent (False) for all segments, and 665 teams managed to score above this threshold. The best submission achieved a F1-score of 0.8527 (public 0.9128) and the top 10 best performing systems were within only 7% difference in score. The vast majority of approaches were based on convolutional neural network ensembles and mostly differed in pre- and post-processing and neural network backbone. Interestingly, few-shot learning techniques were vastly underrepresented despite the fact that some target species only had a handful of training samples. Participants employed various sophisticated post-processing schemes, most notably a percentile based thresholding approach that was established during the 2021 edition [28]. Some participants experimented with different loss functions, especially focal loss being the most notable. However, results were inconsistent across teams. Some teams used audio transformers, but again, results were inconsistent and led to discussions about whether these methods were appropriate for the task of bird call identification.

4   GeoLifeCLEF Challenge: Predicting Species Presence from Multi-modal Remote Sensing, Bioclimatic and Pedologic Images

A detailed description of the challenge and a more complete discussion of the results can be found in the dedicated working note [57].

4.1   Objective

Automatic prediction of the list of species most likely to be present at a given location is useful for many scenarios related to biodiversity management and conservation. First, it can improve species identification tools (whether automatic, semi-automatic or based on traditional field guides) by reducing the list of candidate species observable at a given site. Moreover, it can facilitate decision making related to land use and land management with regard to biodiversity conservation obligations (e.g., to determine new constructible areas or new natural areas to be protected). Last but not least, it can be used in the context of educational and citizen science initiatives, e.g., to determine regions of interest with a high species richness or vulnerable habitats to be monitored carefully.

4.2   Data Set and Evaluation Protocol

Data Collection. The data for this year's challenge is a cleaned-up version of the data from previous years, essentially removing species integrated by error and those observed less than 3 times. A detailed description of the GeoLifeCLEF 2020 dataset is provided in [9] and a complete changelog of the cleaning process is available on the Kaggle page[14]. In a nutshell, the dataset consists of over 1.6 million observations covering 17,037 plant and animal species distributed across US and France (as shown in Fig. 4). Each species observation is paired with high-resolution covariates (RGB-NIR imagery, land cover and altitude data) as illustrated in Fig. 5. These high-resolution covariates are resampled to a spatial resolution of 1 m per pixel and provided as 256×256 images covering a 256 m × 256 m

---

[14] https://www.kaggle.com/c/geolifeclef-2022-lifeclef-2022-fgvc9/data.

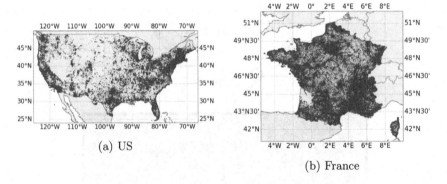

(a) US

(b) France

Fig. 4. Observations distribution over the US and France in GeoLifeCLEF 2022. Blue dots represent training data, red dots represent test data. (Color figure online)

square centered on each observation. RGB-NIR imagery come from the 2009–2011 cycle of the National Agriculture Imagery Program (NAIP) for the US[15], and from the BD-ORTHO® 2.0 and ORTHO-HR® 1.0 databases from the IGN for France[16]. Land cover data originates from the National Land Cover Database (NLCD) [30] for the U.S. and from CESBIO[17] for France. All elevation data comes from the NASA Shuttle Radar Topography Mission (SRTM)[18]. In addition, the dataset also includes traditional coarser resolution covariates: bio-climatic rasters ($1 \mathrm{km}^2$/pixel, from WorldClim [29]) and pedologic rasters ($250 \mathrm{m}^2$/pixel, from SoilGrids [27]).

Train-Test Split. The full set of occurrences is split in a training and testing set using a spatial block holdout procedure to limit the effect of *spatial auto-correlation* in the data [70]. Using this splitting procedure, a model cannot achieve a high performance by simply interpolating between training samples. The split was based on a global grid of 5km × 5km quadrats. 2.5% of these quadrats were randomly sampled and the observations falling in those formed the test set. 10% of those observations were used for the public leaderboard on Kaggle while the remaining 90% allowed to compute the private leaderboard providing the final results of the challenge. Similarly, another 2.5% of the quadrats were randomly sampled to provide an official validation set. The remaining quadrats and their associated observations were assigned to the training set.

Evaluation Metric. For each occurrence in the test set, the goal of the task was to return a candidate set of species likely to be present at that location. To measure the precision of the predicted sets, top-30 error rate was chosen as the main evaluation criterion. Each observation i is associated with a single ground-truth label y_i corresponding to the observed species. For each observation, the

---

[15] https://www.fsa.usda.gov.

[16] https://geoservices.ign.fr.

[17] http://osr-cesbio.ups-tlse.fr/~oso/posts/2017-03-30-carte-s2-2016/.

[18] https://lpdaac.usgs.gov/products/srtmgl1v003/.

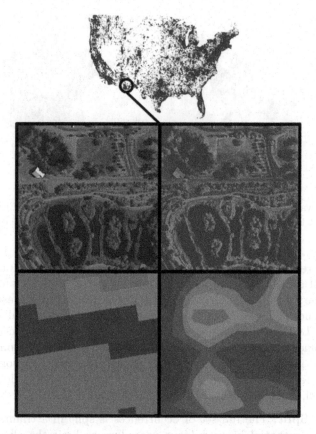

Fig. 5. In the GeoLifeCLEF dataset, each species observation is paired with high-resolution covariates (clockwise from top left: RGB imagery, IR imagery, altitude, land cover). (Color figure online)

submissions provided 30 candidate labels $\hat{y}_{i,1}, \hat{y}_{i,2}, \ldots, \hat{y}_{i,30}$. The top-30 error rate was then computed using

$$\text{Top-30 error rate} = \frac{1}{N} \sum_{i=1}^{N} e_i, \tag{2}$$

where

$$e_i = \begin{cases} 1 & \text{if } \forall k \in \{1, \ldots, 30\}, \ \hat{y}_{i,k} \neq y_i \\ 0 & \text{otherwise} \end{cases}. \tag{3}$$

Note that this evaluation metric does not try to correct the sampling bias inherent to present-only observation data (linked to the density of population, etc.). The absolute value of the resulting figures should thus be taken with care. Nevertheless, this metric does allow to compare the different approaches and

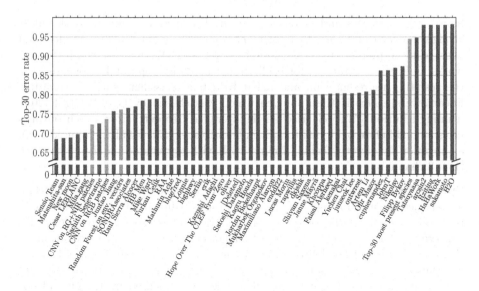

Fig. 6. Results of the GeoLifeCLEF 2022 task. The top-30 error rates of the best submission of each participant are shown in blue. The provided baselines are shown in orange. (Color figure online)

to determine which type of input data and of models are useful for the species presence detection task.

4.3 Participants and Results

52 teams participated and submitted at least one prediction file through the Kaggle[19] page of the GeoLifeCLEF 2022 challenge for a total number submission in the course of the competition of 261. Out of these teams, 7 managed to beat the weakest (non-constant) baseline provided and 5 the strongest one. Details of the baselines provided and of the methods used in the submitted runs are synthesized in the overview working note paper for this task [57]. The runs of 5 of those participants are further developed in their individual working notes [31, 47,49,75,88]. Figure 6 shows the final standings given by the Kaggle's private leaderboard. We briefly highlight the main methods used by the participants.

Multi-modal Data. The main challenge of this competition is to find a proper way to aggregate the heterogeneous sources of data and to deal with their respective characteristics: while RGB and NIR patches are standard images, other data is not directly provided in this format. For instance, altitude can not be casted in `uint8` without loss of information, land cover data is a categorical variable, bioclimatic and pedologic data have a resolution and range of their own, and, localisation (GPS coordinates) is a punctual information. Interestingly, the participants did try different means of aggregating this heterogeneous data with

---

[19] https://www.kaggle.com/c/geolifeclef-2022-lifeclef-2022-fgvc9.

more or less success and conflicting result. For instance, [49] tried the most straight-forward and easy to implement approach: train separate models and average their predictions. The winning team and [47,75] used complete networks as feature extractors for each chosen modality separately, concatenated the resulting representation and fed it to a final classifier (single or multiple linear layers). This is the approach which was chosen by GeoLifeCLEF 2021 winning solution [72]. [88] used single-layer features extractors which outputs are summed before being fed to a Swin transformer [55]. Finally, [75] used early aggregation by directly feeding the network with aggregated patches with more than 3 channels.

Species Imbalance. Another important trait of the dataset is its imbalance: a few species account for most of the observations, while a lot of them have only been observed a handful of times. [31,47] tried to use specialized method for this type of data such as focal loss [52], balanced softmax [69] or more advanced methods. These did not help improve their scores, most likely because the test set shares the same imbalance as the training set and the evaluation metric did take it into account (the fixed list of metrics implemented by Kaggle did not allow us to use a class-averaged top-30 error rate).

Presence-Only Observation Data. One last major characteristic of the dataset is that the observation data provided is presence-only data: at a given location, we only know that one species is present and do not have access of the complete list of species present nor the ones absent. The winning team and [47] tried to address this by using a grid of squared cells to aggregate the species observed into each cell. They then used this information in a different manner. The winning team tried to map the 30 species closest to each training point falling into its cell and used this list as the new label. Unfortunately, in the given time, this approach only resulted in overfitting. On the other hand, [47] successfully used the aggregated observations as a regularization method by replacing the label assigned to each training observation by another species from its cell 10% of the time.

Other methods were also tried out such as different architectures, different approaches for model pretraining (no pretraining, pretraining on ImageNet, on another dataset closer to GeoLifeCLEF 2022, etc.), multi-task learning, and a lot more. These are more exhaustively listed in the GeoLifeCLEF 2022 overview working note paper [57] along with a more detailed description of the methods presented above and further analyses.

5   SnakeCLEF Challenge: Automated Snake Species Identification on a Global Scale

A detailed description of the challenge and a more complete discussion of the results can be found in the dedicated overview paper [64].

5.1 Objective

Building an automatic and robust image-based system for snake species identification is an important goal for biodiversity, conservation, and global health. With over half a million victims of death and disability from venomous snakebite annually, such a system could significantly improve eco-epidemiological data and treatment outcomes (e.g. based on the specific use of antivenoms) [2,6]. Importantly, most herpetological expertise and most snake images are concentrated in developed countries in areas of the world where snake diversity is relatively low and snakebite is not a major public health concern. In contrast, remote parts of developing countries tend to lack expertise and images, even in areas where snake diversity is high and snakebites are common [15]. Thus, snake species identification assistance has a bigger potential to save lives in areas with the least information.

A primary difficulty of snake species identification lies in the high intra-class and low inter-class variance in appearance, which may depend on geographic location, color morph, sex, or age. At the same time, many species are visually similar to other species – mimicry (Fig. 7). Furthermore, our knowledge of which snake species occur in which countries is incomplete, and it is common that most or all images of a given snake species might originate from a small handful of countries or even a single country. Furthermore, many snake species resemble species found on other continents, with which they are entirely allopatric. Incorporating metadata on the geographic origin of an unidentified snake can narrow down the possible correct identifications considerably because only about 125 of the approximately 3,900 snake species co-occur in any given location [71]. It is known that more widespread species with more images are over-predicted relative to rare species with few images [16], and this can be a particularly vexing problem when trying to predict the identity of species that are widespread across areas of the world with few images.

The main goal of the SnakeCLEF 2022 competition was to provide a reliable evaluation ground for automatic snake species recognition. Like other LifeCLEF competitions, the SnakeCLEF 2022 competition was hosted on Kaggle[20] primarily to attract machine learning experts to participate and present their ideas.

5.2 Dataset and Evaluation Protocol

For this year, the dataset used in previous editions [62,65] has been extended with new and rare species. The number of species was doubled and the number of images from remote geographic areas with none or just a few samples was increased considerably, i.e., the uneven species distributions across all the countries was straightened. The SnakeCLEF 2022 dataset is based on 187,129 snake observations – multiple images of the same individual (refer to Fig. 8) – with 318,532 photographs belonging to 1,572 snake species and observed in 208 countries. The dataset has a heavy long-tailed class distribution, where the most

---

[20] https://www.kaggle.com/competitions/fungiclef2022.

Fig. 7. Harmless mimic species *Cemophora coccinea ssp. coccinea* (top row) and poisonous lookalike species. *Micrurus pyrrhocryptus, Micrurus ibiboboca,* and *Micrurus nigrocinctus* (left to right, bot. row). ©*roadmom*–iNaturalist, ©*Anthony Damiani*–iNaturalist, ©*Adam Cushen*–iNaturalist, ©*Alexander Guiñazu*–iNaturalist, ©*Tarik Câmara*–iNaturalist, *and* ©*Cristhian Banegas*–iNaturalist.

frequent species (*Natrix natrix*) is represented by 6,472 images and the least frequent species just by 5 samples. The difference in the number of images between the species with the most and fewest was reduced by an order of magnitude relative to SnakeCLEF2021. All the data was gathered from the online biodiversity platform – iNaturalist[21].

For testing, two sets were created: (i) the full test set for a machine evaluation, with 48,280 images from 28,431 observations, and (ii) the subset from the full test set with 150 observations, tailored for the human performance evaluation. Unlike in other LifeCLEF competitions, where the final testing set remained undisclosed, we provided the test data without labels to the participants. To prevent over fitting to the leaderboard, the evaluation method was composed of two stages; the first being the public leaderboard where the user scores were calculated on an unknown 20% of the test set, and the second a private leaderboard where participants were scored on the remaining part of the test set. In addition to image data, we provide:

---

[21] https://www.inaturalist.com/.

Fig. 8. Two snake observations from SnakeCLEF2022 dataset – three images for each individual. ©*André Giraldi* – iNaturalist, ©*Harshad Sharma* – iNaturalist.

- human verified species labels that allows up-scaling to higher taxonomic ranks,
- the country-species mapping file describing species-country presence to allow better regularization towards all geographical locations, based on The Reptile Database [78], and
- information about endemic species – species that occur only in one geographical region, e.g., Australia or Madagascar.

The geographical information, e.g., state and country labels, was included for approximately 95% of the training and test images. Additionally, we provide a mapping matrix (MM_{cs}) describing country-species presence to allow better worldwide regularization.

$$MM_{cs} = \begin{cases} 1 & \text{if species } S \in \text{country } C, \\ 0 & \text{otherwise.} \end{cases} \tag{4}$$

Unlike last year's dataset, where the vast majority (77%) of all images came from the United States and Canada, the SnakeCLEF 2022 dataset includes just a fraction of the data (28.3%) from the United States and Canada. The rest of the data is distributed across remaining regions, e.g., Europe, Asia, Africa, Australia and Oceania.

Evaluation: The main goal of this challenge was to build a system that is capable of recognizing 1,572 snake species based on the given snake observation – unseen set of images – and relevant geographical location. As a main metric, we use the macro F1 score (F_1^m). The F_1^m is defined as the mean of class-wise F1 scores:

$$F_1^m = \frac{1}{N} \sum_{s=0}^{N} F_{1_s}, \qquad F_{1_s} = 2 \times \frac{P_s \times R_s}{P_s + R_s}, \qquad (5)$$

where s is species index, N equals to the number of classes in a training set. The F1 score for each class represents harmonic mean of the class precision P_s and recall R_s.

5.3   Participants and Results

A total of 29 teams participated in the SnakeCLEF 2022 challenge and contributed with 648 submissions. Everyone who submitted a solution better than baseline submission, i.e., random predictions, was considered a participant. The number of participants quadrupled since last year, primarily as Kaggle was used as an evaluation platform. The best performing team achieved F_1^m of 86.47% on a private part of a test set and 94.01% accuracy on the full test set. On the expert set, the best performing team achieved an F_1^m of 90.28%. The performance evaluation for top-20 Teams is provided in Fig. 9. At the time of writing, the organisers could not reproduce any score from the leaderboard, even though most teams provided code.

Details of the best submitted methods and systems are synthesized in the overview working notes paper [64] and further developed in the individual working notes. The main outcomes we can derive from the achieved results are as follows:

Transformer-Based Architectures Outperformed CNNs. This year various deep neural network architectures – Convolutional Neural Networks and Transformers – were evaluated; ConvNext [56], EfficientNet [74], Vision Transformer [14], Swin Transformer [55], and MetaFormer [13]. Unlike last year, where the CNN architectures overwhelmed the performance, Vision Transformer architectures were a vital asset for most methods submitted this year. The second best method with F_1^m score of 84.56% was based on an ensemble of exclusively ViT models and performed slightly worst (−0.9%) than the best performing system that used a combination of Transformer and CNN models. An ensemble of MetaFormer models achieved the third-best score of 82.36%. It seems that Transformers and CNNs benefit from each other in an ensemble, while a standalone Transformer ensemble performs better than a pure CNN ensemble which achieved an F_1^m score of "only" 70.8%

Loss Function Matters. Several loss functions were evaluated: Label Aware Smoothing [89], (modified) Categorical Cross-Entropy, and Seesaw [83]. Overall, any Loss function if used is better than standard CrossEntropy. The wining team used Label Aware Smoothing. The runner-up used an Effective Logit Adjustment Loss and showed an improvement of around 2% of F_1^m score when compared to Cross Entropy, reducing the error rate by 15%. The third team used Logit adjustment to outperform the Seesaw loss from an F_1^m score of 76.5% to 78.6%.

Self-Supervision has Potential. Adding unlabeled data to the train set is a welcome option when not many observations of a species are available. The third team used the SimCLR [7] method with InfoNCE [61] loss function to increase the F_1^m score from 63.76% to 68.83% when compared to an ImageNet-1k pretrained models.

Geographical Metadata Improves Classification Performance. Most teams report accuracy improvement when adding the metadata into the learning process. The second team achieved an improvement of 10.9% in terms of the F_1^m score using a simple location filtering approach. The third team described an absolute improvement of 7.5% when adding the metadata into the MetaFormer.

Ensemble Helps, but at What Cost? Most teams used ensembling to increase the accuracy of classification. The standard approach was to compute an average of the individual models' decisions. Some teams used a late fusion of deep features by concatenation as an ensemble technique. Even though the improvement in accuracy is observable (around 1% point of F_1^m across the board), it would be interesting to measure the added computational complexity vs the added accuracy. In the case of snakebite, the system's inference time plays a crucial role.

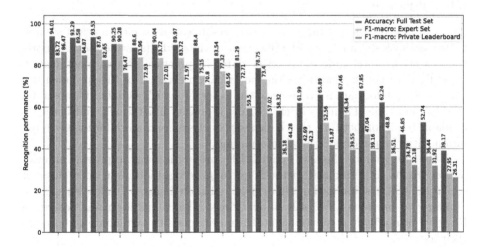

Fig. 9. SnakeCLEF 2022 competition: Top20 teams Performance. Accuracy on Full Test set, and Macro F_1 score on private part of the test set and Expert set. Sorted by performance on the private leaderboard.

6 FungiCLEF Challenge: Fungi Recognition as an Open Set Classification Problem

A detailed description of the challenge and a more complete discussion of the results can be found in the dedicated overview paper [66].

Fig. 10. Two fungi specimen observations from the Danish Fungi 2020 dataset. Atlas of Danish Fungi: ©*Jan Riis-Hansen and* ©*Arne Pedersen.*

6.1 Objective

Automatic recognition of fungi species assists mycologists, citizen scientists and nature enthusiasts in species identification in the wild. Its availability supports the collection of valuable biodiversity data. In practice, species identification typically does not depend solely on the visual observation of the specimen but also on other information available to the observer – such as habitat, substrate, location and time. Thanks to rich metadata, precise annotations, and baselines available to all competitors, the challenge provides a benchmark for image recognition with the use of additional information.

The main goal for the new FungiCLEF competition was to provide an evaluation ground for automatic methods for fungi recognition in an open class set scenario, i.e., the submitted methods have to handle images of unknown species. Similarly to previous LifeCLEF competitions, The competition was hosted on Kaggle[22] primarily to attract machine learning experts to participate and present their ideas.

6.2 Dataset and Evaluation Protocol

Data Collection: The FungiCLEF 2022 dataset is based on data collected through the Atlas of Danish Fungi mobile (iOS[23] and Android[24]) and Web[25] applications.

---

[22] https://www.kaggle.com/competitions/fungiclef2022.
[23] https://apps.apple.com/us/app/atlas-of-danish-fungi/id1467728588.
[24] https://play.google.com/store/apps/details?id=com.noque.svampeatlas.
[25] https://svampe.databasen.org/.

The Atlas of Danish Fungi is a citizen science platform with more than 4,000 actively contributing volunteers and with more than 1 million content-checked observations of approximately 8,650 fungi species.

For training, the competitors were provided with the DanishFungi 2020 (DF20) dataset [63]. DF20 contains 295,938 images – 266,344 for training and 29,594 for validation – belonging to 1,604 species. All training samples passed an expert validation process, guaranteeing high quality labels. Furthermore, rich observation metadata about habitat, substrate, time, location, EXIF etc. are provided.

The test dataset is constructed from all observations submitted in 2021, for which expert-verified species labels are available. It includes observations collected across all substrate and habitat types. The test set contains 59,420 observations with 118,676 images belonging to 3,134 species: 1,165 known from the training set and 1,969 unknown species covering approximately 30% of the test observations. The test set was further split into public (20%) and private (80%) subsets – a common practice for Kaggle competitions to prevent participants from overfitting to the leaderboard.

Task Description: The goal of the task is to return the correct species (or "unknown") for each test observation, consisting from a set of images and metadata. Photographs of unknown fungi species should be classified into an "unknown" class with label id -1. A baseline procedure to include meta-data in the decision problem and baseline pre-trained image classifiers were provided as part of the task description to all participants.

Evaluation Protocol: The evaluation process consisted of two stages: (i) a public evaluation, which was available during the whole competition with a limit of two submissions a day, and (ii) a private evaluation used for the final leaderboard. The main evaluation metric for the competition was the F_1^m, defined as the mean of class-wise F_1 scores:

$$F_1^m = \frac{1}{N} \sum_{s=1}^{N} F_{1_s}, \tag{6}$$

where N represents the number of classes – in case of the Kaggle evaluation, $N = 1,165$ (#classes in the test set) – and s is the species index. The F_1 score for each class is calculated as a harmonic mean of the class precision P_S and recall R_S:

$$F_{1_s} = 2 \times \frac{P_s \times R_s}{P_s + R_s}, \quad P_s = \frac{tp_s}{tp_s + fp_s}, \quad R_s = \frac{tp_s}{tp_s + fn_s} \tag{7}$$

In single-label multi-class classification, the True Positives (tp) of a species represents the number of correct Top1 predictions of that species, False Positive (fp) denotes how many times was different species predicted instead of the (tp), and False Negatives (fn) indicates how many images of species s have been wrongly classified.

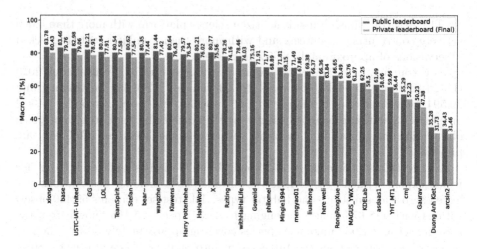

Fig. 11. Results of the FungiCLEF 2022 competition on Kaggle, sorted by performance on the final (private) test set.

6.3 Participants and Results

In total, 38 teams contributed with 701 valid submissions to the challenge evaluation on Kaggle. A detailed description of the methods used in the submitted runs is available in the overview working note paper [66] and further developed in the individual working notes. The results on the public and private test sets (leaderboards) are displayed in Fig. 11.

All submissions that shared their working notes were based on modern Convolutional Neural Network (CNN) or transformer-inspired architectures, such as Metaformer [13], Swin Transformer [55], and BEiT [1]. The best performing teams used ensembles of both CNNs and Transformers. The winning team [85] achieved 80.43% accuracy with a combination of ConvNext-large [56] and MetaFormer [13]. The results were often improved by combining predictions belonging to the same observation and by both training-time and test-time data augmentations.

Participants experimented with a number of different training losses to battle the long tail distribution and fine-grained classification with small inter-class differences and large intra-class differences: besides the standard Cross Entropy loss function, we have seen successful applications of the Seesaw loss [83], Focal loss [52], Arcface loss [11], Sub-Center loss [10] and Adaptive Margin [53].

We were happy to see the participants experimented with different use of the provided observation metadata, which often lead to improvements in the recognition scores. Besides the probabilistic baseline published with the dataset [63], we have seen hand-crafted encoding of the metadata into feature vectors, as well as encoding of the meta-data with a multilingual BERT model [12] and RoBERTa [54]. The meta-data were then combined with image features extracted from a CNN or Transformer image classifier, or directly used as an input to Metaformer [13].

7   Conclusions and Perspectives

The main outcome of this collaborative evaluation is a new snapshot of the performance of state-of-the-art computer vision, bio-acoustic and machine learning techniques towards building real-world biodiversity monitoring systems. This study shows that recent deep learning techniques still allow some consistent progress for most of the evaluated tasks. One of the main new insights of this edition of LifeCLEF is that vision transformers performed better than CNNs in some tasks, in particular in the PlantCLEF task for which the best model is a vision transformer whose training was not yet completed at the time of the challenge closure. This shows the potential of these techniques on huge datasets such as the one of PlantCLEF (4M images of 80K species). However, training those models requires more computational resources that only participants with access to large computational clusters can afford. In the other challenges, what seems to best explain the best performances is the model selection methodology employed given the time constraints and the available computational resources. Participants must carefully prioritize the approaches they want to test with a compromise between novelty and efficiency. New methods are typically more risky than that well-known recipes. However, when they work they can make a real difference to the other participants. The challenge where there were the most methodological novelty is probably the GeoLifeCLEF challenge. It is indeed quite unusual due to its multi-modal nature (mixing very different types) and the originality of the task itself (set-valued classification based on presence-only data). The way all the modalities were combined was clearly one of the main driver of success. Moreover, the set-valued classification problem has encouraged the implementation of an original label swapping strategy that has proven to be effective. In the FungiCLEF challenge, several participants utilized the provided metadata in the decision process of a fine-grained image classification task – either by combining image and metadata embeddings in a classifier, or by directly feeding the image and the metadata in a transformer/MetaFormer [13] architecture. Finally, the long-tail distribution problem (common to all tasks) has also been one of the most explored research topics through the different challenges (in particular the SnakeCLEF and FungiCLEF challenges). While it is difficult to draw a simple conclusion about the superiority of some approaches over others, many participants showed that substantial gains could be made by taking the long tail problem into account (including alternative loss functions to cross-entropy or self-supervision on unlabeled data).

Acknowledgements. This project has received funding from the European Union's Horizon 2020 research and innovation programme under grant agreement No° 863463 (Cos4Cloud project), and the support of #DigitAG.

References

1. Bao, H., Dong, L., Wei, F.: BEiT: BERT pre-training of image transformers. arXiv preprint arXiv:2106.08254 (2021)

2. Bolon, I., et al.: Identifying the snake: first scoping review on practices of communities and healthcare providers confronted with snakebite across the world. PLoS one **15**(3), e0229989 (2020)

3. Bonnet, P., et al.: Plant identification: experts vs. machines in the Era of deep learning. In: Joly, A., Vrochidis, S., Karatzas, K., Karppinen, A., Bonnet, P. (eds.) Multimedia Tools and Applications for Environmental and Biodiversity Informatics. MSA, pp. 131–149. Springer, Cham (2018). https://doi.org/10.1007/978-3-319-76445-0_8

4. Cai, J., Ee, D., Pham, B., Roe, P., Zhang, J.: Sensor network for the monitoring of ecosystem: Bird species recognition. In: 3rd International Conference on Intelligent Sensors, Sensor Networks and Information, 2007. ISSNIP 2007 (2007). https://doi.org/10.1109/ISSNIP.2007.4496859

5. Carranza-Rojas, J., Gonzalez-Villanueva, R., Jimenez-Morales, K., Quesada-Montero, K., Esquivel-Barboza, E., Carvajal-Barboza, N.: Extreme automatic plant identification under constrained resources. In: Working Notes of CLEF 2022 - Conference and Labs of the Evaluation Forum (2022)

6. de Castañeda, R.R., et al.: Snakebite and snake identification: empowering neglected communities and health-care providers with AI. Lancet Digit. Health **1**(5), e202–e203 (2019)

7. Chen, T., Kornblith, S., Norouzi, M., Hinton, G.: A simple framework for contrastive learning of visual representations. In: International Conference on Machine Learning, pp. 1597–1607. PMLR (2020)

8. Chulif, S., Lee, S.H., Chang, Y.L.: A global-scale plant identification using deep learning: Neuon submission to PlantCLEF 2022. In: Working Notes of CLEF 2022 - Conference and Labs of the Evaluation Forum (2022)

9. Cole, E., et al.: The GeoLifeCLEF 2020 dataset. arXiv preprint arXiv:2004.04192 (2020)

10. Deng, J., Guo, J., Liu, T., Gong, M., Zafeiriou, S.: Sub-center ArcFace: boosting face recognition by large-scale noisy web faces. In: Vedaldi, A., Bischof, H., Brox, T., Frahm, J.-M. (eds.) ECCV 2020. LNCS, vol. 12356, pp. 741–757. Springer, Cham (2020). https://doi.org/10.1007/978-3-030-58621-8_43

11. Deng, J., Guo, J., Xue, N., Zafeiriou, S.: Arcface: additive angular margin loss for deep face recognition. In: Proceedings of the IEEE/CVF Conference on Computer Vision and Pattern Recognition, pp. 4690–4699 (2019)

12. Devlin, J., Chang, M.W., Lee, K., Toutanova, K.: BERT: pre-training of deep bidirectional transformers for language understanding. arXiv preprint arXiv:1810.04805 (2018)

13. Diao, Q., Jiang, Y., Wen, B., Sun, J., Yuan, Z.: MetaFormer: a unified meta framework for fine-grained recognition. arXiv preprint arXiv:2203.02751 (2022)

14. Dosovitskiy, A., et al.: An image is worth 16×16 words: transformers for image recognition at scale. arXiv preprint arXiv:2010.11929 (2020)

15. Durso, A.M., et al.: Citizen science and online data: opportunities and challenges for snake ecology and action against snakebite. Toxicon: X **9**, 100071 (2021)

16. Durso, A.M., Moorthy, G.K., Mohanty, S.P., Bolon, I., Salathé, M., Ruiz De Castañeda, R.: Supervised learning computer vision benchmark for snake species identification from photographs: implications for herpetology and global health. Front. Artif. Intell. **4**, 17 (2021)

17. Gaston, K.J., O'Neill, M.A.: Automated species identification: why not? Philos. Trans. Roy. Soc. London B: Biol. Sci. **359**(1444), 655–667 (2004)

18. Ghazi, M.M., Yanikoglu, B., Aptoula, E.: Plant identification using deep neural networks via optimization of transfer learning parameters. Neurocomputing **235**, 228–235 (2017)

19. Glotin, H., Clark, C., LeCun, Y., Dugan, P., Halkias, X., Sueur, J.: Proceedings of the 1st Workshop on Machine Learning for Bioacoustics - ICML4B. ICML, Atlanta USA (2013). http://sabiod.org/ICML4B2013_book.pdf

20. Glotin, H., LeCun, Y., Artières, T., Mallat, S., Tchernichovski, O., Halkias, X.: Neural Information Processing Scaled for Bioacoustics, from Neurons to Big Data. In: Proceedings of the NIPS International Conference (2013). http://sabiod.org/nips4b

21. Goëau, H., Bonnet, P., Joly, A.: Overview of PlantCLEF 2022: image-based plant identification at global scale. In: Working Notes of CLEF 2022 - Conference and Labs of the Evaluation Forum (2022)

22. Goëau, H., Bonnet, P., Joly, A., Bakic, V., Barthélémy, D., Boujemaa, N., Molino, J.F.: The imageclef 2013 plant identification task. In: CLEF task overview 2013, CLEF: Conference and Labs of the Evaluation Forum, September 2013, Valencia, Spain. Valencia (2013)

23. Goëau, H., et al.: The imageclef 2011 plant images classification task. In: CLEF task overview 2011, CLEF: Conference and Labs of the Evaluation Forum, September 2011, Amsterdam, Netherlands (2011)

24. Goëau, H., et al.: Imageclef 2012 plant images identification task. In: CLEF task overview 2012, CLEF: Conference and Labs of the Evaluation Forum, September 2012, Rome, Italy. Rome (2012)

25. Goëau, H., Glotin, H., Planqué, R., Vellinga, W.P., Stefan, K., Joly, A.: Overview of BirdCLEF 2018: monophone vs. soundscape bird identification. In: CLEF task overview 2018, CLEF: Conference and Labs of the Evaluation Forum, September 2018, Avignon, France (2018)

26. Grill, T., Schlüter, J.: Two convolutional neural networks for bird detection in audio signals. In: 2017 25th European Signal Processing Conference (EUSIPCO), pp. 1764–1768, August 2017. https://doi.org/10.23919/EUSIPCO.2017.8081512

27. Hengl, T., et al.: SoilGrids250m: global gridded soil information based on machine learning. PLoS one **12**(2), e0169748 (2017)

28. Henkel, C., Pfeiffer, P., Singer, P.: Recognizing bird species in diverse soundscapes under weak supervision. In: CLEF Working Notes 2021, CLEF: Conference and Labs of the Evaluation Forum, September 2021, Bucharest, Romania (2021)

29. Hijmans, R.J., Cameron, S.E., Parra, J.L., Jones, P.G., Jarvis, A.: Very high resolution interpolated climate surfaces for global land areas. Int. J. Climatol. J. Roy. Meteorol. Soc. **25**(15), 1965–1978 (2005)

30. Homer, C., Dewitz, J., Yang, L., Jin, S., Danielson, P., Xian, G., Coulston, J., Herold, N., Wickham, J., Megown, K.: Completion of the 2011 national land cover database for the conterminous united states - representing a decade of land cover change information. Photogram. Eng. Remote Sens. **81**(5), 345–354 (2015)

31. Jiang, J.: Localization of plant and animal species prediction with convolutional neural networks. In: Working Notes of CLEF 2022 - Conference and Labs of the Evaluation Forum (2022)

32. Joly, A., et al.: Interactive plant identification based on social image data. Eco. Inform. **23**, 22–34 (2014)

33. Joly, A., et al.: Overview of LifeCLEF 2018: a large-scale evaluation of species identification and recommendation algorithms in the Era of AI. In: Bellot, P., et al. (eds.) CLEF 2018. LNCS, vol. 11018, pp. 247–266. Springer, Cham (2018). https://doi.org/10.1007/978-3-319-98932-7_24

34. Joly, A., et al.: Overview of LifeCLEF 2019: identification of Amazonian plants, South and North American birds, and Niche prediction. In: Crestani, F., et al. (eds.) CLEF 2019. LNCS, vol. 11696, pp. 387–401. Springer, Cham (2019). https://doi.org/10.1007/978-3-030-28577-7_29

35. Joly, A., et al.: LifeCLEF 2016: multimedia life species identification challenges. In: Fuhr, N., et al. (eds.) CLEF 2016. LNCS, vol. 9822, pp. 286–310. Springer, Cham (2016). https://doi.org/10.1007/978-3-319-44564-9_26

36. Joly, A., et al.: LifeCLEF 2017 lab overview: multimedia species identification challenges. In: Jones, G.J.F., et al. (eds.) CLEF 2017. LNCS, vol. 10456, pp. 255–274. Springer, Cham (2017). https://doi.org/10.1007/978-3-319-65813-1_24

37. Joly, A., et al.: LifeCLEF 2014: multimedia life species identification challenges. In: Kanoulas, E., et al. (eds.) CLEF 2014. LNCS, vol. 8685, pp. 229–249. Springer, Cham (2014). https://doi.org/10.1007/978-3-319-11382-1_20

38. Joly, A., et al.: LifeCLEF 2015: multimedia life species identification challenges. In: Mothe, J., et al. (eds.) CLEF 2015. LNCS, vol. 9283, pp. 462–483. Springer, Cham (2015). https://doi.org/10.1007/978-3-319-24027-5_46

39. Joly, A., et al.: Overview of LifeCLEF 2020: a system-oriented evaluation of automated species identification and species distribution prediction. In: Arampatzis, A., et al. (eds.) CLEF 2020. LNCS, vol. 12260, pp. 342–363. Springer, Cham (2020). https://doi.org/10.1007/978-3-030-58219-7_23

40. Joly, A., et al.: Overview of LifeCLEF 2021: an evaluation of machine-learning based species identification and species distribution prediction. In: Candan, K.S., et al. (eds.) CLEF 2021. LNCS, vol. 12880, pp. 371–393. Springer, Cham (2021). https://doi.org/10.1007/978-3-030-85251-1_24

41. Kahl, S., Clapp, M., Hopping, A., Goëau, H., Glotin, H., Planqué, R., Vellinga, W.P., Joly, A.: Overview of BirdCLEF 2020: bird sound recognition in complex acoustic environments. In: CLEF task overview 2020, CLEF: Conference and Labs of the Evaluation Forum, September 2020, Thessaloniki, Greece (2020)

42. Kahl, S., Denton, T., Klinck, H., Glotin, H., Goëau, H., Vellinga, W.P., Planqué, R., Joly, A.: Overview of BirdCLEF 2021: bird call identification in soundscape recordings. In: Working Notes of CLEF 2021 - Conference and Labs of the Evaluation Forum (2021)

43. Kahl, S., et al.: Overview of BirdCLEF 2022: Endangered bird species recognition in soundscape recordings. In: Working Notes of CLEF 2022 - Conference and Labs of the Evaluation Forum (2022)

44. Kahl, S., Stöter, F.R., Glotin, H., Planqué, R., Vellinga, W.P., Joly, A.: Overview of Birdclef 2019: large-scale bird recognition in soundscapes. In: CLEF task overview 2019, CLEF: Conference and Labs of the Evaluation Forum, September 2019, Lugano, Switzerland (2019)

45. Kahl, S., Wood, C.M., Eibl, M., Klinck, H.: Birdnet: a deep learning solution for avian diversity monitoring. Eco. Inform. 61, 101236 (2021)

46. Karun, A., Divyasri, K., Balasundaram, P., Sella Veluswami, J.R.: Plant species identification using probability tree approach of deep learning models. In: Working Notes of CLEF 2022 - Conference and Labs of the Evaluation Forum (2022)

47. Kellenberger, B., Devis, T.: Block label swap for species distribution modelling. In: Working Notes of CLEF 2022 - Conference and Labs of the Evaluation Forum (2022)

48. Lasseck, M.: Audio-based bird species identification with deep convolutional neural networks. In: CLEF Working Notes 2018, CLEF: Conference and Labs of the Evaluation Forum, September 2018, Avignon, France (2018)

49. Leblanc, C., Lorieul, T., Servajean, M., Bonnet, P., Joly, A.: Species distribution modeling based on aerial images and environmental features with convolutional neural networks. In: Working Notes of CLEF 2022 - Conference and Labs of the Evaluation Forum (2022)
50. Lee, D.J., Schoenberger, R.B., Shiozawa, D., Xu, X., Zhan, P.: Contour matching for a fish recognition and migration-monitoring system. In: Optics East, pp. 37–48. International Society for Optics and Photonics (2004)
51. Lee, S.H., Chan, C.S., Remagnino, P.: Multi-organ plant classification based on convolutional and recurrent neural networks. IEEE Trans. Image Process. **27**(9), 4287–4301 (2018)
52. Lin, T.Y., Goyal, P., Girshick, R., He, K., Dollár, P.: Focal loss for dense object detection. In: Proceedings of the IEEE International Conference on Computer Vision, pp. 2980–2988 (2017)
53. Liu, H., Zhu, X., Lei, Z., Li, S.Z.: Adaptiveface: adaptive margin and sampling for face recognition. In: Proceedings of the IEEE/CVF Conference on Computer Vision and Pattern Recognition, pp. 11947–11956 (2019)
54. Liu, Y., et al.: Roberta: a robustly optimized Bert pretraining approach. arXiv preprint arXiv:1907.11692 (2019)
55. Liu, Z., et al.: Swin transformer: hierarchical vision transformer using shifted windows. In: Proceedings of the IEEE/CVF International Conference on Computer Vision, pp. 10012–10022 (2021)
56. Liu, Z., Mao, H., Wu, C.Y., Feichtenhofer, C., Darrell, T., Xie, S.: A convnet for the 2020s. arXiv preprint arXiv:2201.03545 (2022)
57. Lorieul, T., Cole, E., Deneu, B., Servajean, M., Joly, A.: Overview of GeoLifeCLEF 2022: Predicting species presence from multi-modal remote sensing, bioclimatic and pedologic data. In: Working Notes of CLEF 2022 - Conference and Labs of the Evaluation Forum (2022)
58. Ong, J.M., Yang, S.J., Ng, K.W., Chan, C.S.: Image-based plant identification with taxonomy aware architecture. In: Working Notes of CLEF 2022 - Conference and Labs of the Evaluation Forum (2022)
59. Mühling, M., Franz, J., Korfhage, N., Freisleben, B.: Bird species recognition via neural architecture search. In: CLEF working notes 2020, CLEF: Conference and Labs of the Evaluation Forum, September 2020, Thessaloniki, Greece (2020)
60. Norouzzadeh, M.S., Morris, D., Beery, S., Joshi, N., Jojic, N., Clune, J.: A deep active learning system for species identification and counting in camera trap images. Methods Ecol. Evol. **12**(1), 150–161 (2021)
61. Van den Oord, A., Li, Y., Vinyals, O.: Representation learning with contrastive predictive coding. arXiv e-prints pp. arXiv-1807 (2018)
62. Picek, L., Ruiz De Castañeda, R., Durso, A.M., Sharada, P.M.: Overview of the snakeclef 2020: Automatic snake species identification challenge. In: CLEF task overview 2020, CLEF: Conference and Labs of the Evaluation Forum, September 2020, Thessaloniki, Greece (2020)
63. Picek, L., et al.: Danish fungi 2020-not just another image recognition dataset. In: Proceedings of the IEEE/CVF Winter Conference on Applications of Computer Vision, pp. 1525–1535 (2022)
64. Picek, L., Durso, A.M., Hrúz, M., Bolon, I.: Overview of SnakeCLEF 2022: automated snake species identification on a global scale. In: Working Notes of CLEF 2022 - Conference and Labs of the Evaluation Forum (2022)
65. Picek, L., Durso, A.M., Ruiz De Castañeda, R., Bolon, I.: Overview of SnakeCLEF 2021: automatic snake species identification with country-level focus. In: Working Notes of CLEF 2021 - Conference and Labs of the Evaluation Forum (2021)

66. Picek, L., Šulc, M., Heilmann-Clausen, J., Matas, J.: Overview of FungiCLEF 2022: fungi recognition as an open set classification problem. In: Working Notes of CLEF 2022 - Conference and Labs of the Evaluation Forum (2022)

67. Pitman, N.C., et al.: Identifying gaps in the photographic record of the vascular plant flora of the Americas. Nature plants **7**(8), 1010–1014 (2021)

68. Pravinkrishnan, K., Sivakumar, N., Balasundaram, P., Kalinathan, L.: Classification of plant species using Alexnet architecture. In: Working Notes of CLEF 2022 - Conference and Labs of the Evaluation Forum (2022)

69. Ren, J., Yu, C., Ma, X., Zhao, H., Yi, S., et al.: Balanced meta-softmax for long-tailed visual recognition. Adv. Neural. Inf. Process. Syst. **33**, 4175–4186 (2020)

70. Roberts, D.R., et al.: Cross-validation strategies for data with temporal, spatial, hierarchical, or phylogenetic structure. Ecography **40**(8), 913–929 (2017)

71. Roll, U., et al.: The global distribution of Tetrapods reveals a need for targeted reptile conservation. Nature Ecol. Evol. **1**(11), 1677–1682 (2017)

72. Seneviratne, S.: Contrastive representation learning for natural world imagery: habitat prediction for 30,000 species. In: Working Notes of CLEF 2021 - Conference and Labs of the Evaluation Forum (2021)

73. Shiu, Y., et al.: Deep neural networks for automated detection of marine mammal species. Sci. Rep. **10**(1), 1–12 (2020)

74. Tan, M., Le, Q.: Efficientnet: rethinking model scaling for convolutional neural networks. In: International Conference on Machine Learning, pp. 6105–6114. PMLR (2019)

75. Teng, M., Elkafrawy, S.: Participation to the GeoLifeCLEF challenge 2022 working notes. In: Working Notes of CLEF 2022 - Conference and Labs of the Evaluation Forum (2022)

76. Towsey, M., Planitz, B., Nantes, A., Wimmer, J., Roe, P.: A toolbox for animal call recognition. Bioacoustics **21**(2), 107–125 (2012)

77. Trifa, V.M., Kirschel, A.N., Taylor, C.E., Vallejo, E.E.: Automated species recognition of antbirds in a Mexican rainforest using hidden Markov models. J. Acoust. Soc. Am. **123**, 2424 (2008)

78. Uetz, P., Freed, P., Hošek, J., et al.: The reptile database (2020). https://reptile-database.reptarium.cz/advanced_search

79. Van Horn, G., et al.: The iNaturalist species classification and detection dataset. In: CVPR (2018)

80. Villon, S., Mouillot, D., Chaumont, M., Subsol, G., Claverie, T., Villéger, S.: A new method to control error rates in automated species identification with deep learning algorithms. Sci. Rep. **10**(1), 1–13 (2020)

81. Wäldchen, J., Mäder, P.: Machine learning for image based species identification. Methods Ecol. Evol. **9**(11), 2216–2225 (2018)

82. Wäldchen, J., Rzanny, M., Seeland, M., Mäder, P.: Automated plant species identification-trends and future directions. PLoS Comput. Biol. **14**(4), e1005993 (2018)

83. Wang, J., et al.: Seesaw loss for long-tailed instance segmentation. In: Proceedings of the IEEE/CVF Conference on Computer Vision and Pattern Recognition, pp. 9695–9704 (2021)

84. Wood, C.M., Kahl, S., Chaon, P., Peery, M.Z., Klinck, H.: Survey coverage, recording duration and community composition affect observed species richness in passive acoustic surveys. Methods Ecol. Evol. **12**(5), 885–896 (2021)

85. Xiong, Z., et al.: An empirical study for fine-grained fungi recognition with transformer and convnet. In: Working Notes of CLEF 2022 - Conference and Labs of the Evaluation Forum (2022)

86. Xu, M., Yoon, S., Lee, J., Park, D.S.: Vision transformer-based unsupervised transfer learning for large scale plant identification. In: Working Notes of CLEF 2022 - Conference and Labs of the Evaluation Forum (2022)
87. Yu, X., Wang, J., Kays, R., Jansen, P.A., Wang, T., Huang, T.: Automated identification of animal species in camera trap images. EURASIP J. Image Video Process. (2013)
88. Zhang, X., Zhou, Y.: A multimodal model for predict the localization of plant and animal species. In: Working Notes of CLEF 2022 - Conference and Labs of the Evaluation Forum (2022)
89. Zhong, Z., Cui, J., Liu, S., Jia, J.: Improving calibration for long-tailed recognition. In: Proceedings of the IEEE/CVF Conference on Computer Vision and Pattern Recognition, pp. 16489–16498 (2021)

Overview of ARQMath-3 (2022): Third CLEF Lab on Answer Retrieval for Questions on Math

Behrooz Mansouri[1(✉)], Vít Novotný[2], Anurag Agarwal[1], Douglas W. Oard[3], and Richard Zanibbi[1]

[1] Rochester Institute of Technology, Rochester, NY, USA
{bm3302,axasma,rxzvcs}@rit.edu
[2] Faculty of Informatics, Masaryk University, Brno, Czech Republic
witiko@mail.muni.cz
[3] University of Maryland, College Park, MD, USA
oard@umd.edu

Abstract. This paper provides an overview of the third and final year of the Answer Retrieval for Questions on Math (ARQMath-3) lab, run as part of CLEF 2022. ARQMath has aimed to introduce test collections for math-aware information retrieval. ARQMath-3 has two main tasks, Answer Retrieval (Task 1) and Formula Search (Task 2), along with a new pilot task Open Domain Question Answering (Task 3). Nine teams participated in ARQMath-3, submitting 33 runs for Task 1, 19 runs for Task 2, and 13 runs for Task 3. Tasks, topics, evaluation protocols, and results for each task are presented in this lab overview.

Keywords: Community Question Answering · Open Domain Question Answering · Mathematical Information Retrieval · Math-aware Search · Math Formula Search

1 Introduction

Math information retrieval (Math IR) aims at facilitating the access, retrieval and discovery of math resources, and is needed in many scenarios [12]. For example, many traditional courses and Massive Open Online Courses (MOOCs) release their resources (books, lecture notes and exercises, etc.) as digital files in HTML or XML. However, due to the specific characteristics of math formulae, classic search engines do not work well for indexing and retrieving math.

Math-aware search systems can be beneficial for learning activities. Students can search for references to help solve problems, increase knowledge, reduce doubts, and clarify concepts. Instructors can also benefit from these systems by creating learning communities within a classroom. For example, a teacher can pool different digital resources to create the subject matter and then let students search through them for mathematical notation and terminology. Math-aware

© The Author(s), under exclusive license to Springer Nature Switzerland AG 2022
A. Barrón-Cedeño et al. (Eds.): CLEF 2022, LNCS 13390, pp. 286–310, 2022.
https://doi.org/10.1007/978-3-031-13643-6_20

search engines can also help researchers identify potentially useful systems, fields, and collaborators. Good examples of this interdisciplinary approach benefiting physics include the AdS/CFT correspondence and holographic duality theories.

A key focus of mathematical searching is formulae. In contrast to simple words or other objects, a formula can have a well defined set of properties, relations, applications, and often also a 'result'. There are many (mathematically) equivalent formulae which are structurally quite different. For example, it is of fundamental importance to ask what information a user wants when searching for $x^2 + y^2 = 1$: is it the value of the variables x and y that satisfy this equation, all indexed objects that contain this formula, all indexed objects containing $a^2 + b^2 = 1$, or the geometric figure that is represented by this equation?

This third Answer Retrieval for Questions on Math (ARQMath-3) lab at the Conference and Labs of the Evaluation Forum (CLEF) completes our development of test collections for Math IR from content found on Math Stack Exchange,[1] a Community Question Answering (CQA) forum. This year, ARQMath continues its two main tasks: Answer Retrieval for Math Questions (Task 1) and Formula Search (Task 2). We also introduce a new pilot task, Open Domain Question Answering (Task 3).

Using the question posts from Math Stack Exchange, participating systems are given a question (in Tasks 1 and 3) or a formula from a question (in Task 2), and asked to return a ranked list of either potential answers to the question (Task 1) or potentially useful formulae (Task 2). For Task 3, given the same questions as Task 1, the participating systems also provide an answer, but are not limited to searching the ARQMath collection to find that answer. Relevance is determined by the expected utility of each returned item. These tasks allow participating teams to explore leveraging math notation together with text to improve the quality of retrieval results.

2 Related Work

Prior to ARQMath, three test collections were developed over a period of five years at the NII Testbeds and Community for Information Access Research (NTCIR) shared task evaluations. To the best of our knowledge, NTCIR-10 [1] was the first shared task on Math IR, considering three scenarios for searching:

- Formula Search: find similar formulae for the given formula query.
- Formula+Text Search: search the documents in the collection with a combination of keywords and formula queries.
- Open Information Retrieval: search the collection using text queries.

NTCIR-11 [2] considered the formula+keyword search task as the main task and introduced an additional Wikipedia open subtask, using the same set of topics with a different collection and different evaluation methods. Finally, in NTCIR-12 [19], the main task was formula+text search on two different collections. A second task was Wikipedia Formula Browsing (WFB), focusing on

---

[1] https://math.stackexchange.com/.

formula search. Formula similarity search (the *simto* task) was a third, where the goal was to find formulae 'similar' (not identical) to the formula query.

An earlier effort to develop a test collection started with the Mathematical REtrieval Collection (MREC) [8], a set of 439,423 scientific documents that contained more than 158 million formulae. This was initially only a collection, with no shared relevance judgments (although the effectiveness of individual systems was measured by manually assessing a set of topics). The Cambridge University MathIR Test Collection (CUMTC) [18] subsequently built on MREC, adding 160 test queries derived from 120 MathOverflow discussion threads (although not all queries contained math). CUMTC relevance judgments were constructed using citations to MREC documents cited in MathOverflow answers.

To the best of our knowledge, ARQMath's Task 1 is the first Math IR test collection to focus directly on answer retrieval. ARQMath's Task 2 (formula search) extends earlier work on formula search, with several improvements:

- **Scale.** ARQMath has an order of magnitude more assessed topics than prior formula search test collections. There are 22 topics in NTCIR-10, and 20 in NTCIR-12 WFB (+20 variants with wildcards).
- **Contextual Relevance.** In the NTCIR-12 WFB task [19], there was less attention to context. ARQMath Task 2, by contrast, has evolved as a contextualized formula search task, where relevance is defined both by the query and retrieved formulae and also the contexts in which those formulae appear.
- **Deduplication.** NTCIR collections measured effectiveness using formula instances. In ARQMath we clustered visually identical formulae to avoid rewarding retrieval of multiple instances of the same formula.
- **Balance.** ARQMath balances formula query complexity, whereas prior collections were less balanced (reannotation shows low complexity topics dominate NTCIR-10 and high complexity topics dominate NTCIR-12 WFB [10]).

In ARQMath-3, we introduced a new pilot task, Open Domain Question Answering. The most similar prior work is the SemEval 2019 [7] math question answering task, which used question sets from Math SAT practice exams in three categories: Closed Algebra, Open Algebra and Geometry. A majority of the Math SAT questions were multiple choice, with some having numeric answers.

While we have focused on search and question answering tasks in ARQ-Math, there are other math information processing tasks that can be considered for future work. For example, extracting definitions for identifiers, math word problem solving, and informal theorem proving are active areas of research: for a survey of recent work in these areas, see Meadows and Ferentes [14]. Summarization of mathematical texts, text/formula co-referencing, and the multimodal representation and linking of information in documents are some other examples.

3   The ARQMath Stack Exchange Collection

For ARQMath-3, we reused the collection[2] from ARQMath-1 and -2.[3] The collection was constructed using the March 1st, 2020 Math Stack Exchange snapshot from the Internet Archive.[4] Questions and answers from 2010–2018 are included in the collection. The ARQMath test collection contains roughly 1 million questions and 28 million formulae. Formulae in the collection are annotated using `<span>` XML elements with the class attribute `math-container`, and a unique integer identifier given in the `id` attribute. Formulae are also provided separately in three index files for different formula representations (LaTeX, Presentation MathML, and Content MathML), which we describe in more detail below.

During ARQMath-2021, participants identified three issues with the ARQMath collection that had not been noticed and corrected earlier. In 2022, we have made the following improvements to the collection:

1. **Formula Representations.** We found and corrected 65,681 formulae with incorrect Symbol Layout Tree (SLT) and Operator Tree (OPT) representations. This resulted from incorrect handling of errors generated by the LaTeXML tool that had been used for generating those representations.
2. **Clustering Visually Distinct Formulae.** Correcting SLT representations resulted in a need to adjust the clustering of formula instances. Each cluster of visually identical formulae was assigned a unique 'Visual ID'. Clustering had been performed using SLT where possible, and LaTeX otherwise. To correct the clustering, we split any cluster that now included formulae with different representations. In such cases, the partition with the largest number of instances retained its Visual ID; remaining formulae were assigned to another existing Visual ID (with the same SLT or LaTeX) or, if necessary, to a new Visual ID. To break ties, the partition with the largest cumulative ARQMath-2 relevance score retained its Visual ID or, failing that, choosing the partition with the lowest Formula ID. 29,750 new Visual IDs resulted.
3. **XML Errors.** In the XML files for posts and comments, the LaTeX for each formula is encoded as a `<span>` XML element with the class attribute `math-container`. We found and corrected 108,242 formulae that had not been encoded in that way.
4. **Spurious Formula Identifiers.** The ARQMath collection includes an index file that includes Formula ID, Visual ID, Post ID, SLT, OPT, and LaTeX for each formula instance. However, there were also formulae in the index file that did not actually occur in any post or comment in the collection. This happened because formula extraction was initially done on the Post History

---

[2] By *collection* we mean the content to be searched. That content together with topics and relevance judgments is a *test collection*. There is only one ARQMath *collection*.

[3] ARQMath-1 was built for CLEF 2020, ARQMath-2 was built for CLEF 2021. We refer to submitted runs or evaluation results by year, as AQRMath-2020 or ARQMath-2021. This distinction is important because ARQMath-2022 participants also submitted runs for both the ARQMath-1 and -2 test collections.

[4] https://archive.org/download/stackexchange.

file, which also contained some content that had later been removed. We added a new annotation to the formula index file to mark such cases.

The Math Stack Exchange collection was distributed to participants as XML files on Google Drive.[5] To facilitate local processing, the organizers provided python code on GitHub[6] for reading and iterating over the XML data, and for generating the HTML question threads. All of the code to generate the corrected ARQMath collection is available in the that same GitHub repository.

4   Task 1: Answer Retrieval

The goal of Task 1 is to find and rank relevant answers to math questions. Topics are constructed from questions posted to Math Stack Exchange in 2021, and the collection to be searched is only the answers to earlier questions (from 2010–2018) in the ARQMath collection. System results ('runs') are evaluated using measures that characterize the extent to which answers judged by relevance assessors as having higher relevance come before answers with lower relevance in the system results (e.g., using nDCG'). In this section, we describe the Task 1 search topics, participant runs, baselines, pooling, relevance assessment, and evaluation measures, and we briefly summarize the results.

4.1   Topics

ARQMath-3 Task 1 topics were selected from questions posted to Math Stack Exchange posted in 2021. There were two strict criteria for selecting candidate topics: (1) any candidate must have at least one formula in the title or the body of the question, (2) any candidate must have at least one known duplicate question (from 2010 to 2018) in the ARQMath collection. Duplicates had been annotated by Math Stack Exchange moderators as part of their ongoing work, and we chose to limit our candidates to topics for which a known duplicated existed in the collection in order to limit the potential for allocating assessment effort to topics that had no relevant answers in the collection. In ARQMath-2 we had included 11 topics for which there were no known duplicates on an experimental basis. Of those 11, 9 had turned out to have no relevant answers found by any participating system or baseline.

We selected 139 candidate topics from among the 3313 questions that satisfied both of our strict criteria by applying additional soft criteria based on the number of terms and formulae in the title and body of the question, the question score that Math Stack Exchange users had assigned to the question, and the number of answers, comments, and views for the question. From those 139, we manually selected 100 topics in a way that balanced three desiderata: (1) A similar topic should not already be present in the ARQMath-1 or ARQMath-2 test collections, (2) we expected that our assessors would have (or be able to

---

[5] https://drive.google.com/drive/folders/1ZPKIWDnhMGRaPNVLi1reQxZWTfH2R 4u3.

[6] https://github.com/ARQMath/ARQMathCode.

easily acquire) the expertise to judge relevance to the topic, and (3) the set of topics maximized diversity across four dimensions (question type, difficulty, dependence, and complexity).

In prior years, we had manually categorized topic type as *computation, concept* or *proof* and we did so again for ARQMath-3. A disproportionately large fraction of Math Stack Exchange questions ask for proofs, so we sought to stratify the ARQMath-3 topics in a way that was somewhat better balanced. Of the 100 ARQMath-3 topics, 49 are categorized as *proof*, 28 as *computation*, and 23 as *concept*. Question difficulty also benefited from restratification. Our insistence that topics have at least one duplicate question in the collection injects a bias in favor of easier questions, and such a bias is indeed evident in the ARQMath-1 and ARQMath-2 test collections. We made an effort to better balance (manually estimated) topic difficulty for the ARQMath-3 test collection, ultimately resulting in 24 topics categorized as hard, 55 as medium, and 21 as easy. We also paid attention to the (manually estimated) dependency of topics on text, formulae, or both, but we did not restratify on that factor. Of the 100 ARQMath-3 topics, 12 are categorized as dependent to text, 28 on formulae, and 60 on both. New this year, we also paid attention to whether a topic actually asks several questions rather than just one. For these multi-part topics, our relevance criteria require that a highly relevant answer provide relevant information for all parts of the question. Among ARQMath-3 topics, 14 are categorized as multi-part questions.

The topics were published in the XML file format illustrated in Fig. 1. Each topic has a unique Topic ID, a Title, a Question (which is the body of the question post), and Tags provided by the asker of the question on the Math Stack Exchange. Notably, links to duplicate or related questions are <u>not</u> included. To facilitate system development, we provided python code that participants could use to load the topics. As in the collection, the formulae in the topic file are placed in `<span>` XML elements, with each formula instance represented by a unique identifier and its LaTeX representation. Similar to the collection, there are three Tab Separated Value (TSV) files, for the LaTeX, OPT and SLT representations of the formulae, in the same format as the collection's TSV files. The Topic IDs in ARQMath-3 start from 301 and continue to 400. In ARQMath-1, Topic IDs were numbered from 1 to 200, and in ARQMath-2, from 201 to 300.

4.2 Participant Runs

ARQMath Participants submitted their runs on Google Drive. As in previous years, we expect all runs to be publicly available.[7] A total of 33 runs were received from 7 teams. Of these, 28 runs were declared to be automatic, with no human intervention at any stage of generating the ranked list for each query. The remaining 5 runs were declared to be manual, meaning that there was some type of human involvement in at least one stage of retrieving answers. Manual runs were invited in ARQMath to increase the quality and diversity of the pool

---

[7] https://drive.google.com/drive/u/1/folders/1l1c2O06gfCk2jWOixgBXI9hAlATy
bxKv.

TASK 1: QUESTION ANSWERING

```
<Topics>
  . . .
  <Topic number="A.384">
    <Title>What does this bracket notation mean?</Title>
    <Question>
      I am currently taking MIT6.006 and I came across this problem on the
      problem set. Despite the fact I have learned Discrete Mathematics
      before, I have never seen such notation before, and I would like to
      know what it means and how it works, Thank you:
      <span class=''math-container'' id=''q`898''>
        $$f`3(n) = "binom n2$$
      </span>
    </Question>
    <Tags>discrete-mathematics, algorithms</Tags>
  </Topic>
  . . .
</Topics>
```

TASK 2: FORMULA RETRIEVAL

```
<Topics>
  . . .
  <Topic number="B.384">
    <Formula`Id>q`898</Formula`Id>
    <Latex>f`3(n) = "binom n2</Latex>
    <Title>What does this bracket notation mean?</Title>
    <Question>
      I am currently taking MIT6.006 and I came across this problem on the
      problem set. Despite the fact I have learned Discrete Mathematics
      before, I have never seen such notation before, and I would like to
      know what it means and how it works, Thank you:
      <span class=''math-container'' id=''q`898''>
        $$f`3(n) = "binom n2$$
      </span>
    </Question>
    <Tags>discrete-mathematics, algorithms</Tags>
  </Topic>
  . . .
</Topics>
```

Fig. 1. Example XML Topic Files. Formula queries in Task 2 are taken from questions for Task 1. Here, ARQMath-3 formula topic B.384 is a copy of ARQMath-3 question topic A.384 with two additional fields for the query formula (1) identifier and (2) LaTeX.

of documents that are judged for relevance, but it important to note that they might not be fairly compared to automatic runs. The teams and submissions are shown in Table 1. For the details of each run, please see the participant papers in the working notes.

4.3   Baseline Runs

For Task 1, five baseline systems were provided by the organizers.[8] This year, the organizers included a new baseline system using PyTerrier [9] for the TF-IDF

---

[8] Source code and instructions for running the baselines are available from Git-Lab (Tangent-S: https://gitlab.com/dprl/tangent-s, PyTerrier: https://gitlab.com/dprl/pt-arqmath/) and GoogleDrive (Terrier: https://drive.google.com/drive/u/0/folders/1YQsFSNoPAFHefweaN01Sy2ryJjb7XnKF).

Table 1. ARQMath-3: Submitted Runs. Baselines for Task 1 (5), Task 2 (1) and Task 3 (1) were generated by the organizers. Primary and alternate runs were pooled to different depths, as described in Sect. 4.4.

	Automatic		Manual	
	Primary	Alternate	Primary	Alternate
Task 1: Answer Retrieval				
Baselines	2	3		
Approach0			1	4
DPRL	1	4		
MathDowsers	1	2		
MIRMU	1	4		
MSM	1	4		
SCM	1	4		
TU_DBS	1	4		
Totals (38 runs)	*8*	*25*	*1*	*4*
Task 2: Formula Retrieval				
Baseline	1			
Approach0			1	4
DPRL	1	4		
MathDowsers	1	2		
JU_NITS	1	2		
XY_PHOC_DPRL	1	2		
Totals (20 runs)	*5*	*11*	*1*	*3*
Task 3: Open Domain QA				
Baseline	1			
Approach0			1	4
DPRL	1	3		
TU_DBS	1	3		
Totals (14 runs)	*3*	*6*	*1*	*4*

model. The other baselines were also run for ARQMath 2020 and 2021. Here is a description of our baseline runs.

1. **TF-IDF.** We provided two TF-IDF baselines. The first uses Terrier [15] with default parameters and raw LaTeX strings, as in prior years of the lab. One problem with this baseline is that Terrier removes some LaTeX symbols during tokenization. The second uses PyTerrier [9], with symbols in LaTeX strings first mapped to English words to avoid tokenization problems.
2. **Tangent-S**. This baseline is an isolated formula search engine that uses both SLT and OPT representations [5]. The target formula was selected from the question title if at least one existed, otherwise from the question body. If

there were multiple formulae in the field, a formula with the largest number of symbols (nodes) in its SLT representation was chosen; if more than one had the largest number of symbols, we chose randomly between them.

3. **TF-IDF + Tangent-S.** Averaging normalized similarity scores from the TF-IDF (only from PyTerrier) and Tangent-S baselines. The relevance scores from both systems were normalized in [0,1] using min-max normalization, and then combined using an unweighted average.

4. **Linked Math Stack Exchange Posts.** Using duplicate post links from 2021 in Math Stack Exchange, this oracle system returns a list of answers from posts in the ARQMath collection that had been given to questions marked in Math Stack Exchange as duplicates to ARQMath-3 topics. These answers are ranked by descending order of their vote scores. Note that the links to duplicate questions were not available to the participants.

4.4   Relevance Assessment

Relevance judgments for Tasks 1 and 3 were performed together, with the results for the two tasks intermixed in the judgment pools.

Pooling. For each topic, participants were asked to rank up to 1,000 answer posts. We created pools for relevance judgments by taking the top-k retrieved answer posts from every participating system or baseline in Tasks 1 or 3. For Task 1 primary runs, the top 45 answer posts were included; for alternate runs the top 20 were included. These pooling depths were chosen based on assessment capacity, with the goal of identifying as many relevant answer posts as possible. Two Task 1 baseline runs, PyTerrier TF-IDF+Tangent-S. and Linked Math Stack Exchange Posts, were pooled as primary runs (i.e., to depth 45); other baselines were pooled as alternate runs (i.e., to depth 20). All Task 3 run results (each of which is a single answer; see Sect. 5.6) were also included in the pools. After merging these top-ranked results, duplicate posts were deleted and the resulting pools were sorted randomly for display to assessors. On average, the judgment pools for Tasks 1 and 3 contain 464 answer posts per topic.

Relevance Definition. The relevance definitions were the same those defined for ARQMath-1 and -2. The assessors were asked to consider an expert (modeling a math professor) judging the relevance of each answer to the topics. This was intended to avoid the ambiguity that might result from guessing the level of math knowledge of the actual posters of the original Math Stack Exchange question. The definitions of the four levels of relevance are shown in Table 2. In judging relevance, ARQMath assessors were asked not to consider any link outside the ARQMath collection. For example, if there is a link to a Wikipedia page, which provides relevant information, the information in the Wikipedia page should not be considered to be a part of the answer.

4.5   Assessor Selection

Paid ARQMath-3 assessors were recruited over email at the Rochester Institute of Technology. 44 students expressed interest, 11 were invited to perform 3 sam-

Table 2. Relevance Assessment Criteria for Tasks 1 and 2.

SCORE	RATING	DEFINITION
Task 1: Answer Retrieval		
3	High	Sufficient to answer the complete question on its own
2	Medium	Provides some path towards the solution. This path might come from clarifying the question, or identifying steps towards a solution
1	Low	Provides information that could be useful for finding or interpreting an answer, or interpreting the question
0	Not Relevant	Provides no information pertinent to the question or its answers. A post that restates the question without providing any new information is considered non-relevant
Task 2: Formula Retrieval		
3	High	Just as good as finding an exact match to the query formula would be
2	Medium	Useful but not as good as the original formula would be
1	Low	There is some chance of finding something useful
0	Not Relevant	Not expected to be useful

ple assessment tasks, and 9 students specializing in mathematics or computer science were then selected, based on an evaluation of their judgments by an expert mathematician. Of those, 6 were assigned to Tasks 1 and 3; the others performed assessment for Task 2.

Assessment Tool. As with ARQMath-1 and ARQMath-2, we used Turkle, a system similar to Amazon Mechanical Turk. As shown in Fig. 2, there are two panes, one having the question topic (left pane) and the other having a candidate answer from the judgment pool (right panel). For each topic, the title

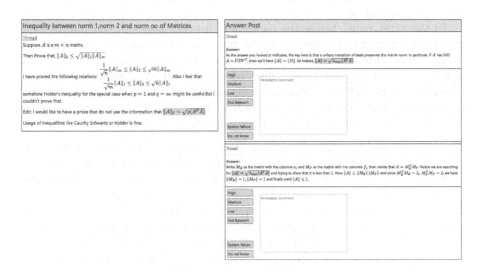

Fig. 2. Turkle Assessment Interface. Shown are hits for Formula Retrieval (Task 2). In the left pane, the formula query is highlighted. In the right pane, two answer posts containing the same retrieved formula are shown. For Task 1, the same interface was used, but without formula highlighting, and presenting only one answer post at a time.

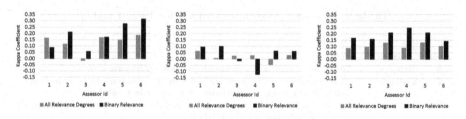

Fig. 3. Inter-annotator agreement for 6 assessors during training sessions for Task 1 (mean Cohen's kappa), with four-way classification in gray, and two-way classification (H+M binarized) in black. Left-to-right: agreements for rounds 1, 2, and 3.

and question body are provided for the assessors. To familiarize themselves with the topic question, assessors can click on the Thread link for the question, which shows the question and the answers given to it (i.e., answers posted in 2021, which were not available to task participants), along with other information such as tags and comments. Another Thread link is also available for the answer post being assessed. By clicking on that link, the assessor can see a copy of the original question thread on Math Stack Exchange in which the candidate answer was given, as recorded in the March 2020 snapshot used for the ARQMath test collection.

Note that these Thread links are provided to help the assessors gain just-in-time knowledge that they might need for unfamiliar concepts, but the content of the threads is neither a part of the topic nor of the answer being assessed, and thus it should have no has no effect on their judgement beyond serving as reference information.

In the right pane, below the candidate answer, assessors can indicate the relevance degree. In addition to four relevance degrees, there are two additional choices: 'System failure' to indicate system issues such as unintelligible rendering of formulae, and 'Do not know' which can be used if after possibly consulting external sources such as Wikipedia or viewing the Threads the assessor is simply not able to decide the relevance degree. We asked the assessors to leave a comment in the event of a 'System failure' or 'Do not know' selection.

Assessor Training. All training was done remotely, over Zoom, in four sessions, with some individual assessment practice between each Zoom session. As in ARQMath-1 and -2, in the first session the task and relevance criteria were explained. A few examples were then shown to the assessors and they were asked for their opinions on relevance, which were then discussed with an expert assessor (a math professor). Then, three rounds of training were conducted, with each round consisting of assessment of small judgment pools for four sample topics from ARQMath-2. For each topic, 5–6 answers with different ground truth relevance degrees (from the ARQMath-2 qrels) were chosen. After each round, we held a Zoom session to discuss their relevance judgements, with the specific goal of clarifying their understanding of the relevance criteria. The assessors discussed the reasoning for their choices, with organizers (always including the math professor) sharing their own judgments and their supporting reasoning. The primary goal of training was to help assessors make self-consistent annotations, as topic interpretations will vary across individuals. Some of the topics involve issues that are not typically covered in regular undergraduate courses, and some such cases required the assessors to get a basic understanding of those issue before they could do the assessment. The assessors found the Threads made available in the Turkle interface helpful in this regard (see Fig. 2).

Figure 3 shows average Cohen's kappa coefficients for agreement between each assessor and all others during training. Collapsing relevance to binary by considering only high and medium as relevant (henceforth "H+M binarization") yielded better agreement among the assessors.[9] The agreement values in the second round are unusually low, but the third round agreement is in line with what we had seen at the end of training in prior years.

Assessment Results. Among 80 topics assessed, two (A.335 and A.367) had only one answer assessed as high or medium; these two topics were removed from the collection as score quantization for MAP′ can be quite substantial when only a single relevant document contributes to the computation. For the remaining 78 topics, an average of 446.8 answers were assessed, with an average assessment time of 44.1 s per answer post. The average number of answers labeled with any degree of relevance (high, medium, or low; henceforth "H+M+L binarization") over those 78 topics was 100.8 per question (twice as high as that seen in ARQMath-2), with the highest number being 295 (for topic A.317) and the lowest being 11 (for topic A.385).

4.6 Evaluation Measures

While this is the third year of the ARQMath lab, with several relatively mature systems participating, it is still possible that many relevant answers may remain undiscovered. To support fair comparisons with future systems that may find different documents, we have adopted evaluation measures that ignore unjudged answers, rather than adopting the more traditional convention of treating unjudged answers as not relevant. Specifically, the primary evaluation measure for Task 1 is the nDCG′ (read as "nDCG-prime") introduced by Sakai and Kando [17]. nDCG′ is simply the nDCG@1000 that would be computed after removing unjudged documents from the ranked list. This measure has shown better discriminative power and somewhat better system ranking stability (with judgement ablation) compared to the bpref [4] measure that had been adopted in the NTCIR Math IR evaluation for similar reasons [11]. Moreover, nDCG′ yields a single-valued measure with graded relevance, whereas bpref, Precision@k, and Mean Average Precision (MAP) all require binarized

---

[9] H+M binarization corresponds to the definition of relevance usually used in the Text Retrieval Conference (TREC).

298 B. Mansouri et al.

Table 3. ARQMath 2022 Task 1 (CQA) results. **P**: primary run, **M**: manual run, (✓): baseline pooled as a primary run. For MAP′ and P′@10, H+M binarization was used. (D)ata indicates use of (T)ext, (M)ath, (B)oth text and math, or link structure (*L).

Run	D	P	M	ARQMATH-1 77 TOPICS NDCG′	MAP′	P′@10	ARQMATH-2 71 TOPICS NDCG′	MAP′	P′@10	ARQMATH-3 78 TOPICS NDCG′	MAP′	P′@10
Baselines												
TF-IDF(Terrier)	B			0.204	0.049	0.073	0.185	0.046	0.063	0.272	0.064	0.124
TF-IDF(PyTerrier) +Tangent-S	B	(✓)		0.249	0.059	0.081	0.158	0.035	0.072	0.229	0.045	0.097
TF-IDF(PyTerrier)	B			0.218	0.079	0.127	0.120	0.029	0.055	0.190	0.035	0.065
Tangent-S	M			0.158	0.033	0.051	0.111	0.027	0.052	0.159	0.039	0.086
Linked MSE posts	*L	(✓)		0.279	0.194	0.384	0.203	0.120	0.282	0.106	0.051	0.168
approach0												
fusion_alpha05	B	✓	✓	0.462	0.244	0.321	0.460	0.226	0.296	**0.508**	**0.216**	**0.345**
fusion_alpha03	B		✓	0.460	0.246	0.312	0.450	0.221	0.278	0.495	0.203	0.317
fusion_alpha02	B		✓	0.455	0.243	0.309	0.443	0.217	0.266	0.483	0.195	0.305
rerank_nostemer	B		✓	0.382	0.205	0.322	0.385	0.187	0.276	0.418	0.172	0.309
a0porter	B		✓	0.373	0.204	0.270	0.383	0.185	0.241	0.397	0.159	0.271
MSM												
Ensemble_RRF	B	✓		0.422	0.172	0.197	0.381	0.119	0.152	0.504	0.157	0.241
BM25_system	B			0.332	0.123	0.168	0.285	0.082	0.116	0.396	0.122	0.194
BM25_TfIdf _system	B			0.332	0.123	0.168	0.286	0.083	0.116	0.396	0.122	0.194
TF-IDF	B			0.238	0.074	0.117	0.169	0.040	0.076	0.280	0.064	0.081
CompuBERT22	B			0.115	0.038	0.099	0.098	0.030	0.090	0.130	0.025	0.059
MIRMU												
MiniLM+RoBERTa	B	✓		0.466	0.246	0.339	0.487	0.233	0.316	0.498	0.184	0.267
MiniLM +MathRoBERTa	B			0.466	0.246	0.339	0.484	0.227	0.310	0.496	0.181	0.273
MiniLM_tuned +MathRoBERTa	B			0.470	0.240	0.335	0.472	0.221	0.309	0.494	0.178	0.262
MiniLM_tuned +RoBERTa	B			0.466	0.246	0.339	0.487	0.233	0.316	0.472	0.165	0.244
MiniLM+RoBERTa	T			0.298	0.124	0.201	0.277	0.104	0.180	0.350	0.107	0.159
MathDowsers												
L8_a018	B	✓		0.511	0.261	0.307	0.510	0.223	0.265	0.474	0.164	0.247
L8_a014	B			0.513	0.257	0.313	0.504	0.220	0.265	0.468	0.155	0.237
L1on8_a030	B			0.482	0.241	0.281	0.507	0.224	0.282	0.467	0.159	0.236
TU_DBS												
math_10	B	✓		0.446	0.268	0.392	0.454	0.228	0.321	0.436	0.158	0.263
Khan_SE_10	B			0.437	0.254	0.357	0.437	0.214	0.309	0.426	0.154	0.236
base_10	B			0.438	0.252	0.369	0.434	0.209	0.299	0.423	0.154	0.228
roberta_10	B			0.438	0.254	0.372	0.446	0.224	0.309	0.413	0.150	0.226
math_10_add	B			0.421	0.264	0.405	0.566	0.445	0.589	0.379	0.149	0.278
DPRL												
SVM-Rank	B	✓		0.508	0.467	0.604	0.533	0.460	0.596	0.283	0.067	0.101
RRF-AMR-SVM	B			**0.587**	**0.519**	**0.625**	**0.582**	**0.490**	**0.618**	0.274	0.054	0.022
QQ-QA-RawText	B			0.511	0.467	0.604	0.532	0.460	0.597	0.245	0.054	0.099
QQ-QA-AMR	B			0.276	0.180	0.295	0.186	0.103	0.237	0.185	0.040	0.091
QQ-MathSE-AMR	B			0.231	0.114	0.218	0.187	0.069	0.138	0.178	0.039	0.081
SCM												
interpolated_text +positional_word 2vec_tangentl	B	✓		0.254	0.102	0.182	0.197	0.059	0.149	0.257	0.060	0.119
joint_word2vec	B			0.247	0.105	0.187	0.183	0.047	0.106	0.249	0.059	0.106
joint_tuned _roberta	B			0.248	0.104	0.187	0.184	0.047	0.109	0.249	0.059	0.105
joint_positional _word2vec	B			0.247	0.105	0.190	0.184	0.047	0.109	0.248	0.059	0.105
joint_roberta_base	T			0.135	0.048	0.101	0.099	0.023	0.060	0.188	0.040	0.077

relevance judgments. As secondary measures, we compute Mean Average Precision (MAP@1000) with unjudged posts removed (MAP') and Precision at 10 with unjudged posts removed (P'@10). For MAP' and P'@10 we used H+M binarization. Note that the answers assessed as "System failure" or "Do not know" were not considered for evaluation, thus can be viewed as answers that are not assessed.

4.7 Results

Progress Testing. In addition to their submissions on the ARQMath-3 topics, we asked each participating team to also submit results from exactly the same systems on ARQMath-1 and ARQMath-2 topics for progress testing. Note, however, that ARQMath-3 systems could be trained on topics from ARQMath-1 and -2; Together, there were 158 topics (77 from ARQMath-1, 81 from ARQMath-2) that could be used for training. The progress test results thus need to be interpreted with this train-on-test potential in mind. Progress test results are provided in Table 3.

ARQMath-3 Results. Table 3 also shows results for ARQMath-3 Task 1. This table shows baselines first, followed by teams, and within teams their systems, ranked by nDCG'. As seen in the table, the manual primary run of the approach0 team achieved the best results, with 0.508 nDCG'. Among automatic runs, nDCG', 0.504, was achieved by the MSM team. Note that the highest possible nDCG' and MAP' values are 1.0, but that because fewer than 10 assessed relevant answers (with H+M binarization) were found in the pools for some topics, the highest P'@10 value in ARQMath-3 Task 1 is 0.95.

5 Task 2: Formula Search

The goal of the formula search task is to find a ranked list of formula instances from both questions and answers in the collection that are relevant to a formula query. The formula queries are selected from the questions in Task 1. One formula was selected from each Task 1 question topic to produce Task 2 topics. For cases in which suitable formulae were present in both the title and the body of the Task 1 question, we selected the Task 2 formula query from the title. For each query, a ranked list of 1,000 formulae instances were returned by their identifiers in the <span> XML elements and the accompanying TSV LaTeX formula index file, along with their associated post identifiers.

While in Task 1, the goal was to find relevant answers for the questions, in Task 2, the goal is to find relevant formulae that are associated with information that can help to satisfy an information need. The post in which a formula is found need not be relevant to the question post in which the formula query originally appeared for a formula to be relevant to a formula query, but those post contexts inform the interpretation of each formula (e.g., by defining operations and identifying variable types). A second difference is that the retrieved formulae instances in Task 2 can be found in either question posts or answer posts, whereas in Task 1, only answer posts were retrieved.

Finally, in Task 2, we distinguish visually distinct formulae from instances of those formulae, and systems are evaluated by the ranking of the visually distinct formulae they return. The same formula can appear in different posts, we call these *formula instances*. By a *visually distinct formula* we mean a set of formula instances that are visually identical when viewed in isolation. For example, x^2 is a formula, $x \cdot x$ is a different (i.e., visually distinct) formula, and each time x^2 appears, it is an instance of the visually distinct formula x^2. Although systems in Task 2 rank formula instances in order to support the relevance judgment process, the evaluation measure for Task 2 is based on the ranking of visually distinct formulae. As shown by Mansouri et al. (2021) [10], using visually-distinct formulae for evaluation can result in a different preference order between systems than would evaluation on formula instances.

5.1   Topics

Each formula query was selected from a Task 1 topic. Similarly to Task 1, Task 2 topics were provided in XML in the format shown in Fig. 1. Differences are:

1. **Topic Id.** Task 2 topic ids are in the form "B.x" where x is the topic number. There is a correspondence between topic id in tasks 1 and 2. For instance, topic id "B.384" indicates the formula is selected from topic "A.384" in Task 1, and both topics include the same question post (see Fig. 1).
2. **Formula Id.** This added field specifies the unique identifier for the query formula instance. There may be other formulae in the Title or Body of the same question post, but the formula query is only the formula instance specified by this Formula_Id.
3. **LaTeX.** This added field is the LaTeX representation of the query formula instance, as found in the question post.

As the query formulae are selected from Task 1 questions, the same LaTeX, SLT and OPT TSV files that were provided for the Task 1 topics can be used when SLT or OPT representations for a query formula are needed.

Formulae for Task 2 were manually selected using a heuristic approach to stratified sampling over two criteria: complexity and elements. Formula complexity was labeled low, medium or high by the third author. For example, $[x, y] = x$ is low complexity, $\int \frac{1}{(x^2+1)^n} dx$ is medium complexity, and $\frac{\sqrt{1-p^2}}{2\pi(1-2p\sin(\varphi)\cos(\varphi))}$ is high complexity. These annotations, available in an auxiliary file, can be useful as a basis for fine-grained result analysis, since formula queries of differing complexity may result in different preference orders between systems [13]. For elements, our intuition was to make sure that we have formula queries that contain different elements and math phenomena such as integral, limit, and matrices.

5.2 Participant Runs

A total of 19 runs were received for Task 2 from a total of five teams, as shown in Table 1. Among the participating runs, 5 were annotated as manual and the others were automatic. Each run retrieved up to 1,000 formula instances for each formula query, ranked by relevance to that query. For each retrieved formula instance, participating teams provided the `formula_id` and the associated `post_id` for that formula. Please see the participant papers in the working notes for descriptions of the systems that generated these runs.

5.3 Baseline Run: Tangent-S

Tangent-S [5] is the baseline system for ARQMath-3 Task 2. That system accepts a formula query without using any associated text context from its associated question post. Since a single formula is specified for each Task 2 query, the formula selection step in the Task 1 Tangent-S baseline is not needed for Task 2. Timing was similar to that of Tangent-S in ARQMath-1 and -2.

5.4 Assessment

Pooling. For each topic, participants were asked to rank up to 1,000 formula instances. However, the pooling was done using visually distinct formulae. The visual ids, which were provided beforehand for the participants, were used for clustering formula instances. Pooling was done by going down in each ranked list until k visually distinct formulae were found. For primary runs (and the baseline system), the first 25 visually distinct formulae were pooled; for alternate runs, the first 15 visually distinct formulae were pooled.

The visual Ids used for clustering retrieval results were determined by the SLT representation when possible, and the LATEX representation otherwise. When SLT was available, we used Tangent-S [5] to create a string representation using a depth-first traversal of the SLT, with each SLT node and edge generating a single item in the SLT string. Formula instances with identical SLT strings were then considered to be the same formula. For formula instances with no Tangent-S SLT string available, we removed the white space from their LATEX strings and grouped formula instances with identical LATEX strings. This process is simple and appears to be reasonably robust, but it is possible that some visually identical formula instances were not captured due to LATEXML conversion failures, or where different LATEX strings produce visually identical formulae (e.g., if subscripts and superscripts appear in a different order in LATEX).

Task 2 assessment was done on formula instances. For each visually distinct formula at most five instances were selected for assessment. As in ARQMath-2 Task 2, formula instances to be assessed were chosen in a way that prefers highly-ranked instances and that prefers instances returned in multiple runs. This was done using a simple voting protocol, where each instance votes by the sum of its reciprocal ranks within each run, breaking ties randomly. For each query, on

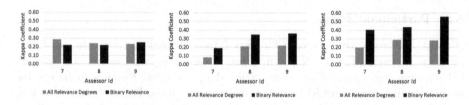

Fig. 4. Annotator agreement for 3 assessors during training for Task 2 (mean Cohen's kappa). Four-way classification is shown in gray, and two-way (H+M binarized) classification in black. Left-to-right: agreements for rounds 1, 2, and 3.

average there were 154.35 visually distinct formulae to be assessed, and only 6% of visually distinct formulae had more than 5 instances.

Relevance Definition. To distinguish between different relevance degrees, we relied on the definitions in Table 2. The usefulness is defined as the likelihood of the candidate formula being associated with information (text) that can help a searcher to accomplish their task. In our case, the task is answering the question from which a query formula is taken.

To judge the relevance of a candidate formula instance, the assessor was given the candidate formula (highlighted) along with the (question or answer) post in which it had appeared. They were then asked to decide on relevance by considering the definitions provided. For each visually distinct formula, up to 5 instances were shown to assessors and they would assess the instances individually. For assessment, they could look at the formula's associated post in an effort to understand factors such as variable types, the interpretation of specific operators, and the area of mathematics it concerns. As in Task 1, assessors could also follow Thread links to increase their knowledge by examining the thread in which the query formula had appeared, or in which a candidate formula had appeared.

Assessment Tool. As in Task 1, we used Turkle for the Task 2 assessment process, as illustrated in Fig. 2. There are two panes, the left pane showing the formula query ($\|A\|_2 = \sqrt{\rho(A^T A)}$ in this case) highlighted in yellow inside its question post, and the right pane showing the (in this case, two) candidate formula instances of a single visually distinct formula. For each topic, the title and question body are provided for the assessors. Thread links can be used by the assessors just for learning more about mathematical concepts in the posts. For each formula instance, the assessment is done separately. As in Task 1, the assessors can choose between different relevance degrees, they can choose 'System failure' for issues with Turkle, or they can choose 'Do not know' if they are not able to decide on a relevance degree.

Assessor Training. Three paid undergraduate and graduate mathematics and computer science students from RIT were selected to perform relevance judgments. As in Task 1, all training sessions were done remotely, over Zoom.

There were four Task 2 training sessions. In the first meeting, the task and relevance criteria were explained to assessors and then a few examples were shown, followed by discussion about relevance level choices. In each subsequent training round, assessors were asked to first assess four ARQMath-2 Task 2 topics, each with 5–6 visually distinct formula candidates with a variety of relevance degrees. Organizers then met with the assessors to discuss their choices and clarify relevance criteria. Figure 4 shows the average agreement (kappa) of each assessor with the others during training. As can be seen, agreement had improved considerably by round three, reaching levels comparable to that seen in prior years of ARQMath.

Assessment Results. Among 76 assessed topics, all have at least two relevant visually distinct formulae with H+M binarization, so all 76 topics were retained in the ARQMath-3 Task 2 test collection. An average of 152.3 visually distinct formulae were assessed per topic, with an average assessment time of 26.6 s per formula instance. The average number of visually distinct formulae with H+M+L binarization was 63.2 per query, with the highest number being 143 (topic B.305) and the lowest being 2 (topic B.333).

5.5   Evaluation Measures

As in Task 1, the primary evaluation measure for Task 2 is nDCG$'$, with MAP$'$ and P$'$@10 also reported. Participants submitted ranked lists of formula instances used for pooling, but with evaluation measures computed over visually distinct formulae. The ARQMath-2 Task 2 evaluation script replaces each formula instance with its associated visually distinct formula, and then deduplicates from the top of the list downward, producing a ranked list of visually distinct formulae, from which our "prime" evaluation measures are then computed using `trec_eval`, after removing unjudged visually distinct formulae. For the visually distinct formulae with multiple instances, the maximum relevance score of any judged instance was used as the relevance visually distinct formula's relevance score. This reflects a goal of having at least one instance that provides useful information. Similar to Task 1, formulas assessed as "System failure" or "Do not know" were treated as not being assessed.

Table 4. Task 2 (Formula Retrieval) results. **P**: primary run, **M**: manual run, (✓): baseline pooled as a primary run. MAP′ and P′@10 use H+M binarization. Baseline results in parentheses. DATA indicates sources used by systems: (M)ath, or (B)oth math and text.

		TYPE		ARQMATH-1 45 TOPICS			ARQMATH-2 58 TOPICS			ARQMATH-3 76 TOPICS		
RUN	DATA	P	M	NDCG′	MAP′	P′@10	NDCG′	MAP′	P′@10	NDCG′	MAP′	P′@10
Baselines												
Tangent-S	M	(✓)		0.691	0.446	0.453	0.492	0.272	0.419	0.540	0.336	0.511
approach0												
fusion_alph05	M	✓	✓	0.647	0.507	0.529	**0.652**	0.471	0.612	**0.720**	**0.568**	**0.688**
fusion_alph03	M		✓	0.644	0.513	0.520	0.649	0.470	0.603	**0.720**	0.565	0.665
fusion_alph02	M		✓	0.633	0.502	0.513	0.646	0.469	0.597	0.715	0.558	0.659
a0	M		✓	0.582	0.446	0.477	0.573	0.420	0.588	0.639	0.501	0.615
fusion02_ctx	B		✓	0.575	0.448	0.496	0.575	0.417	0.590	0.631	0.490	0.611
DPRL												
TangentCFT2ED	M	✓		0.648	0.480	0.502	0.569	0.368	0.541	0.694	0.480	0.611
TangentCFT2	M			0.607	0.438	0.482	0.552	0.350	0.510	0.641	0.419	0.534
T-CFT2TED+MathAMR	B			0.667	0.526	**0.569**	0.630	**0.483**	**0.662**	0.640	0.388	0.478
LTR	M			**0.733**	**0.532**	0.518	0.550	0.333	0.491	0.575	0.377	0.566
MathAMR	B			0.651	0.512	0.567	0.623	0.482	0.660	0.316	0.160	0.253
MathDowsers												
latex_L8_a040	M			0.657	0.460	0.516	0.624	0.412	0.524	0.640	0.451	0.549
latex_L8_a035	M			0.659	0.461	0.516	0.619	0.410	0.522	0.640	0.450	0.549
L8	M	✓		0.646	0.454	0.509	0.617	0.409	0.510	0.633	0.445	0.549
XYPhoc												
xy7o4	M			0.492	0.316	0.433	0.448	0.250	0.435	0.472	0.309	0.563
xy5	M			0.419	0.263	0.403	0.328	0.168	0.391	0.369	0.211	0.518
xy5IDF	M	✓		0.379	0.241	0.374	0.317	0.156	0.391	0.322	0.180	0.461
JU_NITS												
formulaL	M	✓		0.238	0.151	0.208	0.178	0.078	0.221	0.161	0.059	0.125
formulaO	M			0.007	0.001	0.009	0.182	0.101	0.367	0.016	0.008	0.001
formulaS	M			0.000	0.000	0.000	0.142	0.070	0.159	0.000	0.000	0.000

5.6 Results

Progress Testing. As with Task 1, we asked Task 2 teams to run their ARQMath-3 systems on ARQMath-1 and -2 Topics for progress testing (see Table 4). Some progress test results may represent a train-on-test condition: there were 70 topics from ARQMath-2 and 74 topics from ARQMath-1 available for training. Note also that while the relevance definition stayed the same for ARQMath-1, -2, and -3, the assessors were instructed differently in ARQMath-1 on how to handle the specific case in which two formulae were visually identical. In ARQMath-1 assessors were told such cases are always highly relevant, whereas ARQMath-2 and ARQMath-3 assessors were told that from context they might recognize cases in which a visually identical formula would be less relevant, or not relevant at all (e.g., where identical notation is used with very different meaning). Assessor instruction did not change between ARQMath-2 and -3.

ARQMath-3 Results. Table 4 also shows results for ARQMath-3 Task 2. In that table, the baseline is shown first, followed by teams and then their systems ranked by nDCG′ on ARQMath-3 Task 2 topics. As shown, the highest nDCG′ was achieved by the manual primary run from the approach0 team, with an nDCG$'$ value of 0.720. Among automatic runs, the highest nDCG$'$ value was the DPRL primary run, with an NDCG$'$ of 0.694. Note that 1.0 is a possible score for nDCG$'$ and MAP$'$, but that the highest possible P$'$@10 value is 0.93 because (with H+M binarization) 10 visually distinct formulae were not found in the pools for some topics.

6   Task 3: Open Domain Question Answering

The new pilot task developed for ARQMath-3 (Task 3) is Open Domain Question Answering. Unlike Task 1, system answers are not limited to content from any specific source. Rather, answers can be *extracted* from anywhere, automatically *generated*, or even written by a person. For example, suppose that we ask a Task 3 system the question "What does it mean for a matrix to be Hermitian?" An extractive system might first retrieve an article about Hermitian matrices from Wikipedia and then extract the following excerpt as the answer: "In mathematics, a Hermitian matrix (or self-adjoint matrix) is a complex square matrix that is equal to its own conjugate transpose". By contrast, a generative system such as GPT-3 can directly construct an answer such as: "A matrix is Hermitian if it is equal to its transpose conjugate." For a survey of open-domain question answering, see Zhu et al. [21]. In this section, we describe the Task 3 search topics, runs from participant and baseline systems, and the assessment and evaluation procedures used. Then, we provide a summary of the results.

6.1   Topics and Participant Runs

The topics for Task 3 are the Task 1 topics, with the same content provided (title, question body, and tags). A total of 13 runs were received from 3 teams. Each run consists of a single result for each topic. 9 runs from the TU_DBS and DPRL teams were declared to be automatic and 5 runs from the approach0 team were declared as manual. The 4 automatic runs from the TU_DBS team used generative systems, whereas the remaining 9 runs from the DPRL and approach0 teams used extractive systems. The teams and their submissions are listed in Table 1.

6.2   Baseline Run: GPT-3

The ARQMath organizers provided one baseline run for this task using GPT-3.[10] This baseline system uses the *text-davinci-002* model of GPT-3 [3] from OpenAI. First, the system prompts the model with the text Q: followed by the text and the LaTeX formulae of the question, two newline characters, and the text A: as follows:

---

[10] Source code and instructions for running the GPT-3 baseline is available from GitHub: https://github.com/witiko/arqmath3-openqa-tools.

```
Q: What does it mean for a matrix to be Hermitian?

A:
```

Then, GPT-3 completes the text and produces an answer of up to 570 tokens:

```
Q: What does it mean for a matrix to be Hermitian?

A: A matrix is Hermitian if it is equal to its transpose conjugate.
```

If the answer is longer than the maximum of 1,200 Unicode characters, the system retries until the model has produced a sufficiently short answer.

To provide control over how creative an answer is, GPT-3 resmooths the output layer L using the temperature τ as follows: $\text{softmax}(L/\tau)$ [6]. A temperature close to zero ensures deterministic outputs on repeated prompts, whereas higher temperatures allow the model's decoder to consider many different answers. Our system uses the default temperature $\tau = 0.7$.

6.3 Assessment

The answers for Task 3 were assessed together with the Task 1 results, using the same relevance definitions. After the publication of this paper, we plan to provide a random sample of Task 1 and Task 3 answers to assessors and ask them:

1. Whether the answers were computer-generated
2. Whether the answers contained information unrelated to the questions

These assessments will allow us to determine whether there exist qualitative differences between the answers retrieved by Task 1 systems and the answers produced by extractive and generative Task 3 systems.

6.4 Evaluation Measures

In this section, we first describe the evaluation measures that we used to evaluate participating systems. Then, we describe additional evaluation measures that we have developed with the goal of providing a fair comparison between participating systems and future systems that return answers from outside Math Stack Exchange, or that are generated.

Manual Evaluation Measures. As described in Sect. 4.4, the assessors produced a relevance score between 0 and 3 for most answers from each participating system. The exceptions were 'System failure' and 'Do not know' assessments, which we interpreted as relevance score 0 ('Not relevant') in our evaluation of Task 3. To evaluate participating systems, we report the Average Relevance

(AR) score and Precision@1 (P@1). AR is equivalent to the unnormalized Discounted Cumulative Gain at position 1 (DCG@1).[11] P@1 is computed using H+M binarization.

Task 1 systems approximate a restricted class of Task 3 systems. For this reason, in our working notes paper we plan to also report AR and P@1 for ARQMath-3 Task 1 systems in order to extend the range of system comparisons that can be made. To do this, we will truncate the Task 1 result lists after the first result. Note, however, that Task 3 answers were limited to a maximum of 1,200 Unicode characters, whereas Task 1 systems had no such limitation. Approximately 15% of all answer posts in the collection are longer than 1,200 Unicode characters (when represented as text and LaTeX). Therefore, the Task 3 measures that we plan to report for Task 1 systems should be treated as somewhat optimistic estimates of what might have been achieved by an extractive system that was limited to the ARQMath collection.

Table 5. Task 3 (Open Domain QA) ARQMath-3 results. **P**: primary run, **M**: manual run, **G**: generative system, (✓): baseline pooled as primary run. All runs use (B)oth math and text. P@1 uses H+M binarization. AR: Average Relevance. Task 3 topics are the same as Task 1 topics. Baseline results are in parentheses.

		TYPE			78 Topics	
RUN	DATA	P	M	G	AR	P@1
Baselines						
GPT-3	B	(✓)		✓	(1.346)	(0.500)
approach0						
run1	B		✓		**1.282**	**0.436**
run4	B		✓		1.231	0.397
run3	B		✓		1.179	0.372
run2	B		✓		1.115	0.321
run5	B	✓		✓	0.949	0.282
DPRL						
SBERT-SVMRank	B				0.462	0.154
BERT-SVMRank	B	✓			0.449	0.154
SBERT-QQ-AMR	B				0.423	0.128
BERT-QQ-AMR	B				0.385	0.103
TU_DBS						
amps3_sel_hints	B			✓	0.325	0.078
se3_len_pen_10	B			✓	0.244	0.064
amps3_sel_len_pen_20_sample_hint	B			✓	0.231	0.051
shortest	B	✓		✓	0.205	0.026

---

[11] For ranked lists of depth 1 there is no discounting or accumulation, and in ARQMath the relevance value is used directly as the gain.

Automatic Evaluation Measures. In Task 1, systems pick answers from a fixed collection of potential answers. When evaluated with measures that differentiate between relevant, non-relevant, and unjudged answers, reasonable comparisons can be made between participating systems that contributed to the judgement pools and future systems that did not. By contrast, the open-ended nature of Task 3 means that relevance judgements on results from participating systems can not be used in the same way to evaluate future systems that might (and hopefully will!) generate different answers.

The problem lies in the way AR and P@1 are defined; they rely on our ability to match new answers with judged answers. For future systems, however, the best we might reasonably hope for is similarity between the new answers and the judged answers. If we are to avoid the need to keep assessors around forever, we will need automatic evaluation measures that can be used to compare participating Task 3 systems with future Task 3 systems. With that goal in mind, we expect to also report Task 3 results using the following evaluation measures in the working notes paper:

1. **Lexical Overlap.** Following SQuAD and CoQA [16, Sect. 6.1], we represent answers as a bag of tokens, where tokens are words for text and SLT nodes for math. For every topic, we compute the token F_1 score between the system's answer and each known relevant Task 1 and Task 3 answer (using H+M binarization). The score for a topic is the maximum across these F_1 scores. The final score is the average score for a topic.

2. **Contextual Similarity.** Although lexical overlap can account for answers with high surface similarity, it cannot recognize answers that use different tokens with similar meaning. For context similarity, we use BERTScore [20] with the MathBERTa[12] language model. As with our computation of lexical overlap, for BERTScore we also compute a token F_1 score, but instead of exact matches, we match tokens with the most similar contextual embeddings and interpret their similarity as fractional membership. For every topic, we compute BERTScore between the system's answer and each known relevant answer (with H+M binarization). The score for a topic is the maximum across these F_1 scores. The final score is the average score for a topic.

To determine the suitability of our automatic evaluation measures for the evaluation of future systems, we also expect to report the correlation between these automatic measures and our manual measures. When computing the automatic measures for a participating system, we plan to exclude relevant answers uniquely contributed to the pools by systems from the same team. This ablation avoids the perfect overlap scores that systems contributing to the pools would otherwise get from matching their own results.

---

[12] https://huggingface.co/witiko/mathberta.

6.5   Results

Manual Evaluation Measures. Table 5 shows results for ARQMath-3 Task 3. This table shows baselines first, followed by teams, and within teams their systems ranked by Average Recall (AR). As seen in the table, the automatic generative baseline run using GPT-3 achieved the best results, with 1.346 AR. Note that uniquely among ARQMath evaluation measures, AR is not bounded between 0 and 1; rather, it is bounded between 0 and 3. Among manual extractive non-baseline runs, the highest AR was achieved by a run from the approach0 team, with 1.282 AR. Among automatic extractive non-baseline runs, the highest AR was achieved by a run from the DPRL team, with 0.462 AR. Among automatic generative non-baseline runs, the highest AR was achieved by the TU_DBS team, with 0.325 AR. No manual generative non-baseline runs were submitted to ARQMath-3 Task 3.

7   Conclusion

Over the course of three years, ARQMath has created test collections for three tasks that together include relevance judgments for hundreds of topics for two of those tasks, and 78 topics for the third. Coming as it did at the dawn of the neural age in information retrieval, considerable innovation in methods has been evident throughout the three years of the lab. ARQMath has included substantial innovation in evaluation design as well, including better contextualized definitions for graded relevance, and piloting a new task on open domain question answering. Having achieved our twin goals of building a new test collection from Math Stack Exchange posts and bringing together a research community around that test collection, the time as now come to end this lab at CLEF. We expect, however, that both that collection and that community will continue to contribute to advancing the state of the art in Math IR for years to come.

Acknowledgements. We thank our student assessors from RIT: Duncan Brickner, Jill Conti, James Hanby, Gursimran Lnu, Megan Marra, Gregory Mockler, Tolu Olatunbosun, and Samson Zhang. This material is based upon work supported by the National Science Foundation (USA) under Grant No. IIS-1717997 and the Alfred P. Sloan Foundation under Grant No. G-2017-9827.

References

1. Aizawa, A., Kohlhase, M., Ounis, I.: NTCIR-10 math pilot task overview. In: Proceedings of the 10th NTCIR (2013)
2. Aizawa, A., Kohlhase, M., Ounis, I.: NTCIR-11 math-2 task overview. In: Proceedings of the 11th NTCIR (2014)
3. Brown, T., et al.: Language models are few-shot learners (2020)
4. Buckley, C., Voorhees, E.M.: Retrieval evaluation with incomplete information. In: Proceedings of the 27th Annual International ACM SIGIR Conference on Research and Development in Information Retrieval (2004)

5. Davila, K., Zanibbi, R.: Layout and semantics: combining representations for mathematical formula search. In: Proceedings of the 40th International ACM SIGIR Conference on Research and Development in Information Retrieval (2017)
6. Ficler, J., Goldberg, Y.: Controlling linguistic style aspects in neural language generation. In: Proceedings of the Workshop on Stylistic Variation. Association for Computational Linguistics, September 2017
7. Hopkins, M., Le Bras, R., Petrescu-Prahova, C., Stanovsky, G., Hajishirzi, H., Koncel-Kedziorski, R.: SemEval-2019 task 10: math question answering. In: Proceedings of the 13th International Workshop on Semantic Evaluation (2019)
8. Líška, M., Sojka, P., Růžička, M., Mravec, P.: Web interface and collection for mathematical retrieval WebMIaS and MREC (2011)
9. Macdonald, C., Tonellotto, N.: Declarative experimentation in information retrieval using PyTerrier. In: Proceedings of the 2020 ACM SIGIR on International Conference on Theory of Information Retrieval (2020)
10. Mansouri, B., Oard, D.W., Agarwal, A., Zanibbi, R.: Effects of context, complexity, and clustering on evaluation for math formula retrieval. arXiv preprint arXiv:2111.10504 (2021)
11. Mansouri, B., Rohatgi, S., Oard, D.W., Wu, J., Giles, C.L., Zanibbi, R.: Tangent-CFT: an embedding model for mathematical formulas. In: Proceedings of the 2019 ACM SIGIR International Conference on Theory of Information Retrieval (ICTIR) (2019)
12. Mansouri, B., Zanibbi, R., Oard, D.W.: Characterizing Searches for Mathematical Concepts. In: Joint Conference on Digital Libraries (JCDL) (2019)
13. Mansouri, B., Zanibbi, R., Oard, D.W.: Learning to Rank for Mathematical Formula Retrieval. In: Proceedings of the 44th International ACM SIGIR Conference on Research and Development in Information Retrieval (2021)
14. Meadows, J., Freitas, A.: A survey in mathematical language processing. arXiv preprint arXiv:2205.15231 (2022)
15. Ounis, I., Amati, G., Plachouras, V., He, B., Macdonald, C., Johnson, D.: Terrier information retrieval platform. In: Losada, D.E., Fernández-Luna, J.M. (eds.) ECIR 2005. LNCS, vol. 3408, pp. 517–519. Springer, Heidelberg (2005). https://doi.org/10.1007/978-3-540-31865-1_37
16. Reddy, S., Chen, D., Manning, C.D.: CoQA: a conversational question answering challenge. Trans. Assoc. Comput. Linguist. **7**, 249–266 (2019)
17. Sakai, T., Kando, N.: On information retrieval metrics designed for evaluation with incomplete relevance assessments. Information Retrieval **11**, 447–470 (2008)
18. Stathopoulos, Y., Teufel, S.: Retrieval of research-level mathematical information needs: a test collection and technical terminology experiment. In: Proceedings of the 53rd Annual Meeting of the Association for Computational Linguistics and the 7th International Joint Conference on Natural Language Processing (Volume 2: Short Papers) (2015)
19. Zanibbi, R., Aizawa, A., Kohlhase, M., Ounis, I., Topic, G., Davila, K.: NTCIR-12 MathIR task overview. In: Proceedings of the 12th NTCIR (2016)
20. Zhang, T., Kishore, V., Wu, F., Weinberger, K.Q., Artzi, Y.: BERTScore: evaluating text generation with BERT. arXiv preprint arXiv:1904.09675 (2019)
21. Zhu, F., Lei, W., Wang, C., Zheng, J., Poria, S., Chua, T.S.: Retrieving and reading: a comprehensive survey on open-domain question answering. arXiv preprint arXiv:2101.00774v3 (2021)

Overview of Touché 2022: Argument Retrieval

Alexander Bondarenko[1(✉)], Maik Fröbe[1], Johannes Kiesel[2], Shahbaz Syed[3], Timon Gurcke[4], Meriem Beloucif[5], Alexander Panchenko[6], Chris Biemann[7], Benno Stein[2], Henning Wachsmuth[4], Martin Potthast[3], and Matthias Hagen[1]

[1] Martin-Luther-Universität Halle-Wittenberg, Halle, Germany
touche@webis.de
[2] Bauhaus-Universität Weimar, Weimar, Germany
[3] Leipzig University, Leipzig, Germany
[4] Paderborn University, Paderborn, Germany
[5] Uppsala University, Uppsala, Sweden
[6] Skolkovo Institute of Science and Technology, Moscow, Russia
[7] Universität Hamburg, Hamburg, Germany
https://touche.webis.de/

Abstract. This paper is a condensed report on the third year of the Touché lab on argument retrieval held at CLEF 2022. With the goal to foster and support the development of technologies for argument mining and argument analysis, we organized three shared tasks in the third edition of Touché: (a) argument retrieval for controversial topics, where participants retrieve a gist of arguments from a collection of online debates, (b) argument retrieval for comparative questions, where participants retrieve argumentative passages from a generic web crawl, and (c) image retrieval for arguments, where participants retrieve images from a focused web crawl that show support or opposition to some stance.

1 Introduction

Decision making and opinion formation are routine human tasks that often involve weighing pro and con arguments. Since the Web is full of argumentative texts on almost any topic, everybody has, in principle, the chance to acquire knowledge to come to informed decisions or opinions by simply using a search engine. However, large amounts of the arguments accessible easily may be of low quality. For example, they may be irrelevant, contain incoherent logic, provide insufficient support, or use foul language. Such arguments should rather remain "invisible" in search results which implies several retrieval challenges—regardless of whether a query is about socially important topics or "only" about personal decisions. The challenges range from assessing an argument's relevance to a query and estimating how well an implied stance is justified, to identifying what is the main "gist" of an argument's reasoning as well as finding images that help to illustrate some stance. Still, today's popular web search engines do not

A. Barrón-Cedeño et al. (Eds.): CLEF 2022, LNCS 13390, pp. 311–336, 2022.
https://doi.org/10.1007/978-3-031-13643-6_21

really address these challenges, thus lacking a sophisticated support for searchers in argument retrieval scenarios—a gap we aim to close with the Touché labs.[1]

In the spirit of the two successful Touché labs on argument retrieval at CLEF 2020 and 2021 [13,16], we organized a third lab edition to again bring together researchers from the fields of information retrieval and natural language processing who work on argumentation. At Touché 2022, we organized the following three shared tasks, the last of which being fully new to this edition:

1. Argumentative sentence retrieval from a focused collection (crawled from debate portals) to support argumentative conversations on controversial topics.
2. Argument retrieval from a large collection of text passages to support answering comparative questions in the scenario of personal decision making.
3. Image retrieval to corroborate and strengthen textual arguments and to provide a quick overview of public opinions on controversial topics.

In the Touché lab, we followed the classic TREC-style[2] methodology: documents and topics were provided to the participants who then submitted their ranked results (up to five runs) for every topic to be judged by human assessors. While the first two Touché editions focused on retrieving complete arguments and documents, the third edition focused on more refined problems. Three shared tasks explored whether argument retrieval can support decision making and opinion formation more directly by extracting the argumentative gist from documents, by classifying their stance as pro or con towards the issue in question, and by retrieving images that show support or opposition to some stance.

The teams that participated in the third year of Touché were able to use the topics as well as the relevance and argument quality judgments from the previous lab editions to improve their approaches. Only a few decided to train and optimize their pipelines using the judgments provided, though. Alongside dense retrieval models like BM25 [70], this year approaches focus on more recent Transformer-based models, such as T5 [67] and T0 [76] in zero-shot settings, to predict relevance, argument quality, and stance. Also many re-ranking methods are proposed based on a wide range of diverse characteristics including a word mover's distance, linguistic properties of documents, as well as document "argumentativeness" and argument quality. A more comprehensive overview of all submitted approaches is covered in the extended overview [15].

2   Related Work

Queries in argument retrieval often may be phrases that describe a controversial topic, questions that ask to compare two options, or even statements that capture complete claims or short arguments [85]. In the Touché lab, we address the first two types in three different shared tasks. Here, we briefly summarize the related work for all three tasks.

[1] 'touché' is commonly "used to acknowledge a hit in fencing or the success or appropriateness of an argument" [https://merriam-webster.com/dictionary/touche]
[2] https://trec.nist.gov/tracks.html

2.1   Argument Retrieval

The goal of argument retrieval is to deliver arguments to support users in making a decision or in persuading an audience of a specific point of view. An argument is usually modeled as a conclusion with one or more supporting or attacking premises [83]. While a conclusion is a statement that can be accepted or rejected, a premise is a more grounded statement (e.g., statistical evidence).

The development of an argument search engine is faced with challenges that range from identifying argumentative queries [2] to mining arguments from unstructured text to assessing their relevance and quality [83]. Argument retrieval follows several paradigms that start from different sources and perform argument mining and retrieval tasks in different orders [3]. Wachsmuth et al. [83], for instance, extract arguments offline using heuristics that are tailored to online debate portals. Their argument search engine args.me uses BM25F [71] to rank the indexed arguments, giving conclusions more weight than premises. Also Levy et al. [48] use distant supervision to mine arguments offline for a set of topics from Wikipedia before ranking. Following a different paradigm, Stab et al. [79] retrieve documents from the Common Crawl[3] in an online fashion (no prior offline argument mining) and use a topic-dependent neural network to extract arguments from the retrieved documents at query time. With the three Touché tasks, we address the paradigms of Wachsmuth et al. [83] (Task 1) and Stab et al. [79] (Tasks 2 and 3), respectively.

Argument retrieval should rank arguments according to their topical relevance but also to their quality. What makes a good argument has been studied since the time of Aristotle [6]. Wachsmuth et al. [81] categorized the different aspects of argument quality into a taxonomy that covers three dimensions: logic, rhetoric, and dialectic. Logic concerns the strength of the internal structure of an argument, i.e., the conclusion and the premises along with their relations. Rhetoric covers the effectiveness of the argument in persuading an audience with its conclusion. Dialectic, finally, addresses the relations of an argument to other arguments on the topic. For example, an argument attacked by many others may be rather vulnerable in a debate. The relevance of an argument to a query's topic is categorized under dialectical quality [81].

Researchers assess argument relevance by measuring an argument's similarity to a query's topic or by incorporating its support and attack relations to other arguments. Potthast et al. [63] evaluate four standard retrieval models for ranking arguments with regard to four quality dimensions: relevance, logic, rhetoric, and dialectic. One of the main findings is that DirichletLM is better at ranking arguments than BM25, DPH, and TF-IDF. Gienapp et al. [32] extend this work by proposing a pairwise strategy that reduces the costs of crowdsourcing argument retrieval annotations in a pairwise fashion by 93% (i.e., annotating only a small subset of argument pairs).

Wachsmuth et al. [84] create a graph of arguments by connecting two arguments when one uses the other's conclusion as a premise. They exploit this

---

[3] http://commoncrawl.org

structure to rank the arguments in the graph using PageRank scores [58]. This method is shown to outperform baselines that only consider the content of the argument and its internal structure (conclusion and premises). Dumani et al. [25] introduce a probabilistic framework that operates on semantically similar claims and premises and that utilizes support and attack relations between clusters of premises and claims as well as between clusters of claims and a query. It is found to outperform BM25 in ranking arguments. Later, Dumani and Schenkel [26] also proposed an extension of the framework to include the quality of a premise as a probability by using the fraction of premises which are worse with regard to three quality dimensions: cogency, reasonableness, and effectiveness. Using a pairwise quality estimator trained on the Dagstuhl-15512 ArgQuality Corpus [82], their probabilistic framework with the argument quality component outperformed the one without on the 50 Task 1 topics of Touché 2020.

2.2 Retrieval for Comparisons

Comparative information needs in web search have first been addressed by basic interfaces where two products to be compared are entered separately in a left and a right search box [55,80]. Comparative sentences are then identified and mined from product reviews in favor or against one or the other product using opinion mining approaches [39,40,42]. Recently, the identification of the comparison preference (the "winning" entity) in comparative sentences has been tackled in a more broad domain (not just product reviews) by applying feature-based and neural classifiers [52,60]. Such preference classification forms the basis of the comparative argumentation machine CAM [77] that takes two entities and some comparison aspect(s) as input, retrieves comparative sentences in favor of one or the other entity using BM25, and then classifies their preference for a final merged result table presentation. A proper argument ranking, however, is still missing in CAM. Chekalina et al. [18] later extend the system to accept comparative questions as input and to return a natural language answer to the user. A comparative question is parsed by identifying the comparison objects, aspect(s), and predicate. The system's answer is either generated directly based on Transformers [22] or by retrieval from an index of comparative sentences. Identifying comparative information needs in question queries is proposed by Bondarenko et al. [12] and Bondarenko et al. [11] who study such information needs in a search engine log, propose a cascading ensemble of classifiers (rule-based, feature-based, and neural models) that identifies comparative questions, and label a respective dataset. They also propose an approach to identify entities of interest such as comparison objects, aspects, and predicates in comparative questions and to detect the stance of potential answers towards the comparison objects. The respective stance dataset is provided for Touché Task 2 participants to train their approaches for the stance classification of retrieved passages.

2.3   Image Retrieval

Images can provide contextual information and express, underline, or popularize an opinion [24], thereby taking the form of subjective statements [27]. Some images express both a premise and a conclusion, making them full arguments [35, 73]. Other images may provide contextual information only and have to be combined with a textual conclusion to form a complete argument. In this regard, a recent SemEval task distinguished a total of 22 persuasion techniques in memes alone [23]. Moreover, argument quality dimensions like acceptability, credibility, emotional appeal, and sufficiency [82] all apply to arguments that include images as well.

Keyword-based image search by analyzing the content of images or videos has been studied for decades [1], pre-dated only by approaches relying on metadata and similarity measures [17]. Early approaches exploited keyword-based web search (e.g., by Yanai [88]). In a recent survey, Latif et al. [46] categorize image features into color, texture, shape, and spatial features. Current commercial search engines also index text found in images, surrounding text, alternative texts displayed when an image is unavailable, and their URLs [34, 87]. As for the retrieval of argumentative images, a closely related concept is "emotional images", which is based on image features like color and composition [78, 86]. Since argumentation goes hand in hand with emotions, those emotional features may be promising for retrieving images for arguments in the future. To retrieve images for arguments is a relatively new task that has been recently proposed by Kiesel et al. [44], which forms the basis of the Touché Task 3.

3   Lab Overview and Statistics

In this year, we received 58 registrations in total, doubling the number of registered participants in the previous year (29 registrations in 2021). We received 17 registrations for Task 1, 10 for Task 2, and 4 for Task 3 (the new task this year); 27 teams registered for more than one task. The majority of registrations came from Germany and Italy (13 each), followed by 12 from India, 3 from the United States, 2 from the Netherlands, France, Switzerland, and Bangladesh, and one each from Pakistan, Portugal, United Kingdom, Indonesia, China, Russian Federation, Bulgaria, Nigeria, and Lebanon. Aligned with the lab's fencing-related title, participants selected a real or a fictional fencer or swordsman character (e.g., Zorro) as their team name upon registration.

Out of 58 registered teams, 23 actively participated in the tasks and submitted their results (27 teams submitted in 2021 and 17 teams in 2020).[4] Using the setup of the previous Touché editions, we encouraged the participants to submit software in TIRA [64] to improve the reproducibility of the developed approaches. TIRA is an integrated cloud-based Evaluation-as-a-Service research architecture where shared task participants can install their software on a dedicated virtual machine to which they have a full administrative access. By default,

---

[4] Three teams declined to proceed in the task after submitting the results

the virtual machines run the server version of Ubuntu 20.04 with one CPU (Intel Xeon E5-2620), 4 GB RAM, and 16 GB HDD. However, we customized the resources as needed to meet participants' requirements. We pre-installed the latest versions of reasonable software in the virtual machines (e.g., Docker and Python) to simplify the deployment of the approaches within TIRA.

We allowed participants to submit software submissions and run file submissions in TIRA. For software submissions, participants created the run files with their software using the web UI of TIRA. The process for software submissions ensured that the software is fully installed in the virtual machine: the respective virtual machine is shut down, disconnected from the internet, powered on again in a sandbox mode, mounting the test datasets for the respective tasks. The interruption of the internet connection ensured that the participants' software worked without external web services that may disappear or become incompatible, which could reduce reproducibility (i.e., downloading additional external code or models during the execution is not possible). We offered support in case of problems during deployment. Later, we archived the virtual machines that the participants used for their submissions such that the respective systems can be re-evaluated or applied to new datasets.

Overall, 9 of the 23 teams submitted traditional run files instead of software in TIRA. We allowed each team to submit up to 5 runs that should follow the standard TREC-style format.[5] We checked the validity of all submitted run files, asking participants to resubmit their run files (or software) if there were any problems—again, also offering our support in case of problems. All 23 teams submitted valid runs, resulting in 84 valid runs.

4   Task 1: Argument Retrieval for Controversial Questions

The goal of the Touché 2022 lab's first task was to support individuals who search for opinions and arguments on socially important controversial topics like "Are social networking sites good for our society?". Such scenarios benefit from obtaining the gists of various web resources that briefly summarize different standpoints (pro or con) on controversial topics. The task we considered in this regard followed the idea of extractive argument summarization [5].

4.1   Task Definition

Given a controversial topic and a collection of arguments, the task was to retrieve sentence pairs that represent the gist of their corresponding arguments (e.g., the main claim and premise). Sentences in a pair may not contradict each other and ideally build upon each other in a logical manner comprising a coherent text.

4.2   Data Description

Topics. We used 50 controversial topics from the previous iterations of Touché. Each topic is formulated as a question that the user might pose as a query to the

---

[5] The expected format of submissions was also described at https://touche.webis.de

Table 1. Example topic for Task 1: Argument Retrieval for Controversial Questions.

Number	34
Title	Are social networking sites good for our society?
Description	Democracy may be in the process of being disrupted by social media, with the potential creation of individual filter bubbles. So a user wonders if social networking sites should be allowed, regulated, or even banned.
Narrative	Highly relevant arguments discuss social networking in general or particular networking sites, and its/their positive or negative effects on society. Relevant arguments discuss how social networking affects people, without explicit reference to society.

search engine, accompanied by a description summarizing the information need and the search scenario, along with a narrative to guide assessors in recognizing relevant results (see Table 1).

Document Collection. The document collection for Task 1 was based on the args.me corpus [3] that contains about 400,000 structured arguments (from debatewise.org, idebate.org, debatepedia.org, and debate.org). It is freely available for download[6] and can also be accessed through the args.me API.[7] To account for this year's changes in the task definition (the focus on gists), a pre-processed version of the corpus was created. Pre-processing steps included sentence splitting, and removing premises and conclusions shorter than two words, resulting in 5,690,642 unique sentences with 64,633 claims and 5,626,509 premises.

4.3 Participant Approaches

This year's approaches included standard retrieval models such as TF-IDF, BM25, DirichletLM, and DPH. Participants also used multiple existing toolkits, such as the Project Debater API [7] for stance and evidence detection in arguments, Apache OpenNLP[8] for language detection, and classifiers proposed by Gienapp et al. [32] and Reimers et al. [69] trained on the IBM Rank 30K corpus [36] for argument quality detection. Additionally, semantic similarity of word and sentence embeddings based on doc2vec [47] and SBERT [68] was employed for retrieving coherent sentence pairs as required by the task definition. One team leveraged the text generation capabilities of GPT-2 [66] to find subsequent sentences while another team similarly used the next sentence prediction (NSP) of BERT [22] for this. These toolkits augmented the document pre-processing and re-ranking of the retrieved results.

---

[6] https://webis.de/data.html#args-me-corpus
[7] https://www.args.me/api-en.html
[8] https://opennlp.apache.org/

4.4   Task Evaluation

Participants submitted their rankings as classical TREC-style runs where document IDs are sorted by descending relevance score for each search topic (i.e., the most relevant argument occurs at Rank 1). Given the large number of runs and the possibility of retrieving up to 1000 documents (in our case, these are sentence pairs) per topic in a run, we created the pools using a top-5 pooling strategy for judgments with TrecTools [59], resulting in 6,930 unique documents for manual assessment of relevance, quality (argumentativeness), and textual coherence. Relevance was judged on a three-point scale: 0 (not relevant), 1 (relevant), and 2 (highly relevant). For quality, annotators assessed whether a retrieved pair of sentences are rhetorically well-written on a three-point scale: 0 (low quality/non-argumentative), 1 (average quality), and 2 (high quality). Finally, textual coherence (if the two sentences in a pair logically build upon each other) was also judged on a three-point scale: 0 (unrelated/contradicting), 1 (average coherence), and 2 (high coherence).

4.5   Task Results

We used nDCG@5 for evaluation of relevance, quality, and coherence. Table 2 shows the results of the best run per team. For all evaluation categories at least eight out of ten teams managed to beat the provided baseline. Similar to previous years' results, quality appeared to be the evaluation category which is covered best by the approaches followed by relevance and the newly added coherence. A more comprehensive discussion including all teams' approaches is covered in the extended lab overview [15].

In terms of relevance Team *Porthos* achieved the highest results followed by Team *Daario Naharis* with nDCG@5 scores of 0.742 and 0.683 respectively. For quality and coherence Team *Daario Naharis* obtained the highest scores (0.913 and 0.458) followed by Team *Porthos* (0.873 and 0.429). The two-stage re-ranking employed by Team *Daario Naharis* improved coherence and quality in comparison to other approaches. They first ensured that retrieved pairs were relevant to their context in the argument alongside the topic that also boosted quality (argumentativeness). Then, a second re-ranking based on stance to determine the final pairing of the retrieved sentences boosted coherence. Below, we briefly describe our baseline and summarize the submitted approaches.

Our baseline *Swordsman* employed a graph-based approach that ranks argument's sentences by their centrality in the argument graph as proposed by Alshomary et al. [5]. The top two sentences are then retrieved as the final pair.

Team *Bruce Banner* employed BM25 retrieval model provided by the Pyserini toolkit [49][9] with its default parameters. Two query variants were used: standalone query and an expanded query (narrative and description appended). Likewise two variants of the sentence pairs were indexed: standalone pair and pair with the topic appended.

---

[9] https://pypi.org/project/pyserini/

Table 2. Results for Task 1 Argument Retrieval for Controversial Questions. Table shows the evaluation score of a team's best run for the three dimensions of relevance, quality, and coherence of the retrieved sentence pairs. Best scores per dimension are in bold. Team names are sorted alphabetically; the baseline Swordsman is emphasized.

Team	nDCG@5		
	Relevance	Quality	Coherence
Bruce Banner	0.651	0.772	0.378
D'Artagnan	0.642	0.733	0.378
Daario Naharis	0.683	**0.913**	**0.458**
Gamora	0.616	0.785	0.285
General Grevious	0.403	0.517	0.231
Gorgon	0.408	0.742	0.282
Hit Girl	0.588	0.776	0.377
Korg	0.252	0.453	0.168
Pearl	0.481	0.678	0.398
Porthos	**0.742**	0.873	0.429
Swordsman	*0.356*	*0.608*	*0.248*

Team *D'Artagnan* combined sparse retrieval with multiple text preprocessing and query expansion approaches. They used different combinations of retrieval models such as BM25 and DirichletLM, preprocessing steps, for instance, stemming, n-grams, and stopword removal, and query expansion with synonyms using WordNet [54] and word2vec [53]. Relevance judgments from the previous year were used for optimizing parameter values. Specifically, they used word and character n-grams (bi-grams and tri-grams) and built five different vocabularies for the word2vec model.

Team *Daario Naharis* developed a standard retrieval system using the Lucene TF-IDF implementation. Additionally, they introduced a new coefficient for scoring the discriminant power of a term. Re-ranking was performed based on stance detection using the Project Debater API. The highest nDCG@5 scores were achieved with a combination of the following components: Letter Tokenizer, English Stemmer, No Stop-List, POS Tag, WordNet, Evidence Detection, ICoefficient, and LMDirichlet Similarity.

Team *Gamora* developed Lucene-based approaches using deduplication and contextual feature-enriched indexing, adding the title of a discussion and the stance on the topic, to obtain document-level relevance and quality scores following the approach used in previous Touché editions [16]. To find relevant sentence pairs rather than relevant documents, these results were used to limit the number of documents by creating a new index for only the sentences of relevant documents (double indexing) or creating all possible sentence combinations and ranking them based on a weighted average of the argumentative quality (using an SVR) of the pair and its source document. BM25 and DirichletLM were used for document similarity and SBERT [68] and TF-IDF for sentence agreement. The best approach is based on double indexing and a combination of query reduction, query boosting, query decorators, query expansion with respect to important keywords and synonyms, and using the EnglishPossessiveFilter, LengthFilter and the Krovetz stemmer.

Team *General Grevious* used a conventional IR pipeline based on Lucene, extended with a LowerCaseFilter, an EnglishPossessiveFilter (removes possessive words (trailing 's) from words), and a LengthFilter (retains tokens between 3 and 20 characters in length and removes the others). BM25 and Dirichlet-based document relevance and sentence relevance were used for retrieval along with Rapid Automatic Keyword Extraction (RAKE) [74] query expansion. Sentiment analysis and readability analysis were used for re-ranking. However, their best model does not include re-ranking, but relies solely on query expansion.

Team *Gorgon* used the Lucene project for document retrieval and compared BM25 and LMDirichlet similarity measures, developing four different analyzers with combinations of the following components: LowercaseFilter, Krovetz stemmer, EnglishPossessiveFilter, StopwordFilter. Sentence pairs were created by creating all combinations within a single document before indexing. The best approach is a combination of the LowercaseFilter, EnglishPossessiveFilter and the similarity measure BM25.

Team *Hit Girl* proposed a two-stage retrieval pipeline that combines semantic search and re-ranking via argument quality agnostic models. Internal evaluation results showed that while re-ranking improved the argument quality to varying degrees, it affected the relevance. Additionally, they proposed a novel re-ranking method called structural distance which employs a fuzzy matching between query and the sentences based on part of speech tags. This performed best in comparison to standard methods such as maximal marginal relevance and word mover's distance.

Team *Korg* proposed to first use Elasticsearch[10] with the LM-Dirichlet similarity measure to find the best matching argumentative sentences for a query. Then, either doc2vec [47], trained on all sentences in the argsme corpus, or GPT-2 [66] was used to find similar sentences by direct comparison and by generation, respectively. AsciifoldingFilter and LowercaseFilter were used together with the Krovetz stemmer and a user-defined stopword list to preprocess the sentences. Their best approach is based on doc2vec's similarity calculation.

Team *Pearl* also proposed a two-stage retrieval pipeline using DirichletLM and DPH models to retrieve argumentative sentences. Argument quality scores were used as a pre-processing step to remove noisy examples. First, a vertical prototype was developed as a baseline model for revealing the weakness of the DPH model. Specifically, they found that this model assigns high relevance to sentences even if their terms are a part of a URL, or other sources in the text and is susceptible to homonyms thus negatively affecting the retrieval performance. To account for this, a refined prototype was developed that combines an argument quality prediction model and query expansion.

---

[10] https://www.elastic.co/

Team *Porthos* used Elasticsearch with DirichletLM and BM25 for retrieval after removing sentence duplicates and filtering irrelevant sentences by removing sentences in incorrect language based on POS heuristics and their argumentativeness using the support vector machine (SVM) of [32] and the BERT approach of [69]. The approaches are based on a search term as a composition of single terms and Boolean queries together with [69] to reorder the retrieved sentences according to their argumentative quality. The sentences are paired with SBERT [68] and BERT [22] trained for the next sentence prediction task (NSP). The best approach is based on DirichletLM, NSP, using the sentence classifier in preprocessing, Boolean query with Noun Chunking for retrieval, and the BERT approach of [69] for re-ranking.

5   Task 2: Argument Retrieval for Comparative Questions

The goal of the Touché 2022 lab's second task was to support individuals in coming to informed decisions in more "everyday" or personal comparison situations—for questions like "Should I major in philosophy or psychology?". Decision making in such situations benefits from finding balanced grounds for choosing one option over the other, for instance, in the form of opinions and arguments.

5.1   Task Definition

Given a comparison search topic with two comparison objects and a collection of text passages, the task was to retrieve relevant argumentative passages for one or both objects, and to detect the passages' stances with respect to the objects.

5.2   Data Description

Topics. For the task on comparative questions, we provided 50 search topics that described scenarios of personal decision making (cf. Table 3). Each of these topics had a *title* in terms of a comparative question, *comparison objects* for the stance detection of the retrieved passages, a *description* specifying the particular search scenario, and a *narrative* that served as a guideline for the assessors.

Document Collection. The collection for Task 2 was a focused collection of 868,655 passages extracted from the ClueWeb12[11] for the 50 search topics of the task. We constructed this passage corpus from 37,248 documents in the top-100 pools for all runs submitted in the previous Touché editions. Using the TREC CAsT tools[12]), we split the documents at the sentence boundary into fixed-length passages of approximately 250 terms since working with fixed-length passages

---

[11] https://lemurproject.org/clueweb12/index.php
[12] https://github.com/grill-lab/trec-cast-tools

Table 3. Example topic for Task 2: Argument Retrieval for Comparative Questions.

Number	88
Title	Should I major in philosophy or psychology?
Objects	major in philosophy, psychology
Description	A soon-to-be high-school graduate finds themself at a crossroad in their life. Based on their interests, majoring in philosophy or in psychology are the potential options and the graduate is searching for information about the differences and similarities, as well as advantages and disadvantages of majoring in either of them (e.g., with respect to career opportunities or gained skills).
Narrative	Relevant documents will overview one of the two majors in terms of career prospects or developed new skills, or they will provide a list of reasons to major in one or the other. Highly relevant documents will compare the two majors side-by-side and help to decide which should be preferred in what context. Not relevant are study program and university advertisements or general descriptions of the disciplines that do not mention benefits, advantages, or pros/cons.

is more effective than variable-length original passages [41]. From the initial 1,286,977 passages we removed near-duplicates with CopyCat [29] to mitigate negative impacts [30,31], resulting in the final collection of 868,655 passages.

To lower the entry barrier of this task, we also provided the participants with a number of previously compiled resources. These included the document-level relevance and argument quality judgments from the previous Touché editions as well as, for passage-level relevance judgments, a subset of MS MARCO [56] with comparative questions identified by our ALBERT-based [45] classifier (about 40,000 questions are comparative) [11]. Each comparative question in MS MARCO contains 10 text passages with relevance labels. For stance detection, a dataset comprising 950 comparative questions and answers extracted from Stack Exchange was provided [11]. For the identification of claims and premises, the participants could use any existing argument tagging tool, such as the API[13] of TARGER [19] hosted on our own servers, or develop an own method if necessary. Additionally, we provided the collection of 868,655 passages expanded with queries generated using the docT5query model [57].

5.3 Participant Approaches

For Task 2, seven teams submitted their results (25 valid runs). Interestingly, only two participants decided to use the relevance judgments from the previous Touché editions to fine-tune models or to optimize parameters. The others

---

[13] Also available as a Python library: https://pypi.org/project/targer-api/

preferred to manually label a sample of retrieved documents themselves for the intermediate evaluation or relied on the zero-shot approaches such as the Transformer model T0++ [76]. Two teams also used the document collection expanded with docT5query [57] as a retrieval collection. Overall, the main trend of this year was using Transformer-based models for ranking and re-ranking such as ColBERT [43] and MonoT5 and DuoT5 [65]. The baseline retrieval approach was BM25. Five out of seven participants also submitted the results for stance detection for retrieved passages (additional task). They either trained their own classifiers on the provided stance dataset, fine-tuned pre-trained language models or directly used pre-trained models as zero-shot classifiers. The baseline stance detector simply output 'no stance' for all text passages.

5.4   Task Evaluation

Similar to Task 1, our volunteer human assessors labeled the relevance to a respective topic with three labels: 0 (not relevant), 1 (relevant), and 2 (highly relevant), and they assessed whether arguments are present in a result and whether they are rhetorically well-written [82] with three labels: 0 (low quality, or no arguments in a document), 1 (sufficient quality), and 2 (high quality). Additionally, we asked the assessors to label documents with respect to the comparison objects given in search topics as (a) pro first object (expresses a stronger positive attitude towards the first object), (b) pro second object (positive attitude towards the second object), (c) neutral (both comparison objects are equally good or bad), and (d) no stance (no attitude/opinion/argument towards the objects entailed). Following the strategy from Task 1, we pooled the top-5 documents from the runs resulting in 2,107 unique documents that were manually judged.

5.5   Task Results

For Task2, we used nDCG@5 to evaluate submitted rankings based on the relevance and argument quality judgments. The effectiveness of the stance detection approaches was evaluated using a macro-averaged F_1 score. Table 4 shows the results for the most effective runs of the participated teams based on the relevance and argument quality. For the stance detection (additional task) we evaluated all documents across all runs for each team that appeared in the top-5 pooling. A more comprehensive discussion including all teams' approaches is covered in the extended lab overview [15].

Team *Captian Levi* (submitted the relevance-wise most effective run) first retrieved 2,000 documents using Pyserini's BM25 [49] ($k_1 = 1.2$ and $b = 0.68$) by combining the top-1000 results for the original query (topic title) with the results for modified queries, where they (1) only removed stopwords (using the NLTK [9] stopword list), (2) replaced comparative adjectives with synonyms and antonyms found in WordNet [54], (3) added extra terms using pseudo-relevance feedback, and (4) used queries expanded with the docT5query model [57] provided by the Touché organizers. Queries and corpus were also processed by using stopwords and punctuation removal and lemmatization (with the WordNet lemmatizer).

Table 4. Results for Task 2 Argument Retrieval for Comparative Questions. The left part (a) shows the evaluation results of a team's best run according to the results' relevance, while the middle part (b) shows the best runs according to the results' quality, and the right part (c) shows the stance detection results (the teams' ordering is the same as in the part (b)). An asterisk (*) indicates that the runs with the best relevance and the best quality differ for a team. The baseline BM25 ranking is shown in bold; the baseline stance detector always predicts 'no stance'.

(a) Best relevance score per team			(b) Best quality score per team			(c) Stance
	nDCG@5			nDCG@5		F_1 macro
Team	Rel.	Qual.	Team	Qual.	Rel.	
Captain Levi	0.758	0.744	Aldo Nadi*	0.774	0.695	–
Aldo Nadi*	0.709	0.748	Captain Levi	0.744	0.758	0.261
Katana*	0.618	0.643	Katana*	0.644	0.601	0.220
Captain Tempesta*	0.574	0.589	Captain Tempesta*	0.597	0.557	–
Olivier Armstrong	0.492	0.582	Olivier Armstrong	0.582	0.492	0.191
Puss in Boots	**0.469**	**0.476**	**Puss in Boots**	**0.476**	**0.469**	**0.158**
Grimjack	0.422	0.403	Grimjack	0.403	0.422	0.235
Asuna	0.263	0.332	Asuna	0.332	0.263	0.106

The initially retrieved results were re-ranked using monoT5 and duoT5 [65]. Additionally ColBERT [43] also was used for initial ranking. The team Captain Levi submitted in total 5 runs that differ in strategies of modifying queries, initial ranking models, and final re-ranking models. Finally, the stance was detected using the pre-trained RoBERTA-Large-MNLI language model [50] without fine-tuning in two steps: by first detecting if the document has a stance and after that for documents that were not classified as 'neutral' or 'no stance' detecting which comparison object the document favors. This stance detector achieved the highest macro-averaged F_1 score across all teams.

Team *Aldo Nadi* (submitted the quality-wise most effective run) re-ranked passages that were initially retrieved with BM25F [71] (default Lucene implementation with $k_1 = 1.2$ and $b = 0.75$) on two fields: the text of the original passages, and the passages expanded with docT5query. All texts were processed with the Porter stemmer [61], removing stopwords using different lists such as Snowball [62], a default Lucene stopword list, a custom list containing the 400 most frequent terms in the retrieval collection excluding the comparison objects contained in the 50 search topics, etc. Queries were expanded using a relevance feedback method that is based on the Rocchio Algorithm [72]. For the final ranking, the team experimented with re-ranking (up to top-1000 documents from the initial ranking) based on the argument quality by multiplying the relevance and the quality scores and Reciprocal Ranking Fusion [20]. The quality scores were predicted using the IBM Project Debater API [7]. Aldo Nadi submitted 5 runs, which vary by different combinations of the proposed methods, e.g., testing different stopword lists, using the quality-based re-ranking or fusion, etc. The team did not detect the stance.

Team *Katana* submitted in total 3 runs that all used different variants of Col-BERT [43]: (1) pre-trained on MS MARCO [56] by the University on Glasgow,[14] (2) pre-trained by Katana from scratch on MS MARCO replacing a cosine similarity between a query and a document representation with L2 distance, and (3) the latter model fine-tuned on the relevance and quality judgments from the previous Touché editions. As queries the team used topic titles without additional processing. For the stance detection Katana used a pre-trained XGBoost-based classifier that is part of Comparative Argumentation Machine [60,77].

Team *Captain Tempesta* used linguistic properties of text such as non-informative symbol frequency (hashtags, emojis, etc.), the difference between the short word (less or equal than 4 characters) frequency and the long word (more than 4 characters) frequency, and adjective and comparative adjective frequencies. Based on these properties for each document in the retrieval corpus, the quality score was computed as a weighted sum (weights were assigned manually). At a query time, the relevance score of Lucene BM25 ($k_1 = 1.2$ and $b = 0.75$) was multiplied with the quality score; the final ranking was created by sorting documents by the descending final scores. Search queries were created by removing stopwords (Lucene default list) from topic titles and lowercasing query terms except for the brand names,[15] query terms were stemmed using Lovins stemmer [51]. The team's 5 submitted runs differ in the weights manually assigned for the different quality properties. They did not detect the stance.

Team *Olivier Armstrong* submitted one run. They first identified the compared objects, aspects, and predicates in queries (topic titles) using a RoBERTa-based classifier fine-tuned on the provided stance dataset. After removing stopwords, queries were expanded with synonyms found with WordNet. Then 100 documents were retrieved using Elasticsearch BM25 ($k_1 = 1.2$ and $b = 0.75$) as initial ranking. Using the DistilBERT-based classifier [75] fine-tuned by Alhamzeh et al. [4] (Touché 2021 participant), Olivier Armstrong identified premises and claims in the retrieved documents. Before the final ranking the following scores were calculated for each candidate document: (1) arg-BM25 score by querying the new re-indexed corpus (only premises and claims are kept) using the original queries, (2) argument support score, i.e., the ratio of premises and claims in the document, (3) similarity score, i.e., the averaged cosine similarity between the original query and every argumentative sentence in the document, both represented using the SBERT embeddings [68]. The final score was obtained by summing up the normalized individual scores. The final ranking included 25 documents sorted by the descending score. For the stance detection, the team used an LSTM-based neural network with one hidden layer that was pre-trained on the provided stance dataset.

Team *Puss in Boots* was our baseline retrieval model that used a BM25 implementation in Pyserini [49] with default parameters ($k_1 = 0.9$ and $b = 0.4$) and original topic titles as queries. The baseline stance detector simply assigned 'no stance' to all documents in the ranked list.

---

[14] http://www.dcs.gla.ac.uk/~craigm/colbert.dnn.zip
[15] https://github.com/MatthiasWinkelmann/english-words-names-brands-places

Team *Grimjack* submitted 5 runs using query expansion and query reformulation to increase recall followed by a re-ranking step to improve precision and balance the stance distribution. For the first result they simply retrieved 100 passages ranked with the query likelihood with Dirichlet smoothing ($\mu = 1000$) using the original, unmodified queries (topic titles). Another approach re-ranks the top-10 of the initially retrieved passages using (1) argumentative axioms [8,14] that are based on premises and claims in documents that were identified using TARGER [19], (2) newly proposed comparative axioms that "prefer" more comparative objects or earlier occurrence of comparative objects premises and claims, and (3) argument quality axioms that rank higher documents with a higher argument quality score; the quality scores were calculated using the IBM Project Debater API [7]. Next result ranking is based on the previous one, where the document positions are changed based on the predicted stance such as the 'pro first object' document is followed by the 'pro send object' followed by 'neutral' stance; the steps are then repeated. The document stance was predicted using the IBM Project Debater API [7]. The last two results used T0++ [76] to expand queries, e.g., by combining original queries with newly generated, where T0++ received topic descriptions as input, to assess the argument quality, and to detect the stance in zero-shot settings. The runs differed in whether the re-ranking that balanced the stance classes distribution was used.

Team *Asuna* proposed a three-step approach that consisted of preprocessing, search, and re-ranking. For each document (text passage) in the retrieval corpus the following 3 components were computed: one-sentence extractive summary using LexRank [28], premises and claims were identified with TARGER [19], and spam scores were found in the Waterloo Spam Rankings dataset [21].[16] Initial retrieval of top-40 documents was performed with a Pyserini [49] implementation of BM25F with default parameters ($k_1 = 0.9$ and $b = 0.4$). Queries (topic titles) were lemmatized and stopwords were removed using NLTK and extended with the most frequent terms coming from the topics modeled using LDA [10] generated for the initially retrieved documents. The extended queries were used to again retrieve top-40 passages with BM25F. Finally, team Asuna re-ranked the initially retrieved documents with the Random Forests classifier [37] fed with the following features: BM25F score, number of times the document was retrieved for different queries (original, three LDA topics from documents, and one LDA topic from the task topics' descriptions), number of tokens in documents, number of sentences in documents, number of premises in documents, number of claims in documents, spam-scores, predicted argument quality scores, and predicted stances. The classifier was trained on the Touché 2020 and 2021 relevance judgments. The argument quality was predicted using DistilBERT fine-tuned on the Webis-ArgQuality-20 corpus [33]. The stance was also predicted using DistilBERT fine-tuned on the provided stance dataset [11].

---

[16] https://lemurproject.org/clueweb12/related-data.php

6 Task 3: Image Retrieval for Arguments

The goal of the Touché 2022 lab's third task was to provide argumentation support through image search. The retrieval of relevant images should provide both a quick visual overview of frequent arguments on some topic and for compelling images to support one's argumentation. To this end, the goal of the third task was to retrieve images that indicate an agreement or disagreement to some stance on a given topic as two separate lists similar to textual argument search.

6.1 Task Definition

Given a controversial topic, the task was to retrieve images (from web pages) for each stance (pro and con) that show support for that stance.

6.2 Data Description

Topics. Task 3 employs the same 50 controversial topics as Task 1 (cf. Sect. 4).

Document Collection. This task's document collection stems from a focused crawl of 23,841 images and associated web pages from late 2021. For each of the 50 topics, we issued 11 queries (with different filter words like "good", "meme", "stats", "reasons", or "effects") to Google's image search and downloaded the top 100 images and associated web pages. 868 duplicate images were identified and removed using pHash[17] and manual checks. The dataset contains various resources for each image, including the associated page for which it was retrieved as HTML page and as detailed web archive,[18] and information on how the image was ranked by Google. The full dataset is 368 GB large. To kickstart machine learning approaches, we provided 334 relevance judgments from [44].

6.3 Participant Approaches

In total, 3 teams submitted 12 runs to this task. The teams pursued quite different approaches. However, all participants employed OCR (Tesseract[19]) to extract image text. The teams Boromir and Jester also used the associated web page's text, but Team Jester restricted to text close to the image on the web page. Each team used sentiment or emotion features: based on image colors (Aramis), faces in the images (Jester), image text (all), and the web page text (Boromir, Jester). Team Boromir used the ranking information for internal evaluation.

---

[17] https://www.phash.org/
[18] Archived using https://github.com/webis-de/scriptor
[19] https://github.com/tesseract-ocr/tesseract

6.4   Task Evaluation

We employed crowdsourcing on Amazon Mechanical Turk[20] to evaluate the topical relevance, argumentativeness, and stance of the 6,607 images that the approaches retrieved, employing 5 independent annotators each. Specifically, we asked for each topic for which an image was retrieved: (1) Is the image in some manner related to the topic? (2) Do you think most people would say that, if someone shares this image without further comment, they want to show they approve of the pro-side to the topic? (3) Or do you think most people would rather say the one who shares this image does so to show they disapprove? We described each topic using the topic's title, modified as necessary to convey the description and narrative (cf. Table 1) and to clarify which stance is approve (pro) and disapprove (con). We then employed MACE [38] to identify images with high disagreement (confidence ≤ 0.55) and re-judged them ourselves (2,056 images).

6.5   Task Results

We used Precision@10 for evaluation: the ratio of relevant images among 10 retrieved images for each topic and stance. Table 5 shows the results of each team's most effective run. For each team, the same run performed best across all three measures. A more comprehensive discussion including all teams' approaches is covered in the extended lab overview [15].

Table 5. Results for Task 3 Image Retrieval for Arguments in terms of Precision@10 (per stance) for topic relevance, argumentativeness, and stance relevance. The table shows the best run for each team across all three measures.

Team	Run	Precision@10		
		Topic	Arg.	Stance
Boromir	BERT, OCR, query-processing	0.878	0.768	0.425
Minsc	Baseline	0.736	0.686	0.407
Aramis	Argumentativeness:formula, stance:formula	0.701	0.634	0.381
Jester	With emotion detection	0.696	0.647	0.350

We provided one tough baseline for comparison, called *Minsc*, which ranks images according to the ranking from our original Google queries that included the filter words "good" (for pro) and "anti" (for con). Indeed, only team Boromir was able to beat this tough baseline. Remarkably, they did so especially for on-topic relevance, which is the closest to classical information retrieval.

Team *Aramis* focused on image features. They tested the use of hand-crafted formula vs. fully-connected neural network classifiers for both argumentativeness and stance detection. Features were based on OCR, image color, image category

---

[20] https://www.mturk.com

(graphic vs. photo; diagram-likeness), and query–text similarity. In our evaluation, the hand-crafted formula performed better that the neural approaches, maybe due to differences in the annotation procedure of the training set. However, the performance drop was not large, with their worst runs still achieving a Precision@10 of 0.664 (−0.037), 0.609 (−0.025), and 0.344 (−0.037).

Team *Boromir* indexed both image text (boosted 5-fold) and web page text, using stopword lists, min-frequency filtering, and lemmatization. They clustered images and manually assigned retrieval boosts per cluster to favor more argumentative images, especially diagrams. They employed textual sentiment detection for stance detection, using either a dictionary (AFINN) or a BERT classifier. Their approach performed best and convincingly improved over the baseline. In our evaluation, the BERT classifier improves over the dictionary and the image clustering had negative effects, as it seems to introduce more off-topic images into the ranking: the same setup as the best run but using image clusters achieved a Precision@10 of 0.822 (−0.056), 0.728 (−0.040), and 0.411 (−0.014).

Team *Jester* focused on emotion-based image retrieval per facial image recognition,[21] image text, and the associated web page's text that is close to the image in the HTML source code. They assign positive leaning images to the pro-stance and negative leaning images to the con-stance. For comparison, they submitted a second run without emotion features (thus plain retrieval), which achieved a lower Precision@10: 0.671 (−0.025), 0.618 (−0.029), and 0.336 (−0.014). Thus emotion features seem helpful but insufficient when taken alone.

7   Conclusion

In this paper, we report on the third year of the Touché lab at CLEF 2022 and its three shared tasks: (1) argument retrieval for controversial questions, (2) argument retrieval for comparative questions, and (3) image retrieval for arguments. In the third Touché edition, the units of retrieval were different to the previous editions, including relevant argumentative sentences, passages, and images as well as their stance detection (our previous tasks focused on the retrieval of entire documents). From 58 registered teams, 23 participated in the tasks and submitted at least one valid run. Along with various query processing, query reformulation and expansion methods, and sparse retrieval models, the approaches had an increased focus on Transformer models and diverse re-ranking techniques. Not only the quality of documents and arguments was estimated, but also the predicted stance was considered for creating a final ranking. All evaluation resources developed at Touché are shared freely, including search queries (topics), the assembled manual relevance and argument quality judgments (qrels), and the ranked result lists submitted by the participants (runs). A comprehensive survey of developed approaches is included in the extended lab overview [15].

We plan to continue our activities for establishing a collaborative platform for researchers in the area of argument retrieval by providing submission and evaluation tools as well as by organizing collaborative events such as workshops,

---

[21] https://github.com/justinshenk/fer

fostering the accumulation of knowledge and the development of new approaches in the field. For the next iteration of the Touché lab, we plan to expand current test collections with manual judgments, to extend evaluation with other argument quality dimensions and deeper document pooling.

Acknowledgments. We are very grateful to the CLEF 2022 organizers and the Touché participants, who allowed this lab to happen. We also want to thank our volunteer annotators who helped to create the relevance and argument quality assessments and our reviewers for their valuable feedback on the participants' notebooks.

This work was partially supported by the Deutsche Forschungsgemeinschaft (DFG) through the projects "ACQuA 2.0" (Answering Comparative Questions with Arguments; project number 376430233) and "OASiS: Objective Argument Summarization in Search" (grant WA 4591/3-1), all part of the priority program "RATIO: Robust Argumentation Machines" (SPP 1999), and the German Ministry for Science and Education (BMBF) through the project "Shared Tasks as an Innovative Approach to Implement AI and Big Data-based Applications within Universities (SharKI)" (grant FKZ 16DHB4021). We are also grateful to Jan Heinrich Reimer for developing the TARGER Python library and Erik Reuter for expanding a document collection for Task 2 with docT5query.

References

1. Aigrain, P., Zhang, H., Petkovic, D.: Content-based representation and retrieval of visual media: a state-of-the-art review. Multimed. Tools Appl. **3**(3), 179–202 (1996). https://doi.org/10.1007/BF00393937
2. Ajjour, Y., Braslavski, P., Bondarenko, A., Stein, B.: Identifying argumentative questions in web search logs. In: 45th International ACM Conference on Research and Development in Information Retrieval (SIGIR 2022). ACM, July 2022. https://doi.org/10.1145/3477495.3531864
3. Ajjour, Y., Wachsmuth, H., Kiesel, J., Potthast, M., Hagen, M., Stein, B.: Data acquisition for argument search: the args.me corpus. In: Benzmüller, C., Stuckenschmidt, H. (eds.) KI 2019. LNCS (LNAI), vol. 11793, pp. 48–59. Springer, Cham (2019). https://doi.org/10.1007/978-3-030-30179-8_4
4. Alhamzeh, A., Bouhaouel, M., Egyed-Zsigmond, E., Mitrovic, J.: DistilBERT-based argumentation retrieval for answering comparative questions. In: Proceedings of the Working Notes of CLEF 2021 - Conference and Labs of the Evaluation Forum, CEUR Workshop Proceedings, vol. 2936, pp. 2319–2330, CEUR-WS.org (2021). http://ceur-ws.org/Vol-2936/paper-209.pdf
5. Alshomary, M., Düsterhus, N., Wachsmuth, H.: Extractive snippet generation for arguments. In: Proceedings of the 43nd International ACM Conference on Research and Development in Information Retrieval, SIGIR 2020, pp. 1969–1972. ACM (2020). https://doi.org/10.1145/3397271.3401186
6. Aristotle, Kennedy, G.A.: On Rhetoric: A Theory of Civic Discourse. Oxford University Press, Oxford (2006)
7. Bar-Haim, R., Kantor, Y., Venezian, E., Katz, Y., Slonim, N.: Project debater APIs: decomposing the AI grand challenge. In: Proceedings of the 2021 Conference on Empirical Methods in Natural Language Processing: System Demonstrations, EMNLP 2021, Online and Punta Cana, Dominican Republic, 7–11 November 2021, pp. 267–274. Association for Computational Linguistics (2021). https://doi.org/10.18653/v1/2021.emnlp-demo.31

8. Bevendorff, J., et al.: Webis at TREC 2020: Health Misinformation track. In: Voorhees, E., Ellis, A. (eds.) Proceedings of the 29th International Text Retrieval Conference, TREC 2020, NIST, November 2020

9. Bird, S., Klein, E., Loper, E.: Natural Language Processing with Python. O'Reilly (2009). ISBN 978-0-596-51649-9. http://www.oreilly.de/catalog/9780596516499/index.html

10. Blei, D.M., Ng, A.Y., Jordan, M.I.: Latent Dirichlet allocation. J. Mach. Learn. Res. **3**, 993–1022 (2003). http://jmlr.org/papers/v3/blei03a.html

11. Bondarenko, A., Ajjour, Y., Dittmar, V., Homann, N., Braslavski, P., Hagen, M.: Towards understanding and answering comparative questions. In: Proceedings of the 15th ACM International Conference on Web Search and Data Mining, WSDM 2022. ACM (2022). https://doi.org/10.1145/3488560.3498534

12. Bondarenko, A., et al.: Comparative web search questions. In: Proceedings of the 13th ACM International Conference on Web Search and Data Mining, WSDM 2020, pp. 52–60. ACM (2020). https://dl.acm.org/doi/abs/10.1145/3336191.3371848

13. Bondarenko, A., et al.: Overview of Touché 2020: argument retrieval. In: Working Notes Papers of the CLEF 2020 Evaluation Labs. CEUR Workshop Proceedings, vol. 2696 (2020). http://ceur-ws.org/Vol-2696/

14. Bondarenko, A., Fröbe, M., Kasturia, V., Völske, M., Stein, B., Hagen, M.: Webis at TREC 2019: decision track. In: Voorhees, E., Ellis, A. (eds.) Proceedings of the 28th International Text Retrieval Conference, TREC 2019, NIST, November 2019

15. Bondarenko, A., et al.: Overview of touché 2022: argument retrieval. In: Working Notes of CLEF 2022 - Conference and Labs of the Evaluation Forum. CEUR Workshop Proceedings. CEUR-WS.org, Berlin, Heidelberg, New York (2022, to appear)

16. Bondarenko, A., et al.: Overview of touché 2021: argument retrieval. In: Working Notes Papers of the CLEF 2021 Evaluation Labs. CEUR Workshop Proceedings, vol. 2936 (2021). http://ceur-ws.org/Vol-2936/

17. Chang, N., Fu, K.: Query-by-pictorial-example. IEEE Trans. Softw. Eng. **6**(6), 519–524 (1980). https://doi.org/10.1109/TSE.1980.230801

18. Chekalina, V., Bondarenko, A., Biemann, C., Beloucif, M., Logacheva, V., Panchenko, A.: Which is better for deep learning: Python or MATLAB? Answering comparative questions in natural language. In: Proceedings of the 16th Conference of the European Chapter of the Association for Computational Linguistics: System Demonstrations, EACL 2021, pp. 302–311. Association for Computational Linguistics (2021). https://www.aclweb.org/anthology/2021.eacl-demos.36/

19. Chernodub, A., et al.: TARGER: neural argument mining at your fingertips. In: Proceedings of the 57th Annual Meeting of the Association for Computational Linguistics, ACL 2019, pp. 195–200. ACL (2019). https://doi.org/10.18653/v1/p19-3031

20. Cormack, G.V., Clarke, C.L.A., Büttcher, S.: Reciprocal rank fusion outperforms condorcet and individual rank learning methods. In: Proceedings of the 32nd Annual International ACM SIGIR Conference on Research and Development in Information Retrieval, SIGIR 2009, pp. 758–759. ACM (2009). https://doi.org/10.1145/1571941.1572114

21. Cormack, G.V., Smucker, M.D., Clarke, C.L.A.: Efficient and effective spam filtering and re-ranking for large web datasets. Inf. Retr. **14**(5), 441–465 (2011). https://doi.org/10.1007/s10791-011-9162-z

22. Devlin, J., Chang, M., Lee, K., Toutanova, K.: BERT: pre-training of deep bidirectional transformers for language understanding. In: Proceedings of the 2019 Conference of the North American Chapter of the Association for Computational Linguistics: Human Language Technologies, NAACL-HLT 2019, pp. 4171–4186. ACL (2019). https://doi.org/10.18653/v1/n19-1423
23. Dimitrov, D., et al.: SemEval-2021 task 6: detection of persuasion techniques in texts and images. In: 15th International Workshop on Semantic Evaluation (SemEval 2021), pp. 70–98. Association for Computational Linguistics, August 2021. https://doi.org/10.18653/v1/2021.semeval-1.7, https://aclanthology.org/2021.semeval-1.7
24. Dove, I.J.: On images as evidence and arguments. In: van Eemeren, F.H., Garssen, B. (eds.) Topical Themes in Argumentation Theory: Twenty Exploratory Studies. Argumentation Library, pp. 223–238. Springer, Dordrecht (2012). https://doi.org/10.1007/978-94-007-4041-9_15
25. Dumani, L., Neumann, P.J., Schenkel, R.: A framework for argument retrieval - ranking argument clusters by frequency and specificity. In: Jose, J.M., et al. (eds.) ECIR 2020. LNCS, vol. 12035, pp. 431–445. Springer, Cham (2020). https://doi.org/10.1007/978-3-030-45439-5_29
26. Dumani, L., Schenkel, R.: Quality aware ranking of arguments. In: Proceedings of the 29th ACM International Conference on Information & Knowledge Management, CIKM 2020, pp. 335–344. Association for Computing Machinery (2020). https://doi.org/10.1007/978-3-030-45439-5_29
27. Dunaway, F.: Images, emotions, politics. Mod. Am. Hist. 1(3), 369–376 (2018). https://doi.org/10.1017/mah.2018.17. ISSN 2515-0456, 2397-1851
28. Erkan, G., Radev, D.R.: LexRank: graph-based lexical centrality as salience in text summarization. J. Artif. Intell. Res. 22, 457–479 (2004). https://doi.org/10.1613/jair.1523
29. Fröbe, M., et al.: CopyCat: near-duplicates within and between the ClueWeb and the common crawl. In: Proceedings of the 44th International ACM Conference on Research and Development in Information Retrieval, SIGIR 2021, pp. 2398–2404. ACM (2021). https://dl.acm.org/doi/10.1145/3404835.3463246
30. Fröbe, M., Bevendorff, J., Reimer, J., Potthast, M., Hagen, M.: Sampling bias due to near-duplicates in learning to rank. In: Proceedings of the 43rd International ACM Conference on Research and Development in Information Retrieval, SIGIR 2020. ACM (2020). https://dl.acm.org/doi/10.1145/3397271.3401212
31. Fröbe, M., Bittner, J.P., Potthast, M., Hagen, M.: The effect of content-equivalent near-duplicates on the evaluation of search engines. In: Jose, J.M., et al. (eds.) ECIR 2020. LNCS, vol. 12036, pp. 12–19. Springer, Cham (2020). https://doi.org/10.1007/978-3-030-45442-5_2
32. Gienapp, L., Stein, B., Hagen, M., Potthast, M.: Efficient pairwise annotation of argument quality. In: Proceedings of the 58th Annual Meeting of the Association for Computational Linguistics, pp. 5772–5781. Association for Computational Linguistics, July 2020. https://doi.org/10.18653/v1/2020.acl-main.511, https://aclanthology.org/2020.acl-main.511
33. Gienapp, L., Stein, B., Hagen, M., Potthast, M.: Efficient pairwise annotation of argument quality. In: Proceedings of the 58th Annual Meeting of the Association for Computational Linguistics, ACL 2020, pp. 5772–5781. Association for Computational Linguistics (2020). https://www.aclweb.org/anthology/2020.acl-main.511/
34. Google: Google images best practices. Google Developers (2021). https://support.google.com/webmasters/answer/114016

35. Grancea, I.: Types of visual arguments. Argumentum. J. Seminar Discursive Log. Argument. Theory Rhetoric **15**(2), 16–34 (2017)
36. Gretz, S., et al.: A large-scale dataset for argument quality ranking: construction and analysis. In: The Thirty-Fourth AAAI Conference on Artificial Intelligence, AAAI 2020, The Thirty-Second Innovative Applications of Artificial Intelligence Conference, IAAI 2020, The Tenth AAAI Symposium on Educational Advances in Artificial Intelligence, EAAI 2020, pp. 7805–7813. AAAI Press (2020). https://ojs. aaai.org/index.php/AAAI/article/view/6285
37. Ho, T.K.: Random decision forests. In: Third International Conference on Document Analysis and Recognition, ICDAR 1995, 14–15 August 1995, Montreal, Canada, vol. I, pp. 278–282. IEEE Computer Society (1995). https://doi.org/10. 1109/ICDAR.1995.598994
38. Hovy, D., Berg-Kirkpatrick, T., Vaswani, A., Hovy, E.: Learning whom to trust with MACE. In: Proceedings of the Conference of the North American Chapter of the Association for Computational Linguistics: Human Language Technologies (NAACL-HTL 2013), pp. 1120–1130. Association for Computational Linguistics, Atlanta, June 2013. https://aclanthology.org/N13-1132
39. Jindal, N., Liu, B.: Identifying comparative sentences in text documents. In: Proceedings of the 29th Annual International Conference on Research and Development in Information Retrieval, SIGIR 2006, pp. 244–251. ACM (2006). https:// doi.org/10.1145/1148170.1148215
40. Jindal, N., Liu, B.: Mining comparative sentences and relations. In: Proceedings of the 21st National Conference on Artificial Intelligence and the 18th Innovative Applications of Artificial Intelligence Conference, AAAI 2006, pp. 1331–1336. AAAI Press (2006). http://www.aaai.org/Library/AAAI/2006/aaai06-209.php
41. Kaszkiel, M., Zobel, J.: Passage retrieval revisited. In: Belkin, N.J., Narasimhalu, A.D., Willett, P., Hersh, W.R., Can, F., Voorhees, E.M. (eds.) Proceedings of the 20th Annual International ACM SIGIR Conference on Research and Development in Information Retrieval, SIGIR 1997, Philadelphia, PA, USA, 27–31 July 1997, pp. 178–185. ACM (1997). https://doi.org/10.1145/258525.258561
42. Kessler, W., Kuhn, J.: A corpus of comparisons in product reviews. In: Proceedings of the 9th International Conference on Language Resources and Evaluation, LREC 2014, pp. 2242–2248. European Language Resources Association (ELRA) (2014). http://www.lrec-conf.org/proceedings/lrec2014/summaries/1001.html
43. Khattab, O., Zaharia, M.: ColBERT: efficient and effective passage search via contextualized late interaction over BERT. In: Huang, J., et al. (eds.) Proceedings of the 43rd International ACM SIGIR conference on research and development in Information Retrieval, SIGIR 2020, pp. 39–48. ACM (2020). https://doi.org/10. 1145/3397271.3401075
44. Kiesel, J., Reichenbach, N., Stein, B., Potthast, M.: Image retrieval for arguments using stance-aware query expansion. In: Proceedings of the 8th Workshop on Argument Mining, ArgMining 2021 at EMNLP, pp. 36–45. ACL (2021)
45. Lan, Z., Chen, M., Goodman, S., Gimpel, K., Sharma, P., Soricut, R.: ALBERT: a lite BERT for self-supervised learning of language representations. In: Proceedings of the 8th International Conference on Learning Representations, ICLR 2020. OpenReview.net (2020). https://openreview.net/forum?id=H1eA7AEtvS
46. Latif, A., et al.: Content-based image retrieval and feature extraction: A comprehensive review. Math. Probl. Eng. **2019**, 21 (2019). https://doi.org/10.1155/2019/ 9658350
47. Le, Q., Mikolov, T.: Distributed representations of sentences and documents. In: International Conference on Machine Learning, pp. 1188–1196. PMLR (2014)

48. Levy, R., Bogin, B., Gretz, S., Aharonov, R., Slonim, N.: Towards an argumentative content search engine using weak supervision. In: Proceedings of the 27th International Conference on Computational Linguistics, COLING 2018, pp. 2066–2081. Association for Computational Linguistics (2018). https://www.aclweb.org/anthology/C18-1176/

49. Lin, J., Ma, X., Lin, S., Yang, J., Pradeep, R., Nogueira, R.: Pyserini: A python toolkit for reproducible information retrieval research with sparse and dense representations. In: Proceedings of the 44th International ACM SIGIR Conference on Research and Development in Information Retrieval, SIGIR 2021, pp. 2356–2362. ACM (2021). https://doi.org/10.1145/3404835.3463238

50. Liu, Y., et al.: RoBERTa: a robustly optimized BERT pretraining approach. CoRR abs/1907.11692 (2019). http://arxiv.org/abs/1907.11692

51. Lovins, J.B.: Development of a stemming algorithm. Mech. Transl. Comput. Linguist. **11**(1–2), 22–31 (1968). http://www.mt-archive.info/MT-1968-Lovins.pdf

52. Ma, N., Mazumder, S., Wang, H., Liu, B.: Entity-aware dependency-based deep graph attention network for comparative preference classification. In: Proceedings of the 58th Annual Meeting of the Association for Computational Linguistics, ACL 2020, pp. 5782–5788. Association for Computational Linguistics (2020). https://www.aclweb.org/anthology/2020.acl-main.512/

53. Mikolov, T., Chen, K., Corrado, G., Dean, J.: Efficient estimation of word representations in vector space. In: Proceedings of the 1st International Conference on Learning Representations, ICLR 2013 (2013). http://arxiv.org/abs/1301.3781

54. Miller, G.A.: WordNet: a lexical database for English. Commun. ACM **38**(11), 39–41 (1995)

55. Nadamoto, A., Tanaka, K.: A comparative web browser (CWB) for browsing and comparing web pages. In: Proceedings of the 12th International World Wide Web Conference, WWW 2003, pp. 727–735. ACM (2003). https://doi.org/10.1145/775152.775254

56. Nguyen, T., et al.: MS MARCO: a human generated MAchine reading COmprehension dataset. In: Proceedings of the Workshop on Cognitive Computation: Integrating Neural and Symbolic Approaches 2016 at NIPS, CEUR Workshop Proceedings, vol. 1773. CEUR-WS.org (2016). http://ceur-ws.org/Vol-1773/CoCoNIPS_2016_paper9.pdf

57. Nogueira, R., Lin, J., Epistemic, A.: From doc2query to docTTTTTquery. Online preprint (2019). https://cs.uwaterloo.ca/~jimmylin/publications/Nogueira_Lin_2019_docTTTTTquery-v2.pdf

58. Page, L., Brin, S., Motwani, R., Winograd, T.: The PageRank citation ranking: bringing order to the web. Technical report 1999–66, Stanford InfoLab (1999). http://ilpubs.stanford.edu:8090/422/

59. Palotti, J.R.M., Scells, H., Zuccon, G.: TrecTools: an open-source Python library for information retrieval practitioners involved in TREC-like campaigns. In: Proceedings of the 42nd International Conference on Research and Development in Information Retrieval, SIGIR 2019, pp. 1325–1328. ACM (2019). https://doi.org/10.1145/3331184.3331399

60. Panchenko, A., Bondarenko, A., Franzek, M., Hagen, M., Biemann, C.: Categorizing comparative sentences. In: Proceedings of the 6th Workshop on Argument Mining, ArgMining@ACL 2019, pp. 136–145. Association for Computational Linguistics (2019). https://doi.org/10.18653/v1/w19-4516

61. Porter, M.F.: An algorithm for suffix stripping. Program **14**(3), 130–137 (1980). https://doi.org/10.1108/eb046814

62. Porter, M.F.: Snowball: a language for stemming algorithms (2001). http://snowball.tartarus.org/texts/introduction.html
63. Potthast, M., et al.: Argument search: assessing argument relevance. In: Proceedings of the 42nd International Conference on Research and Development in Information Retrieval, SIGIR 2019, pp. 1117–1120. ACM (2019). https://doi.org/10.1145/3331184.3331327
64. Potthast, M., Gollub, T., Wiegmann, M., Stein, B.: TIRA integrated research architecture. In: Information Retrieval Evaluation in a Changing World. TIRS, vol. 41, pp. 123–160. Springer, Cham (2019). https://doi.org/10.1007/978-3-030-22948-1_5
65. Pradeep, R., Nogueira, R., Lin, J.: The expando-mono-duo design pattern for text ranking with pretrained sequence-to-sequence models. CoRR abs/2101.05667 (2021). https://arxiv.org/abs/2101.05667
66. Radford, A., et al.: Language models are unsupervised multitask learners. OpenAI Blog 1(8), 9 (2019)
67. Raffel, C., et al.: Exploring the limits of transfer learning with a unified text-to-text transformer. J. Mach. Learn. Res. 21, 140:1–140:67 (2020). http://jmlr.org/papers/v21/20-074.html
68. Reimers, N., Gurevych, I.: Sentence-BERT: sentence embeddings using Siamese BERT-networks. In: Proceedings of the 2019 Conference on Empirical Methods in Natural Language Processing and the 9th International Joint Conference on Natural Language Processing, EMNLP-IJCNLP 2019, pp. 3980–3990. Association for Computational Linguistics (2019). https://doi.org/10.18653/v1/D19-1410
69. Reimers, N., Schiller, B., Beck, T., Daxenberger, J., Stab, C., Gurevych, I.: Classification and clustering of arguments with contextualized word embeddings. In: Proceedings of the 57th Annual Meeting of the Association for Computational Linguistics, pp. 567–578. Association for Computational Linguistics, Florence, July 2019. https://doi.org/10.18653/v1/P19-1054, https://aclanthology.org/P19-1054
70. Robertson, S.E., Walker, S., Jones, S., Hancock-Beaulieu, M., Gatford, M.: Okapi at TREC-3. In: Proceedings of The Third Text REtrieval Conference, TREC 1994, NIST Special Publication, vol. 500-225, pp. 109–126. NIST (1994). https://trec.nist.gov/pubs/trec3/papers/city.ps.gz
71. Robertson, S.E., Zaragoza, H., Taylor, M.J.: Simple BM25 extension to multiple weighted fields. In: Proceedings of the 13th International Conference on Information and Knowledge Management, CIKM 2004, pp. 42–49. ACM (2004). https://doi.org/10.1145/1031171.1031181
72. Rocchio, J.: Relevance feedback in information retrieval. In: The Smart Retrieval System-Experiments in Automatic Document Processing, pp. 313–323 (1971)
73. Roque, G.: Visual argumentation: a further reappraisal. In: van Eemeren, F.H., Garssen, B. (eds.) Topical Themes in Argumentation Theory. Argumentation Library, vol. 22, pp. 273–288. Springer, Dordrecht (2012). https://doi.org/10.1007/978-94-007-4041-9_18. ISBN 978-94-007-4040-2
74. Rose, S., Engel, D., Cramer, N., Cowley, W.: Automatic keyword extraction from individual documents. Text Min.: Appl. Theory 1(1–20), 10–1002 (2010)
75. Sanh, V., Debut, L., Chaumond, J., Wolf, T.: DistilBERT, a distilled version of BERT: smaller, faster, cheaper and lighter. CoRR abs/1910.01108 (2019). http://arxiv.org/abs/1910.01108
76. Sanh, V., et al.: Multitask prompted training enables zero-shot task generalization. CoRR abs/2110.08207 (2021). https://arxiv.org/abs/2110.08207

77. Schildwächter, M., Bondarenko, A., Zenker, J., Hagen, M., Biemann, C., Panchenko, A.: Answering comparative questions: better than ten-blue-links? In: Proceedings of the 2019 Conference on Human Information Interaction and Retrieval, CHIIR 2019, pp. 361–365. ACM (2019). https://doi.org/10.1145/3295750.3298916

78. Solli, M., Lenz, R.: Color emotions for multi-colored images. Color Res. Appl. **36**(3), 210–221 (2011). https://doi.org/10.1002/col.20604

79. Stab, C., et al.: ArgumenText: searching for arguments in heterogeneous sources. In: Proceedings of the 2018 Conference of the North American Chapter of the Association for Computational Linguistics, NAACL 2018, pp. 21–25. Association for Computational Linguistics (2018). https://www.aclweb.org/anthology/N18-5005

80. Sun, J., Wang, X., Shen, D., Zeng, H., Chen, Z.: CWS: a comparative web search system. In: Proceedings of the 15th International Conference on World Wide Web, WWW 2006, pp. 467–476. ACM (2006). https://doi.org/10.1145/1135777.1135846

81. Wachsmuth, H., et al.: Argumentation quality assessment: theory vs. practice. In: Proceedings of the 55th Annual Meeting of the Association for Computational Linguistics, ACL 2017, pp. 250–255. Association for Computational Linguistics (2017). https://doi.org/10.18653/v1/P17-2039

82. Wachsmuth, H., et al.: Computational argumentation quality assessment in natural language. In: Proceedings of the 15th Conference of the European Chapter of the Association for Computational Linguistics, EACL 2017, pp. 176–187 (2017). http://aclweb.org/anthology/E17-1017

83. Wachsmuth, H., et al.: Building an argument search engine for the web. In: Proceedings of the Fourth Workshop on Argument Mining (ArgMining), pp. 49–59. Association for Computational Linguistics (2017). https://doi.org/10.18653/v1/w17-5106

84. Wachsmuth, H., Stein, B., Ajjour, Y.: "PageRank" for argument relevance. In: Proceedings of the 15th Conference of the European Chapter of the Association for Computational Linguistics, EACL 2017, pp. 1117–1127. Association for Computational Linguistics (2017). https://doi.org/10.18653/v1/e17-1105

85. Wachsmuth, H., Syed, S., Stein, B.: Retrieval of the best counterargument without prior topic knowledge. In: Proceedings of the 56th Annual Meeting of the Association for Computational Linguistics, ACL 2018, pp. 241–251. Association for Computational Linguistics (2018). https://www.aclweb.org/anthology/P18-1023/

86. Wang, W., He, Q.: A survey on emotional semantic image retrieval. In: International Conference on Image Processing (ICIP 2008), pp. 117–120. IEEE (2008). https://doi.org/10.1109/ICIP.2008.4711705

87. Wu, A.: Learn more about what you see on google images. Google Blog (2020). https://support.google.com/webmasters/answer/114016

88. Yanai, K.: Image collector: an image-gathering system from the world-wide web employing keyword-based search engines. In: International Conference on Multimedia and Expo (ICME 2001). IEEE (2001). https://doi.org/10.1109/ICME.2001.1237772

Overview of BioASQ 2022: The Tenth BioASQ Challenge on Large-Scale Biomedical Semantic Indexing and Question Answering

Anastasios Nentidis[1,2]([✉]), Georgios Katsimpras[1], Eirini Vandorou[1], Anastasia Krithara[1], Antonio Miranda-Escalada[3], Luis Gasco[3], Martin Krallinger[3], and Georgios Paliouras[1]

[1] National Center for Scientific Research "Demokritos", Athens, Greece
{tasosnent,gkatsibras,evandorou,akrithara,paliourg}@iit.demokritos.gr
[2] Aristotle University of Thessaloniki, Thessaloniki, Greece
[3] Barcelona Supercomputing Center, Barcelona, Spain
{antonio.miranda,lgasco,martin.krallinger}@bsc.es

Abstract. This paper presents an overview of the tenth edition of the BioASQ challenge in the context of the Conference and Labs of the Evaluation Forum (CLEF) 2022. BioASQ is an ongoing series of challenges that promotes advances in the domain of large-scale biomedical semantic indexing and question answering. In this edition, the challenge was composed of the three established tasks a, b and Synergy, and a new task named DisTEMIST for automatic semantic annotation and grounding of diseases from clinical content in Spanish, a key concept for semantic indexing and search engines of literature and clinical records. This year, BioASQ received more than 170 distinct systems from 38 teams in total for the four different tasks of the challenge. As in previous years, the majority of the competing systems outperformed the strong baselines, indicating the continuous advancement of the state-of-the-art in this domain.

Keywords: Biomedical knowledge · Semantic Indexing · Question Answering

1 Introduction

Advancing the state-of-the-art in large-scale biomedical semantic indexing and question answering has been the main focus of the BioASQ challenge for more than 10 years. To this end, respective tasks are organized annually, where different teams develop systems that are evaluated on the same benchmark datasets that represent the real information needs of experts in the biomedical domain. Many research teams have participated over the years in these challenges or have profited by its publicly available datasets.

In this paper, we present the shared tasks and the datasets of the tenth BioASQ challenge in 2022, as well as an overview of the participating systems

A. Barrón-Cedeño et al. (Eds.): CLEF 2022, LNCS 13390, pp. 337–361, 2022.
https://doi.org/10.1007/978-3-031-13643-6_22

and their performance. The remainder of this paper is organized as follows. Section 2 presents a general description of the shared tasks, which took place from December 2021 to May 2022, and the corresponding datasets developed for the challenge. Followed by this is Sect. 3 that provides a brief overview of the systems developed by the participating teams for the different tasks. Detailed descriptions for some of the systems are available in the proceedings of the lab. Then, in Sect. 4, we focus on evaluating the performance of the systems for each task and sub-task, using state-of-the-art evaluation measures or manual assessment. The final section concludes the paper by giving some conclusions regarding the 2022 BioASQ challenge.

2   Overview of the Tasks

The tenth edition of the BioASQ challenge consisted of four tasks: (1) a large-scale biomedical semantic indexing task (task 10a), (2) a biomedical question answering task (task 10b), (3) task on biomedical question answering on the developing problem of COVID-19 (task Synergy), all considering documents in English, and (4) a new medical semantic annotation and concept normalization task in Spanish (DisTEMIST). In this section, we first describe the two established tasks 10a and 10b with focus on differences from previous versions of the challenge [25]. For a more detailed description of these tasks the readers can refer to [32]. Additionally, we discuss this year's version of the Synergy task and also present the new DisTEMIST task on medical semantic annotation.

2.1   Large-Scale Semantic Indexing - Task 10a

In task 10a, participants are asked to classify articles from the PubMed/MED-LINE[1] digital library into concepts of the MeSH hierarchy. Specifically, new PubMed articles that are not yet annotated by the indexers in the National Library of Medicine (NLM) are collected to build the test sets for the evaluation of the competing systems. However, NLM scaled-up its policy of fully automated indexing to all MEDLINE citations by mid-2022[2]. In response to this change, the schedule of task 10a was shifted a few weeks earlier in the year and the task was completed in fewer rounds compared to previous years. The details of each test set are shown in Table 1. In consequence, we believe that, ten years after its initial introduction, task a full-filled its goal in facilitating the advancement of biomedical semantic indexing research and no new editions of this task are planned in the context of the BioASQ challenge.

The task was designed into three independent batches of 5 weekly test sets each. However, due to the early adoption of the new NLM policy the third batch finally consists of a single test set. A second test set was also initially released in the context of the third batch, but due to the fully automated annotation of all its articles by NLM, it was disregarded and no results will be released for it. Overall,

---

[1] https://pubmed.ncbi.nlm.nih.gov/.
[2] https://www.nlm.nih.gov/pubs/techbull/nd21/nd21_medline_2022.html.

Table 1. Statistics on test datasets for task 10a. Due to the early adoption of a new NLM policy for fully automated indexing, the third batch finally consists of a single test set.

Batch	Articles	Annotated Articles	Labels per Article
	9659	9450	13.03
	4531	4512	12.00
1	4291	4269	13.04
	4256	4192	12.81
	4862	4802	12.75
Total	27599	27225	12.72
	8874	8818	12.70
	4071	3858	12.38
2	4108	4049	12.60
	3193	3045	11.74
	3078	2916	12.07
Total	23324	22686	12.29
3	2376	1870	12.31
	28	0	-
Total	2404	1870	12.31

two scenarios are provided in this task: i) on-line and ii) large-scale. The test sets contain new articles from all available journals. Similar to previous versions of the task [6], standard flat and hierarchical information retrieval measures were used to evaluate the competing systems as soon as the annotations from the NLM indexers were available. Moreover, for each test set, participants had to submit their answers in 21 h. Additionally, a training dataset that consists of 16,218,838 articles with 12.68 labels per article, on average, and covering 29,681 distinct MeSH labels in total was provided for task 10a.

2.2 Biomedical Semantic QA - Task 10b

Task 10b consists of a large-scale question answering challenge in which participants have to develop systems for all the stages of question answering in the biomedical domain. As in previous editions, the task examines four types of questions: "yes/no", "factoid", "list" and "summary" questions [6]. In this edition, the available training dataset, which the competing teams had to use to develop their systems, contains 4,234 questions that are annotated with relevant golden elements and answers from previous versions of the task. Table 2 shows the details of both training and testing sets for task 10b.

Differently from previous challenges, task 10b was split into six independent bi-weekly batches. These include five official batches, as in previous versions of the task, and an additional sixth batch with questions posed by new biomedical experts. The motivation for this additional batch was to investigate how

Table 2. Statistics on the training and test datasets of task 10b. The numbers for the documents and snippets refer to averages per question.

Batch	Size	Yes/No	List	Factoid	Summary	Documents	Snippets
Train	4,234	1148	816	1252	1018	9.22	12.24
Test 1	90	23	14	34	19	3.22	4.06
Test 2	90	18	15	34	23	3.13	3.79
Test 3	90	25	11	32	22	2.76	3.33
Test 4	90	24	12	31	23	2.77	3.51
Test 5	90	28	18	29	15	3.01	3.60
Test 6	37	6	15	6	10	3.35	4.78
Total	**4,721**	1272	901	1418	1130	3.92	5.04

interesting could be the responses of the systems for biomedical experts that are not familiar with BioASQ. In particular, a collaborative schema was adopted for this additional batch, where the new experts posed their questions in the field of biomedicine and the experienced BioASQ expert team reviewed these questions to guarantee their quality. The test set of the sixth batch contains 37 questions developed by eight new experts.

Task 10b is also divided into two phases: (phase A) the retrieval of the required information and (phase B) answering the question, which run during two consecutive days for each batch. In each phase, the participants receive the corresponding test set and have 24 h to submit the answers of their systems. This year, a test set of 90 questions, written in English, was released for phase A and the participants were expected to identify and submit relevant elements from designated resources, including PubMed/MEDLINE articles and snippets extracted from these articles. Then, the manually selected relevant articles and snippets for these 90 questions were also released in phase B and the participating systems were asked to respond with *exact answers*, that is entity names or short phrases, and *ideal answers*, that is, natural language summaries of the requested information.

2.3 Task Synergy

In order to make the advancements of biomedical information retrieval and questions answering available for the study of developing problems, we aim at a synergy between the biomedical experts and the automated question answering systems. So that the experts receive and assess the systems' responses and their assessment is fed back to the systems in order to help improving them, in a continuous iterative process. In this direction, last year we introduced the BioASQ Synergy task [25] envisioning a continuous dialog between the experts and the systems. In this model, the experts pose open questions and the systems provide relevant material and answers for these questions. Then, the experts assess the submitted material (documents and snippets) and answers, and provide feedback to the systems, so that they can improve their responses. This process proceeds with new feedback and new predictions from the systems in an iterative way.

This year, task Synergy took place in four rounds, focusing on unanswered questions for the developing problem of the COVID-19 disease. In each round the systems responses and expert feedback refer to the same questions, although some new questions or new modified versions of some questions could be added into the test sets. Table 3 shows the details of the datasets used in task Synergy.

Table 3. Statistics on the datasets of task Synergy. "Answer" stands for questions marked as having enough relevant material from previous rounds to be answered. "Feedback" stands for questions that already have some expert feedback from previous rounds.

Round	Size	Yes/No	List	Factoid	Summary	Answer	Feedback
1	72	21	20	13	18	13	26
2	70	20	19	13	18	25	70
3	70	20	19	13	18	41	70
4	64	18	19	10	17	47	64

Contrary to the task B, this task was not structured into phases, but both relevant material and answers were received together. However, for new questions only relevant material (documents and snippets) is required until the expert considers that enough material has been gathered during the previous round and mark the questions as "ready to answer". When a question receives a satisfactory answer that is not expected to change, the expert can mark the question as "closed", indicating that no more material and answers are needed for it.

In order to reflect the rapid developments in the field, each round of this task utilizes material from the current version of the COVID-19 Open Research Dataset (CORD-19) [35]. This year the time interval between two successive rounds was extended into three weeks, from two weeks in BioASQ9, to keep up with the release of new CORD-19 versions that were less frequent compared to the previous version of the task. In addition, apart from PubMed documents of the current CORD-19, CORD-19 documents from PubMed Central and ArXiv were also considered as additional resources of knowledge. Similar to task b, four types of questions are examined in the Synergy task: yes/no, factoid, list, and summary, and two types of answers, exact and ideal. Moreover, the assessment of the systems' performance is based on the evaluation measures used in task 10b.

2.4   Medical Semantic Annotation in Spanish - DisTEMIST

The DisTEMIST track [21] tries to overcome the lack of resources for indexing disease information content in languages other than English, moreover harmonizing concept mentions to controlled vocabularies. SNOMED CT was explicitly chosen to normalize disease mentions for DisTEMIST, because it is a comprehensive, multilingual and widely used clinical terminology [7].

Over the last years, scientific production has increased significantly. And, especially with the COVID-19 health crisis, it has become evident that it is necessary to integrate information from multiple data sources, including biomedical literature and clinical records. Therefore, semantic indexing tools need to efficiently work with heterogeneous data sources to achieve that information integration. But they also need to work beyond data in English, in particular considering publications like clinical case reports, as well as electronic medical records, which are generated in the native language of the healthcare professional/system [4].

In semantic indexing, certain types of concepts or entities are of particular relevance for researches, clinicians as well as patients alike. For instance, more than 20% of PubMed search queries are related to diseases, disorders, and anomalies [15], representing the second most used search type after authors. Some efforts were made to extract diseases from text using data in English, like the 2010 i2b2 corpus [34] and NCBI-Disease corpus [16]. Few resource are available for non-English content, particularly with the purpose to process diverse data sources.

DisTEMIST is promoted by the Spanish Plan for the Advancement of Language Technology (Plan TL)[3] and organized by the Barcelona Supercomputing Center (BSC) in collaboration with BioASQ. Besides, the extraction of disease mentions is of direct relevance for many use cases such as study of safety issues of biomaterials and implants, or occupational health (associating diseases to professions and occupations).

Figure 1 provides an overview of the DisTEMIST shared task setting. Using the generated DisTEMIST resources, participants create their automatic systems and generate predictions. These predictions are later evaluated, and systems are ranked according to their performance. It is structured into two independent sub-tasks (participants may choose to participate in the first, the second, or both), each taking into account a critical scenario:

- *DisTEMIST-entities subtask.* It required automatically finding disease mentions in clinical cases. All disease mentions are defined by their corresponding character offsets (start character and end character) in UTF-8 plain text.
- *DisTEMIST-linking subtask.* It is a two-step subtask. It required, first, automatically detection of disease mentions, and then they had to assign, to each mention, its corresponding SNOMED CT concept identifier.

To enable the development of disease recognition and linking systems, we have generated the DisTEMIST Gold Standard corpus. It is a collection of 1000 carefully selected clinical cases written in Spanish, that were manually annotated with disease mentions by clinical experts. All mentions were exhaustively revised to mapped them to their corresponding SNOMED CT concept identifier. The manual annotation and code assignment were done following strict annotation guidelines (see the DisTEMIST annotation guidelines in Zenodo[4]), and quality checks were implemented. The inter-annotator agreement for the disease mention annotation was 82.3% (computed as the pairwise agreement between two independent annotators with the 10% of the corpus).

---

[3] https://plantl.mineco.gob.es.
[4] https://doi.org/10.5281/zenodo.6458078.

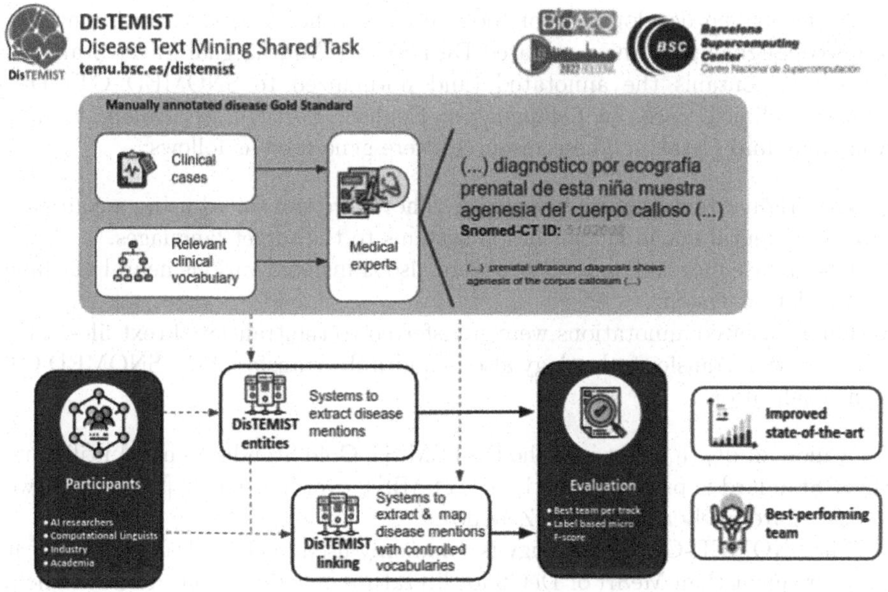

Fig. 1. Overview of the DisTEMIST Shared Task.

The corpus was randomly split into training (750 clinical cases) and test (250). Participants used the training set annotations and SNOMED CT assignments for developing their systems, while generating predictions for the test set for evaluation purposes. Table 4 shows the overview statistics of the DisTEMIST Gold Standard.

Table 4. DisTEMIST Gold Standard corpus statistics

	Documents	Annotations	Unique codes	Sentences	Tokens
Training	750	8,066	4,819	12,499	305,166
Test	250	2,599	2,484	4,179	101,152
Total	1,000	10,665	7,303	16,678	406,318

A large number or medical literature is written in languages different from English, this is particularly true for clinical case reports or publications of relevance for health-aspects specific to a certain region or country. For instance, the Scielo repository[5] contains 6741148 references in Portuguese and 388528 in Spanish. And relevant non-English-language studies are being published in languages such as Chinese, French, German or Portuguese [4].

---

[5] https://scielo.org/.

To foster the development of tools also for other languages including low resource languages, we have released DisTEMIST Multilingual Silver Standard corpus. It contains the annotated (and normalized to SNOMED CT) Dis-TEMIST clinical cases in 6 languages: English, Portuguese, Italian, French, Romanian and Catalan. These resources were generated as follows:

1. The original text files are in Spanish. They were translated using a combination of neural machine translation systems to the target languages.
2. The disease mention annotations were also translated various neural machine translation system.
3. The translated annotations were transferred to the translated text files. This annotation transfer technology also includes the transfer of the SNOMED CT normalization.

A more in-depth analysis of the DisTEMIST Gold Standard and Multilingual Silver Standard is presented in the DisTEMIST overview paper [21]. These two resources are freely available at Zenodo[6].

The SNOMED-CT terminology is commonly used in clinical scenarios, but it is less frequent than MeSH or DeCS for literature indexing applications. To help participants used to working with other terminologies, in addition to the manual mappings to SNOMED-CT, we generated cross-mappings to MeSH, ICD-10, HPO, and OMIM through the UMLS Metathesaurus.

Finally, we have generated the DisTEMIST gazetteer, containing official terms and synonyms from the relevant branches of SNOMED CT for the grounding of disease mentions. This was done because SNOMED CT cover different types of information that need to be recorded in clinical records, not just diseases. Indeed, the July 31, 2021 release of the SNOMED CT International Edition included more than 350,000 concepts. In the evaluation phase, mentions whose assigned SNOMED CT term is not included in the versions 1.0, and 2.0 of the DisTEMIST gazetteer were not considered. This resource is accessible at Zenodo[7].

3   Overview of Participation

3.1   Task 10a

This year, 8 teams participated with a total of 21 different systems in this task. Below, we provide a brief overview of those systems for which a description was available, stressing their key characteristics. The participating systems along with their corresponding approaches are listed in Table 5.

The team of Wellcome participated in task 10a with two different systems ("*xlinear*" and "*bertMesh*"). In particular, the "*xlinear*" model is a linear model

---

[6] https://doi.org/10.5281/zenodo.6408476.
[7] https://doi.org/10.5281/zenodo.6458114.

Table 5. Systems and approaches for task 10a. Systems for which no description was available at the time of writing are omitted.

System	Approach
xlinear, bertMesh	pecos, tf-idf, linear model, BertMesh, PubMedBERT, multilabel attention head
NLM	SentencePiece, CNN, embeddings, ensembles, PubMedBERT
dmiip_fdu	BertMesh, PubMedBERT, BioBERT, LTR, SVM
D2V_scalar	Doc2Vec, scalar product projection

that uses tf-idf features and it is heavily optimised for fast training and inference, while the *"bertMesh"*[8] model is a custom implementation based on the BertMesh that utilizes a multilabel attention head and PubMedBERT. The Institut de Recherche en Informatique de Toulouse (IRIT) team participated with one system, *"D2V_scalar"*, which uses Doc2Vec to map textual information into vectors and then applies a scoring mechanism to filter the results. This year, the National Library of Medicine (NLM) team competed with five systems that followed the same approaches used by the systems in previous versions of the task [29]. Finally, the Fudan University team (*"dmiip_fdu"*) also relied upon existing systems that already participated in the previous version of the task. Their systems are based on a learning to rank approach, where the component methods include both the deep learning based method BERTMeSH, which extends AttentionXML with BioBERT, and traditional SVM based methods.

As in previous versions of the challenge, two systems, developed by NLM to facilitate the annotation of articles by indexers in MEDLINE/PubMed, were available as baselines for the semantic indexing task. The first system is MTI [23] as enhanced in [38] and the second is an extension of it based on features suggested by the winners of the first version of the task [33].

3.2   Task 10b

In task 10b, 20 teams competed this year with a total of 70 different systems for both phases A and B. In particular, 10 teams with 35 systems participated in phase A, while in phase B, the number of participants and systems were 16 and 49 respectively. Six teams engaged in both phases. An overview of the technologies employed by the teams is provided in Table 6 for the systems for which a description was available. Detailed descriptions for some of the systems are available at the proceedings of the workshop.

The *"UCSD"* team competed in both phases of the task with four systems (*"bio-answerfinder"*). For both phases their systems were built upon previously developed systems [28]. In phase A, apart from improving tokenization and morphological query expansion facilities, they introduced a relaxation of the greedy ranked keyword based iterative document retrieval for cases where there were no

---

[8] https://huggingface.co/Wellcome/WellcomeBertMesh.

Table 6. Systems and approaches for task 10b. Systems for which no information was available at the time of writing are omitted.

Systems	Phase	Approach
bio-answerfinder	A, B	Bio-AnswerFinder, ElasticSearch, Bio-ELECTRA, ELECTRA, BioBERT, SQuAD, wRWMD, BM25, LSTM, T5
bioinfo	A, B	BM25, ElasticSearch, distant learning, DeepRank, universal weighting passage mechanism (UPWM), PARADE-CNN, PubMedBERT
LaRSA	A, B	ElasticSearch, BM25, SQuAD, Marco Passage Ranking, BioBERT, BoolQA, BART
ELECTROBERT	A, B	ELECTRA, ALBERT, BioELECTRA, BERT
RYGH	A	BM25, BioBERT, PubMedBERT, T5, BERTMeSH, SciBERT
gsl	A	BM25, BERT, dual-encoder
BioNIR	A	sBERT, distance metrics
KU-systems	B	BioBERT, data augmentation
MQ	B	tf-idf, sBERT, DistilBERT
lr_sys	B	BERT, SQuAD1.0, SpanBERT, XLNet, PubMedBERT, BioELECTRA, BioALBERT, BART
UDEL-LAB	B	BioM-ALBERT, BioM-ELECTRA, SQuAD
MQU	B	BART, summarization
NCU-IISR/AS-GIS	B	BioBERT, BERTScore, SQuAD, logistic-regression

or very few documents, and combined it with a BM25 based retrieval approach on selected keywords. The keywords are ranked with a cascade of LSTM layers. For phase B, their systems used a T5 based abstractive summarization system instead of the default extractive summarization subsystem.

Another team participating in both phases is the team from the University of Aveiro. Their systems ("*bioinfo*") relied on their previous transformer-UPWM model [2] and they also experimented with the PARADE-CNN model [19]. In both systems, they used a fixed PubMedBERT transformer model. Regarding phase B, they tried to answer the yes or no questions by using a simple classifier over a fixed PubMedBERT transformer model.

The team from Mohamed I Uni participated in both phases with the system "*LaRSA*". In phase A, they used ElasticSearch with BM25 as a retriever, Roberta-base-fine tuned on SQuAD as a reader, along with a cross-encoder based re-ranker trained on MS Marco Passage Ranking task. In phase B, they used a BioBERT model fine-tuned on SQuAD for both factoid and list questions, while they used a BioBERT fine-tuned on BoolQA and PubMed QA datasets for yes/no questions. For ideal answers they used a BART model fine-tuned on the CNN dataset and the ebmsum corpus.

The BSRC Alexander Fleming team also participated in both phases with four systems in total. Their systems ("*ELECTROBERT*") are based on a transformer model that combines the replaced token prediction of the ELECTRA system [12] with the sentence order prediction used in the ALBERT system [18], and are pre-trained on the 2022 baseline set of all PubMed abstracts provided by the National Library of Medicine and fine-tuned using pairs of relevant and non-relevant question-abstract pairs generated using the BioASQ9 dataset [25].

In phase A, the *"RYGH"* team participated with five systems. They adopted a multi-stage information retrieval system that utilized the BM25 along with several pre-trained models including BioBERT, PubMedBERT and SciBERT. The Google team competed also in phase A with five systems (*"gsl"*). Their systems are based on a zero-shot hybrid model consisting of two stages: retrieval and re-ranking. The retrieval model is a hybrid of BM25 and a dual-encoder model while the re-ranking is a cross-attention model with ranking loss and is trained using the output of the retrieval model. The TU Kaiserslautern team participated with five systems (*"BioNIR"*) in phase A. Their systems are based on a sBERT sentence transformer which encodes the query and each abstract, sentence by sentence, and it is trained using the BioASQ 10 dataset. They also apply different distance metrics to score and rank the sentences accordingly.

In phase B, the *"KU-systems"* team participated with five systems. Their systems are based on a BioBERT backbone architecture that involves also a data augmentation method which relies on a question generation technique. There were two teams from the Macquarie University. The first team participated with two systems (*"MQ"*) in phase B and focused on finding the ideal answers. Their systems used DistilBERT and were trained on the BioASQ10 dataset. The second team competed with two systems (*"MQU"*) which utilized a BART-based abstractive summarization system.

The Fudan University team participated with four systems (*"Ir_sys"*) in all four types of question answering tasks in phase B. For Yes/no questions, the employed BERT as their backbone and initialized its weights with BioBERT. For Factoid/List questions, they also used a BERT-based model fine-tuned with SQuAD1.0 and BioASQ 10b Factoid/List training datasets. For Summary questions, they adopted BART as the backbone of their model.

The University of Delaware team participated with five systems (*"UDEL-LAB"*) which are based on BioM-Transformers models [3]. In particular, they used both BioM-ALBERT and BioM-ELECTRA, and this year they investigated three main areas: optimizing the hyper-parameters settings, merging both List and Factoid questions to address the limited size of the Factoid training dataset, and finally, investigating the randomness with Transformers-based models by submitting two identical models with the same hyper-parameters.

The National Central Uni team competed with four systems *"NCU-IISR/AS-GIS"* in phase B. For exact answers, they used a pre-trained BioBERT model and took the possible answer list combined with the snippets score generated by Linear Regression model [39]. For yes/no questions, they used a BioBERT-MNLI model. For factoid and list type, the used a BioBERT - SQuAD model. For ideal answers, they relied on their previous BERT-based model [39]. To improve their results they replaced ROUGE-SU4 with BERTScore.

As in previous editions of the challenge, a baseline was provided for phase B exact answers, based on the open source OAQA system[37]. This system that relies on more traditional NLP and Machine Learning approaches, used to achieve top performance in older editions of the challenge and now serves a baseline. The system is developed based on the UIMA framework. In particular, question and snippet parsing is done with ClearNLP. Then, MetaMap, TmTool [36], C-Value and LingPipe [5] are employed for identifying concept that are retrieved from the UMLS Terminology Services (UTS). Finally, the relevance of concepts, documents and snippets is identified based on some classifier components and some scoring and ranking techniques are also employed.

3.3 Task Synergy

In this edition of the task Synergy 6 teams participated submitting the results from 22 distinct systems. An overview of systems and approaches employed in this task is provided in Table 7, for the systems for which a description was available. More detailed descriptions for some of the systems are available at the proceedings of the workshop.

Table 7. Systems and their approaches for task Synergy. Systems for which no description was available at the time of writing are omitted.

System	Approach
RYGH	BM25, BioBERT, PubMedBERT, T5, BERTMeSH, SciBERT
PSBST	BERT, SQuAD1.0, SpanBERT, XLNet, PubMedBERT, BioELECTRA, BioALBERT, BART
bio-answerfinder	Bio-ELECTRA++, BERT, weighted relaxed word mover's distance (wRWMD), pyserini with MonoT5, SQuAD, GloVe
MQ	tf-idf, sBERT, DistilBERT
bioinfo	BM25, ElasticSearch, distant learning, DeepRank, universal weighting passage mechanism (UPWM), BERT

The Fudan University (*"RYGH"*, *"PSBST"*) competed in task Synergy with the same models they used for task 10b. Additionally, they applied a query expansion technique in the preliminary retrieval stage and they used the Feedback data to further fine-tune the model.

The *"UCSD"* team competed in task Synergy with three systems. Their systems (*"bio-answerfinder"*) used the Bio-AnswerFinder end-to-end QA system they had previously developed [28] with few improvements.

The Macquarie University team participated with four systems. Their systems (*"MQ"*) retrieved the documents by sending the unmodified question to the search API provided by BioASQ. Then, the snippets were obtained by reranking the document sentences based on cosine similarity with the query using two variants: tf-idf, and sBERT. The ideal answers were obtained by sending the top snippets to a re-ranker based on DistilBERT and trained on the BioASQ9b training data.

The University of Aveiro team participated with five systems. Their systems (*"bioinfo"*) are based on their implementation [1] for the previous edition of

Synergy, which employs a relevant feedback approach. Their approach creates a strong baseline using a simple relevance feedback technique, based on tf-idf and the BM25 algorithm.

3.4 Task DisTEMIST

The DisTEMIST track received significant interest from a heterogeneous public. 159 teams registered for the task, and 9 of them submitted their predictions from countries such as Mexico, Germany, Spain, Italy, and Argentina. These teams provided 19 systems for DisTEMIST-entities and 15 for DisTEMIST-linking during the task period. Besides, 6 extra systems were submitted post-workshop.

Table 8. Systems and approaches for task DisTEMIST. Systems for which no description was available at the time of writing are omitted.

Team	Ref	Approach
PICUSLab	[24]	Entities: fine-tuning pre-trained biomedical language model. Linking: pre-trained biomedical language model embeddings similarity
HPI-DHC	[9]	Entities: based on Spanish Clinical Roberta. Linking: ensemble of a TF-IDF / character-n-gram based approach + multilingual embeddings (SapBERT)
SINAI	[11]	Entities: fine-tuning two different RoBERTA-based models. Linking: biomedical RoBERTa embeddings cosine similarity
Better Innovations Lab & Norwegian Centre for E-health Research	[8]	Entities: fine-tuning Spanish transformer model. Linking: FastText model embeddings Approximate Nearest Neighbour similarity
NLP-CIC-WFU	[31]	Entities: fine-tuned multilingual BERT for token classification and applied simple post-processing to deal with subword tokenization and some punctuation marks
PU++	[30]	Entities: fine-tuning multilingual BERT. Linking: FastText embeddings cosine similarity
Terminología	[10]	Use of terminology resources, NLP preprocess & lookup
iREL	-	Entities: fine-tuning BiLSTM-CRF with Spanish medical embeddings
Unicage	[27]	Entities: dictionary lookup from several ontologies

Table 8 describes the general methods used by the participants. Most teams treated DisTEMIST-entities as a NER problem and used pre-trained language models. For DisTEMIST-linking, the most common approach was generating embeddings with the test entities and the ontology terms and applying vector similarity measures, being cosine distance the most popular one. There are interesting variations, such as combining different language models [11] or similarity measures [8], adding post-processing rules [31], or using dictionary lookup methods [10, 27].

For benchmarking purposes, we introduced two baselines systems for the DisTEMIST-entities subtask, (1) DiseaseTagIt-VT: a vocabulary transfer method based on Levenshtein distance, and (2) DiseaseTagIt-Base: a modified BiLSTM-CRF architecture. To create the DisTEMIST-linking baseline, the output from these two systems was fed into a string matching engine to look for similar terminology entries to the entities. DiseaseTagIt-VT obtained a 0.2262 and 0.124 f1-score in DisTEMIST-entities and DisTEMIST-linking, respectively. DiseaseTagIt-Base reached 0.6935 f1-score in DisTEMIST-entities and 0.2642 in DisTEMIST-linking.

4   Results

4.1   Task 10a

Table 9. Average system ranks across the batches of the task 10a. The ranking for Batch 3 is based on the single test set of this batch. A hyphenation symbol (-) indicates insufficient participation in a batch, that is less than four test sets for Batch1 and Batch2. Systems with insufficient participation in all three batches are omitted.

System	Batch 1		Batch 2		Batch 3	
	MiF	LCA-F	MiF	LCA-F	MiF	LCA-F
NLM System 2	**1.5**	**1.5**	4	4	7	7
NLM System 1	2	2	7.5	6.625	8	8
attention_dmiip_fdu	3.5	4	2.25	**2**	2	3
deepmesh_dmiip_fdu	4.25	4.75	**2**	2.5	1	2
NLM CNN	5.5	6.5	9.75	10.375	12	12
MTI First Line Index	6.75	5.5	10.5	10.25	11	11
Default MTI	7.25	6.5	10	9.75	10	10
XLinear model	8.75	8.75	13.75	13.75	16	15
Dexstr system	10	10	-	-	18	17
Plain dict match	12.5	12.5	-	-	-	-
deepmesh_dmiip_fdu_	-	-	3	2.5	3	1
dmiip_fdu	-	-	4	4.25	4	5
NLM System 4	-	-	4.75	4.75	6	6
NLM System 3	-	-	6.25	6.625	9	9
coomat inference	-	-	7	6.5	5	4
similar to BertMesh	-	-	12.25	12.25	13	13
BioASQ Filtering	-	-	13.75	14.5	15	16
ediranknn	-	-	-	-	14	14
svm_baseline	-	-	-	-	17	18

In task 10a, each of the three batches were independently evaluated as presented in Table 9. As in previous editions of the task, the classification performance of the systems was measured with standard evaluation measures [6], both hierarchical and flat. In particular, the official measures for identifying the winners of each batch were the Lowest Common Ancestor F-measure (LCA-F) and the micro F-measure (MiF) [17].

As each batch consists of five test sets, we compare the participating systems based on their average rank across all multiple datasets, as suggested by Demšar [13]. Based on the rules of the challenge, the average rank of each system for a batch is the average of the four best ranks of the system in the five test sets of the batch. However, for the third batch of task 10a, where no multiple test sets are available, the ranking is based in the single available test set. In particular, the system with the best performance in a test set gets rank 1.0 for this test set, the second best rank 2.0 and so on. In case two or more systems tie, they all receive the average rank. The average rank of each system, based on both the flat MiF and the hierarchical LCA-F scores, for the three batches of the task are presented in Table 9.

The results of task 10a reveal that several participating systems manage to outperform the strong baselines in all test batches and considering either the flat or the hierarchical measures. Namely, the "NLM" systems and the "*dmiip_fdu*" systems from the Fudan University team achieve the best performance in all three batches of the task. More detailed results can be found in the online results page[9]. Figure 2 presents the improvement of the MiF scores achieved by both the MTI baseline and the top performing participant systems through the ten years of the BioASQ challenge.

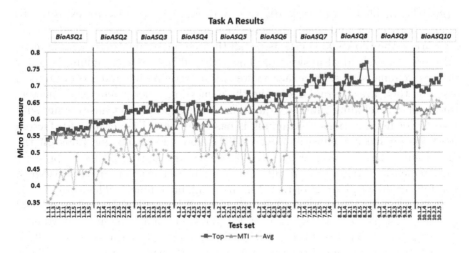

Fig. 2. The micro f-measure (MiF) achieved by systems across different years of the BioASQ challenge. For each test set the MiF score is presented for the best performing system (Top) and the MTI, as well as the average micro f-measure of all the participating systems (Avg).

---

[9] http://participants-area.bioasq.org/results/10a/.

4.2   Task 10b

Phase A: In phase A of task 10b, the evaluation of system performance on document retrieval is based on the Mean Average Precision (MAP) measure. For snippets, on the other hand, interpreting the MAP which is based on the number of relevant elements, is more complicated as the same golden snippet may overlap with several distinct submitted snippets. Therefore, the official ranking of the systems in snippet retrieval is based on the F-measure, which is calculated based on character overlaps[10].

Since BioASQ8, a modified version of Average Precision (AP) is adopted for MAP calculation. In brief, since BioASQ3, the participant systems are allowed to return up to 10 relevant items (e.g. documents or snippets), and the calculation of AP was modified to reflect this change. However, some questions with fewer than 10 golden relevant items have been observed in the last years, resulting to relatively small AP values even for submissions with all the golden elements. Therefore, the AP calculation was modified to consider both the limit of 10 elements and the actual number of golden elements [26].

Tables 10 and 11 present some indicative preliminary results for the retrieval of documents and snippets in batch 4. The full results are available online in the result page of task 10b, phase A[11]. These results are currently preliminary, as the manual assessment of the system responses by the BioASQ team of biomedical experts is still in progress and the final results for the task 10b will be available after its completion.

Phase B: In phase B of task 10b, the participating systems are expected to submit both exact and ideal answers. Regarding the sub-task of ideal answer generation, the BioASQ experts assess all the systems responses, assigning manual scores to each ideal answer [6]. Then, the official system ranking is based on these manual scores. For exact answers, the participating systems are ranked based on their average ranking in the three question types where exact answers are required, excluding summary questions for which no exact answers are submitted. For list questions the ranking is based on mean F1-measure, for factoid questions on mean reciprocal rank (MRR), and for yes/no questions on the F1-measure, macro-averaged over the classes of yes and no. Table 12 presents some indicative preliminary results on exact answer extraction from batch 4. The full results of phase B of task 10b are available online[12]. These results are preliminary, as the final results for task 10b will be available after the manual assessment of the system responses by the BioASQ team of biomedical experts.

The top performance of the participating systems in exact answer generation for each type of question during the ten years of BioASQ is presented in Fig. 3. These preliminary results reveal that the participating systems keep improving in all types of questions. In batch 4, for instance, presented in Table 12, in yes/no questions several systems manage to answer correctly for all yes/no questions.

---

[10] http://participants-area.bioasq.org/Tasks/b/eval_meas_2021/.

[11] http://participants-area.bioasq.org/results/10b/phaseA/.

[12] http://participants-area.bioasq.org/results/10b/phaseB/.

Table 10. Preliminary results for document retrieval in batch 4 of phase A of task 10b. Only the top-10 systems are presented, based on MAP.

System	Mean Precision	Mean Recall	Mean F-measure	MAP	GMAP
RYGH-3	**0.1091**	0.5478	0.1703	**0.4058**	0.0169
RYGH-1	0.1091	**0.5496**	**0.1704**	0.4040	**0.0183**
RYGH	0.1080	0.5381	0.1684	0.3925	0.0138
gsl_zs_rrf2	0.1011	0.5024	0.1574	0.3913	0.0083
gsl_zs_hybrid	0.1011	0.5015	0.1573	0.3904	0.0084
RYGH-4	0.1111	0.5424	0.1720	0.3883	0.0166
RYGH-5	0.1100	0.5387	0.1703	0.3873	0.0152
gsl_zs_rrf1	0.0989	0.4960	0.1541	0.3829	0.0082
gsl_zs_rrf3	0.1000	0.4997	0.1558	0.3778	0.0089
bioinfo-3	0.1133	0.5116	0.1728	0.3613	0.0117

Table 11. Preliminary results for snippet retrieval in batch 4 of phase A of task 10b. Only the top-10 systems are presented, based on F-measure.

System	Mean Precision	Mean Recall	Mean F-measure	MAP	GMAP
bio-answerfinder	**0.1270**	0.2790	**0.1619**	0.4905	0.0047
RYGH-5	0.1126	0.3292	0.1578	0.6596	0.0036
RYGH-4	0.1119	**0.3333**	0.1577	**0.6606**	0.0036
bio-answerfinder-3	0.1114	0.2672	0.1463	0.4456	0.0031
RYGH-3	0.0859	0.2862	0.1257	0.3669	**0.0067**
RYGH-1	0.0845	0.2801	0.1235	0.3620	0.0059
RYGH	0.0836	0.2747	0.1215	0.3523	0.0049
bio-answerfinder-4	0.0887	0.2342	0.1197	0.2973	0.0031
Basic e2e mid speed	0.0887	0.2146	0.1184	0.3321	0.0019
bio-answerfinder-2	0.0878	0.2301	0.1182	0.2949	0.0031

Table 12. Results for batch 4 for exact answers in phase B of task 10b. Only the top-10 systems based on Yes/No F1 and the BioASQ Baseline are presented.

System	Yes/No		Factoid			List		
	F1	Acc.	Str. Acc.	Len. Acc.	MRR	Prec.	Rec.	F1
UDEL-LAB3	**1.0000**	**1.0000**	0.5161	0.6129	0.5484	0.5584	0.4438	0.4501
UDEL-LAB4	**1.0000**	**1.0000**	0.5484	0.6129	0.5613	**0.6162**	0.4753	0.4752
UDEL-LAB5	**1.0000**	**1.0000**	0.5161	0.5806	0.5484	0.6132	0.4426	0.4434
lr_sys1	**1.0000**	**1.0000**	0.4839	0.6452	0.5495	0.4444	0.2410	0.2747
lr_sys2	**1.0000**	**1.0000**	0.4516	0.5161	0.4839	0.3889	0.2847	0.2718
lalala	**1.0000**	**1.0000**	**0.5806**	0.6452	**0.5995**	0.4089	0.4507	0.3835
UDEL-LAB1	0.9515	0.9583	0.4839	0.6129	0.5387	0.5799	0.5017	0.4950
UDEL-LAB2	0.9515	0.9583	0.4839	0.6129	0.5484	0.5834	**0.5844**	**0.5386**
bio-answerfinder	0.9473	0.9583	0.3548	0.4194	0.3871	0.3727	0.2701	0.2733
BioASQ_Baseline	0.2804	0.2917	0.1613	0.3226	0.2177	0.2163	0.4035	0.2582

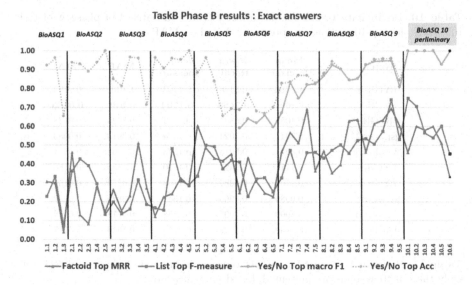

Fig. 3. The evaluation scores of the best performing systems in task B, Phase B, for exact answers, across the ten years of the BioASQ challenge. Since BioASQ6 the official measure for Yes/No questions is the macro-averaged F1 score (macro F1), but accuracy (Acc) is also presented as the former official measure. The black dots in 10.6 highlight that these scores are for the additional batch with questions from new experts.

Some improvements are also observed in the preliminary results for list questions compared to the previous years, but there is still more room for improvement, as dose for factoid questions where the preliminary performance is comparable to the one of the previous year. The performance of the best systems in the additional collaborative batch (10.6 in Fig. 3), although clearly decreased compared to the other batches, it is still comparable to them. This suggests that the responses of the participating systems could be useful to biomedical experts that are not necessarily familiarized with the BioASQ framework and did not contribute with any questions to the development of the training dataset.

4.3   Task Synergy

In task Synergy the participating systems were expected to retrieve documents and snippets, as in phase A of task 10b, and, at the same time, provide answers for some of these questions, as in phase B of task 10b. In contrast to task 10b, it is possible that no answer exists for some questions. Therefore only some of the questions provided in each test set, that were indicated to have enough relevant material gathered from previous rounds, require the submission of exact and ideal answers. Also in contrast to task B, for new questions no golden documents and snippets were provided, while for questions from previous rounds a separate file with feedback from the experts, based on the previously submitted responses, was provided.

The feedback concept was introduced in this task to further assist the collaboration between the systems and the BioASQ team of biomedical experts. The feedback includes the already judged documentation and answers along with their evaluated relevancy to the question. The documents and snippets included in the feedback are not considered valid for submission in the following rounds, and even if accidentally submitted, they were not be taken into account for the evaluation of that round. The evaluation measures for the retrieval of documents and snippets are the MAP and F-measure respectively, as in phase A of task 10b.

Regarding the ideal answers, the systems are ranked according to manual scores assigned to them by the BioASQ experts during the assessment of systems responses as in phase B of task B [6]. For the exact answers, which are required for all questions except the summary ones, the measure considered for ranking the participating systems depends on the question type. For the yes/no questions, the systems were ranked according to the macro-averaged F1-measure on prediction of no and yes answer. For factoid questions, the ranking was based on mean reciprocal rank (MRR) and for list questions on mean F1-measure.

Table 13. Results for document retrieval of the third round of the Synergy 10 task. Only the top-10 systems are presented.

System	Mean precision	Mean Recall	Mean F-Measure	MAP	GMAP
bio-answerfinder-3	**0.3063**	**0.2095**	**0.1970**	**0.2622**	**0.0184**
RYGH	0.2500	0.2017	0.1733	0.2125	0.0157
RYGH-3	0.2361	0.1699	0.1551	0.2019	0.0079
RYGH-1	0.2267	0.1677	0.1522	0.1944	0.0116
RYGH-4	0.2125	0.1642	0.1450	0.1797	0.0105
bio-answerfinder-2	0.2484	0.1204	0.1382	0.1736	0.0036
bio-answerfinder	0.2402	0.1187	0.1334	0.1669	0.0031
PSBST2	0.1844	0.1578	0.1327	0.1511	0.0131
RYGH-5	0.1891	0.1615	0.1353	0.1413	0.0108
bioinfo-3	0.1798	0.1060	0.1010	0.1158	0.0025

Some indicative results for the Synergy task are presented in Table 13. The full results of Synergy task are available online[13]. Although the scores on information retrieval and extraction of exact answers are quite moderate, compared to task 10b, where the questions are not open nether for a developing issue, the experts did found the submissions of the participants useful, as most of them stated they would be interested in using a tool following the BioASQ Synergy process to identify interesting material and answers for their research.

---

[13] http://participants-area.bioasq.org/results/synergy_v2022/.

4.4 Task DisTEMIST

The performance range of DisTEMIST participants varies depending on the method employed, the subtask (DisTEMIST-entities vs. DisTEMIST-linking), and even within the same team. The highest micro-average F1-score in DisTEMIST-entities is 0.777, and it is 0.5657 in DisTEMIST-linking.

Table 14. Results of DisTEMIST systems. The best run per team and subtask is shown. For full results, see the DisTEMIST overview paper [21]. MiP, MiR and MiF stands for micro-averaged Precision, Recall and F1-score. DisTEMIST-e stands for DisTEMIST-entities and DisTEMIST-l stands for DisTEMIST-linking.

Team	System	DisTEMIST-e			DisTEMIST-l		
		MiP	MiR	MiF	MiP	MiR	MiF
PICUSLab	NER_results	0.7915	0.7629	0.777			
	EL_results				0.2814	0.2748	0.278
HPI-DHC	3-r.c.e.-linear-lr-pp	0.7434	0.7483	0.7458			
	5-ensemble-reranking-pp.				0.6207	0.5196	0.5657
SINAI	run2-biomedical_model	0.752	0.7259	0.7387			
	run1-clinical_model				0.4163	0.4081	0.4122
Better Innovations Lab & Norwegian.	run1-ner	0.7724	0.6925	0.7303			
	run1-snomed				0.5478	0.4577	0.4987
NLP-CIC-WFU	System_mBERT	0.6095	0.4938	0.5456			
PU++	run2_mbertM5	0.601	0.4488	0.5139			
	run2-scieloBERT				0.2754	0.1494	0.1937
Terminologa	distemist-subtrack1	0.5622	0.3772	0.4515			
	distemist-subtrack2				0.4795	0.2292	0.3102
iREL	iREL	0.4984	0.3576	0.4164			
Unicage	XL_LEX_3spc	0.2486	0.3303	0.2836			
BSC baselines	DiseaseTagIt-VT	0.1568	0.4057	0.2262	0.1003	0.1621	0.124
	DiseaseTagIt-Base	0.7146	0.6736	0.6935	0.3041	0.2336	0.2642

As shown in Table 14, the top performer in DisTEMIST-entities was the NER system of PICUSLab, based on the fine-tuning of a pre-trained biomedical transformer language model. In the case of DisTEMIST-linking, the highest micro-average F1-score, precision and recall were obtained by the *ensemble-reranking-postprocess* system from the HPI-DHC team. It is based on an ensemble of a TF-IDF and character-n-gram-based approach with multilingual embeddings. Comparing the participant performances with the baseline, all teams outperformed the vocabulary transfer baseline (DiseaseTagIt-VT) in both subtasks. Based on BiLSTM-CRF architecture (DiseaseTagIt-Base), the competitive baseline ranked 5th in DisTEMIST-entities and 6th in DisTEMIST-linking.

The DisTEMIST-entities shared task results are comparable to the results of previous shared tasks such as PharmaCoNER [14], CodiEsp [22], CANTEMIST [20] and MEDDOCAN. All of them are NER challenges on Spanish clinical

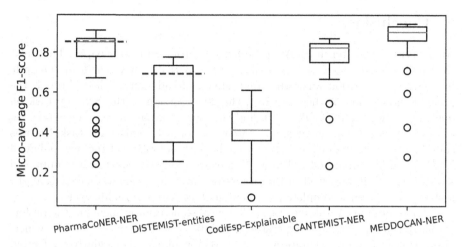

Fig. 4. Micro-average F1-score distribution of PharmaCoNER, DisTEMIST, CodiEsp, CANTEMIST, and MEDDOCAN NER systems. Themicro-average F1-scores of PharmaCoNER and DisTEMIST-entities baseline are shown in blue and red, respectively. (Color figure online)

documents. The results are specially comparable as DisTEMIST-entities, PharmaCoNER's first subtask, and CodiEsp's ExplainableAI subtask used the same corpus of documents, each annotated with different criteria and entity types - DisTEMIST with disease entities, PharmaConer with medication entities, and CodiEsp with Diagnosis and Procedures according to the ICD-10 definitions-. Additionally, the baseline of DisTEMIST and PharmaCoNER used the same architecture.

When comparing the approaches, DisTEMIST NER participants mainly employed combinations of large pre-trained transformer models, and the same is true for highest-scoring CANTEMIST participants. PharmaCoNER took place when these models were unavailable, and the most popular deep learning architectures had Recurrent and Convolutional Neural Networks at their cores. Finally, several successful CodiEsp teams used lexical lookup approaches to match the ICD-10 definitions. PharmaCoNER, CANTEMIST, and MEDDO-CAN results were higher - Fig. 4 compares the distribution of f1-scores in all tasks-.

This performance difference is directly related to the complexity of the target entities. For instance, the average number of characters per annotation in the PharmaCoNER training set was 9.7 and in DisTEMIST-entities was 24.6. The entity complexity influenced the annotation (the inter-annotator agreement of PharmaCoNER corpus was 93%, and it was 82.3% for DisTEMIST) and the system performance. It is remarkable that, given this increase in entity complexity, participants still developed competitive systems in DisTEMIST.

5 Conclusions

An overview of the tenth BioASQ challenge is provided in this paper. This year, the challenge consisted of four tasks: The two tasks on biomedical semantic indexing and question answering in English, already established through the previous nine years of the challenge, the 2022 version of the Synergy task on question answering for COVID-19, and the new task DisTEMIST on retrieving disease information from medical content in Spanish. This year, task 10a was completed earlier than expected, due to the early adoption of the new policy of NLM for fully automated indexing. Although a slight trend towards improved scores can be still observed in the results of this tenth year, we believe that the task has successfully completed its main goal, concluding its life cycle.

The preliminary results for task 10b reveal some improvements in the performance of the top participating systems, mainly for yes/no and list answer generation. However, room for improvement is still available, particularly for factoid and list questions. The introduction of an additional collaborative batch with questions from new biomedical experts that are not familiarized with BioASQ and did not contribute to the development of the BioASQ datasets before, allows for interesting observations. Although the scores for list and factoid answers in this batch are lower compared to the other batches, they are still comparable, highlighting the usefulness of the systems for any biomedical expert. For yes/no questions, in particular, some systems even managed to answer all the questions correctly.

The new task DisTEMIST introduced two new challenging subtasks beyond the one on medical literature. Namely, Named Entity Recognition and Entity Linking of diseases in Spanish clinical documents. Due to the importance of semantic interoperability across data sources, SNOMED CT was the target terminology employed in this task, and multilingual annotated resources have been released. This novel task on disease information indexing in Spanish highlighted the importance of generating resources to develop and evaluate systems that (1) effectively work in multilingual and non-English scenarios and (2) combine heterogeneous data sources.

The second year of the Synergy task in an effort to enable a dialogue between the participating systems with biomedical experts revealed that state-of-the-art systems, despite they still have room for improvement, can be a useful tool for biomedical experts that need specialized information in the context of developing problems such as the COVID-19 pandemic.

As already observed during the last years, the participating systems focus more and more on deep neural approaches. Almost all competing solutions are based on state-of-the-art neural architectures (BERT, PubMedBERT, BioBERT, BART etc.) adapted to the biomedical domain and specifically to the tasks of BioASQ. New promising approaches have been explored this year, especially for the exact answer generation, leading to improved results.

Overall, several systems managed to outperform the strong baselines on the challenging tasks offered in BioASQ, as in previous versions of the challenge, and the top preforming of them were able to improve over the state of the art performance from previous years. BioASQ keeps pushing the research frontier in biomedical semantic indexing and question answering for ten years now, offering both well established and new tasks. Lately is has been extended beyond the English language and biomedical literature, with the tasks MESINESP and DisTEMIST. In addition, BioASQ reaches a more and more broad community of biomedical experts that may benefit from the advancements in the field. This has been done initially through BioASQ Synergy for COVID-19 and this year with the collaborative batch of task 10b. The future plans for the challenge include to further extend of the benchmark data for question answering though a community-driven process, as well as extending the Synergy task into other developing problems beyond COVID-19.

Acknowledgments. Google was a proud sponsor of the BioASQ Challenge in 2021. The tenth edition of BioASQ is also sponsored by the Atypon Systems inc. BioASQ is grateful to NLM for providing the baselines for task 10a and to the CMU team for providing the baselines for task 10b. The DisTEMIST track was supported by the Spanish Plan for advancement of Language Technologies (Plan TL) and the Secretaría de Estado de Digitalización e Inteligencia Artificial (SEDIA), the European Union's Horizon Europe Coordination & Support Action under Grant Agreement No 101058779 and by the AI4PROFHEALTH (PID2020-119266RA-I00).

References

1. Almeida, T., Matos, S.: BioASQ synergy: a strong and simple baseline rooted in relevance feedback. CLEF (Working Notes) (2021)
2. Almeida, T., Matos, S.: Universal passage weighting mechanism (UPWM) in BioASQ 9b. CLEF (Working Notes) (2021)
3. Alrowili, S., Shanker, V.: BioM-transformers: building large biomedical language models with BERT, ALBERT and ELECTRA. In: Proceedings of the 20th Workshop on Biomedical Language Processing, pp. 221–227. Association for Computational Linguistics, June 2021. https://www.aclweb.org/anthology/2021.bionlp-1.24
4. Amano, T., et al.: Tapping into non-English-language science for the conservation of global biodiversity. PLoS Biol. **19**(10), e3001296 (2021)
5. Baldwin, B., Carpenter, B.: Lingpipe (2003). World Wide Web: http://alias-i.com/lingpipe
6. Balikas, G., et al.: Evaluation framework specifications. Project deliverable D4.1, UPMC, May 2013
7. Benson, T.: Principles of Health Interoperability HL7 and SNOMED. Springer, Heidelberg (2012). https://doi.org/10.1007/978-1-4471-2801-4
8. Bernik, M., Tovornik, R., Fabjan, B., Marco-Ruiz, L.: Diagñoza: a natural language processing tool for automatic annotation of clinical free text with SNOMED-CT (2022)
9. Borchert, F., Schapranow, M.P.: Hpi-dhc @ bioasq distemist: Spanish biomedical entity linking with cross-lingual candidate retrieval and rule-based reranking (2022)
10. Castano, J., Gambarte, M.L., Otero, C., Luna, D.: A simple terminology-based approach to clinical entity recognition (2022)

11. Chizhikova, M., Collado-Montañez, J., López-Úbeda, P., Díaz-Galiano, M.C., Ureña-López, L.A., Martín-Valdivia, M.T.: SINAI at CLEF 2022: Leveraging biomedical transformers to detect and normalize disease mentions (2022)
12. Clark, K., Luong, M.T., Le, Q.V., Manning, C.D.: Electra: pre-training text encoders as discriminators rather than generators. arXiv preprint arXiv:2003.10555 (2020)
13. Demsar, J.: Statistical comparisons of classifiers over multiple data sets. J. Mach. Learn. Res. **7**, 1–30 (2006)
14. Gonzalez-Agirre, A., Marimon, M., Intxaurrondo, A., Rabal, O., Villegas, M., Krallinger, M.: Pharmaconer: pharmacological substances, compounds and proteins named entity recognition track. In: Proceedings of the 5th Workshop on BioNLP Open Shared Tasks, pp. 1–10 (2019)
15. Islamaj Dogan, R., Murray, G.C., Névéol, A., Lu, Z.: Understanding pubmed® user search behavior through log analysis. Database 2009 (2009)
16. Islamaj Doğan, R., Leaman, R., Lu, Z.: NCBI disease corpus: a resource for disease name recognition and concept normalization. J. Biomed. Informa. **47**, 1–10 (2014). https://doi.org/10.1016/j.jbi.2013.12.006. https://www.sciencedirect.com/science/article/pii/S1532046413001974
17. Kosmopoulos, A., Partalas, I., Gaussier, E., Paliouras, G., Androutsopoulos, I.: Evaluation measures for hierarchical classification: a unified view and novel approaches. Data Min. Knowl. Disc. **29**(3), 820–865 (2014). https://doi.org/10.1007/s10618-014-0382-x
18. Lan, Z., Chen, M., Goodman, S., Gimpel, K., Sharma, P., Soricut, R.: Albert: a lite BERT for self-supervised learning of language representations. arXiv preprint arXiv:1909.11942 (2019)
19. Li, C., Yates, A., MacAvaney, S., He, B., Sun, Y.: Parade: passage representation aggregation for document reranking. arXiv preprint arXiv:2008.09093 (2020)
20. Miranda-Escalada, A., Farré, E., Krallinger, M.: Named entity recognition, concept normalization and clinical coding: overview of the cantemist track for cancer text mining in Spanish, corpus, guidelines, methods and results. In: Proceedings of the Iberian Languages Evaluation Forum (IberLEF 2020). CEUR Workshop Proceedings (2020)
21. Miranda-Escalada, A., et al.: Overview of DISTEMIST at BioASQ: automatic detection and normalization of diseases from clinical texts: results, methods, evaluation and multilingual resources (2022)
22. Miranda-Escalada, A., Gonzalez-Agirre, A., Armengol-Estapé, J., Krallinger, M.: Overview of automatic clinical coding: annotations, guidelines, and solutions for non-English clinical cases at CodiEsp track of CLEF ehealth 2020. In: Working Notes of Conference and Labs of the Evaluation (CLEF) Forum. CEUR Workshop Proceedings (2020)
23. Mork, J.G., Demner-Fushman, D., Schmidt, S.C., Aronson, A.R.: Recent enhancements to the NLM medical text indexer. In: Proceedings of Question Answering Lab at CLEF (2014)
24. Moscato, V., Postiglione, M., Sperl[í], G.: Biomedical Spanish language models for entity recognition and linking at BioASQ DisTEMIST (2022)
25. Nentidis, A., et al.: Overview of BioASQ 2021: the ninth BioASQ challenge on large-scale biomedical semantic indexing and question answering. In: Candan, K.S., et al. (eds.) CLEF 2021. LNCS, vol. 12880, pp. 239–263. Springer, Cham (2021). https://doi.org/10.1007/978-3-030-85251-1_18

26. Nentidis, A., et al.: Overview of BioASQ 2020: the eighth BioASQ challenge on large-scale biomedical semantic indexing and question answering. In: Arampatzis, A., et al. (eds.) CLEF 2020. LNCS, vol. 12260, pp. 194–214. Springer, Cham (2020). https://doi.org/10.1007/978-3-030-58219-7_16

27. Neves, A.: Unicage at distemist - named entity recognition system using only bash and unicage tools (2022)

28. Ozyurt, I.B.: End-to-end biomedical question answering via bio-answerfinder and discriminative language representation models. CLEF (Working Notes) (2021)

29. Rae, A.R., Mork, J.G., Demner-Fushman, D.: A neural text ranking approach for automatic mesh indexing. In: CLEF (Working Notes), pp. 302–312 (2021)

30. Reyes-Aguillón, J., del Moral, R., Ramos-Flores, O., Gómez-Adorno, H., Bel-Enguix, G.: Clinical named entity recognition and linking using BERT in combination with Spanish medical embeddings (2022)

31. Tamayo, A., Burgos, D.A., Gelbukh, A.: mBERT and simple post-processing: a baseline for disease mention detection in Spanish (2022)

32. Tsatsaronis, G., et al.: An overview of the BioASQ large-scale biomedical semantic indexing and question answering competition. BMC Bioinform. **16**, 138 (2015). https://doi.org/10.1186/s12859-015-0564-6

33. Tsoumakas, G., Laliotis, M., Markontanatos, N., Vlahavas, I.: Large-scale semantic indexing of biomedical publications. In: 1st BioASQ Workshop: A Challenge on Large-Scale Biomedical Semantic Indexing and Question Answering (2013)

34. Uzuner, O., South, B.R., Shen, S., DuVall, S.L.: 2010 i2b2/VA challenge on concepts, assertions, and relations in clinical text. J. Am. Med. Inform. Assoc. **18**(5), 552–556 (2011). https://doi.org/10.1136/amiajnl-2011-000203

35. Wang, L.L., et al.: CORD-19: the COVID-19 open research dataset. ArXiv (2020)

36. Wei, C.H., Leaman, R., Lu, Z.: Beyond accuracy: creating interoperable and scalable text-mining web services. Bioinform. (Oxford, Engl.) **32**(12), 1907–10 (2016). https://doi.org/10.1093/bioinformatics/btv760

37. Yang, Z., Zhou, Y., Eric, N.: Learning to answer biomedical questions: Oaqa at bioasq 4b. ACL **2016**, 23 (2016)

38. Zavorin, I., Mork, J.G., Demner-Fushman, D.: Using learning-to-rank to enhance NLM medical text indexer results. ACL **2016**, 8 (2016)

39. Zhang, Y., Han, J.C., Tsai, R.T.H.: NCU-IISR/AS-GIS: results of various pre-trained biomedical language models and linear regression model in BioASQ task 9b phase b. In: CEUR Workshop Proceedings (2021)

A Concise Overview of LeQua@CLEF 2022: Learning to Quantify

Andrea Esuli, Alejandro Moreo, Fabrizio Sebastiani$^{(\boxtimes)}$, and Gianluca Sperduti

Istituto di Scienza e Tecnologie dell'Informazione,
Consiglio Nazionale delle Ricerche, 56124 Pisa, Italy
{andrea.esuli,alejandro.moreo,fabrizio.sebastiani,
gianluca.sperduti}@isti.cnr.it

Abstract. LeQua 2022 is a new lab for the evaluation of methods for "learning to quantify" in textual datasets, i.e., for training predictors of the relative frequencies of the classes of interest $\mathcal{Y} = \{y_1, ..., y_n\}$ in sets of unlabelled textual documents. While these predictions could be easily achieved by first classifying all documents via a text classifier and then counting the numbers of documents assigned to the classes, a growing body of literature has shown this approach to be suboptimal, and has proposed better methods. The goal of this lab is to provide a setting for the comparative evaluation of methods for learning to quantify, both in the binary setting and in the single-label multiclass setting; this is the first time that an evaluation exercise solely dedicated to quantification is organized. For both the binary setting and the single-label multiclass setting, data were provided to participants both in ready-made vector form and in raw document form. In this overview article we describe the structure of the lab, we report the results obtained by the participants in the four proposed tasks and subtasks, and we comment on the lessons that can be learned from these results.

1 Learning to Quantify

In a number of applications involving classification, the final goal is not determining which class (or classes) individual unlabelled items (e.g., textual documents, images, or other) belong to, but estimating the *prevalence* (or "relative frequency", or "prior probability", or "prior") of each class $y \in \mathcal{Y} = \{y_1, ..., y_n\}$ in the unlabelled data. Estimating class prevalence values for unlabelled data via supervised learning is known as *learning to quantify* (LQ) (or *quantification*, or *supervised prevalence estimation*) [7,20].

LQ has several applications in fields (such as the social sciences, political science, market research, epidemiology, and ecological modelling) which are inherently interested in characterising *aggregations* of individuals, rather than the individuals themselves; disciplines like the ones above are usually *not* interested in finding the needle in the haystack, but in characterising the haystack. For instance, in most applications of tweet sentiment classification we are not concerned with estimating the true class (e.g., Positive, or Negative, or Neutral) of individual tweets. Rather, we are concerned with estimating the relative frequency of these classes in the set of unlabelled tweets under study; or, put in another way, we are interested in estimating as accurately as possible the true distribution of tweets across the classes.

It is by now well known that performing quantification by classifying each unlabelled instance and then counting the instances that have been attributed to the class (the "classify and count" method) usually leads to suboptimal quantification accuracy (see e.g., [2,3,5,15,18,20,22,27,30]); this may be seen as a direct consequence of "Vapnik's principle" [45], which states

> If you possess a restricted amount of information for solving some problem, try to solve the problem directly and never solve a more general problem as an intermediate step. It is possible that the available information is sufficient for a direct solution but is insufficient for solving a more general intermediate problem.

In our case, the problem to be solved directly is quantification, while the more general intermediate problem is classification.

One reason why "classify and count" is suboptimal is that many application scenarios suffer from *distribution shift*, the phenomenon according to which the distribution across the classes $y_1, ..., y_n$ in the sample (i.e., set) σ of *unlabelled* documents may substantially differ from the distribution across the classes in the labelled *training* set L; distribution shift is one example of *dataset shift* [33, 41], the phenomenon according to which the joint distributions $p_L(\mathbf{x}, y)$ and $p_\sigma(\mathbf{x}, y)$ differ. The presence of distribution shift means that the well-known IID assumption, on which most learning algorithms for training classifiers hinge, does not hold. In turn, this means that "classify and count" will perform suboptimally on sets of unlabelled items that exhibit distribution shift with respect to the training set, and that the higher the amount of shift, the worse we can expect "classify and count" to perform.

As a result of the suboptimality of the "classify and count" method, LQ has slowly evolved as a task in its own right, different (in goals, methods, techniques, and evaluation measures) from classification [20]. The research community has investigated methods to correct the biased prevalence estimates of general-purpose classifiers [3,18,27], supervised learning methods specially tailored to quantification [2,5,15,22,30], evaluation measures for quantification [14,43], and protocols for carrying out this evaluation. Specific applications of LQ have also been investigated, such as sentiment quantification [12,13,19,36], quantification in networked environments [31], or quantification for data streams [29]. For the near future it is easy to foresee that the interest in LQ will increase, due (a) to the increased awareness that "classify and count" is a suboptimal solution when it comes to prevalence estimation, and (b) to the fact that, with larger and larger quantities of data becoming available and requiring interpretation, in more and more scenarios we will only be able to afford to analyse these data at the aggregate level rather than individually.

2   The Rationale for LeQua 2022

The LeQua 2022 lab (https://lequa2022.github.io/) at CLEF 2022 has a "shared task" format; it is a new lab, in two important senses:

- No labs on LQ have been organized before at CLEF conferences.
- Even outside the CLEF conference series, quantification has surfaced only episodically in previous shared tasks. The first such shared task was SemEval 2016 Task 4 "Sentiment Analysis in Twitter" [37], which comprised a *binary quantification* subtask and an *ordinal quantification* subtask (these two subtasks were offered again in the 2017 edition). Quantification also featured in the Dialogue Breakdown Detection Challenge [23], in the Dialogue Quality subtasks of the NTCIR-14 Short Text Conversation task [46], and in the NTCIR-15 Dialogue Evaluation task [47]. However, quantification was never the real focus of these tasks. For instance, the real focus of the tasks described in [37] was sentiment analysis on Twitter data, to the point that almost all participants in the quantification subtasks used the trivial "classify and count" method, and focused, instead of optimising the quantification component, on optimising the sentiment analysis component, or on picking the best-performing learner for training the classifiers used by "classify and count". Similar considerations hold for the tasks discussed in [23,46,47].

This is the first time that a shared task whose explicit focus is quantification is organized. A lab on this topic was thus sorely needed, because the topic has great applicative potential, and because a lot of research on this topic has been carried out without the benefit of the systematic experimental comparisons that only shared tasks allow.

We expect the quantification community to benefit significantly from this lab. One of the reasons is that this community is spread across different fields, as also witnessed by the fact that work on LQ has been published in a scattered way across different areas, e.g., information retrieval [5,12,27], data mining [15,18], machine learning [1,10], statistics [25], or in the areas to which these techniques get applied [4,19,24]. In their papers, authors often use as baselines only the algorithms from their own fields; one of the goals of this lab was thus to pull together people from different walks of life, and to generate cross-fertilisation among the respective sub-communities.

While quantification is a general-purpose machine learning/data mining task that can be applied to any type of data, in this lab we focus on its application to data consisting of textual documents.

3   Setting up LeQua 2022

In quantification, a *data item* (usually represented as $\mathbf{x}$) is the individual unit of information; for instance, a textual document, an image, a video, are examples of data items. In LeQua 2022, as data items we use textual *documents* (and, more specifically, product reviews). A document $\mathbf{x}$ has a *label*, i.e., it belongs to a certain class $y \in \mathcal{Y} = \{y_1, ..., y_n\}$; in this case we say that y is the label of $\mathbf{x}$. In LeQua 2022, classes are either merchandise classes for products, or sentiment classes for reviews (see Sect. 3.4 for more).

Some documents are such that their label is known to the quantification algorithm, and are thus called *labelled items*; we typically use them as training examples for the quantifier-training algorithm. Some other documents are such that their label is unknown to the quantifier-training algorithm and to the trained quantifier, and are thus called *unlabelled items*; for testing purposes we use documents whose label we hide to the quantifier. Unlike a classifier, a quantifier must not predict labels for individual documents, but must predict *prevalence values* for *samples* (i.e., sets) of unlabelled documents; a prevalence value for a class y and a sample σ is a number in [0,1] such that the prevalence values for the classes in $\mathcal{Y} = \{y_1, ..., y_n\}$ sum up to 1. Note that when, in the following, we use the term "label", we always refer to the label of an individual document (and not of a sample of documents; samples do not have labels, but prevalence values for classes).

3.1 Tasks

Two tasks (T1 and T2) were offered within LeQua 2022, each admitting two subtasks (A and B).

In Task T1 (the *vector task*) participant teams were provided with vectorial representations of the (training/development/test) documents. This task was offered so as to appeal to those participants who are not into *text* learning, since participants in this task did not need to deal with text preprocessing issues. Additionally, this task allowed the participants to concentrate on optimising their quantification methods, rather than spending time on optimising the process for producing vectorial representations of the documents.

In Task T2 (the *raw documents task*), participant teams were provided with the raw (training/development/test) documents. This task was offered so as to appeal to those participants who wanted to deploy end-to-end systems, or to those who wanted to also optimise the process for producing vectorial representations of the documents (possibly tailored to the quantification task).

The two subtasks of both tasks were the *binary quantification subtask* (T1A and T2A) and the *single-label multiclass quantification subtask* (T1B and T2B); in both subtasks each document belongs to only one of the classes of interest $y_1, ..., y_n$, with $n = 2$ in T1A and T2A and $n > 2$ in T1B and T2B.

For each subtask in { T1A, T1B, T2A, T2B }, participant teams were required not to use (training/development/test) documents other than those provided for that subtask. In particular, participants were explicitly advised against using any document from either T2A or T2B in order to solve either T1A or T1B.

3.2 The Evaluation Protocol

As the protocol for generating the test samples on which the quantifiers will be tested we adopt the so-called *artificial prevalence protocol* (APP), which is by now a standard protocol for generating the datasets to be used in the evaluation of quantifiers. Using the APP consists of taking the test set U of unlabelled data items, and extracting from it a number of subsets (the *test samples*), each characterised by a *predetermined* vector $(p_\sigma(y_1), ..., p_\sigma(y_n))$ of prevalence values, where $y_1, ..., y_n$ are the classes of interest. In other words, for extracting a test sample σ, we generate a vector of prevalence values, and randomly select documents from U accordingly (i.e., by class-conditional random selection of documents until the desired class prevalence values are obtained).[1]

The goal of the APP is to generate samples characterised by widely different vectors of prevalence values; this is meant to test the robustness of a *quantifier* (i.e., of an estimator of class prevalence values) in confronting class prevalence values possibly different (or very different) from the ones of the set it has been trained on. For doing this we draw the vectors of class prevalence values uniformly at random from the set of all legitimate such vectors, i.e., from the *unit $(n-1)$-simplex* of all vectors $(p_\sigma(y_1), ..., p_\sigma(y_n))$ such that $p_\sigma(y_i) \in [0, 1]$ for all $y_i \in \mathcal{Y}$ and $\sum_{y_i \in \mathcal{Y}} p_\sigma(y_i) = 1$.

3.3   The Evaluation Measures

In a recent theoretical study on the adequacy of evaluation measures for the quantification task [43], *relative absolute error* (RAE) and *absolute error* (AE) have been found to be the most satisfactory, and are thus the only measures used in LeQua 2022. RAE and AE are defined as

$$\text{RAE}(p_\sigma, \hat{p}_\sigma) = \frac{1}{n} \sum_{y \in \mathcal{Y}} \frac{|\hat{p}_\sigma(y) - p_\sigma(y)|}{p_\sigma(y)} \qquad (1)$$

$$\text{AE}(p_\sigma, \hat{p}_\sigma) = \frac{1}{n} \sum_{y \in \mathcal{Y}} |\hat{p}_\sigma(y) - p_\sigma(y)| \qquad (2)$$

where p_σ is the true distribution on sample σ, $\hat{p}_\sigma$ is the predicted distribution, $\mathcal{Y}$ is the set of classes of interest, and $n = |\mathcal{Y}|$. Note that RAE is undefined when at least one of the classes $y \in \mathcal{Y}$ is such that its prevalence in the sample σ of unlabelled items is 0. To solve this problem, in computing RAE we smooth all $p_\sigma(y)$'s and $\hat{p}_\sigma(y)$'s via additive smoothing, i.e., we take $\underline{p}_\sigma(y) = (\epsilon + p_\sigma(y))/(\epsilon \cdot n + \sum_{y \in \mathcal{Y}} p_\sigma(y))$, where $\underline{p}_\sigma(y)$ denotes the smoothed version of $p_\sigma(y)$ and the denominator is just a normalising factor (same for the $\underline{\hat{p}}_\sigma(y)$'s); following [18], we use the quantity $\epsilon = 1/(2|\sigma|)$ as the smoothing factor. In Eq. 1 we then use the smoothed versions of $p_\sigma(y)$ and $\hat{p}_\sigma(y)$ in place of their original non-smoothed versions; as a result, RAE is now always defined.

As the official measure according to which systems are ranked, we use RAE; we also compute AE results, but we do not use them for ranking the systems. The official score obtained by a given quantifier is the average value of the official evaluation measure (RAE) across all test samples; for each system we

---

[1] Everything we say here on how we generate the test samples also applies to how we generate the development samples.

also compute and report the value of AE. For each subtask in { T1A, T1B, T2A, T2B } we use a two-tailed t-test on related samples at different confidence levels ($\alpha = 0.05$ and $\alpha = 0.001$) to identify all participant runs that are *not* statistically significantly different from the best run, in terms of RAE and in terms of AE. We also compare all pairs of methods by means of *critical difference diagrams* (CD-diagrams – [8]). We adopt the Nemenyi test and set the confidence level to $\alpha = 0.05$. The test compares the average ranks in terms of RAE and takes into account the sample size $|\sigma|$.

3.4   Data

The data we have used are Amazon product reviews from a large crawl of such reviews. From the result of this crawl we have removed (a) all reviews shorter than 200 characters and (b) all reviews that have not been recognised as "useful" by any users; this has yielded the dataset Ω that we have used for our experimentation. As for the class labels, (i) for the two binary tasks (T1A and T2A) we have used two *sentiment* labels, i.e., Positive (which encompasses 4-stars and 5-stars reviews) and Negative (which encompasses 1-star and 2-stars reviews), while for the two multiclass tasks (T1B and T2B) we have used 28 *topic* labels, representing the merchandise class the product belongs to (e.g., Automotive, Baby, Beauty).[2]

We have used the same data (training/development/test sets) for the binary vector task (T1A) and for the binary raw document task (T2A); i.e., the former are the vectorized (and shuffled) versions of the latter. Same for T1B and T2B. In order to generate the document vectors, we compute the average of the GloVe vectors [38] for the words contained in each document, thus producing 300-dimensional document embeddings. Each of the 300 dimensions of the document embeddings is then (independently) standardized, so that it has zero mean and unit variance.

The L_B (binary) training set and the L_M (multiclass) training set consist of 5,000 documents and 20,000 documents, respectively, sampled from the dataset Ω via *stratified sampling* so as to have "natural" prevalence values for all the class labels. (When doing stratified sampling for the binary "sentiment-based" task, we ignore the "topic" dimension; and when doing stratified sampling for the multiclass "topic-based" task, we ignore the "sentiment" dimension).

The development (validation) sets D_B (binary) and D_M (multiclass) consist of 1,000 development samples of 250 documents each (D_B) and 1,000 development samples of 1,000 documents each (D_M) generated from $\Omega \backslash L_B$ and $\Omega \backslash L_M$ via the Kraemer algorithm.

The test sets U_B and U_M consist of 5,000 test samples of 250 documents each (U_B) and 5,000 test samples of 1,000 documents each (U_M), generated from $\Omega \setminus (L_B \cup D_B)$ and $\Omega \setminus (L_M \cup D_M)$ via the Kraemer algorithm. A submission ("run") for a given subtask consists of prevalence estimations for the relevant

---

[2] The set of 28 topic classes is flat, i.e., there is no hierarchy defined upon it.

classes (the two sentiment classes for the binary subtasks and the 28 topic classes for the multiclass subtasks) for each sample in the test set of that subtask.

3.5 Baselines

In order to set a sufficiently high bar for the participants to overcome, we made them aware of the availability of QuaPy [34], a library of quantification methods that contains, among others, implementations of a number of methods that have performed well in recent comparative evaluations.[3] QuaPy is a publicly available, open-source, Python-based framework that we have recently developed, and that implements not only learning methods, but also evaluation measures, parameter optimisation routines, and evaluation protocols, for LQ.

We used a number of quantification methods, as implemented in QuaPy, as baselines for the participants to overcome.[4] These methods were:

- **Maximum Likelihood Prevalence Estimation** (MLPE): Rather than a true quantification method, this a (more than) trivial baseline, consisting in assuming that the prevalence $p_\sigma(y_i)$ of a class y_i in the test sample σ is the same as the prevalence $p_L(y_i)$ that was observed for that class in the training set L.
- **Classify and Count** (CC): This is the trivial baseline, consisting in training a standard classifier h on the training set L, using it to classify all the data items $\mathbf{x}$ in the sample σ, counting how many such items have been attributed to class y_i, doing this for all classes in $\mathcal{Y}$, and dividing the resulting counts by the cardinality $|\sigma|$ of the sample.
- **Probabilistic Classify and Count** (PCC) [3]: This is a probabilistic variant of CC where the "hard" classifier h is replaced with a "soft" (probabilistic) classifier s, and where counts are replaced with expected counts.
- **Adjusted Classify and Count** (ACC) [16]: This is an "adjusted" variant of CC in which the prevalence values predicted by CC are subsequently corrected by considering the misclassification rates of classifier h, as estimated on a held-out validation set. For our experiments, this held-out set consists of 40% of the training set.
- **Probabilistic Adjusted Classify and Count** (PACC) [3]: This is a probabilistic variant of ACC where the "hard" classifier h is replaced with a "soft" (probabilistic) classifier s, and where counts are replaced with expected counts. Equivalently, it is an "adjusted" variant of PCC in which the prevalence values predicted by PCC are corrected by considering the (probabilistic versions of the) misclassification rates of soft classifier s, as estimated on a held-out validation set. For our experiments, this held-out set consists of 40% of the training set.
- **HDy** [22]: This is a probabilistic binary quantification method that views quantification as the problem of minimising the divergence (measured in terms

---

[3] https://github.com/HLT-ISTI/QuaPy.
[4] Check the branch https://github.com/HLT-ISTI/QuaPy/tree/lequa2022.

of the Hellinger Distance, HD) between two distributions of posterior probabilities returned by the classifier, one coming from the unlabelled examples and the other coming from a validation set consisting of 40% of the training documents. HDy seeks for the mixture parameter $\alpha \in [0,1]$ that minimizes the HD between (a) the mixture distribution of posteriors from the positive class (weighted by α) and from the negative class (weighted by $(1-\alpha)$), and (b) the unlabelled distribution.

- The **Saerens-Latinne-Decaestecker** algorithm (SLD) [42] (see also [11]): This is a method based on Expectation Maximization, whereby the posterior probabilities returned by a soft classifier s for data items in an unlabelled set U, and the class prevalence values for U, are iteratively updated in a mutually recursive fashion. For SLD we calibrate the classifier since, for reasons discussed in [11], this yields an advantage for this method.[5]

- **QuaNet** [12]: This is a deep learning architecture for quantification that predicts class prevalence values by taking as input (i) the class prevalence values as estimated by CC, ACC, PCC, PACC, SLD; (ii) the posterior probabilities $\Pr(y|\mathbf{x})$ for the positive class (since QuaNet is a binary method) for each document $\mathbf{x}$, and (iii) embedded representations of the documents. For task T1A, we directly use the vectorial representations that we have provided to the participants as the document embeddings, while for task T2A we use the RoBERTa embeddings (described below). For training QuaNet, we use the training set L for training the classifier. We then use the validation set for training the network parameters, using 10% of the validation samples for monitoring the validation loss (we apply early stop after 10 epochs that have shown no improvement). Since we devote the validation set to train part of the model, we did not carry out model selection for QuaNet, which was used with default hyperparameters (a learning rate of $1e^{-4}$, 64 dimensions in the LSTM hidden layer, and a drop-out probability of 0.5).

All the above methods (with the exception of MLPE) are described in more detail in [36, §3.3 and §3.4], to which we refer the interested reader; all these methods are well-established, the most recent one (QuaNet) having been published in 2018. For all methods, we have trained the underlying classifiers via logistic regression, as implemented in the `scikit-learn` framework (https://scikit-learn.org/stable/index.html). Note that we have used HDy and QuaNet as baselines only in T1A and T2A, since they are binary-only methods. All other methods are natively multiclass, so we have used them in all four subtasks.

We optimize two hyperparameters of the logistic regression learner by exploring C (the inverse of the regularization strength) in the range $\{10^{-3}, 10^{-2}, \ldots, 10^{+3}\}$ and CLASS_WEIGHT (indicating the relative importance of each class) in $\{\text{"balanced"}, \text{"not-balanced"}\}$. For each quantification method, model selection is carried out by choosing the combination of hyperparameters yielding the lowest average RAE across all validation samples.

---

[5] Calibration does not yield similar improvements for other methods such as PCC, PACC, and QuaNet, though. For this reason, we only calibrate the classifier for SLD.

For the raw documents subtasks (T2A and T2B), for each baseline quantification method we have actually generated *two* quantifiers, using two different methods for turning documents into vectors. (The only two baseline methods for which we do not do this are MLPE, which does not use vectors, and QuaNet, that internally generates its own vectors.) The two methods are

- The standard tfidf term weighting method, expressed as

$$\text{tfidf}(f, \mathbf{x}) = \log \#(f, \mathbf{x}) \times \log \frac{|L|}{|\mathbf{x}' \in L : \#(f, \mathbf{x}') > 0|} \tag{3}$$

where $\#(f, \mathbf{x})$ is the raw number of occurrences of term f in document $\mathbf{x}$; weights are then normalized via cosine normalization, as

$$w(f, \mathbf{x}) = \frac{\text{tfidf}(f, \mathbf{x})}{\sqrt{\sum_{f' \in F} \text{tfidf}(f', \mathbf{x})^2}} \tag{4}$$

where F is the set of all unigrams and bigrams that occur at least 5 times in L.
- The RoBERTa transformer [28], from the Hugging Face hub.[6] In order to use RoBERTa, we truncate the documents to the first 256 tokens, and fine-tune RoBERTa for the task of classification via prompt learning for a maximum of 10 epochs on our training data, thus taking the model parameters from the epoch which yields the best macro F_1 as monitored on a held-out validation set consisting of 10% of the training documents randomly sampled in a stratified way. For training, we set the learning rate to $1e^{-5}$, the weight decay to 0.01, and the batch size to 16, leaving the other hyperparameters at their default values. For each document, we generate features by first applying a forward pass over the fine-tuned network, and then averaging the embeddings produced for the special token [CLS] across all the 12 layers of RoBERTa. (In experiments that we carried out for another project, this latter approach yielded slightly better results than using the [CLS] embedding of the last layer alone.) The embedding size of RoBERTa, and hence the number of dimensions of our vectors, amounts to 768.

4   The Participating Systems

Six teams submitted runs to LeQua 2022. The most popular subtask was, unsurprisingly, T1A (5 teams), while the subtask with the smallest participation was T2B (1 team). We here list the teams in alphabetical order:

- **DortmundAI** [44] submitted a run each for T1A and T1B. Their original goal was to use a modified version of the SLD algorithm described in Sect. 3.5. The modification introduced by DortmundAI consists of the use of

---

[6] https://huggingface.co/docs/transformers/model_doc/roberta.

a regularization technique meant to smooth the estimates that expectation maximization computes for the class prevalence values at each iteration. After extensively applying model selection, though, the team realized that the best configurations of hyperparameters often reduce the strength of such regularization, so as to make the runs produced by their regularized version of SLD almost identical to a version produced by using the "traditional" SLD algorithm. They also found that a thorough optimization of the hyperparameters of the base classifier was instead the key to producing good results.

- **KULeuven** [40] submitted a run each for T1A and T1B. Their system consisted of a robust calibration of the SLD [42] method based on the observations made in [32]. While the authors explored *trainable calibration strategies* (i.e., regularization constraints that modify the training objective of a classifier in favour of better calibrated solutions), the team finally contributed a solution based on the Platt rescaling [39] of the SVM outputs (i.e., a *post-hoc calibration method* that is applied after training the classifier) which they found to perform better in validation. Their solution differs from the version of SLD provided as baseline mainly in the choice of the underlying classifier (the authors chose SVMs while the provided baseline is based on logistic regression) and in the amount of effort devoted to the optimization of the hyperparameters (which was higher in the authors' case).

- **UniLeiden** [26] submitted a run for T1A only. The authors' system is a variant of the Median Sweep (MS) method proposed in [17,18], called *Simplified Continuous Sweep*, which consists of a smooth adaptation of the original method. The main modifications come down to computing the mean (instead of the median) of the class prevalence estimates by integrating over continuous functions (instead of summing across discrete functions) that represent the classification counts and misclassification rates. Since the underlying distributions of these counts and rates are unknown, kernel density estimation is used to approximate them. Although the system did not yield improved results with respect to MS, it paves the way for better understanding the theoretical implications of MS.

- **UniOviedo(Team1)** [21] submitted a run each for all four subtasks. Their system consists of a deep neural network architecture explicitly devised for the quantification task. The learning method is non-aggregative and does not need to know the labels of the training items composing a sample. As the training examples to train the quantifiers that produced the submissions it used the samples with known prevalence from the development sets D_B and D_M (each set is used for its respective task). A generator of additional samples that produces mixtures of pairs of samples of known prevalence is used to increase the number of training examples. Data from training sets L_B and L_M are used only to generate additional training samples when overfitting is observed. Every sample is represented as a set of histograms, each one representing the distribution of values of an input feature. For tasks T1A and T1B, histograms are directly computed on the input vectors. For tasks T2A and T2B, the input text are first converted into dense vectors using a BERT model, for which the histograms are computed. The network uses

372   A. Esuli et al.

RAE as the loss function, modified by the smoothing parameter so as to avoid undefined values when a true prevalence is zero, thus directly optimizing the official evaluation measure.

- **UniOviedo(Team2)** [6] submitted a run each for T1A and T1B. For T1A, this team used a highly optimized version of the HDy system (that was also one of the baseline systems), obtained by optimizing three different parameters (similarity measure used, number of bins used, method used for binning the posteriors returned by the classifier). For T1B, this team used a version of HDy (called EDy) different from the previous one; EDy uses, for the purpose of measuring the distance between two histograms, the "energy distance" in place of the Hellinger Distance.

- **UniPadova** [9] submitted a run for T2A only. Their system consisted of a classify-and-count method in which the underlying classifier is a probabilistic "BM25" classifier. The power of this method thus only derives from the term weighting component, since nothing in the method makes explicit provisions for distribution shift.

5   Results

In this section we discuss the results obtained by our participant teams in the four subtasks we have proposed. The evaluation campaign started on Dec 1, 2021, with the release of the training sets (L_B and L_M) and of the development sets (D_B and D_M); alongside them, the participant teams were provided with a dummy submission, a format checker, and the official evaluation script. The unlabelled test sets (U_B and U_M) were released on Apr 22, 2022; and runs had to be submitted by May 11, 2022. Each team could submit up to two runs per subtask, provided each such run used a truly different method (and not, say, the same method using different parameter values); however, no team decided to take advantage of this, and each team submitted at most one run per subtask. An instantiation of Codalab (https://codalab.org/) was set up in order to allow the teams to submit their runs. The true labels of the unlabelled test sets were released on May 13, 2022, after the submission period was over and the official results had been announced to the participants. In the rest of this section we discuss the results that the participants' systems and the baseline systems have obtained in the vector subtasks (T1A and T1B – Sect. 5.1), in the raw document subtasks (T2A and T2B – Sect. 5.2), in the binary subtasks (T1A and T2A – Sect. 5.3), and in the multiclass subtasks (T1B and T2B – Sect. 5.4).

We report the results of the participants' systems and the baseline systems in Fig. 1 (for subtask T1A), Fig. 2 (T1B), Fig. 3 (T2A), and Fig. 4 (T2B). In each such figure we also display critical-distance diagrams illustrating how the systems rank in terms of RAE and when the difference between the systems is statistically significant.

Rank	Run	RAE	AE
1	KULeuven	**0.10858** ± 0.27476	0.02418 ± 0.01902
2	UniOviedo(Team1)	$0.10897^{\ddagger}$ ± 0.21887	**0.02327** ± 0.01811
3	UniOviedo(Team2)	$0.11130^{\ddagger}$ ± 0.23144	0.02475 ± 0.01908
4	*SLD*	$0.11382^{\ddagger}$ ± 0.26605	0.02518 ± 0.01977
5	UniDortmund	$0.11403^{\dagger}$ ± 0.20345	0.02706 ± 0.02096
6	*HDy*	0.14514 ± 0.45617	0.02814 ± 0.02212
7	*PACC*	0.15218 ± 0.46435	0.02985 ± 0.02258
8	*ACC*	0.17020 ± 0.50795	0.03716 ± 0.02935
9	UniLeiden	0.19624 ± 0.82620	0.03171 ± 0.02424
10	*QuaNet*	0.31764 ± 1.35223	0.03418 ± 0.02527
11	*CC*	1.08400 ± 4.31046	0.09160 ± 0.05539
12	*PCC*	1.39402 ± 5.62067	0.11664 ± 0.06977
13	*MLPE*	3.26692 ± 14.85223	0.32253 ± 0.22961

(a)

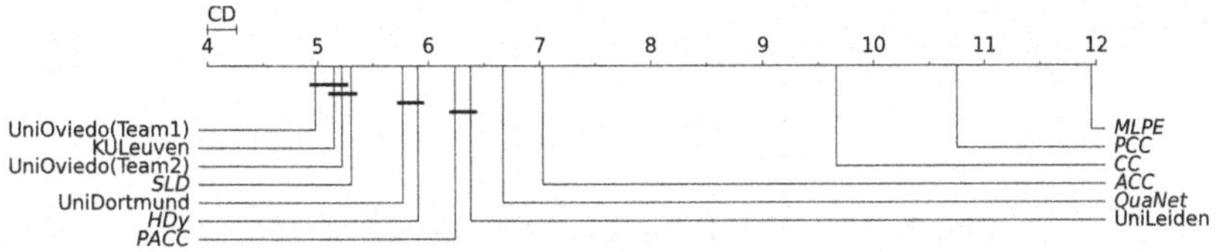

(b)

Fig. 1. Results of Task T1A. Table (a) reports the results of participant teams in terms of RAE (official measure for ranking) and AE, averaged across the 5,000 test samples. **Boldface** indicates the best method for a given evaluation measure. Superscripts † and ‡ denote the methods (if any) whose scores are *not* statistically significantly different from the best one according to a paired sample, two-tailed t-test at different confidence levels: symbol † indicates $0.001 < p$-value < 0.05 while symbol ‡ indicates $0.05 \leq p$-value. The absence of any such symbol indicates p-value ≤ 0.001 (i.e., that the difference in performance between the method and the best one is statistically significant at a high confidence level). Baseline methods are typeset in *italic*. Subfigure (b) reports the CD-diagram for Task T1A for the averaged ranks in terms of RAE.

Interestingly enough, no system (either participants' system or baseline system) was the best performer in more than one subtask, with four different systems (the **KULeuven** system for T1A, the **DortmundAI** system for T1B, the **QuaNet** baseline system for T2A, and the **UniOviedo(Team1)** system for T2B) claiming top spot for the four subtasks. Overall, the performance of UniOviedo(Team1) was especially noteworthy since, aside from topping the rank in T2B, it obtained results not statistically significantly different $(0.05 \leq p$-value) from those of the top-performing team also in T1A and T1B.

Rank	Run	RAE	AE
1	UniDortmund	**0.87987** ± 0.75139	**0.01173** ± 0.00284
2	UniOviedo(Team1)	0.88415[‡] ± 0.45537	0.02799 ± 0.00723
3	UniOviedo(Team2)	1.11395 ± 0.92516	0.01178[‡] ± 0.00329
4	KULeuven	1.17798 ± 1.05501	0.01988 ± 0.00395
5	SLD	1.18207 ± 1.09757	0.01976 ± 0.00399
6	PACC	1.30538 ± 0.98827	0.01578 ± 0.00379
7	ACC	1.42134 ± 1.26958	0.01841 ± 0.00437
8	CC	1.89365 ± 1.18721	0.01406 ± 0.00295
9	PCC	2.26462 ± 1.41613	0.01711 ± 0.00332
10	MLPE	4.57675 ± 4.51384	0.04227 ± 0.00414

(a)

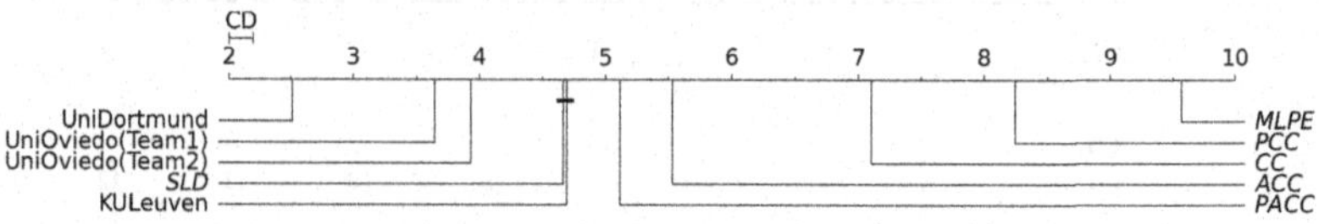

(b)

Fig. 2. As in Fig. 1, but for T1B in place of T1A.

The results allow us to make a number of observations. We organize the discussion of these results in four sections (Sect. 5.1 to Sect. 5.4), one for each of the four dimensions (vectors vs. raw documents, binary vs. multiclass) according to which the four subtasks are structured. However, before doing that, we discuss some conclusions that may be drawn from the results and that affect all four dimensions.

1. MLPE is the worst predictor. This is true in all four subtasks, and was expected, given the fact that the test data are generated by means of the APP, which implies that the test data contain a very high number of samples characterized by substantial distribution shift, and that on these samples MLPE obviously performs badly.
2. CC and PCC obtain very low quantification accuracy; this is the case in all four subtasks, where these two methods are always near the bottom of the ranking. This confirms the fact (already recorded in previous work – see e.g., [34–36]) that they are not good performers when the APP is used for generating the dataset, i.e., they are not good performers when there is substantial distribution shift. Interestingly enough, CC always outperforms PCC, which was somehow unexpected.
3. ACC and PACC are mid-level performers; this holds in all four subtasks, in which both methods are always in the middle portion of the ranking. Interestingly enough, PACC always outperforms ACC, somehow contradicting the impression (see Bullet 2) that "hard" counts are better than expected counts and/or that the calibration routine has not done a good job.
4. SLD is the strongest baseline; this is true in all four subtasks, in which SLD, while never being the best performer, is always in the top ranks. This confirms

Rank	Run	RAE	AE
1	*QuaNet*	**0.07805** ± 0.25437	**0.01306** ± 0.01009
2	*SLD*-tfidf	0.08703[†] ± 0.16721	0.01952 ± 0.01543
3	UniOviedo(Team1)	0.10704 ± 0.27896	0.01916 ± 0.01467
4	*HDy*-tfidf	**0.12198** ± 0.17207	0.02914 ± 0.02266
5	*SLD*-RoBERTa	0.13616 ± 0.45312	0.02208 ± 0.01562
6	*PACC*-tfidf	**0.13804** ± 0.48977	0.02626 ± 0.02080
7	*ACC*-tfidf	**0.16113** ± 0.54750	0.03090 ± 0.02443
8	*HDy*-RoBERTa	**0.16285** ± 0.55900	0.02421 ± 0.01612
9	*PACC*-RoBERTa	0.32902 ± 1.46314	0.03227 ± 0.02381
10	*ACC*-RoBERTa	0.33023 ± 1.49746	0.03374 ± 0.02539
11	*CC*-RoBERTa	0.41222 ± 1.81806	0.04053 ± 0.02976
12	*PCC*-RoBERTa	0.45182 ± 1.92703	0.04077 ± 0.02817
13	*CC*-tfidf	1.06748 ± 4.83335	0.10286 ± 0.07348
14	*PCC*-tfidf	1.36165 ± 6.37488	0.14414 ± 0.10237
15	UniPadova	3.02245 ± 11.99428	0.25067 ± 0.14675
16	*MLPE*	3.26692 ± 14.85223	0.32253 ± 0.22961

(a)

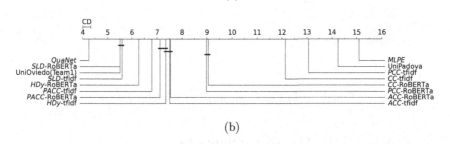

(b)

Fig. 3. As in Fig. 1, but for T2A in place of T1A.

the fact (already recorded in previous work – see e.g., [34–36]) that SLD is a very strong performer when the APP is used for generating the dataset, i.e., when the test data contain many samples characterized by substantial distribution shift.

5. Overall, the ranking MLPE < PCC < CC < ACC < PACC < SLD (where "<" means "performs worse than") clearly emerges from all four tasks.

As it might be expected, not always a good performance according to RAE (our official measure) also corresponds to a good performance on AE (our other measure). Only in 2 subtasks out of 4 (T1B, with the DortmundAI system, and T2A, with the QuaNet baseline system) the system that scores best according to RAE also scores best according to AE; in the other 2 subtasks this is not the case, and in one case (T2B) the system that performs best according to RAE (the UniOviedo(Team1) system) has a very low performance according to AE. This suggests that for some systems, including the UniOviedo(Team1) system, parameter optimization (which, quite naturally, is performed by trying to optimize the official measure) may have played an especially important role.

Rank	Run	RAE		AE	
1	UniOviedo(Team1)	**1.23085**	± 0.72831	0.03208	± 0.00921
2	*SLD*-RoBERTa	1.30978	± 1.61205	0.01552	± 0.00439
3	*SLD*-tfidf	1.31950	± 1.23382	0.01829	± 0.00376
4	*PACC*-RoBERTa	1.45429	± 1.00967	**0.01220**	± 0.00260
5	*ACC*-RoBERTa	1.48661	± 1.07152	0.01310	± 0.00290
6	*PACC*-tfidf	1.53853	± 1.43093	0.01789	± 0.00508
7	*CC*-RoBERTa	1.69071	± 1.15729	0.01367	± 0.00296
8	*PCC*-RoBERTa	1.77143	± 1.15163	0.01328	± 0.00272
9	*ACC*-tfidf	2.01440	± 2.16362	0.01993	± 0.00548
10	*CC*-tfidf	2.24393	± 1.52031	0.01949	± 0.00399
11	*PCC*-tfidf	3.06004	± 2.21288	0.02913	± 0.00469
12	*MLPE*	4.57675	± 4.51384	0.04227	± 0.00414

(a)

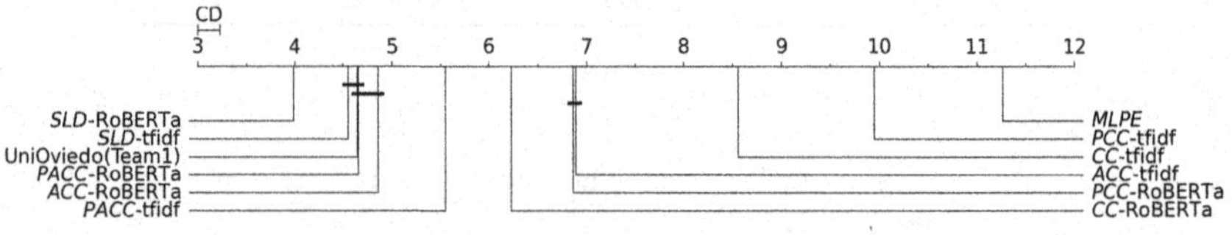

(b)

Fig. 4. As in Fig. 1, but for T2B in place of T1A.

5.1   T1A and T1B: The Vector Subtasks

In the vector subtasks the top-performing systems, KULeuven for T1A and UniDortmund for T1B, both consist of carefully optimized instances of SLD. The KULeuven system outperformed all the baseline systems in both tasks, while the UniDortmund system ranked 5th in T1A, one position below the SLD baseline.

The runs from UniOviedo(Team1) and UniOviedo(Team2) obtained 2nd and 3rd ranks, respectively, in both T1A and T1B. The UniOviedo(Team1) system performed very well in both cases, obtaining RAE scores that, according to the test of statistical significance, are not significantly different from the best result obtained in each of these subtasks. Things are different if we instead look at the AE scores, for which UniOviedo(Team1) obtained the best result in T1A but the second-worst result in T1B.

5.2   T2A and T2B: The Raw Documents Subtasks

In both raw document tasks (T2A and T2B) the best-performing methods is always one based on deep learning (the QuaNet baseline for T2A and the UniOviedo(Team1) system for T2B).

A direct comparison between the UniOviedo(Team1) system and QuaNet in the multiclass case (T2B) is not possible because QuaNet is a binary-only method (see Sect. 3.5) and was thus not used in T2B. A common characteristic between these two methods is that both use (part of the) samples from the validation data not for tuning hyperparameters but for training the model.

Concerning the baseline systems, the results do not give a definitive answer on which between tfidf and RoBERTa is the best method for mapping raw documents into vectors. In fact, out of 9 cases (5 for T2A, 4 for T2B) in which we have generated both variants of the same baseline, the tfidf variant outperforms the RoBERTa variant in 4 cases and is outperformed by it in 5 cases. This was unexpected, since RoBERTa is a way more sophisticated and modern method than the time-worn tfidf. Interestingly (and mysteriously) enough, the tfidf variant is almost always the better performer in the binary case (T2A – 4 cases out of 5), while the RoBERTa variant always outperforms the tfidf variant in the multiclass case (T2B – 4 cases out of 4).

5.3   T1A and T2A: The Binary Subtasks

Concerning T1A and T2A (the binary subtasks), we should first observe that we here use two further baseline systems, namely, HDy and QuaNet; we only use them in the binary subtasks since they are not natively multiclass. HDy performs fairly well in both T1A and T2A, outperforming MLPE, PCC, CC, ACC, and PACC (but not SLD) in both cases. Instead, QuaNet performs less consistently, since it places in the mid-lower ranks of the table in T1A but is no less than the best performer in T2A.

The inconsistent results obtained by QuaNet on binary tasks contrast with those obtained by the UniOviedo(Team1) system, the other method based on deep learning, which ranks among the top positions in both T1A and T2A. This is somehow surprising, given that in T1A (unlike in T2A), the source vectors used by UniOviedo(Team1) and QuaNet methods were exactly the same.

5.4   T1B and T2B: The Multiclass Subtasks

Regarding the multiclass subtasks, the UniOviedo(Team1) system stands out, since it consistently obtained results that either outperform all other methods (T2B) or were not different, in a statistically significant sense, from the best-performing method (T1B). UniOviedo(Team1) was the only team participating in the raw-document multiclass subtask T2B. Although UniOviedo(Team1) beat all other baselines in terms of RAE, it performed comparably worse in terms of AE to most of the baselines (actually, worse than all baselines but MLPE).

6   Final Remarks

Overall, something that we learn from this shared task is that SLD is very hard to beat (thereby confirming recent results reported in [34–36]), and that it tends to fare very well across different settings, including binary and multiclass quantification problems, and including different ways of processing text. This observation is reinforced by the fact that two of the best-performing systems

(KULeuven and UniDortmund, which placed 1st in T1A and T1B, respectively) actually consist of carefully-tuned instances of SLD. Another "classic" method that has also proven to behave well is HDy, a method that forms the basis on which one of the best-performing methods (UniOviedo(Team2)) is built upon. However, the system that has delivered the most consistently competitive results across all tasks (UniOviedo(Team1)) is a "non-classical" one, since it is based on deep-learning technology.

To conclude, we think that LeQua 2022 has proven very useful for the quantification community, since it has confirmed, in a controlled settings, some intuitions about "classic" quantification systems (e.g., SLD) that had already surfaced in the recent literature, but has also shown that there are margins of improvement over them, especially if using "deep" learning approaches (such as QuaNet and the system used by UniOviedo(Team1)).

We plan to propose a LeQua edition for CLEF 2023, so as to allow the LeQua 2022 participants to profit from their 2022 experience in order to consolidate their systems so as to improve on their 2022 performance, and so as to allow prospective participants who could not make it for 2022 to jump in. The experimental setting that we have used for LeQua 2022 will be the starting point, but we might want to incorporate in it possible suggestions that might arise during the LeQua session at the CLEF 2022 conference.

This session will host (a) a keynote talk by George Forman (Amazon Research), (b) a detailed presentation by the organisers, overviewing the lab and the results of the participants, (c) oral presentations by the participating teams, and (d) a final discussion on the takeaway message that LeQua 2022 gives us.

Acknowledgments. This work has been supported by the SoBigData++ project, funded by the European Commission (Grant 871042) under the H2020 Programme INFRAIA-2019-1, and by the AI4Media project, funded by the European Commission (Grant 951911) under the H2020 Programme ICT-48-2020. The authors' opinions do not necessarily reflect those of the European Commission. We thank Alberto Barron Cedeño, Juan José del Coz, Preslav Nakov, and Paolo Rosso, for advice on how to best set up this lab.

References

1. Alaíz-Rodríguez, R., Guerrero-Curieses, A., Cid-Sueiro, J.: Class and subclass probability re-estimation to adapt a classifier in the presence of concept drift. Neurocomputing **74**(16), 2614–2623 (2011)
2. Barranquero, J., Díez, J., del Coz, J.J.: Quantification-oriented learning based on reliable classifiers. Pattern Recognit. **48**(2), 591–604 (2015)
3. Bella, A., Ferri, C., Hernández-Orallo, J., Ramírez-Quintana, M.J.: Quantification via probability estimators. In: Proceedings of the 11th IEEE International Conference on Data Mining (ICDM 2010), Sydney, Australia, pp. 737–742 (2010)
4. Card, D., Smith, N.A.: The importance of calibration for estimating proportions from annotations. In: Proceedings of the 2018 Conference of the North American Chapter of the Association for Computational Linguistics (HLT-NAACL 2018), New Orleans, USA, vol. 1, pp. 1636–1646 (2018)

5. Da San Martino, G., Gao, W., Sebastiani, F.: Ordinal text quantification. In: Proceedings of the 39th ACM Conference on Research and Development in Information Retrieval (SIGIR 2016), Pisa, Italy, pp. 937–940 (2016)
6. del Coz, J.J.: UniOviedo(Team2) at LeQua 2022: comparison of traditional quantifiers and a new method based on energy distance. In: Working Notes of the 2022 Conference and Labs of the Evaluation Forum (CLEF 2022), Bologna, IT (2022)
7. del Coz, J.J., González, P., Moreo, A., Sebastiani, F.: Learning to quantify: methods and applications (LQ 2021). In: Proceedings of the 30th ACM International Conference on Knowledge Management (CIKM 2021), pp. 4874–4875, Gold Coast, AU (2021)
8. Demšar, J.: Statistical comparisons of classifiers over multiple data sets. J. Mach. Learn. Res. **7**, 1–30 (2006)
9. Di Nunzio, G.M.: UniPadova at LeQua 2022: a preliminary study of a Tidyverse approach to quantification. In: Working Notes of the 2022 Conference and Labs of the Evaluation Forum (CLEF 2022), Bologna, Italy (2022)
10. du Plessis, M.C., Niu, G., Sugiyama, M.: Class-prior estimation for learning from positive and unlabeled data. Mach. Learn. **106**(4), 463–492 (2016). https://doi.org/10.1007/s10994-016-5604-6
11. Esuli, A., Molinari, A., Sebastiani, F.: A critical reassessment of the Saerens-Latinne-Decaestecker algorithm for posterior probability adjustment. ACM Trans. Inf. Syst. **39**(2), Article 19 (2021)
12. Esuli, A., Molinari, A., Sebastiani, F.:A recurrent neural network for sentiment quantification. In: Proceedings of the 27th ACM International Conference on Information and Knowledge Management (CIKM 2018), Torino, Italy, pp. 1775–1778 (2018)
13. Esuli, A., Moreo, A., Sebastiani, F.: Cross-lingual sentiment quantification. IEEE Intell. Syst. **35**(3), 106–114 (2020)
14. Esuli, A., Sebastiani, F.: Sentiment quantification. IEEE Intell. Syst. **25**(4), 72–75 (2010)
15. Esuli, A., Sebastiani, F.: Optimizing text quantifiers for multivariate loss functions. ACM Trans. Knowl. Discov. Data **9**(4), Article 27, 1–27 (2015),
16. Forman, G.: Counting positives accurately despite inaccurate classification. In: Gama, J., Camacho, R., Brazdil, P.B., Jorge, A.M., Torgo, L. (eds.) ECML 2005. LNCS (LNAI), vol. 3720, pp. 564–575. Springer, Heidelberg (2005). https://doi.org/10.1007/11564096_55
17. Forman, G.: Quantifying trends accurately despite classifier error and class imbalance. In: Proceedings of the 12th ACM SIGKDD International Conference on Knowledge Discovery and Data Mining (KDD 2006), Philadelphia, USA, pp. 157–166 (2006)
18. Forman, G.: Quantifying counts and costs via classification. Data Min. Knowl. Disc. **17**(2), 164–206 (2008). https://doi.org/10.1007/s10618-008-0097-y
19. Gao, W., Sebastiani, F.: From classification to quantification in tweet sentiment analysis. Soc. Netw. Anal. Min. **6**(1), 1–22 (2016). https://doi.org/10.1007/s13278-016-0327-z
20. González, P., Castaño, A., Chawla, N.V., del Coz, J.J.: A review on quantification learning. ACM Comput. Surv. **50**(5), 74:1–74:40 (2017)
21. González, P.: UniOviedo(Team1) at LeQua 2022: sample-based quantification using deep learning. In: Working Notes of the 2022 Conference and Labs of the Evaluation Forum (CLEF 2022), Bologna, Italy (2022)
22. González-Castro, V., Alaiz-Rodríguez, R., Alegre, E.: Class distribution estimation based on the Hellinger distance. Inf. Sci. **218**, 146–164 (2013)

23. Higashinaka, R., Funakoshi, K., Inaba, M., Tsunomori, Y., Takahashi, T., Kaji, N.: Overview of the 3rd Dialogue Breakdown Detection challenge. In: Proceedings of the 6th Dialog System Technology Challenge, Long Beach, USA (2017)
24. Hopkins, D.J., King, G.: A method of automated nonparametric content analysis for social science. Am. J. Pol. Sci. **54**(1), 229–247 (2010)
25. King, G., Ying, L.: Verbal autopsy methods with multiple causes of death. Stat. Sci. **23**(1), 78–91 (2008)
26. Kloos, K., Meertens, Q.A., Karch, J.D.: UniLeiden at LeQua 2022: the first step in understanding the behaviour of the median sweep quantifier using continuous sweep. In: Working Notes of the 2022 Conference and Labs of the Evaluation Forum (CLEF 2022), Bologna, Italy (2022)
27. Levin, R., Roitman, H.: Enhanced probabilistic classify and count methods for multi-label text quantification. In: Proceedings of the 7th ACM International Conference on the Theory of Information Retrieval (ICTIR 2017), pp. 229–232, Amsterdam, Netherlands (2017)
28. Liu, Y., et al. RoBERTa: a robustly optimized BERT pretraining approach (2019). arXiv:1907.11692
29. Maletzke, A.G., dos Reis, D.M., Batista, G.E.A.P.A.: Combining instance selection and self-training to improve data stream quantification. J. Braz. Comput. Soc. **24**(1), 1–17 (2018). https://doi.org/10.1186/s13173-018-0076-0
30. Milli, L., Monreale, A., Rossetti, G., Giannotti, F., Pedreschi, D., Sebastiani, F.: Quantification trees. In: Proceedings of the 13th IEEE International Conference on Data Mining (ICDM 2013), Dallas, USA, pp. 528–536 (2013)
31. Milli, L., Monreale, A., Rossetti, G., Pedreschi, D., Giannotti, F., Sebastiani, F.: Quantification in social networks. In: Proceedings of the 2nd IEEE International Conference on Data Science and Advanced Analytics (DSAA 2015), Paris, France (2015)
32. Molinari, A., Esuli, A., Sebastiani, F.: Active learning and the Saerens-Latinne-Decaestecker algorithm: an evaluation. In: Proceedings of the 2nd Joint Conference of the Information Retrieval Communities in Europe (CIRCLE 2022), Samatan, France (2022)
33. Moreno-Torres, J.G., Raeder, T., Alaíz-Rodríguez, R., Chawla, N.V., Herrera, F.: A unifying view on dataset shift in classification. Pattern Recogn. **45**(1), 521–530 (2012)
34. Moreo, A., Esuli, A., Sebastiani, F.: QuaPy: a Python-based framework for quantification. In: Proceedings of the 30th ACM International Conference on Knowledge Management (CIKM 2021), Gold Coast, Australia, pp. 4534–4543 (2021)
35. Moreo, A., Sebastiani, F.: Re-assessing the "classify and count" quantification method. In: Proceedings of the 43rd European Conference on Information Retrieval (ECIR 2021), Lucca, Italy, vol. II, pp. 75–91 (2021)
36. Moreo, A., Sebastiani, F.: Tweet sentiment quantification: an experimental re-evaluation. PLoS One (2022, forthcoming)
37. Nakov, P., Ritter, A., Rosenthal, S., Sebastiani, F., Stoyanov, V.: SemEval-2016 task 4: sentiment analysis in Twitter. In: Proceedings of the 10th International Workshop on Semantic Evaluation (SemEval 2016), San Diego, USA, pp. 1–18 (2016)
38. Pennington, J., Socher, R., Manning, C.D.: Glove: global vectors for word representation. In: Proceedings of the 12th Conference on Empirical Methods in Natural Language Processing (EMNLP 2014), Doha, Qatar, pp. 1532–1543 (2014)

39. Platt, J.C.: Probabilistic outputs for support vector machines and comparison to regularized likelihood methods. In: Smola, A., Bartlett, P., Schölkopf, B., Schuurmans, D., (eds.) Advances in Large Margin Classifiers, pp. 61–74. The MIT Press, Cambridge (2000)
40. Popordanoska, T., Blaschko, M.B.: KULeuven at LeQua 2022: model calibration in quantification learning. In: Working Notes of the 2022 Conference and Labs of the Evaluation Forum (CLEF 2022), Bologna, Italy (2022)
41. Quiñonero-Candela, J., Sugiyama, M., Schwaighofer, A., Lawrence, N.D. (eds.): Dataset Shift in Machine Learning. The MIT Press, Cambridge (2009)
42. Saerens, M., Latinne, P., Decaestecker, C.: Adjusting the outputs of a classifier to new a priori probabilities: a simple procedure. Neural Comput. **14**(1), 21–41 (2002)
43. Sebastiani, F.: Evaluation measures for quantification: an axiomatic approach. Inf. Retr. J. **23**(3), 255–288 (2020). https://doi.org/10.1007/s10791-019-09363-y
44. Senz, M., Bunse, M.: DortmundAI at LeQua 2022: regularized SLD. In: Working Notes of the 2022 Conference and Labs of the Evaluation Forum (CLEF 2022), Bologna, Italy (2022)
45. Vapnik, V.: Statistical Learning Theory. Wiley, New York (1998)
46. Zeng, Z., Kato, S., Sakai, T.: Overview of the NTCIR-14 Short Text Conversation task: dialogue quality and nugget detection subtasks. In: Proceedings of the 14th Workshop on NII Testbeds and Community for Information access Research (NTCIR 2019), Tokyo, Japan, pp. 289–315 (2019)
47. Zeng, Z., Kato, S., Sakai, T., Kang, I.: Overview of the NTCIR-15 dialogue evaluation task (DialEval-1). In: Proceedings of the 15th Workshop on NII Testbeds and Community for Information Access Research (NTCIR 2020), Tokyo, Japan, pp. 13–34 (2020)

Overview of PAN 2022: Authorship Verification, Profiling Irony and Stereotype Spreaders, and Style Change Detection

Janek Bevendorff[1], Berta Chulvi[2], Elisabetta Fersini[3], Annina Heini[4], Mike Kestemont[5], Krzysztof Kredens[4], Maximilian Mayerl[6], Reynier Ortega-Bueno[2], Piotr Pęzik[4], Martin Potthast[7], Francisco Rangel[8], Paolo Rosso[2], Efstathios Stamatatos[9], Benno Stein[1], Matti Wiegmann[1], Magdalena Wolska[1(✉)], and Eva Zangerle[6]

[1] Bauhaus-Universität Weimar, Weimar, Germany
pan@webis.de, magdalena.wolska@uni-weimar.de
[2] Universitat Politècnica de València, Valencia, Spain
[3] Universitty Milano-Bicocca, Milan, Italy
[4] Aston University, Birmingham, UK
[5] University of Antwerp, Antwerpen, Belgium
[6] University of Innsbruck, Innsbruck, Austria
[7] Leipzig University, Leipzig, Germany
[8] Symanto Research, Nuremberg, Germany
[9] University of the Aegean, Mitilini, Greece
http://pan.webis.de

Abstract. The paper gives a brief overview of three shared tasks which have been organized at the PAN 2022 lab on digital text forensics and stylometry hosted at the CLEF 2022 conference. The tasks include authorship verification across discourse types, multi-author writing style analysis and author profiling. Some of the tasks continue and advance past editions (authorship verification and multi-author analysis) and some are new (profiling irony and stereotypes spreaders). The general goal of the PAN shared tasks is to advance the state of the art in text forensics and stylometry while ensuring objective evaluation on newly developed benchmark datasets.

1 Introduction

PAN is a workshop series and a networking initiative for stylometry and digital text forensics. The workshop's goal is to bring together scientists and practitioners studying technologies which analyze texts with regard to originality, authorship, trust, and ethicality. Since its inception 15 years back PAN has included shared tasks on specific computational challenges related to authorship analysis, computational ethics, and determining the originality of a piece of writing. Over the years, the respective organizing committees of the 54 shared tasks have

A. Barrón-Cedeño et al. (Eds.): CLEF 2022, LNCS 13390, pp. 382–394, 2022.
https://doi.org/10.1007/978-3-031-13643-6_24

assembled evaluation resources for the aforementioned research disciplines that amount to 51 datasets plus nine datasets contributed by the community.[1] Each new dataset introduced new variants of author verification, profiling, or author obfuscation tasks as well as multi-author analysis and determining the morality, quality, or originality of a text. The 2022 edition of PAN continued in the same vein, introducing new resources as well as previously unconsidered problems to the community. As in earlier editions, PAN is committed to reproducible research in IR and NLP therefore all shared tasks ask for software submissions on our TIRA platform [11]. We briefly outline the 2022 tasks and results in the sections that follow.

2 Authorship Verification

Authorship verification is a fundamental task in author identification and all questioned authorship cases, be it closed-set or open-set scenarios, can be decomposed into a series of verification instances [9]. Previous editions of PAN included across-domain authorship verification tasks where texts of known and unknown authorship come from different domains [2,3,28]. In most of the examined cases, domains corresponded to topics (or thematic areas) and fandoms (non-professional fiction that is nowadays published online in significant quantities by fans of high-popularity authors or works, so-called fanfiction). The obtained results of the latest editions have demonstrated that it is feasible to handle such cases with relatively high performance [2,3]. In addition, at PAN'15, cross-genre authorship verification was partially studied using datasets in Dutch and Spanish covering essays and reviews [28]. However, these are relatively similar genres with respect to communication purpose, intended audience, or level of formality. On the other hand, it is not clear yet how to handle more difficult authorship verification cases where texts of known and unknown authorship belong to different discourse types (DTs), especially when these DTs have few similarities (e.g., argumentative essays vs. text messages to family members). In such cases, it is very challenging to distinguish the authorial characteristics that remain intact along DTs.

In the current edition of the authorship verification task we adopt the simplified version used in the most recent PAN editions [2,3] where text pairs are considered. Formally, one has to approximate the target function $\phi : (d_k, d_u) \rightarrow \{T, F\}$, d_k being a text of known authorship and d_u being a text of unknown or disputed authorship. If $\phi(d_k, d_u) = T$, then the author of d_k is also the author of d_u and if $\phi(d_k, d_u) = F$, then the author of d_k is not the same as the author of d_u. The main novelty of the current edition is that d_k and d_u belong to different discourse types.

Dataset

A new dataset has been created based on the recent Aston 100 Idiolects Corpus in English[2] including a rich set of DTs written by around 100 individuals. We

---

[1] https://pan.webis.de/data.html
[2] https://fold.aston.ac.uk/handle/123456789/17

used the following DTs: emails, essays, text messages, and business memos. All individuals have similar age (18–22) and are native English speakers. The topic of text samples is not restricted while the level of formality can vary within a certain DT (e.g., text messages may be addressed to family members or non-familial acquaintances).

First, we split available individuals into two equal and non-overlapping sets, one to be used for the training dataset and the other for the test dataset. That way, it is ensured that any kind of particularities among the training authors will not affect the performance on the test dataset. In addition, we took advantage of available demographic metadata and used a similar gender distribution of individuals in both training and test datasets.

The dataset comprises a set of text pairs and in each pair the two texts belong to two different DTs. All six combinations of the four available DTs are taken into account. However, the distribution of text pairs over the combination of DTs is not homogeneous since it depends on the available texts belonging to each DT. For example, the corpus comprises only one business memo and multiple email messages per individual. Anyway, the distribution of verification instances per DT combination is similar in both training and test datasets as can be seen in Table 1. Similar, both training and test datasets have balanced distribution of positive/negative verification cases. This is also valid for each combination of DTs (e.g., half of the pairs belonging to the combination essay-email is positive and the other half is negative).

Since the length of texts belonging to certain DTs is very small, we concatenated multiple texts of the same DT to produce longer text samples that are used in the text pairs of authorship verification instances. In more detail, email messages are concatenated so that a text sample of at least 2,000 characters is obtained. The date of email messages is taken into account so that consecutive messages are concatenated. In the case of text messages, we concatenate messages sent either to friends or family so that text samples of at least 500 characters are obtained. The text length information provided in Table 1 for email and text messages refers to text samples produced as explained above.

Evaluation Setup and Results

The evaluation framework is similar to the one used in recent shared tasks at PAN. For each AV instance (a text pair) of the test dataset, participants have to produce a scalar score a_i (in the $[0, 1]$ range) indicating the probability that the pair was written by the same author. It is possible for participants to leave text pairs unanswered by submitting a score of precisely $a_i = 0.5$. As concerns the experimental setup, the set of evaluation measures used in the last edition of PAN is also adopted. These include the area under ROC (AUROC), $c@1$ that rewards unanswered cases over wrong predictions, F_1, $F_{0.5u}$, and the complement of Brier score (so that higher scores correspond to better performance) [2]. The average of these diverse measures is used as final score to rank participants.

Two baseline approaches were made available to the participants: a compression-based approach based on Prediction by Partial Matching

Table 1. Statistics of the new dataset used in the authorship verification task.

	Training	Test
Text pairs		
Positive	6,132 (50.0%)	5,239 (50.0%)
Negative	6,132 (50.0%)	5,239 (50.0%)
Email - Text message	7,484 (61.0%)	6,092 (58.1%)
Essay - Email	1,618 (13.2%)	1,454 (13.9%)
Essay - Text message	1,182 (9.6%)	1,128 (10.8%)
Business memo - Email	1,014 (8.3%)	900 (8.6%)
Business memo - Text message	780 (6.4%)	718 (6.9%)
Essay - Business memo	186 (1.5%)	186 (1.8%)
Text length (avg. chars)		
Essay	11,098	10,117
Email	2,385	2,323
Business memo	1,255	1,042
Text message	611	601

Table 2. Final results for the cross-discourse-type authorship verification task at PAN'22. Submitted systems are ranked by their mean performance across five evaluation metrics. Best result per column is shown in bold.

System	AUROC	c@1	F_1	$F_{0.5u}$	Brier	Overall
BASELINE-cngdist	0.546	0.496	0.669	0.542	0.749	**0.600**
najafi22	**0.598**	**0.571**	0.576	**0.571**	0.618	0.587
galicia22	0.512	0.499	0.628	0.544	0.741	0.585
jinli22	0.577	0.557	0.581	0.563	0.589	0.573
BASELINE-compressor	0.541	0.493	0.570	0.478	**0.750**	0.566
lei22	0.539	0.539	0.399	0.488	0.539	0.501
yihuiye22	0.542	0.526	0.398	0.461	0.565	0.499
huang22	0.519	0.519	0.196	0.328	0.519	0.416
cresposanchez22	0.500	0.500	0	0	0.748	0.350

(PPM) [30] and a naive distance-based character n-gram model [7]. We received 7 submissions and evaluated their performance using the TIRA experimentation framework. The overall results of all participants and the baselines can be found in Table 2.

As can be seen, the general performance of all submissions is quite low reflecting the difficulty of the task. It is surprising that a naive baseline achieved the best overall score despite the fact that most participant models are quite sophisticated. On the other hand, the most effective submitted method (najafi22) outperforms all other submissions and baselines in three out of five evaluation measures indicating a promising potential. More details on the evaluation results and the submissions will be available in the task overview paper [27].

3   Author Profiling

Author profiling is the problem of distinguishing between classes of authors by studying how language is shared by people. This helps in identifying authors' individual characteristics, such as age, gender, or language variety, among others. During the years 2013–2021 we addressed several of these aspects in the shared tasks organised at PAN.[3] In 2013 the aim was to identify gender and age in social media texts for English and Spanish [18]. In 2014 we addressed age identification from a continuous perspective (without gaps between age classes) in the context of several genres, such as blogs, Twitter, and reviews (in Trip Advisor), both in English and Spanish [16]. In 2015, apart from age and gender identification, we addressed also personality recognition on Twitter in English, Spanish, Dutch, and Italian [20]. In 2016, we addressed the problem of cross-genre gender and age identification (training on Twitter data and testing on blogs and social media data) in English, Spanish, and Dutch [21]. In 2017, we addressed gender and language variety identification in Twitter in English, Spanish, Portuguese, and Arabic [19]. In 2018, we investigated gender identification in Twitter from a multimodal perspective, considering also the images linked within tweets; the dataset was composed of English, Spanish, and Arabic tweets [17]. In 2019 the focus was on profiling bots and discriminating bots from humans on the basis of textual data only [15]. We used Twitter data both in English and Spanish. Bots play a key role in spreading inflammatory content and also fake news. Advanced bots that generated human-like language, also with metaphors, were the most difficult to profile. It is interesting to note that when bots were profiled as humans, they were mostly confused with males. In 2020 we focused on profiling fake news spreaders [13]. The easiness of publishing content in social media has led to an increase in the amount of disinformation that is published and shared. The goal was to profile those authors who have shared some fake news in the past. Early identification of possible fake news spreaders on Twitter should be the first step towards preventing fake news from further dissemination. In 2021 the focus was on profiling hate speech spreaders in social media [12]. The goal was to identify Twitter users who can be considered haters, depending on the number of tweets with hateful content that they had spread. The task was set in English and Spanish.

Profiling Irony and Stereotype Spreaders on Twitter (IROSTEREO)

With irony, language is employed in a figurative and subtle way to mean the opposite to what is literally stated [22]. In case of sarcasm, a more aggressive type of irony, the intent is to mock or scorn a victim without excluding the possibility to hurt [6]. Stereotypes are often used, especially in discussions about controversial issues such as immigration [29] or sexism [23] and misogyny [1]. At PAN 2022 we focused on profiling ironic authors in Twitter. Special emphasis was given to those authors that employ irony to spread stereotypes. The goal was to classify authors as ironic or not depending on their number of tweets with ironic content. Among those authors we considered a subset that employs irony to convey stereotypes in order to investigate if state-of-the-art models are

---

[3] To generate the datasets, we have followed a methodology that complies with the EU General Data Protection Regulation [14]

able to distinguish also these cases. Therefore, given authors together with their tweets, the goal was to profile those authors that can be considered as ironic, and among them those that employ irony to convey stereotypical messages. As an evaluation setup, we created a collection that contains tweets posted by users in Twitter. One document consisted of a feed of tweets written by the same user.

Taxonomy of Stereotype Categories

Recently [26] developed the Social Bias Frame, a new conceptual formalism that aims to model the pragmatic frames in which people project social biases and stereotypes onto others. To support this research they developed the Social Bias Inference Corpus (SBIC) with 150K structured annotations of social media posts covering 34k implications about social groups. For example: "If cameras do really add ten pounds, do Africans really exist?". For each post, annotators from Amazon Mechanical Turk indicate whether or not: (i) the post is offensive, (ii) the intent is to offend, and (iii) it contains lewd or sexual content. Only if annotators indicate potential offensiveness they answer the group implication question: who is referred to/targeted by this post? Two possible answers were: (i) yes, this could be offensive to a group and (ii) no, this is just an insult to an individual or a non-identity-related group of people. If the post targets or references a group or demographic, annotators select or write which group is referenced. For each selected group, they then write two to four stereotypes that are used in this post; for the given example, annotators write as stereotype: "Africans are all starving". Finally, workers are asked whether they think the speaker is part of one of the minority groups referenced by the post. From 16,739 instances in SBIC, 8,167 refer to a group of people in the field of "target minority".

To build the IROSTEREO corpus we examine the "target minority" field of SBIC and we identify 600 unique labels that could be considered a social group or a social category. We define a social category following a long tradition of research in Social Psychology [4] which considers that a social group exists when two or more persons define themselves as members of the group and when their existence is recognised by at least one other person. [26] classify the groups referenced in seven categories: (1) body (2) culture (3) disabled (4) gender (5) race (6) social and (7) victims. In order to focus specifically on stereotypes as the expression of a prejudice against certain groups or social categories that are often the object of an ironic and hurtful discourse we create a more granular taxonomy to classify the 600 labels in 17 categories: (1) national majority groups, (2) illness/health groups, (3) age and role family groups, (4) victims, (5) political groups, (6) ethnic/racial minorities, (7) immigration/national minorities (8) professional and class groups, (9) sexual orientation groups, (10) women, (11) physical appearance groups, (12) religious groups, (13) style of life groups, (14) non-normative behaviour groups, (15) man/male groups, (16) minorities expressed in generic terms and (17) white people. As keywords to retrieve the tweets we use the labels associated to groups only from categories 5 to 14 of the taxonomy.

Dataset and Annotation Process

The Twitter API was used to retrieve tweets with two conditions: (i) tweets that contain the hashtag #irony or #sarcasm and at least one of the labels included in categories 5 to 14 of the taxonomy and (ii) the same labels about social groups but without #irony or #sarcasm. Users with more cases in classes 1 and 2 were identified and the tweets that accomplish these two conditions were downloaded. The annotators had to identify ironic tweets and tweets that use stereotypes among this set of users. To identify irony, the annotators were asked to mark the tweets where the user "expresses the opposite of what was saying as a disguised mockery". If a user had more than five ironic tweets it was labelled as ironic.

Positive examples of classes 1 (users that express irony without stereotypes), 2 (non-ironic users that use stereotypes) and 3 (users that express irony and use stereotypes) were selected and 200 tweets from their timeline were downloaded. To find the non-ironic and non-stereotype class (4) the lexicon used in the three previous classes was analysed in order to reduce topic bias. Moreover, tweets should not contain the labels of social groups associated to stereotypes. A second annotation was done to check that class 4 does not contain irony.

Table 3 presents the statistics of the corpus that consists of 600 authors for English language, completely balance between the two classes (ironic and non ironic), and with a 66/33 balance between users using stereotypes or not for each class. For each author, we retrieved via the Twitter API their last 200 tweets. We have split the corpus into training and test sets, following a proportion of 70/30 for training and testing respectively.

Table 3. Number of authors in the PAN-AP-22 corpus distributed between the two classes, Ironic vs Non-Ironic, and within each class, distributed between users who use stereotypes vs. users who do not use stereotypes.

	Ironic			Non Ironic			
Set	Stereotypes	Non stereo.	Total	Stereotypes	Non stereo.	Total	Total
Training	140	70	210	140	70	210	420
Test	60	30	90	60	30	90	180
Total	200	100	300	200	100	300	600

Evaluation Setup

Since the dataset is completely balanced for the two target classes, ironic vs. non ironic, we have used the accuracy measure and ranked the performance of the systems by that metric. More than 60 teams participated in the IROSTEREO author profiling task. At the moment of the writing-up of this overview paper, we are still evaluating the last submissions. The results will be presented in the IROSTEREO overview paper [10].

4   Multi-author Writing Style Analysis

The goal of the style change detection task is to identify—based on an intrinsic style analysis—the text positions at which the author switches in a multi-author document. Style change detection is a crucial part of the authorship identification process and multi-author document analysis. This task has been part of PAN since 2016, with varying task definitions, data sets, and evaluation procedures. In 2016, participants were asked to identify and group fragments of a given document that correspond to individual authors [24]. In 2017, the task was to detect whether a given document is multi-authored. If the document was indeed multi-authored, participants were asked to determine the positions at which authorship changes [31]. Since this task was deemed highly complex, we reduced the complexity of the task in 2018 and asked participants to predict whether a given document is single- or multi-authored [8], which has to lead promising results. In 2019, participants were asked first to detect whether a document was single- or multi-authored and to predict the number of authors if it was indeed written by multiple authors [35]. In 2020, we steered the task back to its original definition, i.e., to find the positions at which authorship changes. We asked participants to first determine whether a document was written by one or by multiple authors and, for multi-author documents, they had to detect between which paragraphs the authors change [34]. Continuing these efforts, in the 2021 edition, we asked participants to first detect whether a document was authored by one or multiple authors. For two-author documents, the task was to find the position of the authorship change and for multi-author documents, the task was to find all positions of authorship change and identify which author wrote any given paragraph [32].

Multi-author Writing Style Analysis at PAN'22

The analysis of author writing styles is the foundation for author identification. In this sense, methods for multi-author writing style analysis can pave the way for authorship attribution at the sub-document level and thus, intrinsic plagiarism detection (i.e., detecting plagiarism without the use of a reference corpus). Given the importance of these tasks, we foster research in this direction through our continued development of benchmarks.

Based on the progress made towards this goal in previous years and to entice novices and experts, we extend the set of challenges. Therefore, the style change detection task at PAN'22 involves three subtasks in increasing difficulty: (1) Style Change Basic (subtask1): for a text written by two authors that contains a single style change only, find the position of this change (i.e., cut the text into the two authors' texts on the paragraph-level), (2) Style Change Advanced (subtask2): for a text written by two or more authors, find all positions of writing style change (i.e., assign all paragraphs of the text uniquely to some author out of the number of authors assumed for the multi-author document), and (3) Style Change Real-World (subtask3): for a text written by two or more authors, find all positions of writing style change, where style changes now not only occur between paragraphs but at the sentence level.

Data Set and Evaluation

The datasets underlying this task were created from posts of the popular Stack-Exchange network of Q&A sites. Based on a dump of questions and answers from the StackExchange network, we extracted a subset of topics (so-called sites)[4]. Initial data cleaning involved removing questions and answers that were edited after they were originally posted and removing images, URLs, code snippets, block quotes, and bullet lists from all questions and answers. The general procedure for generating one of our datasets then works as follows. All questions and answers were split into paragraphs; we removed paragraphs of less than 100 characters. Based on these paragraphs, we create documents by drawing paragraphs from a single question thread to ensure that topic changes cannot be leveraged for detecting style changes. We randomly pick the number of authors per document between one and five. Following that, we randomly choose a corresponding number of authors from the authors who contributed to the question thread we were drawing paragraphs from. In the next step, we take the paragraphs written by the selected authors and shuffle them to obtain the final documents. If a resulting document has fewer than two paragraphs or is fewer than 1,000 or more than 10,000 characters long, we discard it.

We applied this procedure, with slightly different parameters, to generate a separate dataset for each of this year's three subtasks. For the dataset for subtask 1, we ensured that every generated document has exactly one style change in it. For subtask 2, we used the procedure exactly as outlined above. For subtask 3, we changed the procedure to operating on sentences instead of paragraphs. The three datasets we obtained in this way contain a total of 2,000, 10,000, and 10,000 documents, respectively, and were then all split into training, validation, and test sets. The training sets consist of 70% of all generated documents for a given dataset, whereas the test and validation set each consist of 15% of the documents.

The three subtasks are evaluated independently. As primary evaluation metric, we compute the macro-averaged F1-score value across all documents. To add a further perspective on the results obtained, we evaluate two further measures for subtask 2: Diarization Error Rate (DER) [5] and Jaccard Error Rate (JER) [25]. These measures essentially capture the fraction of text that is not correctly attributed to an author and are borrowed from the field of text transcription.

---

[4] The following StackExchange sites were used: Code Review, Computer Graphics, CS Educators, CS Theory, Data Science, DBA, DevOps, GameDev, Network Engineering, Raspberry Pi, Superuser, and Server Fault

Table 4. Overall results for the style change detection task, ranked by average F_1 performance across all three subtasks (ST).

Participant	ST1 F_1	ST2 F_1	ST3 F_1	ST3 DER	ST3 JER
Intrinsic Approaches					
tzumilin22	0.7540	0.5100	0.7156	0.8059	0.6905
xinyin22	0.7346	0.4687	0.6720	0.7620	0.6862
qidilao22	0.7471	0.4170	0.6314	0.7364	0.6359
zhang22	0.7162	0.4174	0.6581	0.7114	0.6444
yang22	0.6690	0.4011	0.6483	0.7036	0.6323
alvi22	0.7052	0.3213	0.5636	0.6076	0.4782
castro22a	0.5661	0.2735	0.5565	0.5965	0.4229
alshmasy22	0.5272	0.2207	0.4995	0.5760	0.3557
Extrinsic Approaches					
graner22	0.9932	0.9855	0.9929	0.9960	0.9960

Results

The style change detection task received nine software submissions, eight of which used intrinsic approaches and one used an extrinsic approach. The individual results achieved by the participants are presented in Table 4. For the intrinsic approaches, the best results were achieved by *tzumilin22*, who obtained the highest score for every subtask and evaluation metric. Further details on the approaches taken can be found in the overview paper [33].

Acknowledgments. The contributions from Bauhaus-Universität Weimar and Leipzig University have been partially funded by the German Ministry for Science and Education (BMBF) project "Shared Tasks as an innovative approach to implement AI and Big Data-based applications within universities (SharKI)" (grant FKZ 16DHB4021). The Cross-DT corpus was developed at the Aston Institute for Forensic Linguistics with funding from Research England's Expanding Excellence in England (E3) Fund. The work of the researchers from the Universitat Politècnica de València was partially funded by the Spanish MICINN under the project MISMIS-FAKEnHATE on MISinformation and MIScommunication in social media: FAKE news and HATE speech (PGC2018-096212-B-C31), and by the Generalitat Valenciana under the project DeepPattern (PROMETEO/2019/121). The work of Francisco Rangel has been partially funded by the Centre for the Development of Industrial Technology (CDTI) of the Spanish Ministry of Science and Innovation under the research project IDI-20210776 on Proactive Profiling of Hate Speech Spreaders - PROHATER (Perfilador Proactivo de Difusores de Mensajes de Odio).

References

1. Anzovino, M., Fersini, E., Rosso, P.: Automatic identification and classification of misogynistic language on Twitter. In: Silberztein, M., Atigui, F., Kornyshova, E., Métais, E., Meziane, F. (eds.) NLDB 2018. LNCS, vol. 10859, pp. 57–64. Springer, Cham (2018). https://doi.org/10.1007/978-3-319-91947-8_6

2. Bevendorff, J., et al.: Overview of PAN 2021: authorship verification, profiling hate speech spreaders on Twitter, and style change detection. In: Candan, K.S., et al. (eds.) CLEF 2021. LNCS, vol. 12880, pp. 419–431. Springer, Cham (2021). https://doi.org/10.1007/978-3-030-85251-1_26

3. Bevendorff, J., et al.: Overview of PAN 2020: authorship verification, celebrity profiling, profiling fake news spreaders on Twitter, and style change detection. In: Arampatzis, A., et al. (eds.) CLEF 2020. LNCS, vol. 12260, pp. 372–383. Springer, Cham (2020). https://doi.org/10.1007/978-3-030-58219-7_25

4. Brown, R.: Prejudice: Its Social Psychology. Wiley, Hoboken (2011)

5. Fiscus, J.G., Ajot, J., Michel, M., Garofolo, J.S.: The rich transcription 2006 spring meeting recognition evaluation. In: Renals, S., Bengio, S., Fiscus, J.G. (eds.) MLMI 2006. LNCS, vol. 4299, pp. 309–322. Springer, Heidelberg (2006). https://doi.org/10.1007/11965152_28

6. Frenda, S., Cignarella, A., Basile, V., Bosco, C., Patti, V., Rosso, P.: The unbearable hurtfulness of sarcasm. In: Expert Systems with Applications (2022). https://doi.org/10.1016/j.eswa.2021.116398

7. Kestemont, M., Stover, J., Koppel, M., Karsdorp, F., Daelemans, W.: Authenticating the writings of Julius Caesar. Expert Syst. Appl. **63**, 86–96 (2016)

8. Kestemont, M., et al.: Overview of the author identification task at PAN 2018: cross-domain authorship attribution and style change detection. In: CLEF 2018 Labs and Workshops, Notebook Papers (2018)

9. Koppel, M., Winter, Y.: Determining if two documents are written by the same author. J. Am. Soc. Inf. Sci. **65**(1), 178–187 (2014)

10. Ortega-Bueno, R., Chulvi, B., Rangel, F., Rosso, P., Fersini, E.: Profiling Irony and stereotype spreaders on Twitter (IROSTEREO) at PAN 2022. In: CLEF 2022 Labs and Workshops, Notebook Papers. CEUR-WS.org (2022)

11. Potthast, M., Gollub, T., Wiegmann, M., Stein, B.: TIRA integrated research architecture. In: Ferro, N., Peters, C. (eds.) Information Retrieval Evaluation in a Changing World. TIRS, vol. 41, pp. 123–160. Springer, Cham (2019). https://doi.org/10.1007/978-3-030-22948-1_5

12. Rangel, F., De-La-Peña-Sarracén, G.L., Chulvi, B., Fersini, E., Rosso, P.: Profiling hate speech spreaders on Twitter task at PAN 2021. In: Faggioli, G., Ferro, N., Joly, A., Maistro, M., Piroi, F. (eds.) CLEF 2021 Labs and Workshops, Notebook Papers. CEUR-WS.org (2021)

13. Rangel, F., Giachanou, A., Ghanem, B., Rosso, P.: Overview of the 8th author profiling task at PAN 2019: profiling fake news spreaders on Twitter. In: CLEF 2020 Labs and Workshops, Notebook Papers. CEUR Workshop Proceedings (2020)

14. Rangel, F., Rosso, P.: On the implications of the general data protection regulation on the organisation of evaluation tasks. Lang. Law/Linguagem e Direito **5**(2), 95–117 (2019)

15. Rangel, F., Rosso, P.: Overview of the 7th author profiling task at PAN 2019: bots and gender profiling. In: CLEF 2019 Labs and Workshops, Notebook Papers (2019)

16. Rangel, F., et al.: Overview of the 2nd author profiling task at PAN 2014. In: CLEF 2014 Labs and Workshops, Notebook Papers (2014)

17. Rangel, F., Rosso, P., Montes-y-Gómez, M., Potthast, M., Stein, B.: Overview of the 6th author profiling task at PAN 2018: multimodal gender identification in Twitter. In: CLEF 2019 Labs and Workshops, Notebook Papers (2018)

18. Rangel, F., Rosso, P., Moshe Koppel, M., Stamatatos, E., Inches, G.: Overview of the author profiling task at PAN 2013. In: CLEF 2013 Labs and Workshops, Notebook Papers (2013)

19. Rangel, F., Rosso, P., Potthast, M., Stein, B.: Overview of the 5th author profiling task at PAN 2017: gender and language variety identification in Twitter. Working Notes Papers of the CLEF (2017)

20. Rangel, F., Rosso, P., Potthast, M., Stein, B., Daelemans, W.: Overview of the 3rd author profiling task at PAN 2015. In: CLEF 2015 Labs and Workshops, Notebook Papers (2015)

21. Rangel, F., Rosso, P., Verhoeven, B., Daelemans, W., Potthast, M., Stein, B.: Overview of the 4th author profiling task at PAN 2016: cross-genre evaluations. In: CLEF 2016 Labs and Workshops, Notebook Papers, September 2016. ISSN 1613-0073

22. Reyes, A., Rosso, P.: On the difficulty of automatically detecting irony: beyond a simple case of negation. Knowl. Inf. Syst. **40**(3), 595–614 (2014)

23. Rodríguez-Sánchez, F., et al.: Overview of exist 2021: sexism identification in social networks. In: Procesamiento del Lenguaje Natural (SEPLN), no. 67, pp. 195–207 (2021)

24. Rosso, P., Rangel, F., Potthast, M., Stamatatos, E., Tschuggnall, M., Stein, B.: Overview of PAN'16–new challenges for authorship analysis: cross-genre profiling, clustering, diarization, and obfuscation. In: Experimental IR Meets Multilinguality, Multimodality, and Interaction, 7th International Conference of the CLEF Initiative (CLEF 16) (2016)

25. Ryant, N., et al.: The second DIHARD diarization challenge: dataset, task, and baselines. arXiv preprint arXiv:1906.07839 (2019)

26. Sap, M., Gabriel, S., Qin, L., Jurafsky, D., Smith, N.A., Choi, Y.: Social bias frames: reasoning about social and power implications of language. In: Proceedings of the 58th Annual Meeting of the Association for Computational Linguistics, pp. 5477–5490. Association for Computational Linguistics, July 2020. https://doi.org/10.18653/v1/2020.acl-main.486. https://aclanthology.org/2020.acl-main.486

27. Stamatatos, E., et al.: Overview of the authorship verification task at PAN 2022. In: CLEF 2022 Labs and Workshops, Notebook Papers. CEUR-WS.org (2022)

28. Stamatatos, E., Potthast, M., Rangel, F., Rosso, P., Stein, B.: Overview of the PAN/CLEF 2015 evaluation lab. In: Mothe, J., et al. (eds.) CLEF 2015. LNCS, vol. 9283, pp. 518–538. Springer, Cham (2015). https://doi.org/10.1007/978-3-319-24027-5_49

29. Sánchez-Junquera, J., Chulvi, B., Rosso, P., Ponzetto, S.P.: How do you speak about immigrants? Taxonomy and stereoimmigrants dataset for identifying stereotypes about immigrants. Appl. Sci. **11**(8), 3610 (2021)

30. Teahan, W.J., Harper, D.J.: Using compression-based language models for text categorization. In: Croft, W.B., Lafferty, J. (eds.) Language Modeling for Information Retrieval, vol. 13, pp. 141–165. Springer, Dordrecht (2003). https://doi.org/10.1007/978-94-017-0171-6_7

31. Tschuggnall, M., et al.: Overview of the author identification task at PAN 2017: style breach detection and author clustering. In: CLEF 2017 Labs and Workshops, Notebook Papers (2017)

32. Zangerle, E., Mayerl, M., Potthast, M., Stein, B.: Overview of the style change detection task at PAN 2021. In: Faggioli, G., Ferro, N., Joly, A., Maistro, M., Piroi, F. (eds.) CLEF 2021 Labs and Workshops, Notebook Papers. CEUR-WS.org (2021)
33. Zangerle, E., Mayerl, M., Potthast, M., Stein, B.: Overview of the style change detection task at PAN 2022. In: CLEF 2022 Labs and Workshops, Notebook Papers. CEUR-WS.org (2022)
34. Zangerle, E., Mayerl, M., Specht, G., Potthast, M., Stein, B.: Overview of the style change detection task at PAN 2020. In: CLEF 2020 Labs and Workshops, Notebook Papers (2020)
35. Zangerle, E., Tschuggnall, M., Specht, G., Stein, B., Potthast, M.: Overview of the style change detection task at PAN 2019. In: CLEF 2019 Labs and Workshops, Notebook Papers (2019)

Intelligent Disease Progression Prediction: Overview of iDPP@CLEF 2022

Alessandro Guazzo[1], Isotta Trescato[1], Enrico Longato[1], Enidia Hazizaj[1], Dennis Dosso[1], Guglielmo Faggioli[1], Giorgio Maria Di Nunzio[1], Gianmaria Silvello[1], Martina Vettoretti[1], Erica Tavazzi[1], Chiara Roversi[1], Piero Fariselli[2], Sara C. Madeira[3], Mamede de Carvalho[3], Marta Gromicho[3], Adriano Chiò[2], Umberto Manera[2], Arianna Dagliati[4], Giovanni Birolo[2], Helena Aidos[3], Barbara Di Camillo[1], and Nicola Ferro[1(✉)]

[1] University of Padua, Padua, Italy
{alessandro.guazzo,isotta.trescato,guglielmo.faggioli}@phd.unipd.it,
{enrico.longato,dennis.dosso,giorgiomaria.dinunzio,gianmaria.silvello,
martina.vettoretti,erica.tavazzi,barbara.dicamillo,
nicola.ferro}@unipd.it, {enidia.hazizaj,chiara.roversi}@studenti.unipd.it
[2] University of Turin, Turin, Italy
{piero.fariselli,adriano.chio,umberto.manera,giovanni.birolo}@unito.it
[3] University of Lisbon, Lisbon, Portugal
{sacmadeira,haidos}@fc.ul.pt, mamedemg@mail.telepac.pt,
mgromichosilva@medicina.ulisboa.pt
[4] University of Pavia, Pavia, Italy
arianna.dagliati@unipv.it

Abstract. *Amyotrophic Lateral Sclerosis (ALS)* is a severe chronic disease characterized by progressive or alternate impairment of neurological functions, characterized by high heterogeneity both in symptoms and disease progression. As a consequence its clinical course is highly uncertain, challenging both patients and clinicians. Indeed, patients have to manage alternated periods in hospital with care at home, experiencing a constant uncertainty regarding the timing of the disease acute phases and facing a considerable psychological and economic burden that also involves their caregivers. Clinicians, on the other hand, need tools able to support them in all the phases of the patient treatment, suggest personalized therapeutic decisions, indicate urgently needed interventions. The goal of iDPP∴CLEF is to design and develop an evaluation infrastructure for AI algorithms able to:

1. better describe disease mechanisms;
2. stratify patients according to their phenotype assessed all over the disease evolution;
3. predict disease progression in a probabilistic, time dependent fashion.

A. Guazzo and I. Trescato—These authors contributed equally.

A. Barrón-Cedeño et al. (Eds.): CLEF 2022, LNCS 13390, pp. 395–422, 2022.
https://doi.org/10.1007/978-3-031-13643-6_25

1   Introduction

Amyotrophic Lateral Sclerosis (ALS) is a neurological disease that causes the progressive degeneration of the motor neurons that control voluntary muscles, resulting in an increasing impairment of motor and vital functions and leading to death usually within 4–5 years from the diagnosis. Likely resulting from a complex interplay of genetic and environmental factors, ALS is characterized by high heterogeneity in both symptoms and disease progression, especially in the early stages of the disease. This heterogeneity is partly responsible for the lack of effective prognostic tools in medical practice, as well as for the current absence of a therapy able to effectively slow down or reverse the disease course. On the one hand, patients need support for facing the psychological and economic burdens deriving from the uncertainty of how the disease will progress; on the other, clinicians require tools that may assist them throughout the patient's care, recommending tailored therapeutic decisions and providing alerts for urgently needed actions.

In order to improve the current diagnostic and prognostic situation, we should design and develop *Artificial Intelligence (AI)* algorithms be able to:

- stratify patients according to their phenotype, assessed all over the disease evolution;
- predict the progression of the disease in a probabilistic, time dependent fashion;
- better describe disease mechanisms.

The *Intelligent Disease Progression Prediction at CLEF (iDPP∵CLEF)* lab[1] aims to design and develop an evaluation infrastructure for driving the development of such AI algorithms. By "evaluation infrastructure", we mean experimental collections, evaluation protocols, evaluation measures, ground-truth creation protocols, and so on. Indeed, in this context, it is fundamental, even if not so common yet, to develop shared approaches, promote the use of common benchmarks, and foster the comparability and replicability of the experiments. Differently from previous challenges in the field, iDPP∵CLEF addresses in a systematic way some issues related to the application of AI in clinical practice in ALS. Therefore, in addition to defining the risk scores based on the probability that an event will occur in the short or long term period, iDPP∵CLEF also addresses the issue of providing information in a more structured and understandable way to clinicians.

The paper is organized as follows: Sect. 2 presents related challenges; Sect. 3 describes its tasks; Sect. 4 discusses the developed dataset; Sect. 5 explains the setup of the lab and introduces the participants; Sect. 6 introduces the evaluation measures adopted to score the runs; Sect. 7 analyzes the experimental results for the different tasks; finally, Sect. 8 draws some conclusions and outlooks some future work.

---

[1] https://brainteaser.health/open-evaluation-challenges/idpp-2022/.

2   Related Challenges

To the best of our knowledge, within CLEF, there have been no other labs on this or similar topics before.

Outside CLEF, there have been a recent challenge on Kaggle[2] in 2021 and some older ones, the DREAM 7 ALS Prediction challenge[3] in 2012 and the DREAM ALS Stratification challenge[4] in 2015.

The Kaggle challenge used a mix of clinical and genomic data to seek insights about the mechanisms of ALS and difference between people with ALS who progress faster versus those who develop it more slowly. The DREAM 7 ALS Prediction challenge [12] asked to use 3 months of ALS clinical trial information (months 0–3) to predict the future progression of the disease (months 3-12), expressed as the slope of change in *ALS Functional Rating Scale Revisited (ALSFRS-R)* [5], a functional scale that ranges between 0 and 40. The DREAM ALS Stratification challenge asked participants to stratify ALS patients into meaningful subgroups, to enable better understanding of patient profiles and application of personalized ALS treatments.

Differently from these previous challenges, iDPP@CLEF focuses on explainable AI and on temporal progression of the disease.

3   Tasks

iDPP@CLEF 2022 is the first edition of the lab and consists of pilot activities aimed both at an initial exploration of ALS progression prediction and at understanding of the challenges and limitations to refine and tune the labs itself for future iterations.

In particular, iDPP@CLEF targets two kinds of activities:

1. preliminary and exploratory pilot tasks on disease progression prediction;
2. position papers on the explainability of the prediction algorithms.

Overall, this mix provides participants with the opportunity to make some hands-on experience with these data and provide feedback about the task design as well as to brainstorm on how to evaluate this kind of algorithms and, in particular, assess their explainability.

3.1   Pilot Task 1: Ranking Risk of Impairment

As shown in Fig. 1, this task focuses on ranking of patients based on the risk of impairment in specific domains. More in detail, we use the ALSFRS-R scale to monitor speech, swallowing, handwriting, dressing/hygiene, walking and respiratory ability in time and ask participants to *rank patients based on time to event risk* of experiencing impairment in each specific domain.

---

[2] https://www.kaggle.com/alsgroup/end-als.

[3] https://dreamchallenges.org/dream-7-phil-bowen-als-prediction-prize4life/.

[4] https://dx.doi.org/10.7303/syn2873386.

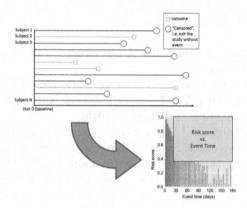

Fig. 1. Task 1: from patients to ranking of patients based on time of event risk.

More in detail, participants are asked to rank subjects based on the risk of early occurrence of

- **Task 1a:** *Non-Invasive Ventilation (NIV)* or (competing event) Death[5], whichever occurs first;
- **Task 1b:** *Percutaneous Endoscopic Gastrostomy (PEG)* or (competing event) Death, whichever occurs first;
- **Task 1c:** Death[6].

For each of these tasks, participants are given a dataset containing 6 months of visits and are asked to rank patients on the risk of occurrence of one of the above events after month 6.

In particular, for each sub-task, we ask for two types of submission from participants:

- submissions using only data available until `Time 0`, i.e. the time of the first ALSFRS-R questionnaire;
- submissions using data available until `Month 6`.

Indeed, from the clinicians point of view, it is of interest to understand what they can say the first time they see the patient (`Time 0`) and what they can say if they collect additional data for the following 6 months.

3.2   Pilot Task 2: Predicting Time of Impairment

As shown in Fig. 2, this task refines Task 1 asking participants to *predict when specific impairments will occur* (i.e. in the correct time-window). In this regard,

---

[5] Death is considered a competing event since a patient might incur death before experiencing the event of interest; the models should account for that.

[6] For the tasks 1c and 2c, death is not a competing event anymore but the focus of the models' predictions.

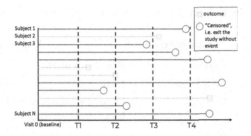

Fig. 2. Task 2: from patients to time of impairment.

we assess model calibration in terms of the ability of the proposed algorithms to estimate a probability of an event close to the true probability within a specified time-window.

In particular, participants are asked to predict the time of the event. Where the event is

- **Task 2a:** NIV or (competing event) Death, whichever occurs first;
- **Task 2b:** PEG or (competing event) Death, whichever occurs first;
- **Task 2c:** Death.

As in the previous case, for each sub-task, we ask two type of submissions from participants:

- submissions using only data available until `Time 0`, i.e. the time of the first ALSFRS-R questionnaire;
- submissions using data available until `Month 6`.

3.3 Position Papers Task 3: Explainability of AI Algorithms

This task is not an evaluation challenge but rather a discussion on how to make these prediction algorithms explainable, also in a visual way.

Therefore, this task called for position papers to start a discussion on AI explainability including proposals on how the single patient data can be visualized in a multivariate fashion contextualizing its dynamic nature and the model predictions together with information on the predictive variables that most influence the prediction. We evaluated proposals of different visualization frameworks able to show the multivariate nature of the data and the model predictions in an explainable, possibly interactive, way.

Even if this task is not an evaluation challenge, authors of the papers were welcome to use the datasets provided by iDPP∵CLEF, if they wished to give examples of their algorithms and solutions, or to explore the submissions made by other participants in iDPP∵CLEF and apply their explainability techniques to them.

Table 1. Main features of the iDPP⋰CLEF dataset.

Section	Sub-section	Variables
Baseline	Patient	Sex, Date of Birth
	ALS Onset	Date, Site
	Diagnosis	Date, Regions affected, Diagnostic Delay, FVC, BMI at diagnosis
Follow-up	Progression scores	ALSFRS-R, Rate of disease progression
	Tests	Hematologic tests, Muscle strength assessed by manual testing, Respiratory function tests
	Therapy	ALS treatments
	Other	Regions affected, Upper and lower motor neuron signs, Cognitive and neurophysiological changes
Clinical Events	History	BMI premorbid, Family history, Comorbidities, Previous surgery and trauma
	Interventions	Date of NIV, Date of PEG, Date of Tracheostomy
	Survival	Date of death
Lifestyle	Lifestyle	Working activity, Physical activity, History of smoking, Marital status, Education level

4   Dataset

iDPP⋰CLEF developed a dataset containing patient records from two clinical institutions in Turin, Italy, and in Lisbon, Portugal.

The dataset is fully anonymized, meaning that all the information which might reveal the identity of a patient, e.g. place of birth or city of residence, are removed; we also avoided absolute dates and made everything relative to Time 0, i.e. the date of the first ALSFRS-R questionnaire [5].

Table 1 summarizes the main features and variables available in the dataset. The following data are available for both the training and the test sets:

- the first available ALSFRS-R questionnaire at Time 0 (both single question scores and total score).

  Thus, for example, time-of-onset and time-of-diagnosis are expressed as relative delta with respect to Time 0 in months (also fractions);
- the slope of the ALSFRS-R score between time-of-onset and Time 0 as:

$$slope = \frac{48 - \text{ALSFRS-R-score}\,(\texttt{Time 0})}{\texttt{Time 0} - \texttt{TimeOnset}}$$

- all the other static data, with a complete list available at http://brainteaser. dei.unipd.it/challenges/idpp2022/assets/other/static-vars.txt

- visits, containing either other ALSFRS-R questionnaires or Spirometry, i.e. *Forced Vital Capacity (FVC)*. The complete list of variables for each visit is available at http://brainteaser.dei.unipd.it/challenges/idpp2022/assets/other/visits.txt.

We ensured that, for each patient, there are 6 months of data, so that predictions can be made using either only data available at `Time 0` or all the data available until month 6.

The following data are available only for the training set:

- Time of event (`NIV`, `PEG`, or `DEATH`); or
- Censoring time, i.e. time of the last available visit if none of the previous events occurs;

according to the following format:

```
0x4bed50627d141453da7499a7f6ae84ab 1 PEG 20.5
0x4d0e8370abe97d0fdedbded6787ebcfc 1 PEG 18.3
0x5bbf2927feefd8617b58b5005f75fc0d 1 DEATH 17.6
0x814ec836b32264453c04bb989f7825d4 0 NONE 37.4
0x71dabb094f55fab5fc719e348dffc85 1 PEG 8.2
...
```

where:

- Columns are separated by a white space;
- The first column is the `patient ID`, a 128 bit hex number (should be considered just as a string);
- The second column indicates whether the one of the above events occurred (`1`) or not (`0`);
- The third column is the occurred event. It comes from a controlled vocabulary and it can be either `NIV`, `PEG`, `DEATH`, or `NONE`;
- The fourth column is the time of the event, or the censoring time, from `Time 0` in months.

Training and test datasets follow a (roughly) 80%–20% proportion; more details about the split into training and test are provided below.

Both Task 1 and Task 2 use the same datasets but we prepared a separate dataset for each of the sub-tasks to make it simpler for participants to focus on a specific event to be predicted. Table 2 provides details about the created datasets.

Creation of the Datasets. The full dataset contained approximately 4,800 records linked to patients, with around 20,000 ALSFRS-R questionnaires in total and 5,500 records concerning spirometries. The original data contain minor inconsistencies and typos. Therefore, we first process the data, removing records that are likely wrong or do not provide essential information to enable prediction. In terms of patient records we removed those presenting an unordered sequence of events (i.e., onset after diagnosis or diagnosis after death). Such event sequences are likely due to typos and other human errors, which result in wrong records that might introduce noise and spurious information in the final dataset.

Table 2. Training and test datasets.

		Training		
Sub-task	Patients	ALSFRS-R	Spirometry	Outcome
Sub-task a	1,454	3,668	1,189	– NIV: 675 patients (46.42%) – DEATH: 636 patients (43.74%) – NONE: 143 patients (9.83%)
Sub-task b	1,715	4,264	1,506	– PEG: 501 patients (29.21%) – DEATH: 969 patients (56.50%) – NONE: 245 patients (14.29%)
Sub-task c	1,756	4,366	1,536	– DEATH: 1,486 patients (84.62%) – NONE: 270 patients (15.38%)
		Test		
Sub-task	Patients	ALSFRS-R	Spirometry	Outcome
Sub-task a	350	872	273	– NIV: 162 patients (46.29%) – DEATH: 152 patients (43.43%) – NONE: 36 patients (10.29%)
Sub-task b	430	1,049	361	– PEG: 120 patients (27.91%) – DEATH: 251 patients (58.37%) – NONE: 59 patients (13.72%)
Sub-task c	494	1,220	414	– DEATH: 417 patients (84.41%) – NONE: 77 patients (15.59%)

Furthermore, a patient record was dropped if one or more of the following pieces of information were absent:

– onset or diagnosis dates;
– death date in records associated with dead patients;
– at least six months of historical ALSFRS-R questionnaires before an event (NIV, PEG, or (competing event) Death).

We adopt the filtering strategy mentioned above to grant that every record in the final dataset contains enough information to allow proper predictions.

Concerning the ALSFRS-R questionnaires, we removed those records that had one or more of the following problems:

- duplicate records;
- missing date;
- one or more of the ALSFRS-R items missing;
- *ALS Functional Rating Scale (ALSFRS)* questionnaires with the old formulation (thus with items from 1 to 9, plus the old 10th item). We include only records referring to ALSFRS-R.

Furthermore, if one or more of the ALSFRS-R sub-scores or the total ALSFRS-R score do not agree with the sum of the associated ALSFRS-R items, we replace the value reported in the original dataset with the sum of the linked items. Finally, regarding the spirometries, we removed duplicated records, records with a missing date, and FVC percentage value.

Figure 3 illustrates a set of - synthetic - patients and their clinical history, describing whether they satisfy the conditions to be inserted into the dataset. By construction, the first ALSFRS visit (blue bullets) is considered as Time 0, while the moment of the previous spirometries (yellow bullets) and subsequent visits is indicated as the difference in months with respect to the reference ALSFRS.

- Patient 1 is inserted into the dataset, having a proper sequence of visits, questionnaires and events (at least six months of information before the first event).
- Patient 2, on the other hand, cannot be included in the dataset since they do not have enough information.
- For Patient 3, we observe that only four months passed between the first ALSFRS and the first event. Thus, even though we have 6 months of overall information (first spirometry to event), we cannot retain the record.
- Patient 4, regardless of the fact that they have a single ALSFRS, can be included in the dataset since the distance between the first ALSFRS and the event is above six months.
- Both patients 5 and 6 need to be excluded from further analyses: the former does not have six months of information before the first event, while the latter does not have enough history, regardless of the spirometry taken before the first ALSFRS.
- Patients 7 and 8, on the other hand, can be considered: the former has a proper clinical history, while the latter, even though he or she have a "censoring" event – marked with a question mark, has more than six months of history.

Split into Training and Test. Each of the three available datasets (sub-task a, b, and c) was split into a training set and a test set, with proportions 80% and 20%, respectively. The data were split stratifying the subjects according to outcome time and to the specific outcome type (*death, NIV, none* for sub-task a, *death, PEG, none* for sub-task b, and *death, none* for sub-task c). Stratifying by these two variables is instrumental to the fairness of the challenge as it forces an equal distribution of their levels across the two subsets. The simplest method to verify whether stratification has been performed correctly is to compare the distribution of the stratification variables (*outcome time* and *outcome type*) in each training/test pair. From the literature, certain variables are known to be

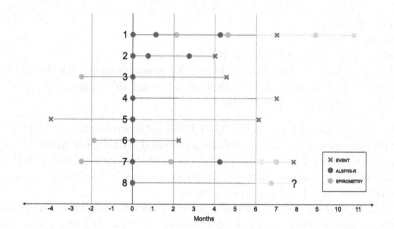

Fig. 3. Sequences of events that allow (or forbid) a patient to be considered as suitable to belong to the dataset. Events grayed out refer to those events happened after another and thus ignored. Visits grayed out refer to visits happened after the first six months. (Color figure online)

particularly relevant in predicting events related to ALS progression [6], therefore, even though they were not included in the stratification criteria, we verified that *sex, age at onset, onset site, ALSFRS-F slope,* and the number of available visits in the first 6 months were also equally represented in the training and test sets. Table 3 reports, as an example, the comparison of the variables' distributions in the training (second column) and test (third column) sets for sub-task a. The comparison for the other two sub-tasks are shown in the extended overview [8]. Since the distributions are similar, we concluded that the training/test split provided to the participants met best-practice quality standards.

5   Lab Setup and Participation

In the remainder of this section, we detail the guidelines the participants had to comply with to submit their runs and the submissions received by iDPP∴CLEF.

5.1   Guidelines

Participating teams were provided with the following guidelines:

- The runs should be submitted in a textual format in the participant repository, both described below;
- Each group can submit a maximum of 5 runs for each sub-task, thus amounting to maximum 15 runs for each of Task 1 and Task 2;
- For each task, participants are asked to submit two types of runs: either using only the information available at Time 0 or using all the information available in the first 6 months.

Table 3. Sub-task a, comparison between training and test populations. Continuous variables are presented as *median [1st–3rd quartiles]*; discrete variables as *count (percentage on sample total)*, for each level.

	Training	Test
Number of subjects	1454	350
Outcome type	Death: 636 (44%) NIV: 675 (46%) Censoring: 143 (10%)	Death: 152 (43%) NIV: 162 (46%) Censoring: 36 (10%)
Outcome time	17.75 [11.14-30.99]	20.72 [11.25-36.76]
Sex	M: 743 (51%) F: 711 (49%)	M: 188 (54%) F: 16 (46%)
Age at onset	64.89 [55.66-70.76]	64.76 [56.66-71.58]
Onset site	Bulbar: 449 (31%) Axial: 3 (0.002%) Generalized: 4 (0.003%) Limbs: 998 (68%)	Bulbar: 105 (30%) Axial: 0 (0%) Generalized: 0 (0%) Limbs: 242 (70%)
ALSFRS-R slope	0.43 [0.24-0.79]	0.41 [0.23-0.80]
Number of available visits	2.00 [2.00-3.00]	3.00 [2.00-3.00]

Runs should be uploaded using the following name convention for their identifiers:

```
<teamname>_T<1|2><a|b|c>_<train>_<freefield>
```

where:

- **teamname** is the name of the participating team;
- **T<1|2><a|b|c>** is the identifier of the task the run is submitted to, e.g. T1b for Task 1, sub-task b;
- **train** is data window used to train the algorithm. It can be either M0, if only the data available at Time 0 have been used, or M6 if all the data available in the first 6 months have been used;
- **freefield** is a free field that participants can use as they prefer.

For example, a complete run identifier may look like

```
upd_T2b_M6_survRF
```

where:

- **upd** is the University of Padua team;
- **T2b** means that the run is submitted for Task 2, sub-task b;
- **M6** means that the algorithm has been trained using all the data available in the first 6 months;
- **survRF** suggests that participants have used survival random forests as a prediction method.

Participant Repository. Participants are provided with an individual git repository for all the tasks they take part in. The repository contains the runs, resources, and possibly the code produced by each participant in order to promote reproducibility and open science. The repository is organised as follows:

- submission: this folder contains the runs submitted for the different tasks.
- score: this folder contains the performance scores of the submitted runs.
- code: this folder contains the source code of the developed system.
- resource: this folder contains any additional resources created during the participation.
- report: this folder contains the template for participant report.

The submission and score folders are organized into sub-folders for each task as follows:

- submission/task1: for the runs submitted to the first task. Similar structure for the other tasks.
- score/task1: for the performance scores of the runs submitted to the first task. Similar structure for the other tasks.

The goal of iDPP⠖CLEF is to speed up the creation of systems and resources for ALS progression prediction as well as openly share these systems and resources as much as possible. Therefore, participants are more than encouraged to share their code and any additional resources they have used or created.

All the contents of these repositories are released under the *Creative Commons Attribution-ShareAlike 4.0 International License*[7].

Task 1 Run Format. Runs had to be submitted as a text file with the following format:

```
0x4bed50627d141453da7499a7f6ae84ab 0.897 0 PEG upd_T1b_M6_survRF
0x4d0e8370abe97d0fdedbded6787ebcfc 0.773 1 PEG upd_T1b_M6_survRF
0x5bbf2927feefd8617b58b5005f75fc0d 0.773 2 DEATH upd_T1b_M6_survRF
0x814ec836b32264453c04bb989f7825d4 0.615 3 NONE upd_T1b_M6_survRF
0x71dabb094f55fab5fc719e348dffc85 0.317 4 PEG upd_T1b_M6_survRF
...
```

where:

- Columns are separated by a white space;
- The first column is the `patient ID`, a 128 bit hex number (should be considered just as a string);
- The second column shows the prediction score that generated the ranking. It is expected to be a floating point number in the range $[0, 1]$. This score must be in descending (non-increasing) order;

---

[7] http://creativecommons.org/licenses/by-sa/4.0/.

- The third column is the rank of the patient by her/his risk of impairment, starting from 0. This is expected to be a strictly increasing integer number. It is important to include the rank so that we can handle tied scores (for a given run) in a uniform fashion;
- The fourth column is the predicted event. It comes from a controlled vocabulary and it can be either NIV, PEG, DEATH, or NONE. Note that, since each sub-task is focused on the prediction of a specific event (NIV, PEG, or DEATH), this column will contain that event or the competing event DEATH or NONE;
- The fifth column is the run identifier, according to the format described above. It must uniquely identify the participating team and the submitted run.

Task 2 Run Format. Runs had to be submitted as a text file with the following format:

```
0x4bed50627d141453da7499a7f6ae84ab 6-12 PEG upd_T2b_M6_survRF
0x4d0e8370abe97d0fdedbded6787ebcfc 18-24 PEG upd_T2b_M6_survRF
0x5bbf2927feefd8617b58b5005f75fc0d 24-30 DEATH upd_T2b_M6_survRF
0x814ec836b32264453c04bb989f7825d4 >36 NONE upd_T2b_M6_survRF
0x71dabb094f55fab5fc719e348dffc85 >36 PEG upd_T2b_M6_survRF
...
```

where:

- Columns are separated by a white space;
- The first column is the `patient ID`, a 128 bit hex number (should be considered just as a string);
- The second column shows the prediction window in months. Possible values are taken from a controlled vocabulary as follows:
  - 6-12: the event will happen in the range of months $(6, 12]$;
  - 12-18: the event will happen in the range of months $(12, 18]$;
  - 18-24: the event will happen in the range of months $(18, 24]$;
  - 24-30: the event will happen in the range of months $(24, 30]$;
  - 30-36: the event will happen in the range of months $(30, 36]$;
  - >36: the event will happen in the range of months $(36, +\infty)$.
- The third column is the rank of the patient by her/his risk of impairment, starting from 0. It is important to include the rank so that we can handle tied scores (for a given run) in a uniform fashion;
- The fourth column is the predicted event. It comes from a controlled vocabulary and it can be either NIV, PEG, DEATH, or NONE. Note that, since each sub-task is focused on the prediction of a specific event (NIV, PEG, or DEATH), this column will contain that event or the competing event DEATH or NONE;
- The fifth column is the run identifier, according to the format described above. It must uniquely identify the participating team and the submitted run.

Table 4. Teams participating in iDPP⁝CLEF 2022.

Team Name	Description	Country	Repository	Paper
BioHIT	National Centre for Scientific Research Demokritos (NCSR Demokritos)	Greece	https://bitbucket.org/brainteaser-health/idpp2022-biohit	–
CompBioMed	Department of Medical Sciences, University of Turin	Italy	https://bitbucket.org/brainteaser-health/idpp2022-compbiomed-unito	Pancotti et al. [16]
FCOOL	Faculty of Sciences of the University of Lisbon	Portugal	https://bitbucket.org/brainteaser-health/idpp2022-fcool	Branco et al. [2] and Nunes et al. [15]
LIG GETALP	Laboratoire d'Informatique de Grenoble, Université Grenoble Alpes	France	https://bitbucket.org/brainteaser-health/idpp2022-lig-getalp	Mannion et al. [14]
SBB	University of Padua	Italy	https://bitbucket.org/brainteaser-health/idpp2022-sbb	Trescato et al. [18]

Table 5. Break-down of the runs submitted by participants for each task and sub-task. Participation in Task 3 does not involve submission of runs and it is marked just with a tick.

Team Name	Total	Task 1			Task 2			Task 3
		a	b	c	a	b	c	
BioHIT	18	3	3	3	3	3	3	–
CompBioMed	40	8	8	6	6	6	6	–
FCOOL	15	–	–	–	5	5	5	✓
LIG GETALP	23	4	4	4	4	4	3	–
SBB	24	4	4	4	4	4	4	–
Total	120	19	19	17	22	22	21	

5.2 Participants

Overall, 43 teams registered for participating in iDPP⁝CLEF but only 5 of them actually managed to submit runs for at least one of the offered tasks. Table 4 reports the details about the participating teams.

Table 5 provides breakdown of the number of runs submitted by each participant for each task and sub-task. Overall, we have received 120 runs which are roughly broken down evenly among the different tasks.

6 Evaluation Measures

iDPP⁝CLEF adopted several state-of-the-art evaluation measures to assess the performance of the prediction algorithms, among which:

- *ROC curve and/or the precision-recall curve (and area under the curve)* to show the trade-off between clinical sensitivity and specificity for every possible cut-off of the risk scores;
- *Concordance Index (C-index)* to summarize how well a predicted risk score describes an observed sequence of events.
- *E/O ratio and Brier Score* to assess whether or not the observed event rates match expected event rates in subgroups of the model population.
- *Specificity and recall* to assess, for each interval, the ability of the models of correctly identify true positives and true negatives.
- *Distance* to assess how far the predicted time interval was from the true time interval.

To ease the computation and reproducibility of the results, scripts for computing the measures are available in the following repository: https://bitbucket. org/brainteaser-health/idpp2022-performance-computation.

The next two sections provide details about the adopted measures for each Task.

6.1   Pilot Task 1: Ranking Risk of Impairment

The runs submitted for Task 1 were evaluated by means of Harrel's concordance index (C-index) [11], area under the receiver operating characteristic curve (AUROC) [10], and the Brier score (BS) [3]. The 95% confidence intervals of the C-index and the AUROC were also considered [17].

The C-index has an advantage over the other considered metrics (i.e., AUROC and BS) in that it can be used to evaluate model discrimination on the test sets regardless of censored data – data for those patients that did not incurred either the relevant event (NIV or PEG) or the competing event (Death). According to the best practices in the field [13], before computing the C-index, a final censoring time equal to the last time-to-event in the training was set on each test set. This ensured consistency between Task 1's final results and those that might have been obtained by the participants during model development.

The AUROC and BS were computed at various prediction horizons (PHs). Specifically, seven clinically relevant PHs were considered, namely: 12, 18, 24, 30, 36, 48, and 60 months after the baseline. For each PH, the corresponding version of the test set comprised: all patients who experienced an event before the PH, and all patients who experienced an event or were censored after the PH as censored patients (and were, thus, censored at that PH). As the status of patients censored before the PH was, by definition, unknown, they were excluded from performance evaluation at that PH.

To contextualize the results obtained by the participants, each run was compared to the empirical lower bound established by the average performance of 100 random classifiers (i.e., such that their output was a random continuous number, uniformly sampled in the range $[0, 1]$).

6.2   Pilot Task 2: Predicting Time of Impairment

To evaluate the predictions of Task 2, the selected evaluation metrics were: the specificity, the recall, and a measure of distance between the predicted and correct time intervals.

Confusion matrices were computed to derive specificity, i.e., the number of correct negative predictions divided by the total number of negatives, and recall, i.e., the ratio of correct positive predictions over the total predicted positives. To do so, the outcome times reported in the column *Time* of the published test sets were mapped to the corresponding interval ("6–12", "12–18", "18–24", "24–30", "30–36", or ">36" months). A conformance check was performed on the participants' predicted times: predictions in the time interval "0–6" were reassigned to the interval "6–12", i.e., the closest allowed interval. The confusion matrices reported the predicted time interval vs the true time interval, independently of the predicted event.

A measure of distance between the predicted and correct time intervals, in months, was also considered (AbsDist). To compute the AbsDist, all the time intervals were replaced with the mean value of each interval (i.e., "6–12" was replaced with 9, "12–18" with 15, "18–24" with 21, "24–30" with 27, "30–36" with 33, and ">36" with 39). The difference between the predicted values and the true values was then computed as $meanValue_{predicted\ time\ interval} - meanValue_{true\ time\ interval}$. The obtained differences were, by construction, in the range $[-36; +36]$ where a smaller modulus corresponds to more accurate predictions. Negative values correspond to a events that occur before the predicted time and positive values to events that occur after. Finally, the AbsDist was obtained by averaging the differences absolute values.

To contextualize the results obtained by the participants, each run was compared to the performance of several synthetic runs, with the following characteristics:

- *min_interval*: a run in which the predicted time intervals are identical for all subjects, and fixed at the first possible time interval, i.e. "6–12";
- *max_interval*: a run in which the predicted time intervals are identical for all subjects, and fixed at the last possible time interval, i.e. ">36";
- *interval_18_24*: a run in which the predicted time intervals are identical for all subjects, and fixed at the time interval "18–24";
- *random_interval*: 100 randomly generated runs, but with the same distribution as the test set distribution (i.e., such that their output was sampled among the labels "6–12", "12–18", "18–24", "24–30", "30–36", ">36" following the same distribution of the true intervals);
- *inverse_distr_interval*: 100 randomly generated runs, but with an inverse distribution compared to the test set distribution (i.e., such that their output was sampled among the labels "6–12", "12–18", "18–24", "24–30", "30–36", ">36" following the inverse distribution of the true outcome);
- *corr_interval*: 100 correlated runs, with correlation coefficient to the true intervals ~0.7.

7 Results

For each task, we report here the analysis of the performance attained by the runs submitted by the Lab's participants according to the metrics described in Sect. 6.

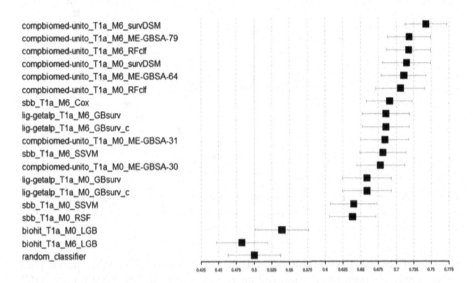

Fig. 4. Sub-task a C-index computed for all submitted runs. The bars in the plot show the 95% confidence intervals. The average C-index of 100 random classifiers is reported in the last row.

7.1 Pilot Task 1: Ranking Risk of Impairment for ALS

Here, only the C-index, the 12-month and 48-month AUROCs, and the 12-month and 48-month BSs obtained for all participants runs submitted for Task 1's sub-task a are shown. Results for all sub-tasks and all PHs are shown in the extended overview [9].

Figure 4 shows the C-index with its 95% confidence intervals computed for all runs submitted for sub-task a and for the 100 random classifiers (last row). As expected, the random classifiers yielded an average C-index of around 0.5. Runs submitted by the BioHit team were comparable to those obtained by the random classifiers. All runs submitted by other participants significantly outperformed the random classifiers (C-index > 0.625) with team CampBioMed leading the pack (C-index > 0.7).

Figure 5 shows the AUROC with its 95% confidence intervals computed for all runs submitted for sub-task a at the 12-month PH. The average 12-month AUROC of the 100 random classifiers is reported in the last row. The 12-month AUROC confirmed the results obtained when considering the C-index. Again, as expected, the random classifiers yielded a 12-month AUROC of around 0.5. Runs submitted by the BioHit team showed a discrimination that was comparable to

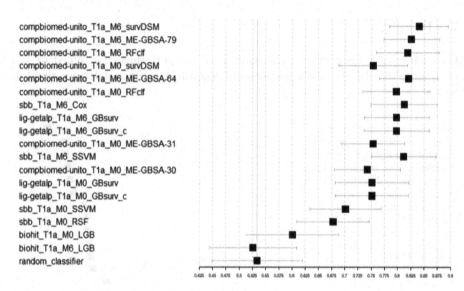

Fig. 5. Sub-task a AUROC computed for all submitted runs with a 12-months PH. The bars in the plot show the 95% confidence intervals. The average 12-months AUROC of 100 random classifiers is reported in the last row.

the one of the random classifiers, and all runs submitted by other participants significantly outperformed the random classifiers (12-month AUROC > 0.675) with some runs of team CampBioMed and team SBB achieving excellent results (12-month AUROC > 0.8) when all the information available in the first 6 months was considered (M6 runs).

Figure 6 shows the BS computed for runs submitted for sub-task a at the 12-month PH. The average 12-month BS of the 100 random classifiers is reported in the last row. The random classifier yield a 12-month BS of around 0.325 as the random probability values were, on average, well distributed in the range [0, 1]. Runs submitted by the CampBioMed team showed the best calibration at this PH (12-month BS < 0.225), while those submitted by the SBB team showed the worst one (12-month BS > 0.675), mainly due to a consistent overestimation of the event probability. Other participants' runs had 12-month BSs comparable with the random classifiers as their models did not correctly predict the event probability but neither showed consistent overestimation trends.

Figure 7 shows the AUROC with its 95% confidence intervals computed for all runs submitted for sub-task a at the 48-month PHs. The average 48-month AUROC of the 100 random classifiers (again, expectedly, around 0.5) is reported in the the last row. The 48-month AUROC confirmed once again the results obtained with the C-index and 12-month AUROC. Runs submitted by the Bio-Hit team had comparable discrimination to the random classifiers, while all runs submitted by other participants significantly outperformed them (48-month AUROC > 0.7). Runs that used all the information available in the first 6 months (M6 runs) submitted by the CampBioMed team were the best performing ones also at this PH (48-month AUROC > 0.8).

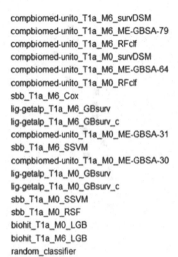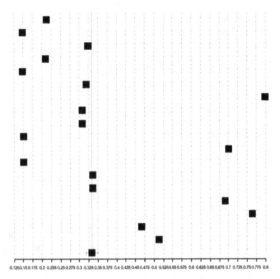

Fig. 6. Sub-task a BS computed for all submitted runs with a 12-months PH. The random classifier average 12-months BS is reported in the last row with its 95% confidence intervals.

Figure 8 shows the BS computed for runs submitted for sub-task a at the 48-month PH. The average 48-month BS of the 100 random classifiers is reported in the last row. The random classifiers yielded a 48-month BS of around 0.325 as the random probability values were, as in the 12-month case, on average, well distributed in the range [0, 1]. All GBSA runs submitted by the CampBioMed team, which had good calibration with a PH of 12 months, led to a poorer calibration at 48 months (48-month BS > 0.75). All other runs submitted by the participants significantly outperformed the random classifiers by showing good calibration at this PH (48-months BS < 0.25).

Overall, for Task 1 sub-task a, runs submitted by the CampBioMed team were the best performing across the board; meanwhile, runs submitted by the BioHit team led to the lowest discrimination, but still yielded acceptable calibration at a long PH (48 months). Finally, the SBB and LIG GETALP teams obtained comparable results when considering runs obtained using all the information available in the first 6 months (M6 runs); meanwhile, when using only the information available at time 0 (M0 runs), runs submitted by the SBB team showed worse discrimination than those submitted by the LIG GETALP teams.

7.2 Pilot Task 2: Predicting Time of Impairment for ALS

Figures 9, 10, and 11 show the specificity-recall plots for three select time intervals ("6–12", "12–18", and "18–24") of Task 2's sub-task a, including all participants' runs and all the synthetic runs. Results for all time intervals and sub-tasks are presented in the extended overview [9]. The graph shows the specificity on

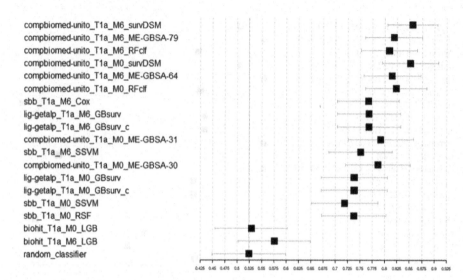

Fig. 7. Sub-task a AUROC computed for all submitted runs with a 48-months PH. The bars in the plot show the 95% confidence intervals. The average 48-months AUROC of 100 random classifiers is reported in the last row.

the x-axis (from 1 to 0, left to right), and the recall on the y-axis (from 0 to 1, bottom to top). The ideal classifier would have specificity = 1 and recall = 1, and would therefore be located in the upper left corner: as a general guidance, the closer a run is to the upper left corner, the better the classification obtained.

In all graphs, the synthetic runs with constant predictions, fixed at the minimum or maximum allowed interval, are located in the two extreme corners of the plot. In detail, the *max_interval* run is located in the lower left corner with specificity = 0 and recall = 1, while the *min_interval* run, in the upper right corner, has specificity = 1 and recall = 0. As expected, the 100 runs with 70% correlation form a cloud in the upper left corner, while the 200 randomly generated runs, 100 with the same distribution and 100 with the inverse distribution always remain in the lower left sector, with $1 > specificity > 0.5$ and $0 > recall > 0.5$.

For the "6–12" interval, represented in Fig. 9, the team with the best classification performance according to specificity and recall was the FCOOL team, whose five submitted runs yielded specificity $\simeq 0.72$ and recall = 0.612. One run from the CompBioMed team also performed well, with specificity = 0.839 and recall = 0.561. In contrast, the other runs submitted by the CompBioMed team were in line with those of the other participants, with rather high specificity but low recall.

Figure 10 reports the results for the "12–18" interval. Again, the CompBioMed team outperformed the other teams with a run with specificity = 0.581 and recall = 0.545. The second best run in this time window was from the LIG GETALP team, with specificity = 0.668 and recall = 0.509. Similar results were

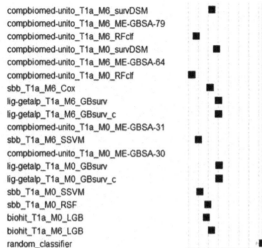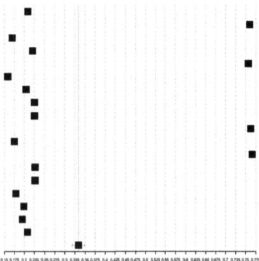

Fig. 8. Sub-task a BS computed for all submitted runs with a 48-months PH. The random classifier average 48-months BS is reported in the last row with its 95% confidence intervals.

obtained by two runs of the SBB team, which reached specificity $\simeq 0.64$ and recall $\simeq 0.44$.

The SBB and LIG GETALP teams obtained the best performance in the interval "18–24", as shown in Fig. 11. Specifically, the SBB team submitted three runs with specificity $\simeq 0.66$ and recall $\simeq 0.47$, while LIG GETALP one with specificity $= 0.697$ and recall $= 0.479$. The other teams, as in the other time intervals, obtained higher specificity scores to the detriment of recall, thus positioning themselves in the lower left quadrant.

Figure 12 shows the AbsDist computed for all runs submitted for sub-task a. The average AbsDist of the synthetic runs is reported as well. As expected, the *max_interval* run led to the worst result (AbsDist > 17 months), as most subjects have a true time interval smaller than the maximum one. Runs *random_interval*, *min_interval*, and *inverse_corr_interval* led to comparable distance values (AbsDist 12–13 months). Runs submitted by the BioHit team had AbsDist values comparable with the synthetic run *interval_18_24* (AbsDist 10–11 months), suggesting that their models might predict the average time interval for most subjects. All runs submitted by the other teams significantly outperformed the aforementioned synthetic runs (AbsDist 7–9 months) with Camp-BioMed team leading the pack. Finally, the *corr_interval* run led to the smallest AbsDist value (AbsDist < 4 months). Note, however, that this run was included only as an arbitrary reference, and its distance value was not strictly expected to be reached by any participant.

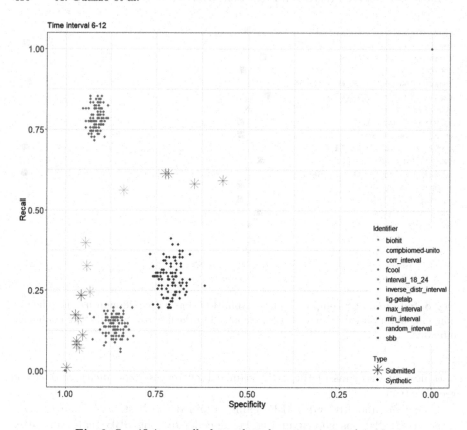

Fig. 9. Specificity-recall plot, sub-task a, time interval 6–12.

Predicting the correct event time interval proved to be a challenge for all teams, especially in terms of recall. However, almost all teams were able to obtain good AbsDist values as, on average, their models, despite not being able to precisely identify the correct time interval, tended to predict an interval that was immediately before or after the true one.

As observed for Task 1, runs performed better when considering all the information available in the first 6 months (M6 runs) rather than only the information available at time 0 (M0 runs).

7.3   Approaches

In this section, we provide a short summary of the approaches adopted by participants in iDPP⁚⋱CLEF. There are two separate sub-sections, one for Task 1 and 2 focused on ALS progression prediction and the other for Task 3, on *eXplainable AI (XAI)* approaches for such kind of algorithms.

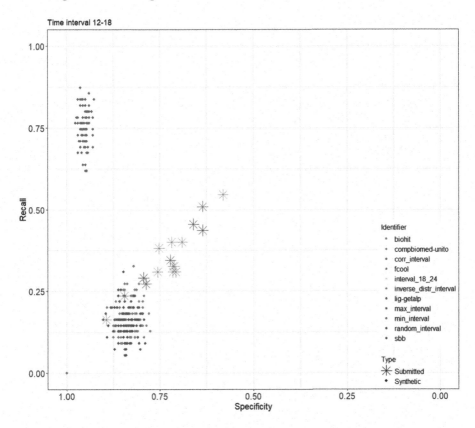

Fig. 10. Specificity-recall plot, sub-task a, time interval 12–18.

Task 1 and 2. BoiHIT explored the use of logistic regression, random forest classifiers, XGBoost, and LightGBM. Decision trees and boosting approaches were preferred due to their ability to deal with both categorical and numerical/continuous features and the interpretability they offer. Even if LightGBM was the model with the best performance, BoiHIT found out that this kind of approaches might not be appropriate for time dependent problems and that time to event analysis methods, such as survival analysis, might yield better results.

CompBioMed [16] considered three main approaches. The simplest one consisted on fitting a standard survival predictor separately for each event as outlined above for independent events, called Naive Multiple Event Survival (NMES). Another was the recently developed Deep Survival Machine (DSM), based on deep learning and capable of handling competing risks. Finally, they also proposed a time-aware classifier ensemble method, that also handles competing risks, called Time-Aware Classifier Ensemble (TACE). All the above approaches achieved comparable performance among them. Only the TACE models appeared to be slightly worse than the rest in when using 6 months of data. Moreover, no clear advantage of the DSM models, that specifically handles competing risks, was observed with respect to the NMES models, which treat all events, as if they were independent.

418   A. Guazzo et al.

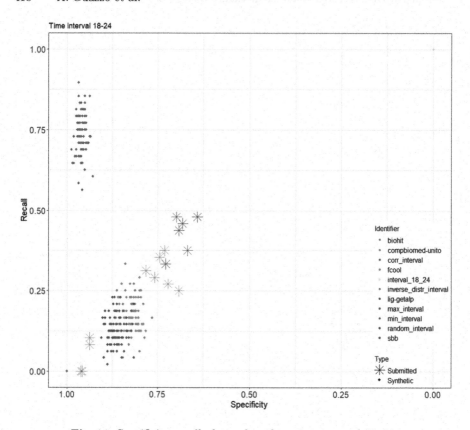

Fig. 11. Specificity-recall plot, sub-task a, time interval 18–24.

FCOOL [2] proposes a hierarchical approach, with a first-stage event prediction, followed by specialized models predicting the time window to a particular
event. The procedure is three-fold: first, it creates patient snapshots based on
clustering with constraints, thus organizing patient records in an efficient manner. Second, it uses a pattern-based approach that incorporates recent advances
on temporal pattern mining to the context of classification. This approach performs end-stage event prediction while allowing the entire patient's medical history to be considered. Finally, exploiting the predictions from the previous step,
specialized models are learned using the original features to predict the time window to an event. This two-stage prediction approach aimed to promote homogeneity and lessen the impact of class imbalance, in comparison to performing
one single multilabel task.

LIG GETALP [14] employed Cox's proportional hazards model to the task
of ranking the risk of impairment, using the gradient boosting learning strategy
The output of the time-independent part of the survival function calculated by
the gradient boosting survival analysis method is then mapped to the interval (0,
1), via a sigmoid function. To estimate the time-to-event, LIG GETALP used a

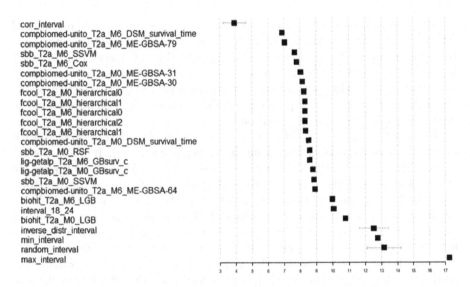

Fig. 12. Sub-task a AbsDist computed for all submitted and synthetic runs. The Abs-Dist of *corr_interval*, *inverse_distr_interval*, and *random_interval* is the average with 95% confidence intervals computed on the corresponding 100 randomly generated runs.

regression model based on Accelerated Gradient Boosting (AGB). This being a standard regression model, it does not take censoring into account and Mannion et al. uses class predictions based on the Task 1 survival model to "censor" the time-to-event predictions.

SBB [18] considered three survival analysis methods, namely: Cox, SSVM, and RSF. They were chosen to represent a broad spectrum of baseline models including parametric (SSVM), semiparametric (Cox), linear (Cox, SSVM), and nonlinear (RSF) models. The Cox model and the RSF can only output risk scores, which can be used to address Task 1 by ranking ALS patients according to their risk of impairment, but do not provide a straightforward solution to predicting Task 2's time of impairment. To extend these approaches to Task 2, the predicted time of impairment for a given patient was selected as the median predicted time to impairment, i.e., the time at which the estimated survival function crossed the 0.5 threshold. Instead, the SSVM can be used either as a ranker or a time regressor depending on how the risk ratio hyperparameter is set during model training. Here, the SSVM was initially trained as a time regressor to address Task 2 directly. Then, its predicted times were converted into risk scores in the range [0–1], as requested by the challenge rules, via Platt scaling.

Task 3. Nunes et al. [15] proposes a novel approach that generates semantic similarity-based explanations for patient-level predictions. The underlying idea is to explain the prediction for one patient by considering aspect-oriented semantic similarity with other relevant patients based on the most important features used by ML approaches or selected by users. To build rich and easy to understand semantic-similarity based explanations, Nunes et al. developed five steps: (1) the enrichment of the Brainteaser Ontology [1] through integration of other biomedical ontologies; (2) the semantic annotation of patients (if not already available); (3) the similarity calculation between patients; (4) selection of the set of patients to explain a specific prediction; and (5) the visualization of the generated similarity-based explanations.

Buonocore et al. [4] trained a set of 4 well-known classifiers to predict death occurrence: Gradient Boosting (using XGB implementation), Random Forest, Logistic Regression and Multilayer perceptron. For the XAI methods Buonocore et al. focused our attention on three different methods for post-hoc, model-agnostic, local explainability, selecting SHAP, LIME and AraucanaXAI. Then, Buonocore et al. evaluated and compared XAI approaches in terms of a set of metrics defined in previous research on XAI in healthcare: *identity*: if there are two identical instances, they must have the same explanations; *fidelity*: concordance of the predictions between the XAI surrogate model and the original ML model; *separability*: if there are 2 dissimilar instances, they must have dissimilar explanations; *time*: average time required by the XAI method to output an explanation across the entire test set. The quantitative evaluation of the three different XAI methods did not reveal definitive superior performance of one of the approaches, albeit SHAP seems to be the better overall performing algorithm. However the explainability evaluation metrics are not all that is needed to thoroughly assess the multifaceted construct of what constitutes a "good" explanation in XAI in healthcare.

8   Conclusions and Future Work

iDPP⁝CLEF is a new pilot activity focusing on predicting the temporal progression of ALS and on the explainability of the AI algorithms for such prediction.

We developed 3 datasets containing anonymized patient data from two medical institutions, one in Turin and the other in Lisbon, for the prediction of NIV, PEG, or death.

Out of 43 registered participants, 5 managed to submit a total of 120 runs, evenly spread across the offered tasks. Participants adopted a range of approaches, including various types of survival analysis, also using deep learning techniques. For the XAI of the prediction algorithms they used both semantic-similarity based techniques and state-of-art post-hoc and model-agnostic XAI approaches.

For this initial iteration of the lab, iDPP⁝CLEF focus on ALS progression prediction. Possible, future cycles will be extended to *Multiple Sclerosis (MS)*, another chronic disease, impairing neurological functions. Moreover, we plan to extend the datasets to also include data from environmental sensor, e.g. concerning pollution.

Acknowledgments. The work reported in this paper has been partially supported by the BRAINTEASER (https://brainteaser.health/) project (contract n. GA101017598), as a part of the European Union's Horizon 2020 research and innovation programme.

References

1. Bettin, M., et al.: Deliverable 9.1 - Project ontology and terminology, including data mapper and RDF graph builder. BRAINTEASER, EU Horizon 2020, Contract N. GA101017598, December 2021. https://brainteaser.health/
2. Branco, R., et al.: Hierarchical modelling for ALS prognosis: predicting the progression towards critical events. In: Faggioli, G., Ferro, N., Hanbury, A., Potthast, M. (eds.) CLEF 2022 Working Notes. CEUR Workshop Proceedings. CEUR-WS.org (2022). ISSN 1613-0073
3. Brier, G.W.: Verification of forecasts expressed in terms of probability. Mon. Weather Rev. **78**(1), 1–3 (1950)
4. Buonocore, T.M., Nicora, G., Dagliati, A., Parimbelli, E.: Evaluation of XAI on ALS 6-months mortality prediction. In: Faggioli, G., Ferro, N., Hanbury, A., Potthast, M. (eds.) CLEF 2022 Working Notes. CEUR Workshop Proceedings. CEUR-WS.org (2022). ISSN 1613-0073
5. Cedarbaum, J.M., et al.: The ALSFRS-R: a revised ALS functional rating scale that incorporates assessments of respiratory function. J. Neurol. Sci. **169**(1–2), 13–21 (1999)
6. Chio, A., et al.: Prognostic factors in ALS: a critical review. Amyotrop. Lateral Sclerosis **10**(5–6), 310–323 (2009)
7. Faggioli, G., Ferro, N., Hanbury, A., Potthast, M. (eds.) CLEF 2022 Working Notes. CEUR Workshop Proceedings. CEUR-WS.org (2022). ISSN 1613-0073
8. Guazzo, A., et al.: Intelligent disease progression prediction: overview of iDPP@CLEF 2022. In: Barrón-Cedeño, A., et al. (eds.) Experimental IR Meets Multilinguality, Multimodality, and Interaction. Proceedings of the Thirteenth International Conference of the CLEF Association (CLEF 2022). LNCS, vol. 13390, pp. 386–413. Springer, Heidelberg (2022)
9. Guazzo, A., et al.: Overview of iDPP@CLEF 2022: the intelligent disease progression prediction challenge. In: Faggioli, G., Ferro, N., Hanbury, A., Potthast, M. (eds.) CLEF 2022 Working Notes. CEUR Workshop Proceedings. CEUR-WS.org (2022). ISSN 1613-0073
10. Hanley, J.A., McNeil, B.J.: The meaning and use of the area under a receiver operating characteristic (ROC) curve. Radiology **143**(1), 29–36 (1982). pMID: 7063747
11. Harrell, F.E., J., Califf, R.M., Pryor, D.B., Lee, K.L., Rosati, R.A.: Evaluating the yield of medical tests. JAMA **247**(18), 2543–2546 (1982). ISSN 0098-7484
12. Küffner, R., et al.: Crowdsourced analysis of clinical trial data to predict amyotrophic lateral sclerosis progression. Nat. Biotechnol. **33**(1), 51–57 (2015)
13. Longato, E., Vettoretti, M., Di Camillo, B.: A practical perspective on the concordance index for the evaluation and selection of prognostic time-to-event models. J. Biomed. Inform. **108**, 103496:1–103496:9 (2020)
14. Mannion, A., Chevalier, T., Schwab, D., Goeuriot, L.: Predicting the Risk of & time to impairment for ALS patients. In: Faggioli, G., Ferro, N., Hanbury, A., Potthast, M. (eds.) CLEF 2022 Working Notes. CEUR Workshop Proceedings. CEUR-WS.org (2022). ISSN 1613-0073

15. Nunes, S., et al.: Explaining artificial intelligence predictions of disease progression with semantic similarity. In: Faggioli, G., Ferro, N., Hanbury, A., Potthast, M. (eds.) CLEF 2022 Working Notes. CEUR Workshop Proceedings. CEUR-WS.org (2022). ISSN 1613-0073
16. Pancotti, C., Birolo, G., Sanavia, T., Rollo, C., Fariselli, P.: Multi-event survival prediction for amyotrophic lateral sclerosis. In: Faggioli, G., Ferro, N., Hanbury, A., Potthast, M. (eds.) CLEF 2022 Working Notes. CEUR Workshop Proceedings. CEUR-WS.org (2022). ISSN 1613-0073
17. Pencina, M.J., D'Agostino, R.B.: Overall C as a measure of discrimination in survival analysis: model specific population value and confidence interval estimation. Stat. Med. **23**(13), 2109–2123 (2004)
18. Trescato, I., et al.: Baseline machine learning approaches to predict amyotrophic lateral sclerosis disease progression. In: Faggioli, G., Ferro, N., Hanbury, A., Potthast, M. (eds.) CLEF 2022 Working Notes. CEUR Workshop Proceedings. CEUR-WS.org (2022). ISSN 1613-0073

Overview of HIPE-2022: Named Entity Recognition and Linking in Multilingual Historical Documents

Maud Ehrmann[1]([✉]) [iD], Matteo Romanello[2] [iD], Sven Najem-Meyer[1] [iD],
Antoine Doucet[3] [iD], and Simon Clematide[4] [iD]

[1] Digital Humanities Laboratory, EPFL, Vaud, Switzerland
{maud.ehrmann,sven.najem-meyer}@epfl.ch
[2] University of Lausanne, Lausanne, Switzerland
matteo.romanello@unil.ch
[3] University of La Rochelle, La Rochelle, France
antoine.doucet@univ-lr.fr
[4] Department of Computational Linguistics, University of Zurich, Zurich, Switzerland
simon.clematide@uzh.ch

Abstract. This paper presents an overview of the second edition of HIPE (Identifying Historical People, Places and other Entities), a shared task on named entity recognition and linking in multilingual historical documents. Following the success of the first CLEF-HIPE-2020 evaluation lab, HIPE-2022 confronts systems with the challenges of dealing with more languages, learning domain-specific entities, and adapting to diverse annotation tag sets. This shared task is part of the ongoing efforts of the natural language processing and digital humanities communities to adapt and develop appropriate technologies to efficiently retrieve and explore information from historical texts. On such material, however, named entity processing techniques face the challenges of domain heterogeneity, input noisiness, dynamics of language, and lack of resources. In this context, the main objective of HIPE-2022, run as an evaluation lab of the CLEF 2022 conference, is to gain new insights into the *transferability* of named entity processing approaches across languages, time periods, document types, and annotation tag sets. Tasks, corpora, and results of participating teams are presented.

Keywords: Named entity recognition and classification · Entity linking · Historical texts · Information extraction · Digitised newspapers · Digital humanities

1 Introduction

Through decades of massive digitisation, an unprecedented amount of historical documents became available in digital format, along with their machine-readable texts. While this represents a major step forward in terms of preservation and

© The Author(s), under exclusive license to Springer Nature Switzerland AG 2022
A. Barrón-Cedeño et al. (Eds.): CLEF 2022, LNCS 13390, pp. 423–446, 2022.
https://doi.org/10.1007/978-3-031-13643-6_26

accessibility, it also bears the potential for new ways to engage with historical documents' contents. The application of machine reading to historical documents is potentially transformative and the next fundamental challenge is to adapt and develop appropriate technologies to efficiently search, retrieve and explore information from this 'big data of the past' [21]. Semantic indexing of historical documents is in great demand among humanities scholars, and the interdisciplinary efforts of the digital humanities (DH), natural language processing (NLP), computer vision and cultural heritage communities are progressively pushing forward the processing of facsimiles, as well as the extraction, linking and representation of the complex information enclosed in transcriptions of digitised collections [28]. In this regard, information extraction techniques, and particularly named entity (NE) processing, can be considered among the first and most crucial processing steps.

Yet, the recognition, classification and disambiguation of NEs in historical texts is not straightforward, and performances are not on par with what is usually observed on contemporary well-edited English news material [8]. In particular, NE processing on historical documents faces the challenges of domain heterogeneity, input noisiness, dynamics of language, and lack of resources [9]. Although some of these issues have already been tackled in isolation in other contexts (with e.g., user-generated text), what makes the task particularly difficult is their simultaneous combination and their magnitude: texts are severely noisy, and domains and time periods are far apart.

Motivation and Objectives. As the first evaluation campaign of its kind on multilingual historical newspaper material, the CLEF-HIPE-2020 edition[1] [13, 14] proposed the tasks of NE recognition and classification (NERC) and entity linking (EL) in ca. 200 years of historical newspapers written in English, French and German. HIPE-2020 brought together 13 teams who submitted a total of 75 runs for 5 different task bundles. The main conclusion of this edition was that neural-based approaches can achieve good performances on historical NERC when provided with enough training data, but that progress is still needed to further improve performances, adequately handle OCR noise and small-data settings, and better address entity linking. HIPE-2022 attempts to drive further progress on these points, and also confront systems with new challenges. An additional point is that in the meantime several European cultural heritage projects have prepared additional NE-annotated text material, thus opening a unique window of opportunity to organize a second edition of the HIPE evaluation lab in 2022.

HIPE-2022[2] shared task focuses on named entity processing in historical documents covering the period from the 18th to the 20th century and featuring several languages. Compared to the first edition, HIPE-2022 introduces several novelties:

---

[1] https://impresso.github.io/CLEF-HIPE-2020.
[2] https://hipe-eval.github.io/HIPE-2022/.

- the addition of a new type of document alongside historical newspapers, namely classical commentaries[3];
- the consideration of a broader language spectrum, with 5 languages for historical newspapers (3 for the previous edition), and 3 for classical commentaries;
- the confrontation with heterogeneous annotation tag sets and guidelines.

Overall, HIPE-2022 confronts participants with the challenges of dealing with more languages, learning domain-specific entities, and adapting to diverse annotation schemas. The objectives of the evaluation lab are to contribute new insights on how best to ensure the transferability of NE processing approaches across languages, time periods, document and annotation types, and to answer the question whether one architecture or model can be optimised to perform well across settings and annotation targets in a cultural heritage context. In particular, the following research questions are addressed:

1. How well can general prior knowledge transfer to historical texts?
2. Are in-domain language representations (i.e. language models learned on the historical document collections) beneficial, and under which conditions?
3. How can systems adapt and integrate training material with different annotations?
4. How can systems, with limited additional in-domain training material, (re)-target models to produce a certain type of annotation?

Recent work on NERC showed encouraging progress on several of these topics: Beryozkin et al. [3] proposed a method to deal with related, but heterogeneous tag sets. Several researchers successfully applied meta-learning strategies to NERC to improve transfer learning: Li et al. [23] improved results for extreme low-resource few-shot settings where only a handful of annotated examples for each entity class are used for training; Wu et al. [36] presented techniques to improve cross-lingual transfer; and Li et al. [24] tackled the problem of domain shifts and heterogeneous label sets using meta-learning, proposing a highly data-efficient domain adaptation approach.

The remainder of this paper is organized as follows. Sections 2 and 3 present the tasks and the material used for the evaluation. Section 4 details the evaluation framework, with evaluation metrics and the organisation of system submissions around tracks and challenges. Section 5 introduces the participating systems, while Sect. 6 presents and discusses their results. Finally, Sect. 7 summarizes the benefits of the task and concludes.[4]

---

[3] Classical commentaries are scholarly publications dedicated to the in-depth analysis and explanation of ancient literary works. As such, they aim to facilitate the reading and understanding of a given literary text.

[4] For space reasons, the discussion of related work is included in the extended version of this overview [15].

Table 1. Overview of HIPE-2022 datasets with an indication of which tasks they are suitable for according to their annotation types.

Dataset alias	Document type	Languages	Suitable for
hipe2020	historical newspapers	de, fr, en	NERC-Coarse, NERC-Fine, EL
newseye	historical newspapers	de, fi, fr, sv	NERC-Coarse, NERC-Fine, EL
sonar	historical newspapers	de	NERC-Coarse, EL
letemps	historical newspapers	fr	NERC-Coarse, NERC-Fine
topres19th	historical newspapers	en	NERC-Coarse, EL
ajmc	classical commentaries	de, fr, en	NERC-Coarse, NERC-Fine, EL

2 Task Description

HIPE-2022 focuses on the same tasks as CLEF-HIPE-2020, namely:

Task 1: Named Entity Recognition and Classification (NERC)

– **Subtask NERC-Coarse:** this task includes the recognition and classification of high-level entity types (person, organisation, location, product and domain-specific entities, e.g. mythological characters or literary works in classical commentaries).
– **Subtask NERC-Fine:** includes the recognition and classification of entity mentions according to fine-grained types, plus the detection and classification of nested entities of depth 1. This subtask is proposed for English, French and German only.

Task 2: Named Entity Linking (EL). This task corresponds to the linking of named entity mentions to a unique item ID in Wikidata, our knowledge base of choice, or to a NIL value if the mention does not have a corresponding item in the knowledge base (KB). We will allow submissions of both end-to-end systems (NERC and EL) and of systems performing exclusively EL on gold entity mentions provided by the organizers (EL-only).

3 Data

HIPE-2022 data consists of six NE-annotated datasets composed of historical newspapers and classic commentaries covering ca. 200 years. Datasets originate from the previous HIPE-2020 campaign, from HIPE organisers' previous research project, and from several European cultural heritage projects which agreed to postpone the publication of 10% to 20% of their annotated material to support HIPE-2022. Original datasets feature several languages and were annotated with different entity tag sets and according to different annotation guidelines. See Table 1 for an overview.

3.1 Original Datasets

Historical Newspapers. The historical newspaper data is composed of several datasets in English, Finnish, French, German and Swedish which originate from various projects and national libraries in Europe:

- **HIPE-2020 data** corresponds to the datasets of the first HIPE-2020 campaign. They are composed of articles from Swiss, Luxembourgish and American newspapers in French, German and English (19C-20C) that were assembled during the *impresso* project[5] [10]. Together, the train, dev and test hipe2020 datasets contain 17,553 linked entity mentions, classified according to a fine-grained tag set, where nested entities, mention components and metonymic senses are also annotated [12].
- **NewsEye data** corresponds to a set of NE-annotated datasets composed of newspaper articles in French, German, Finnish and Swedish (19C-20C) [18]. Built in the context of the *NewsEye* project[6], the newseye train, dev and test sets contain 36,790 linked entity mentions, classified according to a coarse-grained tag set and annotated on the basis of guidelines similar to the ones used for hipe2020. Roughly 20% of the data was retained from the original dataset publication and is published for the first time for HIPE-2022, where it is used as test data (thus the previously published test set became a second dev set in HIPE-2022 data distribution).
- **SoNAR data** is an NE-annotated dataset composed of newspaper articles from the Berlin State library newspaper collections in German (19C-20C), produced in the context of the *SoNAR* project[7]. The sonar dataset contains 1,125 linked entity mentions, classified according to a coarse-grained tag set. It was thoroughly revised and corrected on NE and EL levels by the HIPE-2022 organisers. It is split in a dev and test set – without providing a dedicated train set.
- **Le Temps data:** a previously unpublished, NE-annotated diachronic dataset composed of historical newspaper articles from two Swiss newspapers in French (19C-20C) [8]. This dataset contains 11,045 entity mentions classified according to a fine-grained tag set similar to hipe2020.
- **Living with Machines data** corresponds to an NE-annotated dataset composed of newspaper articles from the British Library newspapers in English (18C-19C) and assembled in the context of the *Living with Machine* project[8]. The topres19th dataset contains 4,601 linked entity mentions, exclusively of geographical types annotated following their own annotation guidelines [5]. Part of this data has been retained from the original dataset publication and is used and released for the first time for HIPE-2022.

---

[5] https://impresso-project.ch.

[6] https://www.newseye.eu/.

[7] https://sonar.fh-potsdam.de/.

[8] https://livingwithmachines.ac.uk/.

Table 2. Entity types used for NERC tasks, per dataset and with information whether nesting and linking apply. ∗: these types are not present in `letemps` data. ∗∗: linking applies, unless the token is flagged as *InSecondaryReference*.

Dataset	Coarse tag set	Fine tag set	Nesting	Linking
hipe2020 letemps	pers	pers.ind	yes	yes
		pers.coll		
		pers.ind.articleauthor		
	org∗	org.adm	yes	yes
		org.ent		
		org.ent.pressagency		
	prod∗	prod.media	no	yes
		prod.doctr		
	time∗	time.date.abs	no	no
	loc	loc.adm.town	yes	yes
		loc.adm.reg		
		loc.adm.nat		
		loc.adm.sup		
		loc.phys.geo	yes	yes
		loc.phys.hydro		
		loc.phys.astro		
		loc.oro	yes	yes
		loc.fac	yes	yes
		loc.add.phys	yes	yes
		loc.add.elec		
		loc.unk	no	no
newseye	pers	pers.articleauthor	yes	yes
	org	-	yes	yes
	humanprod	-	yes	yes
	loc	-	no	yes
topres19th	loc	-	no	yes
	building	-	no	yes
	street	-	no	yes
∗ajmc	pers	pers.author	yes	yes∗∗
		pers.editor		
		pers.myth		
		pers.other		
	work	work.primlit	yes	yes∗∗
		work.seclit		
		work.fragm		
	loc	-	yes	yes∗∗
	object	object.manuscr	yes	no
		object.museum		
	date	-	yes	no
	scope	-	yes	no
sonar	pers	-	no	yes
	loc	-	no	yes
	org	-	no	yes

Historical Commentaries. The classical commentaries data originates from the *Ajax Multi-Commentary* project and is composed of OCRed 19C commentaries published in French, German and English [30], annotated with both universal NEs (person, location, organisation) and domain-specific NEs (bibliographic references to primary and secondary literature). In the field of classical studies, commentaries constitute one of the most important and enduring forms of scholarship, together with critical editions and translations. They are information-rich texts, characterised by a high density of NEs.

These six datasets compose the HIPE-2022 corpus. They underwent several preparation steps, with conversion to the tab-separated HIPE format, correction of data inconsistencies, metadata consolidation, re-annotation of parts of the datasets, deletion of extremely rare entities (esp. for topres19th), and rearrangement or composition of train and dev splits[9].

3.2   Corpora Characteristics

Overall, the HIPE-2022 corpus covers five languages (English, French, Finnish, German and Swedish), with a total of over 2.3 million tokens (2,211,449 for newspapers and 111,218 for classical commentaries) and 78,000 entities classified according to five different entity typologies and linked to Wikidata records. Detailed statistics about the datasets are provided in Table 3 and 4.

The datasets in the corpus are quite heterogeneous in terms of annotation guidelines. Two datasets – hipe2020 and letemps – follow the same guidelines [12,31], and newseye was annotated using a slightly modified version of these guidelines. In the sonar dataset, persons, locations and organisations were annotated, whereas in topres19th only toponyms were considered. Compared to the other datasets, ajmc stands out for having being annotated according to domain-specific guidelines [29], which focus on bibliographic references to primary and secondary literature. This heterogeneity of guidelines leads to a wide variety of entity types and sub-types for the NERC task (see Table 3.1). Among these types, only persons, locations and organisations are found in all datasets (except for topres19th). While nested entities are annotated in all datasets except topres19th and sonar, only hipe2020 and newseye have a sizable number of such entities.

Detailed information about entity mentions that are affected by OCR mistakes is provided in ajmc and hipe2020 (only for the test set for the latter). As OCR noise constitutes one of the main challenges of historical NE processing [9], this information can be extremely useful to explain differences in performance between datasets or between languages in the same dataset. For instance, looking at the percentage of noisy mentions for the different languages in ajmc, we find that it is three times higher in French documents than in the other two languages.

---

[9] Additional information is available online by following the links indicated for each datasets in Table 1.

Table 3. Overview of newspaper corpora statistics (hipe-2022 release v2.1). NIL percentages are computed based on linkable entities (i.e., excluding time entities for hipe2020).

Dataset	Lang.	Fold	Docs	Tokens	Mentions All	Fine	Nested	%noisy	%NIL
hipe2020	de	Train	103	86,446	3,494	3,494	158	-	15.70
		Dev	33	32,672	1,242	1,242	67	-	18.76
		Test	49	30,738	1,147	1,147	73	12.55	17.40
		Total	185	149,856	5,883	5,883	298	-	16.66
	en	Train	-	-	-	-	-	-	-
		Dev	80	29,060	966	-	-	-	44.18
		Test	46	16,635	449	-	-	5.57	40.28
		Total	126	45,695	1,415	-	-	-	42.95
	fr	Train	158	166,218	6,926	6,926	473	-	25.26
		Dev	43	37,953	1,729	1,729	91	-	19.81
		Test	43	40,855	1,600	1,600	82	11.25	20.23
		Total	244	245,026	10,255	10,255	646	-	23.55
Total			555	440,577	17,553	16,138	944	-	22.82
newseye	de	Train	7	374,250	11,381	21	876	-	51.07
		Dev	12	40,046	539	5	27	-	22.08
		Dev2	12	39,450	882	4	64	-	53.74
		Test	13	99,711	2,401	13	89	-	48.52
		Total	44	553,457	15,203	43	1,056	-	49.79
	fi	Train	24	48,223	2,146	15	224	-	40.31
		Dev	24	6,351	223	1	25	-	40.36
		Dev2	21	4,705	203	4	22	-	42.86
		Test	24	14,964	691	7	42	-	47.47
		Total	93	74,243	3,263	27	313	-	41.99
	fr	Train	35	255,138	10,423	99	482	-	42.42
		Dev	35	21,726	752	3	29	-	30.45
		Dev2	35	30,457	1,298	10	63	-	38.91
		Test	35	70,790	2,530	34	131	-	44.82
		Total	140	378,111	15,003	146	705	-	41.92
	sv	Train	21	56,307	2,140	16	110	-	32.38
		Dev	21	6,907	266	1	7	-	25.19
		Dev2	21	6,987	311	1	20	-	37.30
		Test	21	16,163	604	0	26	-	35.43
		Total	84	86,364	3,321	18	163	-	32.82
Total			361	1,092,175	36,790	234	2,237	-	44.36
letemps	fr	Train	414	379,481	9,159	9,159	69	-	-
		Dev	51	38,650	869	869	12	-	-
		Test	51	48,469	1,017	1,017	12	-	-
Total			516	466,600	11,045	11,045	93	-	-
topres19th	en	Train	309	123,977	3,179	-	-	-	18.34
		Dev	34	11,916	236	-	-	-	13.98
		Test	112	43,263	1,186	-	-	-	17.2
		Total	455	179,156	4,601	-	-	-	17.82
Total			455	179,156	4,601	-	-	-	17.82
sonar	de	Train	-	-	-	-	-	-	-
		Dev	10	17,477	654	-	-	-	22.48
		Test	10	15,464	471	-	-	-	33.33
		Total	20	32,941	1,125	-	-	-	27.02
Total			20	32,941	1,125	-	-	-	27.02
Grand Total (newspapers)			**1,907**	**2,211,449**	**71,114**	**27,417**	**3,274**	-	**30.23**

Table 4. Corpus statistics for the `ajmc` dataset (HIPE-2022 release v2.1).

Dataset	Lang.	Fold	Docs	Tokens	Mentions				
					All	Fine	Nested	%noisy	%NIL
`ajmc`	**de**	Train	76	22,694	1,738	1,738	11	13.81	0.92
		Dev	14	4,703	403	403	2	11.41	0.74
		Test	16	4,846	382	382	0	10.99	1.83
	Total		106	32,243	2,523	2,523	13	13.00	1.03
	en	Train	60	30,929	1,823	1,823	4	10.97	1.66
		Dev	14	6,507	416	416	0	16.83	1.70
		Test	13	6,052	348	348	0	10.34	2.61
	Total		87	43,488	2,587	2,587	4	11.83	1.79
	fr	Train	72	24,670	1,621	1,621	9	30.72	0.99
		Dev	17	5,426	391	391	0	36.32	2.56
		Test	15	5,391	360	360	0	27.50	2.80
	Total		104	35,487	2,372	2,372	9	31.16	1.52
Grand Total (ajmc)			**297**	**111,218**	**7,482**	**7,482**	**26**		**1.45**

HIPE-2022 datasets show significant differences in terms of lexical overlap between train, dev and test sets. Following the observations of Augenstein et al. [2] and Taillé et al. [32] on the impact of lexical overlap on NERC performance, we computed the percentage of mention overlap between data folds for each dataset, based on the number of identical entity mentions (in terms of surface form) between train+dev and test sets (see Table 5). Evaluation results obtained on training and test sets with low mention overlap, for example, can be taken as an indicator of the ability of the models to generalise well to unseen mentions. We find that `ajmc`, `letemps` and `topres19th` have a mention overlap which is almost twice that of `hipe2020`, `sonar` and `newseye`.

Finally, regarding entity linking, it is interesting to observe that the percentage of NIL entities (i.e. entities not linked to Wikidata) varies substantially across datasets. The Wikidata coverage is drastically lower for `newseye` than for the other newspaper datasets (44.36%). Conversely, only 1.45% of the entities found in `ajmc` cannot be linked to Wikidata. This fact is not at all surprising considering that commentaries mention mostly mythological figures, scholars of the past and literary works, while newspapers mention many relatively obscure or unknown individuals, for whom no Wikidata entry exists.

3.3   HIPE-2022 Releases

HIPE-2022 data is released as a single package consisting of the neatly structured and homogeneously formatted original datasets. The data is released in IOB format with hierarchical information, similarly to CoNLL-U[10], and consists

---

[10] https://universaldependencies.org/format.html.

Table 5. Overlap of mentions between test and train (plus dev) sets as percentage of the total number of mentions.

Dataset	Lang.	% overlap	Folds
ajmc	de	31.43	train+dev vs test
	en	30.50	train+dev vs test
	fr	27.53	train+dev vs test
	Total	29.87	
hipe2020	de	16.22	train+dev vs test
	en	6.22	dev vs test
	fr	19.14	train+dev vs test
	Total	17.12	
letemps	fr	25.70	train+dev vs test
sonar	de	10.13	dev vs test
newseye	fr	14.79	train+dev vs test
	de	20.77	train+dev vs test
	fi	6.63	train+dev vs test
	sv	10.36	train+dev vs test
	Total	16.18	
topres19th	en	32.33	train+dev vs test

of UTF-8 encoded, tab-separated values (TSV) files containing the necessary information for all tasks (NERC-Coarse, NERC-Fine, and EL). There is one TSV file per dataset, language and split. Original datasets provide different document metadata with different granularity. This information is kept in the files in the form of metadata blocks that encode as much information as necessary to ensure that each document is self-contained with respect to HIPE-2022 settings. Metadata blocks use namespacing to distinguish between mandatory shared task metadata and dataset-specific metadata.

HIPE-2022 data releases are published on the HIPE-eval GitHub organisation repository[11] and on Zenodo[12]. Various licences (of type CC-BY and CC-BY-NC-SA) apply to the original datasets – we refer the reader to the online documentation.

4 Evaluation Framework

4.1 Task Bundles, Tracks and Challenges

To accommodate the different dimensions that characterise the HIPE-2022 shared task (languages, document types, entity tag sets, tasks) and to foster research on transferability, the evaluation lab is organised around **tracks** and **challenges**. Challenges guide participation towards the development of

---

[11] https://github.com/impresso/CLEF-HIPE-2020/tree/master/data.
[12] https://doi.org/10.5281/zenodo.6579950.

approaches that work across settings, e.g. with documents in at least two different languages or annotated according to two different tag sets or guidelines, and provide a well-defined and multi-perspective evaluation frame.

To manage the total combinations of datasets, languages, document types and tasks, we defined the following elements (see also Fig. 1):

- **Task bundle:** a task bundle is a predefined set of tasks as in HIPE-2020 (see bundle table in Fig. 1). Task bundles offer participating teams great flexibility in choosing which tasks to compete for, while maintaining a manageable evaluation frame. Concretely, teams were allowed to submit several 'submission bundles', i.e. a triple composed of dataset/language/taskbundle, with up to 2 runs each.
- **Track:** a track corresponds to a triple composed of dataset/language/task and forms the basic unit for which results are reported.
- **Challenge:** a challenge corresponds to a predefined set of tracks. A challenge can be seen as a kind of tournament composed of tracks.

HIPE-2022 specifically evaluates 3 challenges:

1. **Multilingual Newspaper Challenge** (MNC): This challenge aims at fostering the development of multilingual NE processing approaches on historical newspapers. The requirements for participation in this challenge are that submission bundles consist only of newspaper datasets and include at least two languages for the same task (so teams had to submit a minimum of two submission bundles for this challenge).
2. **Multilingual Classical Commentary Challenge** (MCC): This challenge aims at adapting NE solutions to domain-specific entities in a specific digital humanities text type of classic commentaries. The requirements are that submission bundles consist only of the `ajmc` dataset and include at least three languages for the same task.
3. **Global Adaptation Challenge** (GAC): Finally, the global adaptation challenge aims at assessing how efficiently systems can be retargeted to any language, document type and guidelines. Bundles submitted for this challenge could be the same as those submitted for MNC and MCC challenges. The requirements are that they consist of datasets of both types (commentaries and newspaper) and include at least two languages for the same task.

4.2   Evaluation Measures

As in HIPE-2020, NERC and EL tasks are evaluated in terms of Precision, Recall and F-measure (F1-score) [25]. Evaluation is carried out at entity level according to two computation schemes: micro average, based on true positives, false positives, and false negative figures computed over all documents, and macro average, based on averages of micro figures per document. Our definition of macro differs from the usual one: averaging is done at document level and not at entity type level. This allows to account for variance in document length and

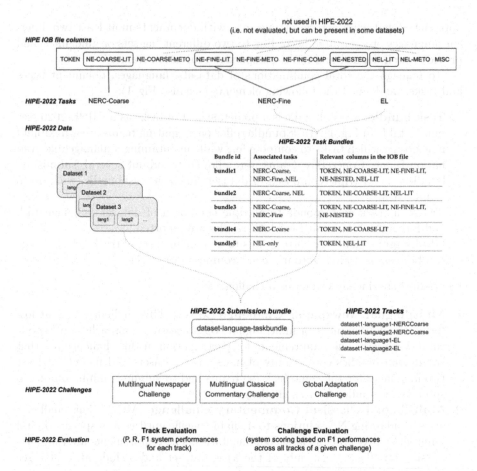

Fig. 1. Overview of HIPE-2022 evaluation setting.

entity distribution within documents and avoids distortions that would occur due to the unevenly distributed entity classes.

Both NERC and EL benefit from strict and fuzzy evaluation regimes, depending on how strictly entity type and boundaries correctness are judged. For NERC (Coarse and Fine), the strict regime corresponds to exact type and boundary matching, and the fuzzy to exact type and overlapping boundaries. It is to be noted that in the strict regime, predicting wrong boundaries leads to a 'double' punishment of one false negative (entity present in the gold standard but not predicted by the system) and one false positive (entity predicted by the system but not present in the gold standard). Although it penalizes harshly, we keep this metric to be consistent with CoNLL and refer to the fuzzy regime when boundaries are of less importance.

The definition of strict and fuzzy regimes differs for entity linking. In terms of boundaries, EL is always evaluated according to overlapping boundaries in both regimes (what is of interest is the capacity to provide the correct link rather than the correct boundaries). EL strict regime considers only the system's top link prediction (NIL or Wikidata QID), while the fuzzy regime expands system predictions with a set of historically related entity QIDs. For example, "Germany" QID is complemented with the QID of the more specific "Confederation of the Rhine" entity and both are considered as valid answers. The resource allowing for such historical normalization was compiled by the task organizers for the entities of the test data sets (for `hipe2020` and `ajmc` datasets), and are released as part of the HIPE-scorer. For this regime, participants were invited to submit more than one link, and F-measure is additionally computed with cut-offs @3 and @5 (meaning, counting a true positive if the ground truth QID can be found within the first 3 or 5 candidates).

4.3   System Evaluation, Scorer and Evaluation Toolkit

Teams were asked to submit system responses based on submission bundles and to specify at least one challenge to which their submitted bundles belong. Micro and macro scores were computed and published for each track, but only micro figures are reported here.

The evaluation of challenges, which corresponds to an aggregation of tracks, was defined as follows: given a specific challenge and the tracks submitted by a team for this challenge, the submitted systems are rewarded points according to their F1-based rank for each track (considering only the best of the submitted runs for a given track). The points obtained are summed over all submitted tracks, and systems/teams are ranked according to their total points. Further details on system submission and evaluation can be found in the HIPE Participation Guidelines [11].

The evaluation is performed using the **HIPE-scorer**[13]. Developed during the first edition of HIPE, the scorer has been improved with minor bug fixes and additional parameterisation (input format, evaluation regimes, HIPE editions). Participants could use the HIPE-scorer when developing their systems. After the evaluation phase, a complete **evaluation toolkit** was also released, including the data used for evaluation (v2.1), the system runs submitted by participating teams, and all the evaluation recipes and resources (e.g. historical mappings) needed to replicate the present evaluation[14].

5   System Descriptions

In this second HIPE edition, 5 teams submitted a total of 103 system runs. Submitted runs do not cover all of the 35 possible tracks (dataset/language/task combinations), nevertheless we received submission for all datasets, with most of them focusing on NERC-Coarse.

---

[13] https://github.com/hipe-eval/HIPE-scorer.
[14] https://github.com/hipe-eval/HIPE-2022-eval.

5.1   Baselines

As a neural baseline (**Neur-bsl**) for NERC-Coarse and NERC-fine, we fine-tuned separately for each HIPE-2022 dataset XLM-R$_{BASE}$, a multilingual transformer-based language representation model pre-trained on 2.5TB of filtered CommonCrawl texts [6]. The models are implemented using HuggingFace[15] [35]. Since transformers rely primarily on subword-based tokenisers, we chose to label only the first subwords. This allows to map the model outputs to the original text more easily. Tokenised texts are split into input segments of length 512. For each HIPE-2022 dataset, fine-tuning is performed on the train set (except for `sonar` and `hipe2020-en` which has only dev sets) for 10 epochs using the default hyperparameters (Adam $\epsilon = 10e-8$, Learning rate $\alpha = 5e-5$). The code of this baseline (configuration files, scripts) is published in a dedicated repository on the HIPE-eval Github organisation[16], and results are published in the evaluation toolkit.

For entity linking in EL-only setting, we provide the NIL baseline (**Nil-bsl**), where each entity link is replaced with the NIL value.

5.2   Participating Systems

The following system descriptions are compiled from information provided by the participants. More details on the implementation and results can be found in the system papers of the participants [16].

Team **L3i**, affiliated with *La Rochelle University* and with the *University of Toulouse*, France, successfully tackled an impressive amount of multilingual newspaper datasets with strong runs for NERC-coarse, NERC-fine and EL. For the classical commentary datasets (`ajmc`) the team had excellent results for NERC[17]. For NERC, L3i – the winning team in HIPE's 2020 edition – builds on their transformer-based approach [4]. Using transformer-based adapters [19], parameter-efficient fine-tuning in a hierarchical multitask setup (NERC-coarse and NERC-fine) has been shown to work well with historical noisy texts [4]. The innovation for this year's submission lies in the addition of context information in the form of external knowledge from two sources (inspired by [34]). First, French and German Wikipedia documents based on dense vector representations computed by a multilingual Sentence-BERT model [27], including a k-Nearest-Neighbor search functionality provided by ElasticSearch framework. Second, English Wikidata knowledge graph (KG) embeddings that are combined with the first paragraph of English Wikipedia pages (Wikidata5m) [33]. For the knowledge graph embeddings, two methods are tested on the HIPE-2022 data: 1) the one-stage KG Embedding Retrieval Module that retrieves top-k KG "documents" (in this context, a document is an ElasticSearch retrieval unit that consists of an entity identifier, an entity description and an entity embedding) via vector similarity on the dense entity embedding vector space; 2) the two-stage

---

[15] https://github.com/huggingface/transformers/.

[16] https://github.com/hipe-eval/HIPE-2022-baseline/.

[17] The EL results for `ajmc` were low, probably due to some processing issues.

KG Embedding Retrieval Module that retrieves the single top similar document first and then in a second retrieval step gets the k most similar documents based on that first entity. All context enrichment techniques work by simply concatenating the original input segment with the retrieved context segments and processing the contextualized segments through their "normal" hierarchical NER architecture. Since the L3i team's internal evaluation on HIPE-2022 data (using a multilingual BERT base pre-trained model) indicated that the two-stage KG retrieval was the best context generator overall, it was used for one of the two officially submitted runs. The other "baseline" run did not use any context enrichment techniques. Both runs additionally used stacked monolingual BERT embeddings for English, French and German, for the latter two languages in the form of Europeana models that were built from digitized historical newspaper text material. Even with improved historical monolingual BERT embeddings, the context-enriched run was consistently better in terms of F1-score in NERC-Coarse and -Fine settings.

Team **histeria**, affiliated with the *Bayerische Staatsbibliothek München*, Germany, the *Digital Philology* department of the University of Vienna, Austria and the *NLP Expert Center, Volkswagen AG* Munich, Germany, focused on the `ajmc` dataset for their NERC-coarse submission (best results for French and English, second best for German), but also provides experimental results for all languages of the `newseye` datasets[18]. Their NER tagging experiments tackle two important questions:

a) How to build an optimal multilingual pre-trained BERT language representation model for historical OCRized documents? They propose and release hmBERT[19], which includes English, Finnish, French, German and Swedish in various model and vocabulary sizes, and specifically apply methods to deal with OCR noise and imbalanced corpus sizes per language. In the end, roughly 27 GB of text per language is used in pre-training.

b) How to fine-tune a multilingual pre-trained model given comparable NER annotations in multiple languages? They compare a single-model approach (training models separately for each language) with a one-model approach (training only one model that covers all languages). The results indicate that, most of the time, the single-model approach works slightly better, but the difference may not be large enough to justify the considerably greater effort to train and apply the models in practice.

HISTeria submitted two runs for each `ajmc` datasets, using careful hyperparameter grid search on the dev sets in the process. Both runs build on the one-model approach in a first multilingual fine-tuning step. Similar to [34], they build monolingual models by further fine-tuning on language-specific training

---

[18] Note that these experiments are evaluated using the officially published *Newseye* test sets [18] (released as dev2 dataset as part of HIPE-2022) and not the HIPE-2022 `newseye` test sets, which were unpublished prior to the HIPE 2022 campaign.

[19] For English data, they used the Digitised Books. c. 1510 - c. 1900, all other languages use Europeana newspaper text data.

data[20]. Run 1 of their submission is based on hmBERT with vocabulary size 32k, while run 2 has a vocabulary size of 64k. Somewhat unexpectedly, the larger vocabulary does not improve the results in general on the development set. For the test set, though, the larger vocabulary model is substantially better overall. Similar to the team L3I, HISTeria also experimented with context enrichment techniques suggested by [34]. However, for the specific domain of classical commentaries, general-purpose knowledge bases such as Wikipedia could not improve the results. Interestingly, L3I also observed much less improvement with Wikipedia context enrichment on `ajmc` in comparison to the `hipe2020` newspaper datasets. In summary, HISTeria outperformed the strong neural baseline by about 10 F1-score percentage points in strict boundary setting, thereby demonstrating the importance of carefully constructed domain-specific pre-trained language representation models.

Team **Aauzh**, affiliated with *University of Zurich*, Switzerland and *University of Milan*, Italy, focused on the multilingual newspaper challenge in NERC-coarse setting and experimented with 21 different monolingual and multilingual, as well as contemporary and historical transformer-based language representation models available on the HuggingFace platform. For fine-tuning, they used the standard token classification head of the transformer library for NER tagging with default hyperparameters and trained each dataset for 3 epochs. In a preprocessing step, token-level NER IOB labels were mapped onto all subtokens. At inference time, a simple but effective summing pooling strategy for NER for aggregating subtoken-level to token-level labels was used [1]. Run 2 of AAUZH are the predictions of the best single model. Run 1 is the result of a hard-label ensembling from different pre-trained models: in case of ties between O and B/I labels, the entity labels were preferred. The performance of the submitted runs varies strongly in comparison with the neural baseline: for German and English it generally beats the baseline clearly for `hipe2020` and `sonar` datasets, but suffers on French `hipe2020` and German/Finnish `newseye` datasets. This again indicates that in transfer learning approaches to historical NER, the selection of pre-trained models has a considerable impact. The team also performed some post-submission experiments to investigate the effect of design choices: Applying soft-label ensembling using averaged token-level probabilities turned out to improve results on the French `newseye` datasets by 1.5% point in micro average and 2.4 points in macro average (F1-score). For all languages of the `newseye`, they also tested a one-model approach with multilingual training. The best multilingual dbmdz Europeana BERT model had a better performance on average (58%) than the best monolingual models (56%). However, several other multilingual pre-trained language models had substantially worse performance, resulting in 57% ensemble F1-score (5 models), which was much lower than 67% achieved by the monolingual ensemble.

Team **Sbb**, affiliated with the *Berlin State Library*, Germany, participated exclusively in the EL-only subtask, but covered all datasets in English, German and French. Their system builds on models and methods developed in the HIPE-

---

[20] This improves the results by 1.2% on average on the HIPE-2022 data.

2020 edition [22]. Their approach uses Wikipedia sentences with an explicit link to a Wikipedia page as textual representations of its connected Wikidata entity. The system makes use of the metadata of the HIPE-2022 documents to exclude entities that were not existing at the time of its publication. Going via Wikipedia reduces the amount of accessible Wikidata IDs, however, for all datasets but `ajmc` the coverage is still 90%. Given the specialised domain of `ajmc`, a coverage of about 55% is to be accepted. The entity linking is done in the following steps: a) A candidate lookup retrieves a given number of candidates (25 for submission run 1, 50 for submission run 2) using a nearest neighbour index based on word embeddings of Wikipedia page titles. An absolute cut-off value is used to limit the retrieval (0.05 for submission 1 and 0.13 for submission 2). b) A probabilistic candidate sentence matching is performed by pairwise comparing the sentence with the mention to link and a knowledge base text snippet. To this end, a BERT model was fine-tuned on the task of whether or not two sentences mention the same entity. c) The final ranking of candidates includes the candidate sentence matching information as well as lookup features from step (a) and more word embedding information from the context. A random forest model calculates the overall probability of a match between the entity mention and an entity linking candidate. If the probability of a candidate is below a given threshold (0.2 for submission run 1 and 2), it is discarded. The random forest model was trained on concatenated training sets of the same language across datasets.

There are no conclusive insights from HIPE-2022 EL-only results whether run 1 or 2 settings are preferable. Post-submission experiments in their system description paper investigate the influence of specific hyperparameter settings on the system performances.

Team **WLV**, University of Wolverhampton, UK, applied classical BERT-based [7] as well as BiLSTM-based NER architectures [20] with a CRF layer to HIPE-2022 English and French newspaper datasets `hipe2020`, `letemps`, and `topres19th` in the NERC-coarse subtask.

6   Results and Discussion

We report results for the best run of each team and consider micro Precision, Recall and F1-score exclusively. Results for NERC-Coarse and NERC-Fine for all languages and datasets according to both evaluation regimes are presented in Table 6 and 7 respectively. Table 8 reports performances for EL-only, with a cut-off @1. We refer the reader to the HIPE-2022 website and the evaluation toolkit for more detailed results[21], and to the extended overview paper for further discussion of the results [15].

---

[21] See https://hipe-eval.github.io/HIPE-2022 and https://github.com/hipe-eval/HIPE-2022-eval.

Table 6. Results for NERC-Coarse (micro P, R and F1-score). Bold font indicates the highest, and underlined font the second-highest value.

	Strict			Fuzzy			Strict			Fuzzy			Strict			Fuzzy		
	P	R	F	P	R	F	P	R	F	P	R	F	P	R	F	P	R	F
								hipe2020										
			French						German							English		
Aauzh	.718	.675	.696	.825	.776	.800	.716	.735	.725	.812	.833	**.822**	.538	490	.513	.726	.661	.692
L3i	.786	.831	**.808**	.883	.933	**.907**	.784	.805	**.794**	.865	.888	**.876**	.624	.617	**.620**	.793	.784	**.788**
Wlv	.640	.712	.674	.767	.853	.808	-	-	-	-	-	-	.400	.430	.414	.582	.626	.603
Neur-bsl	.730	.785	.757	.836	.899	.866	.665	.746	.703	.750	.842	.793	.432	.532	.477	.564	.695	.623
								letemps				**sonar**				**topRes19th**		
			French						German							English		
Aauzh	.589	.710	.644	.642	.773	.701	.512	.548	**.529**	.655	.741	**.695**	.816	.760	**.787**	.869	.810	**.838**
Wlv	.581	.659	.618	.627	.711	.666	-	-	-	-	-	-	.712	.771	.740	.765	.829	.796
Neur-bsl	.595	.744	**.661**	.639	.800	**.711**	.267	.361	.307	.410	.554	.471	.747	.782	.764	.798	.836	.816
								ajmc										
			French						German							English		
HISTeria	.834	.850	**.842**	.874	.903	**.888**	.930	.898	.913	.938	.953	.945	.826	.885	**.854**	.879	.943	**.910**
L3i	.810	.842	.826	.856	.889	.872	.946	.921	**.934**	.965	.940	**.952**	.824	.876	.850	.868	.922	.894
Neur-bsl	.707	.778	.741	.788	.867	.825	.792	.846	.818	.846	.903	.873	.680	.802	.736	.766	.902	.828
								newseye										
			French						German									
Aauzh	.655	.657	**.656**	.785	.787	**.786**	.395	.421	.408	.480	.512	.495						
Neur-bsl	.634	.676	.654	.755	.805	.779	.429	.537	**.477**	.512	.642	**.570**						
			Finnish						Swedish									
Aauzh	.618	.524	.567	.730	.619	.670	.686	.604	.643	.797	.702	.746						
Neur-bsl	.605	.687	**.644**	.715	.812	**.760**	.588	.728	**.651**	.675	.836	**.747**						

Table 7. Results for NERC-Fine and Nested (micro P, R and F1-score).

	French						**German**						**English**					
	Strict			Fuzzy			Strict			Fuzzy			Strict			Fuzzy		
	P	R	F	P	R	F	P	R	F	P	R	F	P	R	F	P	R	F
hipe2020 (Fine)																		
L3i	.702	.782	**.740**	.784	.873	**.826**	.691	.747	**.718**	.776	.840	**.807**						
Neur-bsl	.685	.733	.708	.769	.822	.795	.584	.673	.625	.659	.759	.706						
hipe2020 (Nested)																		
L3i	.390	.366	.377	.416	.390	.403	.714	.411	.522	.738	.425	.539						
ajmc (Fine)																		
L3i	.646	.694	**.669**	.703	.756	**.728**	.915	.898	**.906**	.941	.924	**.933**	.754	.848	**.798**	.801	.899	**.847**
Neur-bsl	.526	.567	.545	.616	.664	.639	.819	.817	.818	.866	.864	.865	.600	.744	.664	.676	.839	.749

Table 8. Results for EL-only (micro P, R and F1-score @1). Bold font indicates the highest value.

	Strict			Fuzzy			Strict			Fuzzy			Strict			Fuzzy		
	P	R	F	P	R	F	P	R	F	P	R	F	P	R	F	P	R	F
							hipe2020											
			French						German						English			
L3i	.602	.602	**.602**	.620	.620	**.620**	.481	.481	.481	.497	.497	.497	.546	.546	**.546**	.546	.546	**.546**
SBB	.707	.515	.596	.730	.532	.616	.603	.435	**.506**	.626	.452	**.525**	.503	.323	.393	.503	.323	.393
							sonar								**topres19th**			
									German						English			
SBB							.616	.446	**.517**	.616	.446	**.517**	.778	.559	.651	.781	.562	.654
N<small>IL-BSL</small>							.333	.333	.333	.333	.333	.333	-	-	-	-	-	-
							newseye											
			French						German									
SBB	.534	.361	.431	.539	.364	.435	.522	.387	.444	.535	.396	.455						
N<small>IL-BSL</small>	.448	.448	**.448**	.448	.448	**.448**	.485	.485	**.485**	.485	.485	**.485**						
							ajmc											
			French						German						English			
SBB	.621	.378	**.470**	.614	.373	**.464**	.712	.389	**.503**	.712	.389	**.503**	.578	.284	**.381**	.578	.284	**.381**
N<small>IL-BSL</small>	.037	.037	.037	.037	.037	.037	.049	.049	.049	.049	.049	.049	.046	.046	.046	.046	.046	.046

General Observations. All systems now use transformer-based approaches with strong pre-trained models. The choice of the pre-trained model – and the corresponding text types used in pre-training – have a strong influence on performance.

The quality of available multilingual pre-trained models for fine-tuning on NER tasks proved to be competitive compared to training individual monolingual models. However, to get the maximum performance out of it, the multilingual fine-tuning in a first phase must be complemented by a monolingual second phase.

NERC. In general, the systems demonstrated a good ability to adapt to heterogeneous annotation guidelines. They achieved their highest F1-scores for the NERC-Coarse task on `ajmc`, a dataset annotated with domain-specific entities and of relatively small size compared to the newspaper datasets, thus confirming the ability of strong pre-trained models to achieve good results when fine-tuned on relatively small datasets. The good results obtained on `ajmc`, however, may be partly due to the relatively high mention overlap between train and test sets (see Sect. 3.2). Moreover, it is worth noting that performances on the French subset of the `ajmc` dataset do not substantially degrade despite the high rate of noisy mentions (three times higher than English and German), which shows a good resilience of transformer-based models to OCR noise on this specific dataset.

EL-Only. Entity linking on already identified mentions appears to be considerably more challenging than NERC, with F1-scores varying considerably across datasets. The linking of toponyms in `topres19th` is where systems achieved the overall best performances. Conversely, EL-only on historical commentaries (`ajmc`) appears to be the most difficult, with the lowest F1-scores compared to the other datasets.

The EL-only performances of the SBB system on the `ajmc` dataset deserve some further considerations, as they are well representative of the challenges faced when applying a generic entity linking system to a domain-specific dataset. Firstly, SBB team reported that `ajmc` is the dataset with the lowest Wikidata coverage: only 57% of the Wikidata IDs in the test set are found in the knowledge base used by their system (a combination of Wikidata record and Wikipedia textual content), whereas the coverage for all other datasets ranges between 86% (`hipe2020`) and 99% (`topres19th`). The reason for the low coverage in `ajmc` is that, when constructing the knowledge base, only Wikidata records describing persons, locations and organisations were kept. In contrast, a substantial number of entities in `ajmc` are literary works, which would have required to retain also records with Wikidata type "literary work" (`Q7725634`) when building the KB.

Secondly, a characteristic of `ajmc` is that both person and work mentions are frequently abbreviated, and these abbreviations tend to be lacking as lexical information in large-scale KBs such as Wikidata. Indeed, an error analysis of SBB's system results shows that only 1.4% of the correctly predicted entity links (true positives) correspond to abbreviated mentions, which nevertheless represent about 47% of all linkable mentions.

Unfortunately, no team has worked on adapting annotation models to be able to use different NER training datasets with sometimes incompatible annotations and benefit from a larger dataset overall. Tackling this challenge remains future work.

7   Conclusion and Perspectives

From the perspective of natural language processing, this second edition of HIPE provided the possibility to test the robustness of existing approaches and to experiment with transfer learning and domain adaptation methods, whose performances could be systematically evaluated and compared on broad historical and multilingual data sets. Besides gaining new insights with respect to domain and language adaptation and advancing the state of the art in semantic indexing of historical material, the lab also contributed an unprecedented set of multilingual and historical NE-annotated datasets that can be used for further experimentation and benchmarking.

From the perspective of digital humanities, the lab's outcomes will help DH practitioners in mapping state-of-the-art solutions for NE processing of historical texts, and in getting a better understanding of what is already possible as opposed to what is still challenging. Most importantly, digital scholars are in need of support to explore the large quantities of digitised text they currently have at hand, and NE processing is high on the agenda. Such processing can support research questions in various domains (e.g. history, political science, literature, historical linguistics) and knowing about their performance is crucial in order to make an informed use of the processed data.

From the perspective of cultural heritage professionals, who increasingly focus on advancing the usage of artificial intelligence methods on cultural heritage text collections [17,26], the HIPE-2022 shared task and datasets represent an excellent opportunity to experiment with multilingual and multi-domain data of various quality and annotation depth, a setting close to the real-world scenarios they are often confronted with.

Overall, HIPE-2022 has contributed to further advance the state of the art in semantic indexing of historical documents. By expanding the language spectrum and document types and integrating datasets with various annotation tag sets, this second edition has set the bar high, and there remains much to explore and experiment.

Acknowledgements. The HIPE-2022 team expresses her greatest appreciation to the HIPE-2022 partnering projects, namely AjMC, impresso-HIPE-2020, Living with Machines, NewsEye, and SoNAR, for contributing (and hiding) their NE-annotated datasets. We particularly thank Mariona Coll-Ardanuy (LwM), Ahmed Hamdi (NewsEye) and Clemens Neudecker (SoNAR) for their support regarding data provision, and the members of the HIPE-2022 advisory board, namely Sally Chambers, Frédéric Kaplan and Clemens Neudecker.

References

1. Ács, J., Kádár, Á., Kornai, A.: Subword pooling makes a difference. In: Proceedings of the 16th Conference of the European Chapter of the Association for Computational Linguistics: Main Volume, pp. 2284–2295. Association for Computational Linguistics, April 2021. https://doi.org/10.18653/v1/2021.eacl-main.194, https://aclanthology.org/2021.eacl-main.194

2. Augenstein, I., Derczynski, L., Bontcheva, K.: Generalisation in named entity recognition: a quantitative analysis. Comput. Speech Lang. **44**, 61–83 (2017). https://doi.org/10.1016/j.csl.2017.01.012, http://www.sciencedirect.com/science/article/pii/S088523081630002X

3. Beryozkin, G., Drori, Y., Gilon, O., Hartman, T., Szpektor, I.: A joint named-entity recognizer for heterogeneous tag-sets using a tag hierarchy. In: Proceedings of the 57th Annual Meeting of the Association for Computational Linguistics, Florence, Italy, pp. 140–150, July 2019. https://aclanthology.org/P19-1014

4. Boros, E., et al.: Alleviating digitization errors in named entity recognition for historical documents. In: Proceedings of the 24th Conference on Computational Natural Language Learning, pp. 431–441. Association for Computational Linguistics, November 2020. https://doi.org/10.18653/v1/2020.conll-1.35

5. Coll Ardanuy, M., Beavan, D., Beelen, K., Hosseini, K., Lawrence, J.: Dataset for toponym resolution in nineteenth-century English newspapers (2021). https://doi.org/10.23636/b1c4-py78

6. Conneau, A., et al.: Unsupervised cross-lingual representation learning at Scale, April 2020

7. Devlin, J., Chang, M.W., Lee, K., Toutanova, K.: BERT: pre-training of deep bidirectional transformers for language understanding. In: Proceedings of the 2019 Conference of the North American Chapter of the Association for Computational Linguistics: Human Language Technologies, Volume 1 (Long and Short Papers), pp. 4171–4186. Minneapolis, Minnesota, June 2019. https://doi.org/10.18653/v1/N19-1423

8. Ehrmann, M., Colavizza, G., Rochat, Y., Kaplan, F.: Diachronic evaluation of NER systems on old newspapers. In: Proceedings of the 13th Conference on Natural Language Processing (KONVENS 2016), pp. 97–107. Bochumer Linguistische Arbeitsberichte, Bochum (2016). https://infoscience.epfl.ch/record/221391

9. Ehrmann, M., Hamdi, A., Pontes, E.L., Romanello, M., Doucet, A.: Named entity recognition and classification on historical documents: a survey. arXiv:2109.11406 [cs], September 2021. (To appear in ACM Journal Computing Surveys in 2022)

10. Ehrmann, M., Romanello, M., Clematide, S., Ströbel, P.B., Barman, R.: Language resources for historical newspapers: the impresso collection. In: Proceedings of the 12th Language Resources and Evaluation Conference, pp. 958–968. European Language Resources Association, Marseille, France, May 2020

11. Ehrmann, M., Romanello, M., Doucet, A., Clematide, S.: HIPE 2022 shared task participation guidelines. Technical report, Zenodo, Feburary 2022. https://doi.org/10.5281/zenodo.6045662, https://zenodo.org/record/6045662

12. Ehrmann, M., Romanello, M., Flückiger, A., Clematide, S.: Impresso named entity annotation guidelines. Annotation guidelines, Ecole Polytechnique Fédérale de Lausanne (EPFL) and Zurich University (UZH), January 2020. https://doi.org/10.5281/zenodo.3604227, https://zenodo.org/record/3585750

13. Ehrmann, M., Romanello, M., Flückiger, A., Clematide, S.: Overview of CLEF HIPE 2020: named entity recognition and linking on historical newspapers. In: Arampatzis, A., et al. (eds.) CLEF 2020. LNCS, vol. 12260, pp. 288–310. Springer, Cham (2020). https://doi.org/10.1007/978-3-030-58219-7_21

14. Ehrmann, M., Romanello, M., Flückiger, A., Clematide, S.: Extended overview of CLEF HIPE 2020: named entity processing on historical newspapers. In: Cappellato, L., Eickhoff, C., Ferro, N., Névéol, A. (eds.) Working Notes of CLEF 2020 - Conference and Labs of the Evaluation Forum, vol. 2696, p. 38. CEUR-WS, Thessaloniki (2020). https://doi.org/10.5281/zenodo.4117566, https://infoscience.epfl.ch/record/281054

15. Ehrmann, M., Romanello, M., Najem-Meyer, S., Doucet, A., Clematide, S.: Extended overview of HIPE-2022: named entity recognition and linking on multilingual historical documents. In: Faggioli, G., Ferro, N., Hanbury, A., Potthast, M. (eds.) Working Notes of CLEF 2022 - Conference and Labs of the Evaluation Forum. CEUR-WS (2022)

16. Faggioli, G., Ferro, N., Hanbury, A., Potthast, M. (eds.): Working Notes of CLEF 2022 - Conference and Labs of the Evaluation Forum. CEUR-WS (2022)

17. Gregory, M., Neudecker, C., Isaac, A., Bergel, G., et al.: AI in relation to GLAMs task FOrce - report and recommendations. Technical report, Europeana Network ASsociation (2021). https://pro.europeana.eu/project/ai-in-relation-to-glams

18. Hamdi, A., et al.: A multilingual dataset for named entity recognition, entity linking and stance detection in historical newspapers. In: Proceedings of the 44th International ACM SIGIR Conference on Research and Development in Information Retrieval, SIGIR 2021, pp. 2328–2334. Association for Computing Machinery, New York, July 2021. https://doi.org/10.1145/3404835.3463255

19. Houlsby, N., et al.: Parameter-efficient transfer learning for NLP. In: Chaudhuri, K., Salakhutdinov, R. (eds.) Proceedings of the 36th International Conference on Machine Learning. Proceedings of Machine Learning Research, vol. 97, pp. 2790–2799. PMLR, 09–15 June 2019. https://proceedings.mlr.press/v97/houlsby19a.html

20. Huang, Z., Xu, W., Yu, K.: Bidirectional LSTM-CRF models for sequence tagging. ArXiv abs/1508.01991 (2015)

21. Kaplan, F., di Lenardo, I.: Big data of the past. Front. Digit. Humanit. **4**, 1–21 (2017)

22. Labusch, K., Neudecker, C.: Named entity disambiguation and linking on historic newspaper OCR with BERT. In: Working Notes of CLEF 2020 - Conference and Labs of the Evaluation Forum. No. 2696 in CEUR Workshop Proceedings, CEUR-WS, September 2020. http://ceur-ws.org/Vol-2696/paper_163.pdf

23. Li, J., Chiu, B., Feng, S., Wang, H.: Few-shot named entity recognition via meta-learning. IEEE Trans. Knowl. Data Eng. 1 (2020). https://doi.org/10.1109/TKDE.2020.3038670. https://ieeexplore.ieee.org/document/9262018

24. Li, J., Shang, S., Shao, L.: MetaNER: named entity recognition with meta-learning. In: Proceedings of The Web Conference 2020, WWW 2020, pp. 429–440. Association for Computing Machinery, New York (2020). https://doi.org/10.1145/3366423.3380127

25. Makhoul, J., Kubala, F., Schwartz, R., Weischedel, R.: Performance measures for information extraction. In: Proceedings of DARPA Broadcast News Workshop, pp. 249–252 (1999)

26. Padilla, T.: Responsible operations: data science, machine learning, and AI in libraries. Technical report, OCLC Research, USA, May 2020. https://doi.org/10.25333/xk7z-9g97

27. Reimers, N., Gurevych, I.: Making monolingual sentence embeddings multilingual using knowledge distillation. In: Proceedings of the 2020 Conference on Empirical Methods in Natural Language Processing (EMNLP), pp. 4512–4525. Association for Computational Linguistics, November 2020. https://doi.org/10.18653/v1/2020.emnlp-main.365, https://aclanthology.org/2020.emnlp-main.365

28. Ridge, M., Colavizza, G., Brake, L., Ehrmann, M., Moreux, J.P., Prescott, A.: The past, present and future of digital scholarship with newspaper collections. In: DH 2019 Book of Abstracts, Utrecht, The Netherlands, pp. 1–9 (2019). http://infoscience.epfl.ch/record/271329

29. Romanello, M., Najem-Meyer, S.: Guidelines for the annotation of named entities in the domain of classics, March 2022. https://doi.org/10.5281/zenodo.6368101

30. Romanello, M., Sven, N.M., Robertson, B.: Optical character recognition of 19th century classical commentaries: the current state of affairs. In: The 6th International Workshop on Historical Document Imaging and Processing (HIP 2021). Association for Computing Machinery, Lausanne, September 2021. https://doi.org/10.1145/3476887.3476911

31. Rosset, S., Grouin, C., Zweigenbaum, P.: Entités nommées structurées: guide d'annotation quaero. Technical report 2011-04, LIMSI-CNRS, Orsay, France (2011)

32. Taillé, B., Guigue, V., Gallinari, P.: Contextualized embeddings in named-entity recognition: an empirical study on generalization. In: Jose, J.M., et al. (eds.) ECIR 2020. LNCS, vol. 12036, pp. 383–391. Springer, Cham (2020). https://doi.org/10.1007/978-3-030-45442-5_48

33. Wang, X., et al.: KEPLER: a unified model for knowledge embedding and pre-trained language representation, November 2019. https://arxiv.org/pdf/1911.06136.pdf

34. Wang, X., et al.: DAMO-NLP at SemEval-2022 task 11: a knowledge-based system for multilingual named entity recognition (2022). https://doi.org/10.48550/ARXIV.2203.00545, https://arxiv.org/abs/2203.00545
35. Wolf, T., et al.: Transformers: state-of-the-art natural language processing. In: Proceedings of the 2020 Conference on Empirical Methods in Natural Language Processing: System Demonstrations, pp. 38–45. Association for Computational Linguistics, October 2020. https://www.aclweb.org/anthology/2020.emnlp-demos.6
36. Wu, Q., et al.: Enhanced meta-learning for cross-lingual named entity recognition with minimal resources. CoRR abs/1911.06161 (2019). http://arxiv.org/abs/1911.06161

Overview of JOKER@CLEF 2022: Automatic Wordplay and Humour Translation Workshop

Liana Ermakova[1,2]([✉]) [ID], Tristan Miller[3] [ID], Fabio Regattin[4] [ID],
Anne-Gwenn Bosser[5], Claudine Borg[6] [ID], Élise Mathurin[1,2], Gaëlle Le Corre[2,7],
Sílvia Araújo[2,8] [ID], Radia Hannachi[2,9], Julien Boccou[1], Albin Digue[1],
Aurianne Damoy[1], and Benoît Jeanjean[1,2]

[1] Université de Bretagne Occidentale, HCTI, 29200 Brest, France
`liana.ermakova@univ-brest.fr`
[2] Maison des sciences de l'homme en Bretagne, 35043 Rennes, France
[3] Austrian Research Institute for Artificial Intelligence, Vienna, Austria
[4] Dipartimento DILL, Universitá degli Studi di Udine, 33100 Udine, Italy
[5] École Nationale d'Ingénieurs de Brest, Lab-STICC CNRS UMR 6285, Brest, France
[6] University of Malta, Msida MSD 2020, Malta
[7] Université de Bretagne Occidentale, CRBC, 29200 Brest, France
[8] Universidade do Minho, CEHUM, 4710-057 Braga, Portugal
[9] Université de Bretagne Sud, HCTI, 56321 Lorient, France

Abstract. While humour and wordplay are among the most intensively studied problems in the field of translation studies, they have been almost completely ignored in machine translation. This is partly because most AI-based translation tools require a quality and quantity of training data (e.g., parallel corpora) that has historically been lacking for humour and wordplay. The goal of the JOKER@CLEF 2022 workshop was to bring together translators and computer scientists to work on an evaluation framework for wordplay, including data and metric development, and to foster work on automatic methods for wordplay translation. To this end, we defined three pilot tasks: (1) classify and explain instances of wordplay, (2) translate single terms containing wordplay, and (3) translate entire phrases containing wordplay (punning jokes). This paper describes and discusses each of these pilot tasks, as well as the participating systems and their results.

Keywords: Machine translation · Humour · Wordplay · Puns · Neologisms · Parallel corpora · Evaluation metrics · Creative language analysis

1 Introduction

Wordplay is a pervasive, highly ingrained, and profoundly meaningful aspect of human experience. We use it to defuse tension in stressful situations, to establish

A. Barrón-Cedeño et al. (Eds.): CLEF 2022, LNCS 13390, pp. 447–469, 2022.
https://doi.org/10.1007/978-3-031-13643-6_27

and enhance our affiliation with social groups, and as a source of entertainment. Wordplay can crop up in almost any type of discourse, and for many of these (including literature, advertising, and social conversations) it is a recurrent and expected feature. It is therefore vitally important that natural language processing applications operating on these discourse types be capable of recognising and appropriately dealing with instances of wordplay. An ideal machine translation system should thus preserve the mirth-provoking effect of comedic texts, if necessary by rewriting puns and similar linguistic oddities for the target culture and language.

However, while humour and wordplay are among the most intensively studied problems in the field of translation studies, they have been almost completely ignored in machine translation. This is partly because AI-based translation tools tend to require a quality and quantity of training data (e.g., parallel corpora) that has historically been lacking of humour and wordplay.

Professional (human) translators employ various strategies for dealing with wordplay, not all of which attempt to retain the wordplay in the target text. However, strategies which do preserve it in some manner can be crucial for maintaining the pragmatic force of the discourse and for avoiding nonsense. Consider the following pun from *Alice's Adventures in Wonderland* by Lewis Carroll, which exploits the homophone pair *lesson* and *lessen* for a humorous effect:

Example 1. 'That's the reason they're called lessons,' the Gryphon remarked: 'because they lessen from day to day.'

Henri Parisot's French translation manages to preserve both the sound and meaning correspondence by using the pair *cours/courts*:

Example 2. «C'est pour cette raison qu'on les appelle des cours: parce qu'ils deviennent chaque jour un peu plus courts.»

By contrast, Google Translate fails to recognise the pun; its translation uses the pair *leçons/diminuent* and the sentence becomes nonsensical:

Example 3. «C'est la raison pour laquelle on les appelle leçons, remarqua le Griffon: parce qu'elles diminuent de jour en jour.»

The goal of the JOKER@CLEF 2022 workshop was to bring together translators, linguists and computer scientists in order to bring us a step closer to the automation of multilingual wordplay analysis and translation. We introduced three pilot tasks making use of a new, multilingual parallel corpus of wordplay and humour that we have produced:

Pilot Task 1 is to classify single terms containing wordplay according to a given typology, and provide lexical-semantic interpretations;
Pilot Task 2 is to translate single terms containing wordplay;
Pilot Task 3 is to translate entire phrases that subsume or contain wordplay.

Forty-nine teams registered for our JOKER track at CLEF 2022. Forty-two users downloaded the data from the server and seven teams submitted 19 runs in total. The statistics for these runs submitted are presented in Table 1.

Table 1. Statistics on submitted runs by pilot task

Team	Task 1	Task 2	Task 3	Total runs
FAST_MT	2	1	1	4
Cecilia	1	1	2	4
Agnieszka	2	1	0	3
eBIHAR	4	0	0	4
TEAM_JOKER	0	1	0	1
LJGG	0	0	2	2
Humorless	0	0	1	1
Total runs	9	4	6	19

2 Background

2.1 Typologies of Wordplay

Wordplay includes a wide variety of phenomena that exploit or subvert the phonological, orthographical, morphological, and semantic conventions of a language [9]. These phenomena include puns; alliteration, assonance, and consonance (repetition of sounds across nearby words); portmanteaux (combining parts of multiple words into a new word); spoonerisms (exchanging the initial sounds of nearby words); anagrams (a word or phrase formed by rearranging the letters of another); and onomatopoeia (a word coined to approximate some non-speech or non-language sound). Our previous annotation work [9] has made it clear that these categories are not mutually exclusive, especially in the case of neologisms (i.e., newly coined words). Furthermore, instances of ambiguity-based wordplay ("punning", broadly construed) can be further subclassified according to their phonological, orthographical, morphological, lexical, or contextual structure, such as whether the two entities forming the ambiguity have the same or different pronunciation (homophony vs. heterophony) or spelling (homography vsheterography), or rely on different morphological analyses of the same word, or arise from different syntactic parses of the subsuming phrase, or exploit both the figurative and literal readings of an idiom [6,16,17,19].

Two longstanding issues with the wordplay typologies used in past work, including our own, are that the typologies tend to be flat rather than hierarchical and do not have clearly defined discrimination criteria; furthermore, application of these typologies in annotation studies does not admit the possibility of a given instance of wordplay meeting the criteria of multiple distinct categories. We introduce a new topology aimed to reduce the drawbacks of existing classification in Sect. 4.

2.2 Translation of Wordplay

Perhaps the most commonly cited typology of wordplay translation strategies is that of Delabastita [5,7]. This typology was developed on the basis of parallel corpus analysis and therefore reflects the techniques used by working translators. And while the typology was developed specifically for puns (a type of wordplay that exploits multiple meanings of a term or of similar-sounding words), many of the strategies are applicable to other forms not based on ambiguity. Delabastita's basic options are the following:

PUN→PUN: The source-text pun is translated by a target-language pun.

PUN→NON-PUN: The pun is translated by a non-punning phrase, which may reproduce all senses of the wordplay or just one of them, but which does not attempt to preserve the level of ambiguity of the original.

PUN→RELATED RHETORICAL DEVICE: The pun is replaced (one could say "compensated for") by some other, rhetorically charged, phrase (involving repetition, alliteration, rhyme, irony, paradox, etc.).

PUN→ZERO: The part of text containing the pun is omitted altogether.

PUN ST=PUN TT: The punning text, and sometimes its immediate environment, is reproduced in the source language, without attempting a target-language rendering.

NON-PUN→PUN: A pun is introduced in the target text where no wordplay was present in the source text.

ZERO→PUN: New textual material involving wordplay is added in the target text, which bears no correspondence whatsoever in the source text.

EDITORIAL TECHNIQUES: All the paratextual strategies involved in explaining, or presenting alternative renderings for, the pun of the source text (footnotes, prefaces, translator's notes, etc.).

Delabastita insists on one further point: the techniques are by no means exclusive. A translator could, for instance, suppress a pun somewhere in their target text (PUN→NON-PUN), explain it in a footnote (EDITORIAL TECHNIQUES), then try to compensate for the loss by adding another pun somewhere else in the text (NON-PUN→PUN or ZERO→PUN).

The very typology of translation strategies drawn by Delbastita directly points to the main reason for the difficulty of conceiving a working model of machine translation (MT) for puns.

2.3 Computational Humour

To date, there have been few studies on the MT of wordplay. One early study [12] proposed a pragmatic-based approach to MT that accounts for the author's locutionary, illocutionary, and perlocutionary intents (that is, the "how", "what", and "why" of the text), and discuss how it might be applied to puns. However, no working system appears to have been implemented. More recent work [30] has proposed an interactive method for the computer-assisted translation of puns, an implementation of which (PunCAT) was later evaluated in a user study [26]. This study was limited to a single language pair (English to German) and translation strategy (namely, the PUN→PUN strategy described above). Furthermore, the tool's functionality is limited to facilitating exploration of the semantic fields corresponding to the two meanings of the pun; actually detecting and interpreting the source-text pun, and devising a complete target-language punning joke, is left to the user.

Numerous studies have been conducted for the related tasks of humour generation and detection. Pun generation systems have often been based on template approaches. One system [38] used lexical constraints to generate adult humour by substituting one word in a pre-existing text; another study [20] trained a system to automatically extract humorous templates which were then used for pun generation. Some current efforts to tackle this difficult problem more generally using neural approaches have been hindered by the lack of a sizable pun corpus [44].

Meanwhile, the recent rise of conversational agents and the need to process large volumes of social media content point to the necessity of automatic humour recognition [32]. Humour and irony studies are now crucial when it comes to social listening [15,24,25,37], dialogue systems (chatbots), recommender systems, reputation monitoring, and the detection of fake news [18] and hate speech [14]. However, the automatic detection, location, and interpretation of humorous wordplay in particular has so far been limited to punning. And while even the earliest such systems have achieved decent performance on the detection and location tasks [31], methods for actually interpreting the double meaning of the pun – a prerequisite for translation – have not been as intensively researched. The first such methods, evaluated at SemEval-2017 [31], achieved an accuracy of 16.0% and 7.7% for homographic and heterographic puns, respectively, but this baseline does not seem to have been improved upon in more recent work [22]. Again, the lack of sufficient training data appears to be a particular stumbling block to further progress, at least for supervised approaches, and especially when processing languages other than English.

A few monolingual humour corpora do exist, including the datasets created for shared tasks of the International Workshop on Semantic Evaluation (SemEval): #HashtagWars: Learning a Sense of Humor [34], Detection and Interpretation of English Puns [31], Assessing Humor in Edited News Headlines [21], and HaHackathon: Detecting and Rating Humor and Offense [28]. Other datasets include one which collects 16 000 humorous sentences and an equal number of negative samples from news titles, proverbs, the British National Corpus, and the Open Mind Common Sense dataset [29], and another which contains 2400 puns and non-puns from news sources, Yahoo! Answers, and proverbs [4,43]. Most datasets are in English, with some notable exceptions for Italian [36], Russian [1,10], and Spanish [3]. To the best of our knowledge, no corpus exists for French or German.

The only parallel corpus of wordplay that we are aware of is the one introduced in our research [9]. We manually collected over a thousand translated examples of wordplay, in English and French, from video games, advertising slogans, literature, and other sources. Each example has been manually classified according to a multi-label inventory of wordplay types and structures, and annotated according to its lexical-semantic or morphosemantic components. The majority of the collected wordplay are puns and *single-term* proper nouns or neologisms based on portmanteau words, the like of which are common in the Asterix and Harry Potter universes.

Large pre-trained AI models, like Jurassic-1 [27], mT5 [42], BERT [8], and GPT [2,35], have outperformed other state-of-the-art models on several natural language processing (NLP) tasks, including MT [39]. Performance of such supervised MT systems depends on the quality and quantity of training data [23]. However, as mentioned above, there exist no large-scale, broad-coverage parallel corpora of wordplay. Such a corpus is a key prerequisite for the training and evaluation of MT models.

Humorous wordplay often exploits the confrontation of similar forms with different meanings, evoking incongruity between expected and presented stimuli. This makes it particularly important in NLP to study the strategies that human translators use for dealing with wordplay [5,41]. This is because MT is generally ignorant of pragmatics and assumes that words in the source text are formed and used in a conventional manner. MT systems fail to recognise the deliberate ambiguity of puns or the unorthodox morphology of neologisms, leaving such terms untranslated or else translating them in ways that lose the humorous aspect [30].

3   Data

We constructed a parallel corpus of wordplay in English and French. Our data is twofold, containing phrase-based wordplay (puns) and term-based wordplay (mainly named entities).

3.1   Parallel Corpus of Puns

Our English corpus of puns is based mainly on that of the SemEval-2017 shared task on pun identification [31]. The original annotated dataset contains 3387 standalone English-language punning jokes, between 2 and 69 words in length, sourced from offline and online joke collections. Roughly half of the puns in the collection are "weakly" homographic (meaning that the lexical units corresponding to the two senses of the pun, disregarding inflections and particles, are spelled identically) while the other half are heterographic (that is, with lemmas spelled differently). The original annotation scheme is rather simple, indicating only the pun's location within the joke, whether it is homographic or heterographic, and the two meanings of the pun (with reference to senses in WordNet [13]).

In order to translate this subcorpus from English into French, we applied a gamification strategy. More precisely, we organised a translation contest[1] which was open to students but also attracted participation from professional and academic translators. The results were submitted via Google Forms. Forty-seven participants submitted 3950 translations of 500 puns from the SemEval-2017 dataset. We took first 250 puns in English from each of the homographic and

---

[1] https://www.joker-project.com/pun-translation-contest/.

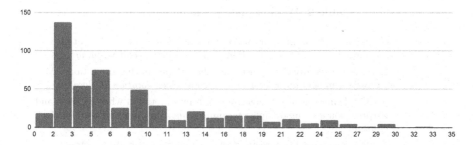

Fig. 1. Histogram of the number of translations per query (all)

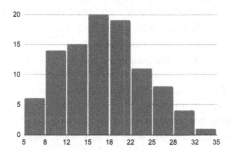

Fig. 2. Histogram of the number of translations per query (first page)

heterographic subsets. In our online submission form, the homographic and heterographic puns were alternated, with 100 puns per page.

Unfortunately, Google Forms does not allow questions to be shuffled for each participant. Thus, we observed a drastic drop in the number of translations per pun starting from the second page. As we had two participants who translated almost all puns, we have a conspicuous peak on the histogram of the number of translations per query (Fig. 1). However, this histogram does not provide a clear idea about the translation difficulty of puns as the vast majority of participants translated only the first page of the form. Figure 2, the number of translations per query on the first page only, perhaps better reflects the translation difficulty distribution.

Besides this SemEval-derived data, we sourced further translation pairs from published literature and from puns translated by Master's students in translation.

We annotated our dataset according to the classification introduced in Sect. 4. The final annotated training set contains a total of 1772 distinct instances in English with 4753 corresponding French translations.

3.2   Parallel Corpus of Term-Based Wordplay

For this part of the corpus, we collected 1409 single terms in English containing wordplay from video games, advertising slogans, literature, and other sources [9]

Table 2. Wordplay interpretation notation

a / b	Distinguishes the location from the interpretation and the different meanings of a wordplay: *meaning 1 (location) / meaning 2 (second meaning)*
a | b	Separates the wordplay instances and their respective interpretations. An expression can contain several wordplay instances: *location 1 | location 2*
a (b)	Specifies definitions or synonyms for each interpretation when location and interpretation are homographs: *meaning (synonym, hyperonym or brief definition)*
a [b]	Specifies comments like foreign language, anagram, palindrome etc.: *interpretation [anagram]*
a { b }	Specifies the frame that activates the ambiguous word when a synonym or a short definition is not available: *meaning {frame activated by meaning}*
< a; · · · >	Groups words from the same lexical field: *<word 1; word 2; word 3>*
"a"	Indicates presence of an idiom: *"idiom"*
a ∼ b	Indicates several possible interpretations for an ambiguous word: *meaning 1 (interpretation 1) ∼ meaning 2 (interpretation 2)*
a + b	Indicates that several words or syllables have been combined: *meaning 1 / meaning 1a + meaning 1b*
A /b	Defines acronyms: *OWL /Ordinary Wizarding Level*
a & b	Shows when the wordplay relies on opposition: *location 1 & location 2*

along with 1420 translations into French. Almost all translations are official ones but we have eleven additional ones proposed by our interns, Master's students in translation.

4 Pilot Task 1: Classify and Explain Instances of Wordplay

Collection and Annotation. For our Task 1, we annotated both phrase-based (puns) and term-based instances of wordplay in English and French (see Sect. 3). Following the SemEval-2017 pun task [31], we annotated each instance of word-play according to its LOCATION and INTERPRETATION. For LOCATION, we mean the precise word(s) in the instance forming the wordplay, such as the ambiguous words of a punning joke.[2] INTERPRETATION means the explanation of the wordplay, which we do, for example, by providing the secondary meaning of a pun. To facilitate preprocessing, we do not use WordNet as in SemEval-2017 but rather introduce the notation described in Table 2.

We further annotated the data according to the following typologies:

– HORIZONTAL/VERTICAL concerns the co-presence of source and target of the wordplay. In horizontal wordplay, both the source and the target of the wordplay are given (Example 1); in vertical wordplay, source and target are collapsed in a single occurrence:

---

[2] Unlike in the SemEval-2017 task, we simply list the word(s) in question rather than indicating their position within the instance.

Example 4. How do you make a cat drink? Easy: put it in a liquidizer.

- MANIPULATION TYPE:
  - Identity: source and target are formally identical, as in Example 4.
  - Similarity: as in Example 1: source and target are not perfectly identical, but the resemblance is obvious.
  - Permutation: the textual material is given a new order, as in anagrams or spoonerisms:

Example 5. Dormitory = dirty room

- Abbreviation: an ad-hoc category for textual material where the initials form another meaning, as in acrostics or "funny" acronyms:

Example 6. BRAINS: Biobehavioral Research Awards for Innovative New Scientists

- Opposition: covers wordplay such as the antonyms *hot & ice | warms & freezing* in the following:

Example 7. Hot ice cream warms you up no end in freezing weather.

- MANIPULATION LEVEL: Most wordplay involves some kind of phonological manipulation, making SOUND our default category. Examples 1 and 4 involve a clear sound similarity or identity, respectively. Only if this category cannot be applied to the wordplay is the instance tagged with another level of manipulation. The next level to be considered is WRITING (as in Examples 5 and 6). If neither SOUND nor WRITING are manipulated, the level of manipulation is specified as OTHER. This level of manipulation may arise, for instance, in chiasmses:

Example 8. We shape our buildings, and afterwards our buildings shape us.

- CULTURAL REFERENCE: This is a binary (true/false) category. In order to understand some instances of wordplay, one has to be aware of some extra-linguistic factors.
- CONVENTIONAL FORM: Another binary category, this time indicating whether the wordplay occurs in a fixed form, such as a Tom Swifty (i.e., wellerism).
- OFFENSIVE: Another binary category, this time indicating whether the wordplay could be considered offensive. (This category was not evaluated in the pilot tasks.)

Statistics on the annotated data are given in Table 3. We furthermore noticed that the LOCATION is usually the last word in wordplay, as evidenced in Fig. 3.

Table 3. Annotation statistics

	English	French
Phrases		
	– 1772 annotated instances	– 4753 annotated instances
	• Vertical 1382	• Vertical 4400
	• Horizontal 212	• Horizontal 320
	– MANIPULATION TYPE	– MANIPULATION TYPE
	• Identity 894	• Identity 2970
	• Similarity 639	• Similarity 1672
	• Opposition 42	• Opposition 51
	• Abbreviation 12	• Permutation 17
	• Permutation 7	• Abbreviation 9
	– MANIPULATION LEVEL	– MANIPULATION LEVEL
	• Sound 1551	• Sound 4540
	• Writing 46	• Writing 179
	• Other 2	• Other 4
	– CULTURAL REFERENCE	– CULTURAL REFERENCE
	• False 1689	• False 4665
	• True 82	• True 88
	– CONVENTIONAL FORM	– CONVENTIONAL FORM
	• False 1604	• False 4665
	• True 167	• True 88
	– OFFENSIVE	– OFFENSIVE
	• Sexist 9	• Sexist 21
	• Possibly 7	• Possibly 6
	• Racist 2	• Racist 4
	• Other 1	• Other 1
Terms		
	– 1409 annotated instances	– 1420 annotated instances
	• Vertical 1408	• Vertical - 1419
	• Horizontal 1	• Horizontal - 1
	– MANIPULATION TYPE	– MANIPULATION TYPE
	• Similarity 606	• Similarity 775
	• Identity 441	• Identity 415
	• Abbreviation 340	• Abbreviation 211
	• Permutation 17	• Permutation 15
	• Opposition 1	• Opposition 1
	– MANIPULATION LEVEL	– MANIPULATION LEVEL
	• Sound 1402	• Sound 1411
	• Writing 7	• Writing 9
	– CULTURAL REFERENCE	– CULTURAL REFERENCE
	• False 1361	• False 1344
	• True 48	• True 76
	– CULTURAL REFERENCE	– CONVENTIONAL FORM
	• NOT APPLICABLE	• NOT APPLICABLE
	– OFFENSIVE	– OFFENSIVE
	• NOT IDENTIFIED	• NOT IDENTIFIED

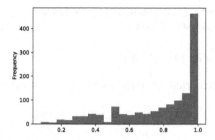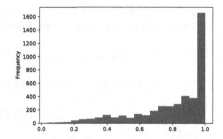

Fig. 3. Wordplay location normalised by text length for English (left); French (right)

```
[{"ID":"noun_1063","WORDPLAY":"Elimentaler","LOCATION":
"Elimentaler","INTERPRETATION":"Emmental (cheese) + Eliminator",
"HORIZONTAL/VERTICAL":"vertical","MANIPULATION_TYPE":"Similarity",
"MANIPULATION_LEVEL":"Sound","CULTURAL_REFERENCE":false,
"CONVENTIONAL_FORM":false,"OFFENSIVE":null},
{"ID":"pun_341","WORDPLAY":"Geologists can be sedimental about
their work.","LOCATION":"sedimental","INTERPRETATION":
"sentimental/sediment","HORIZONTAL/VERTICAL":"vertical",
"MANIPULATION_TYPE":"Similarity","MANIPULATION_LEVEL":"Sound",
"CULTURAL_REFERENCE":false,"CONVENTIONAL_FORM":false,"OFFENSIVE":
null}]
```

Fig. 4. Excerpt of training data (JSON format)

Training Data. Our training data consists of 2078 wordplay instances in English and 2550 in French in the form of a list of translated wordplay instances. This data was provided as a JSON or CSV file with one fields for the unique ID of the instance, one for the text of the instance, and one each for the LOCATION, INTERPRETATION, HORIZONTAL/VERTICAL, MANIPULATION_TYPE, MANIPULATION_LEVEL, and CULTURAL_REFERENCE annotations. Figure 4 shows an excerpt from the JSON file.

Test Data. Our test data contains 3255 instances of wordplay in English from the SemEval-2017 pun task [31] and 4291 instances in French that we did not use for the training set. The test data was provided as a JSON or CSV file with only two fields – one of them a unique ID and the other the text of the instance. Figure 5 shows an excerpt of the JSON test data.

The prescribed output format is similar to the training data format, but with the addition of the fields RUN_ID (to uniquely identify the participating team, pilot task, and run number), MANUAL (to indicate whether the output annotations are produced by a human or a machine), and OFFENSIVE (per our annotation scheme).

[{"ID":"noun_1","WORDPLAY":"Ambipom"},{"ID":"het_1011","WORDPLAY":
"These are my parents, said Einstein relatively"}]

Fig. 5. Excerpt of test data (JSON format)

Table 4. Scores of participants' runs for Pilot Task 1

	LOCATION	MANIP. TYPE	MANIP. LEVEL
FAST_MT		1035	2437
FAST_MT_updated	1455	1667	2437
Cecilia_task_1_run5	1484	1541	2437
Agnieszka_task1_t5	1554		
eBIHAR_en		1392	2437
eBIHAR_en_tfidf_wp		1083	2437
eBIHAR_en_tfidf _wp_preprocessed		536	2437

4.1 Evaluation Metrics

We preprocessed runs to lowercase and trim the values. For the English subcorpus, the labels for LOCATION and INTERPRETATION were provided for puns from the original dataset [31]. All wordplay instances from this dataset were considered to be VERTICAL with manipulation type SOUND. HOMOGRAPHIC puns were attributed the IDENTITY label while HETEROGRAPHIC puns were classified as SIMILARITY manipulation type. We report the absolute values of true labels submitted by the participants.

We discarded all INTERPRETATION values that were equal to the LOCATION fields as we considered this to be insufficient.

In recognition of the fact that there may be slightly different but equally valid INTERPRETATION annotations, for evaluation we retained only the high-level annotation (by removing everything in brackets, parentheses, etc.). We downcased, tokenised, and lemmatised this high-level annotation with the aid of regular expressions and the NLTK WordNetLemmatizer.[3] We then compared the set of lemmas generated by participants with our own annotations.

4.2 Results

All together, four teams submitted eight runs for the English dataset. The eBIHAR team also submitted one run in French. The release of the French dataset was delayed and we also updated the English dataset during the competition. The FAST_MT team submitted runs both for the first release of English dataset and the updated one. The Agnieszka team submitted only partial runs for LOCATION. The results for the participants are given in Table 4.

---

[3] https://www.nltk.org/_modules/nltk/stem/wordnet.html.

All participants, except the Agnieszka team which did not submit predictions for MANIPULATION LEVEL, successfully predicted all classes. However, this success might be explained by the nature of our data, as in the test set the only class was SOUND.

The teams Cecilia, FAST_MT, and Agnieszka demonstrated fairly good results for LOCATION. However, as previously noted, in our dataset the majority of instances had the wordplay located at the last word.

Only the FAST_MT team succeeded in INTERPRETATION prediction for the first data release. For this first run, our annotation coincides with that of the submission in 597 cases; it differs for 61. These differences are, in the majority of cases, not errors but differences in the presentation or human interpretation. The first dataset contained a lot of named entities from popular anime, movies, and video games (e.g., Pokemon), unlike the updated data set. FAST_MT had gathered raw data from various websites explaining puns in Pokemon names and trained their model on it. We should acknowledge that some annotations provided by FAST_MT were more detailed than ours. For the updated dataset, FAST_MT's predictions for LOCATION are identical to those of INTERPRE-TATION. Only one run, Cecilia's run 5, was successful for this dataset with 441 correct results.

We do not provide results for other binary classes; since our data was unbalanced with regard to these categories, the submitted results always provided negative labels.

5   Pilot Task 2: Translate Single Terms Containing Wordplay

For Task 2, participants had to translate single terms containing wordplay from English into French.

5.1   Data

Our training dataset contains 1161 instances of wordplay. Instances were provided as JSON or CSV files with fields denoting the instance's unique ID, the source text in English, and a target text in French. We used 284 wordplay instances for the test data; the format was identical to the training data, minus the field for the target text. The required output format for participants was, as with Task 1, identical to the training data format, but with the addition of the RUN_ID and MANUAL fields.

5.2   Evaluation Metrics

For the wordplay translation tasks (Tasks 2 and 3), there do not yet exist any accepted metrics of translation quality [9]. MT is traditionally measured with the BLEU (Bilingual Evaluation Understudy) metric, which calculates vocabulary overlap between the candidate translation and a reference translation [33].

However, this metric is clearly inappropriate for single-term wordplay translation evaluation, as overlap measures operate only on larger text spans and not on individual words, the morphological analysis of which can be crucial for neologisms [9].

We hypothesised that the majority of proper nouns would not be translated automatically. So we compared the target translation with source wordplay (metric *not translated*).

As our dataset for Task 2 contains "official" translations of wordplay instances coming from various published sources (e.g., Pokemon names), we also tried filtering out these official translations (metric *official*).

We manually evaluated the *non-official* translations according the following metrics:

- *lexical field preservation*: A value of *true* is assigned to translations that preserve the lexical field of the source wordplay (i.e., the translation is close to a literal one).
- *sense preservation*: A value of *true* is assigned to translations that preserve the meaning of the source wordplay.
- *comprehensible terms*: A value of *true* is assigned to translations that do not rely on specialised terminology.
- *wordplay form*: A value of *true* is assigned to translations that employ (as opposed to omit) wordplay.

5.3 Results

Four teams submitted a total of four runs for Task 2. Our initial guess was that the majority of proper nouns would not be translated by machine translation. However, as our dataset contained officially translated named entities (e.g., from Pokemon) that may have been discoverable by participants and large pretrained models, all participants translated all wordplay instances. The results from Table 5 suggest that the majority of translated named entities were indeed the official translations. TEAM_JOKER, however, provided very interesting results, with almost half being non-official translations. Among these, twelve translations were judged as being wordplay. We can also see that among non-official translations, less than 10% are successful in terms of preserving wordplay.

As is evident from Table 6, the majority of non-official translations containing wordplay are accidental, although we observe some francization of English terms.

6   Pilot Task 3: Translate Entire Phrases Containing Wordplay

For Task 3, participants had to translate entire phrases containing wordplay from English into French.

Table 5. Scores of participants' runs for Pilot Task 2

	FAST_MT	TEAM_JOKER	Cecilia	Agnieszka
total	284	284	284	242
not translated	0	0	0	0
official	250	159	216	230
non-official	34	125	68	12
lexical field preservation	16	13	5	0
sense preservation	13	11	5	0
comprehensible terms	26	59	16	2
wordplay form	3	12	3	1

6.1   Data

The dataset for Task 3 is based on that of Task 1. The training set contains 1772 distinct instances in English with 4753 corresponding translations in French, and the test sets contains 2378 instances in English from the SemEval-2017 pun task [31]. The training and test sets use the same formats as in Task 2.

6.2   Evaluation Metrics

As previously stated, the BLEU metric is clearly inappropriate for use with wordplay, where a wide variety of translation strategies (and solutions implementing those strategies) are permissible. And as our *Alice in Wonderland* example from Sect. 1 demonstrates, many of these strategies require metalexical awareness and preservation of features such as lexical ambiguity and phonetic similarity. (Consider how substituting the synonymous *leçons* for *cours* in Parisot's translation would lose the wordplay, and indeed render the translation nonsensical, yet still result in a near-perfect BLEU score with the original translation.)

For our evaluation, participants' runs were pooled together. We filtered out all translations that did not match the regular expression .+[?.!"]\s*$ as we considered these translations to be truncated. Indeed, in some runs (e.g., Cecilia's run 3) the majority of generated translations were too short with regard to the source wordplay and truncated in the middle of the sentence. We refer further the retained translations as *valid*.

We then filtered out French translations identical to the original wordplay in English, as we considered these wordplay instances to be *not translated*.

The pool of *valid* distinct translations into French contains 9513 instances. Three Master's students in translation, French native speakers, manually evaluated each valid translation as follows. We evaluated the following errors:

- *nonsense*: This metric is *true* when the translation contains a nonsensical passage.
- *syntax problem*: This metric is *true* when the translation contains a passage with errors in syntax.

Table 6. List of non-official translations with wordplay

English	French
Orbeetle	Orbétain
Ribombee	Ribombe
Celesteela	Célésteela
Primarina	Primarin
Wimpod	Pompode
Incineroar	Incinéroar
Incineroar	Incinéroque
Toxtricity	Toxtricité
Pyroar	Pyroque
Metallurgix	Métalurgix
Wifix	Ouifix
legilimency	légilimence
butterbeer	bourreau-bourre
Drifdlim	Grodrive
Mismagius	Virgilus
Dwebble	Débébé
Terrible Terror	Terreur terrifiante
Gold Ammolet	Ammolette d'or

- *lexical problem*: This metric is *true* when the translation contains a passage with errors in word choice/use.

An instance was not evaluated for subsequent metrics if one of the above errors was identified. For translations without these errors, we evaluated:

- *lexical field preservation, sense preservation, comprehensible terms, wordplay form*: These four metrics are evaluated as in Task 2.
- *identifiable wordplay*: A value of *true* is assigned to translations that are wordplay and are understandable for general audience. For example, the wordplay *"Je n'abandonnerai jamais mes chiens!" dit Tom cyniquement.* (meaning " 'I'll never abandon my dogs!' Tom said cynically") requires etymological knowledge that is beyond most readers.[4]
- *over-translation*: A value of *true* is assigned to translations that have useless multiple wordplay instances when the source text has just one.
- *style shift*: A value of *true* is assigned to translations that have style shift (e.g., where a vulgarism is present either in the source text or the translation but not in both).
- *humorousness shift*: A value of *true* is assigned to translations that were judged to be much more or much less funnier than the source wordplay.

---

[4] See Sect. 7 for further explanation.

Table 7. Scores of participants' runs for Pilot Task 3

	LJGG DeepL	FAST_MT	LJGG auto	Cecilia run 1	Humorless	Cecilia run 3
total	2378	2378	2378	2378	2378	2378
valid	2324	2120	2264	2343	384	7
not translated	39	103	206	49	22	2
nonsense	59	220	349	51	297	3
syntax problem	17	58	46	41	6	0
lexical problem	25	79	78	52	10	0
lexical field preservation	2184	1739	1595	2155	118	6
sense preservation	1938	1453	1327	1803	100	6
comprehensible terms	1188	867	827	744	56	5
wordplay form	373	345	261	251	19	1
identifiable wordplay	342	318	240	243	16	1
over-translation	3	1	9	13	0	0
style shift	9	12	4	4	0	0
humorousness shift	930	765	838	1427	68	4

Note that the categories *over-translation, style shift* and *humorousness shift* are necessarily subjective.

6.3   Results

Table 7 presents the results of submitted runs for Task 3. We observe that in many cases the successful translations are due to the existence of the same lexical ambiguity (homonymy) in both languages:

Example 9. A train load of paint derailed. Nearby businesses were put in the red.

Un train de peinture a déraillé. Les entreprises voisines ont été mises dans le rouge.

Example 10. An undertaker can be one of your best friends, he is always the last one to let you down.

Un entrepreneur peut être l'un de vos meilleurs amis, il est toujours le dernier á vous laisser tomber.

We also noticed some surprisingly successful translations:[5]

Example 11. Success comes in cans, failure comes in cant's.

Le succès c'est dans les canons, le pétrin c'est dans les canettes.

---

[5] On closer inspection, we determined that Example 12 was very close to an example from a train set.

Example 12. Wal-Mart Is Not the Only Saving Place. Come On In.
Le clerc n'est pas le seul á faire des économies.

Notably, a few successful translations used anglicisms:

Example 13. I used to be addicted to soap, but I'm clean now.
Avant, j'étais accro au savon, mais je suis clean maintenant.

Example 14. When the beekeeper moved into town he created quite a buzz.
Lorsque l'apiculteur s'est installé en ville, il a créé un véritable buzz.

Out of over 1155 translations containing wordplay, only 311 were translations of heterographic puns. This suggests that the state-of-the art machine translation is still unsuitable for translating wordplay, even with a manually annotated training set. The successful machine translations are seemingly accidental, owing to the existences of the same word ambiguity in both languages.

In total only 13% of automatically translated plays on words were successful, compared to the 90% success rate for instances translated by the human participants of our contest.

7   Conclusion

The goal of the JOKER project is to advance the automation of creative-language translation by developing the requisite parallel data and evaluation metrics for detecting, locating, interpreting, and translating wordplay. To this end, we organised the JOKER track at CLEF 2022, consisting of a workshop and associated pilot tasks on automatic wordplay analysis and translation. Seven teams submitted 19 runs for shared tasks.

Participants succeeded in wordplay location but the interpretation tasks raised difficulties. The binary classes HORIZONTAL/VERTICAL, CONVENTIONAL_FORM, CULTURAL_REFERENCE, OFFENSIVE, MANIPULATION_LEVEL were unbalanced, provoking very high but meaningless scores. However, these binary classifications were not the focus of our research.

Looking at the manually constructed data, we noticed that in a few instances, style shift in the translation of the pun could pose an issue. Consider the following pair:

Example 15. I phoned the zoo but the lion was busy.
J'ai appelé le zoo mais on m'a dit phoque you.

The French translation includes a vulgarism, with a pun across languages (fuck/ phoque). This was considered a very successful translation, but would clearly be an inappropriate translation in many contexts. A number of other examples that we could spot introduced strong stereotyping that could be construed as offensive, in contrast to the original.

We decided to annotate the data for those style shifts that introduced in the translation a form of humour relying on vulgarism or stereotyping. In doing so, another issue became evident: an additional bias may be introduced in the data due to the French language. Consider the following pair:

Example 16. Old Quilters never die, they just go under cover.

Les vieilles tricoteuses ne meurent jamais, elles recousent les morceaux.

French is more strongly gendered than English. As many French speakers still consider the use of masculine a default, this translation introduces a stereotype by using a feminine translation for the word *knitter* (tricoteuse). However, using the masculine form only, as a *default gender*, also raises questions in a context where the current evolution of the language seems to go against that usage [40]. In the future, one way forward may be to call for sensitivity readers to annotate the data, in order to provide a varied and unbiased set for ML training. (It also bears mentioning that a few translations used a gender going against the most common stereotype, perhaps deliberately.)

As the majority of translations for Task 2 were official ones, the teams that used large pre-trained models obtained very high scores. However, TEAM_JOKER provided very interesting results different from the official translations, even though they used Google's fine-tuned T5 model. Successful translations of puns in Task 3 are usually accidental as they exploit the ambiguity of the literal translation of the target wordplay term both in English and French. However, some translations are successful due to the right use of anglicism in French.

Although we have gained some insights, we have not done extensive research on the humorousness of wordplay. This aspect was frequently asked about by users of our corpus. The qualitative perception and appreciation of humour is important both for translation and human-computer interaction. (For example, is a given joke funny/offensive in the target language and culture? Should translators strive to preserve the humorousness of jokes, translating weak jokes with weak ones and clever jokes with clever ones, and if so, how can this best be effected? Virtual assistants like Siri and Alexa use predefined jokes in order to limit risks in humour perception, but is "playing it safe" in this manner unduly constraining the humanness of their interactions?) Although cultural background plays an important role in humour perception, other social factors may have a drastic impact on it, as in the case of private jokes. When evaluating translations of wordplay, we noticed that to be considered funny, translations needed to be challenging to some extent, but not too much. Sometimes humour and wordplay might not be understandable even for native speakers. For example, consider the following punning joke:

Example 17. "I'll never abandon my dogs!" Tom said cynically.

Appreciating the humour here requires knowing that the term *cynically* derives from the Greek word for "dog". A subsequent edition of our workshop will allow us to study the knowledge criteria for humorousness and to compare humour perception among native speakers and language learners. We plan to annotate our corpus with these subjective criteria.

Further details on the pilot tasks and the submitted runs can be found in the CLEF CEUR proceedings [11]. Additional information on the track is available on the JOKER website: http://www.joker-project.com/.

Acknowledgments. This work has been funded in part by the National Research Agency under the program *Investissements d'avenir* (Reference ANR-19-GURE-0001) and by the Austrian Science Fund under project M 2625-N31. JOKER is supported by *La Maison des sciences de l'homme en Bretagne*. We thank Orlane Puchalski, Adrien Couaillet, Ludivine Grégoire and Paul Campen for data collection as well as Eric Sanjuan for providing a server. We also thank the PC members: Monika Bokiniec, Ġorġ Mallia, Gordan Matas, Mohamed Saki, Alain Kerhervé, Grigori Sidorov, Victor Manuel Palma Preciado, Fabrice Antoine, and Danica Škara.

References

1. Blinov, V., Bolotova-Baranova, V., Braslavski, P.: Large dataset and language model fun-tuning for humor recognition. In: Proceedings of the 57th Annual Meeting of the Association for Computational Linguistics, pp. 4027–4032. Association for Computational Linguistics (2019). https://doi.org/10.18653/v1/P19-1394

2. Brown, T.B., et al.: Language models are few-shot learners. arXiv preprint arXiv:2005.14165 (2020)

3. Castro, S., Chiruzzo, L., Rosá, A., Garat, D., Moncecchi, G.: A crowd-annotated Spanish corpus for humor analysis. In: Proceedings of the Sixth International Workshop on Natural Language Processing for Social Media, pp. 7–11. Association for Computational Linguistics, July 2018. https://doi.org/10.18653/v1/W18-3502. https://www.aclweb.org/anthology/W18-3502

4. Cattle, A., Ma, X.: Recognizing humour using word associations and humour anchor extraction. In: Proceedings of the 27th International Conference on Computational Linguistics, Association for Computational Linguistics, Santa Fe, New Mexico, USA, pp. 1849–1858 (2018). https://www.aclweb.org/anthology/C18-1157

5. Delabastita, D.: There's a Double Tongue: an Investigation into the Translation of Shakespeare's Wordplay, with Special Reference to Hamlet. Rodopi, Amsterdam (1993)

6. Delabastita, D.: Introduction to the special issue on wordplay and translation. Translator: Stud. Intercultural Commun. **2**(2), 1–22 (1996). https://doi.org/10.1080/13556509.1996.10798970

7. Delabastita, D.: Wordplay as a translation problem: a linguistic perspective. In: Ein internationales Handbuch zur Übersetzungsforschung, vol. 1, pp. 600–606. De Gruyter Mouton, July 2008. https://doi.org/10.1515/9783110137088.1.6.600

8. Devlin, J., Chang, M.W., Lee, K., Toutanova, K.: BERT: pre-training of deep bidirectional transformers for language understanding. In: Proceedings of the 2019 Conference of the North American Chapter of the Association for Computational Linguistics, vol. 1, pp. 4171–4186. Association for Computational Linguistics (2019). https://doi.org/10.18653/v1/n19-1423

9. Ermakova, L., et al.: CLEF workshop JOKER: automatic wordplay and humour translation. In: Hagen, M., et al. (eds.) ECIR 2022. LNCS, vol. 13186, pp. 355–363. Springer, Cham (2022). https://doi.org/10.1007/978-3-030-99739-7_45

10. Ermilov, A., Murashkina, N., Goryacheva, V., Braslavski, P.: Stierlitz meets SVM: humor detection in Russian. In: Ustalov, D., Filchenkov, A., Pivovarova, L., Žižka, J. (eds.) AINL 2018. CCIS, vol. 930, pp. 178–184. Springer, Cham (2018). https://doi.org/10.1007/978-3-030-01204-5_17

11. Faggioli, G., Ferro, N., Hanbury, A., Potthast, M. (eds.): Proceedings of the Working Notes of CLEF 2022: Conference and Labs of the Evaluation Forum. CEUR Workshop Proceedings. CEUR-WS.org (2022)

12. Farwell, D., Helmreich, S.: Pragmatics-based MT and the translation of puns. In: Proceedings of the 11th Annual Conference of the European Association for Machine Translation, pp. 187–194, June 2006. http://www.mt-archive.info/EAMT-2006-Farwell.pdf

13. Fellbaum, C. (ed.): WordNet: An Electronic Lexical Database. MIT Press, Cambridge (1998)

14. Francesconi, C., Bosco, C., Poletto, F., Sanguinetti, M.: Error analysis in a hate speech detection task: the case of HaSpeeDe-TW at EVALITA 2018. In: Bernardi, R., Navigli, R., Semeraro, G. (eds.) Proceedings of the 6th Italian Conference on Computational Linguistics, November 2018. http://ceur-ws.org/Vol-2481/paper32.pdf

15. Ghanem, B., Karoui, J., Benamara, F., Moriceau, V., Rosso, P.: IDAT@FIRE2019: overview of the track on irony detection in Arabic tweets. In: Proceedings of the 11th Forum for Information Retrieval Evaluation, pp. 10–13. Association for Computing Machinery (2019). https://doi.org/10.1145/3368567.3368585

16. Giorgadze, M.: Linguistic features of pun, its typology and classification. Eur. Sci. J. **10**(10) (2014). https://eujournal.org/index.php/esj/article/view/4819

17. Gottlieb, H.: You got the picture? On the polysemiotics of subtitling wordplay. In: Delabastita, D. (ed.) Traductio: Essays on Punning and Translation, pp. 207–232. St. Jerome, Manchester (1997)

18. Guibon, G., Ermakova, L., Seffih, H., Firsov, A., Le Noé-Bienvenu, G.: Multilingual fake news detection with satire. In: CICLing: International Conference on Computational Linguistics and Intelligent Text Processing, La Rochelle, France, April 2019. https://halshs.archives-ouvertes.fr/halshs-02391141

19. Hempelmann, C.F., Miller, T.: Puns: taxonomy and phonology. In: Attardo, S. (ed.) The Routledge Handbook of Language and Humor. Routledge Handbooks in Linguistics, pp. 95–108. Routledge, New York, February 2017. https://doi.org/10.4324/9781315731162-8

20. Hong, B.A., Ong, E.: Automatically extracting word relationships as templates for pun generation. In: Computational Approaches to Linguistic Creativity: Proceedings of the Workshop, pp. 24–31. Association for Computational Linguistics, June 2009

21. Hossain, N., Krumm, J., Gamon, M., Kautz, H.: SemEval-2020 task 7: assessing humor in edited news headlines. In: Proceedings of the Fourteenth Workshop on Semantic Evaluation, pp. 746–758. International Committee for Computational Linguistics, December 2020. https://aclanthology.org/2020.semeval-1.98

22. Jain, A., Yadav, P., Javed, H.: Equivoque: detection and interpretation of English puns. In: Proceedings of the 8th International Conference System Modeling and Advancement in Research Trends, pp. 262–265 (2019). https://doi.org/10.1109/SMART46866.2019.9117433

23. Jiang, C., Maddela, M., Lan, W., Zhong, Y., Xu, W.: Neural CRF model for sentence alignment in text simplification. arXiv:2005.02324 [cs], June 2020

24. Karoui, J., Benamara, F., Moriceau, V., Patti, V., Bosco, C., Aussenac-Gilles, N.: Exploring the impact of pragmatic phenomena on irony detection in tweets: a multilingual corpus study. In: 15th Conference of the European Chapter of the Association for Computational Linguistics, vol. 1, pp. 262–272. Association for Computational Linguistics (2017). https://oatao.univ-toulouse.fr/18921/

25. Karoui, J., Farah, B., Moriceau, V., Aussenac-Gilles, N., Hadrich-Belguith, L.: Towards a contextual pragmatic model to detect irony in tweets. In: Proceedings of the 53rd Annual Meeting of the Association for Computational Linguistics and the 7th International Joint Conference on Natural Language Processing, vol. 2, pp. 644–650. Association for Computational Linguistics (2015). https://doi.org/10.3115/v1/P15-2106. http://aclweb.org/anthology/P15-2106

26. Kolb, W., Miller, T.: Human-computer interaction in pun translation. In: Hadley, J., Taivalkoski-Shilov, K., Teixeira, C.S.C., Toral, A. (eds.) Using Technologies for Creative-Text Translation. Routledge (2022, to appear)

27. Lieber, O., Sharir, O., Lentz, B., Shoham, Y.: Jurassic-1: technical details and evaluation. White paper, AI21 Labs, August 2021. https://uploads-ssl.webflow.com/60fd4503684b466578c0d307/61138924626a6981ee09caf6_jurassic_tech_paper.pdf

28. Meaney, J.A., Wilson, S., Chiruzzo, L., Lopez, A., Magdy, W.: SemEval-2021 task 7: HaHackathon, detecting and rating humor and offense. In: Proceedings of the 15th International Workshop on Semantic Evaluation, pp. 105–119. Association for Computational Linguistics, August 2021. https://doi.org/10.18653/v1/2021.semeval-1.9. https://aclanthology.org/2021.semeval-1.9

29. Mihalcea, R., Strapparava, C.: Making computers laugh: investigations in automatic humor recognition. In: Human Language Technology Conference and Conference on Empirical Methods in Natural Language Processing: Proceedings of the Conference, Stroudsburg, PA, pp. 531–538. Association for Computational Linguistics, October 2005. https://doi.org/10.3115/1220575.1220642. http://www.aclweb.org/anthology/H/H05/H05-1067

30. Miller, T.: The punster's amanuensis: the proper place of humans and machines in the translation of wordplay. In: Proceedings of the Second Workshop on Human-Informed Translation and Interpreting Technology, pp. 57–64, September 2019. https://doi.org/10.26615/issn.2683-0078.2019_007

31. Miller, T., Hempelmann, C.F., Gurevych, I.: SemEval-2017 task 7: detection and interpretation of English puns. In: Proceedings of the 11th International Workshop on Semantic Evaluation, pp. 58–68, August 2017. https://doi.org/10.18653/v1/S17-2005

32. Nijholt, A., Niculescu, A., Valitutti, A., Banchs, R.E.: Humor in human-computer interaction: a short survey. In: Proceedings of INTERACT 2017 (2017)

33. Papineni, K., Roukos, S., Ward, T., Zhu, W.J.: BLEU: a method for automatic evaluation of machine translation. In: Proceedings of the 40th Annual Meeting of the Association for Computational Linguistics, pp. 311–318 (2002). https://doi.org/10.3115/1073083.1073135. https://www.aclweb.org/anthology/P02-1040

34. Potash, P., Romanov, A., Rumshisky, A.: SemEval-2017 task 6: #HashtagWars: learning a sense of humor. In: Proceedings of the 11th International Workshop on Semantic Evaluation, pp. 49–57. Association for Computational Linguistics, August 2017. https://doi.org/10.18653/v1/S17-2004

35. Radford, A., Wu, J., Child, R., Luan, D., Amodei, D., Sutskever, I.: Language models are unsupervised multitask learners. Technical report (2019). https://cdn.openai.com/better-language-models/language_models_are_unsupervised_multitask_learners.pdf

36. Reyes, A., Buscaldi, D., Rosso, P.: An analysis of the impact of ambiguity on automatic humour recognition. In: Matoušek, V., Mautner, P. (eds.) TSD 2009. LNCS (LNAI), vol. 5729, pp. 162–169. Springer, Heidelberg (2009). https://doi.org/10.1007/978-3-642-04208-9_25

37. Reyes, A., Rosso, P., Buscaldi, D.: From humor recognition to irony detection: the figurative language of social media. Data Knowl. Eng. **74**, 1–12 (2012). https://doi.org/10.1016/j.datak.2012.02.005
38. Valitutti, A., Toivonen, H., Doucet, A., Toivanen, J.M.: "Let everything turn well in your wife": generation of adult humor using lexical constraints. In: Proceedings of the 51st Annual Meeting of the Association for Computational Linguistics, vol. 2, pp. 243–248. Association for Computational Linguistics, August 2013. https://aclanthology.org/P13-2044
39. Vaswani, A., et al.: Attention is all you need. arXiv:1706.03762 [cs], December 2017
40. Viennot, E.: Le langage inclusif: pourquoi, comment. Les Éditions iXe (2020)
41. Vrticka, P., Black, J.M., Reiss, A.L.: The neural basis of humour processing. Nat. Rev. Neurosci. **14**(12), 860–868 (2013). https://doi.org/10.1038/nrn3566
42. Xue, L., et al.: mT5: a massively multilingual pre-trained text-to-text transformer. In: Proceedings of the 2021 Conference of the North American Chapter of the Association for Computational Linguistics, pp. 483–498. Association for Computational Linguistics, June 2021. https://doi.org/10.18653/v1/2021.naacl-main.41. https://aclanthology.org/2021.naacl-main.41
43. Yang, D., Lavie, A., Dyer, C., Hovy, E.: Humor recognition and humor anchor extraction. In: Proceedings of the 2015 Conference on Empirical Methods in Natural Language Processing, pp. 2367–2376. Association for Computational Linguistics, September 2015. https://doi.org/10.18653/v1/D15-1284. https://www.aclweb.org/anthology/D15-1284
44. Yu, Z., Tan, J., Wan, X.: A neural approach to pun generation. In: Proceedings of the 56th Annual Meeting of the Association for Computational Linguistics, vol. 1, pp. 1650–1660. Association for Computational Linguistics, July 2018. https://doi.org/10.18653/v1/P18-1153. https://aclanthology.org/P18-1153

Overview of the CLEF 2022 SimpleText Lab: Automatic Simplification of Scientific Texts

Liana Ermakova[1]([✉]), Eric SanJuan[2], Jaap Kamps[3], Stéphane Huet[2], Irina Ovchinnikova[4], Diana Nurbakova[5], Sílvia Araújo[6], Radia Hannachi[7], Elise Mathurin[1], and Patrice Bellot[8]

[1] Université de Bretagne Occidentale, HCTI, Brest, France
liana.ermakova@univ-brest.fr
[2] Avignon Université, LIA, Avignon, France
[3] University of Amsterdam, Amsterdam, The Netherlands
[4] ManPower Language Solution, Tel Aviv, Israel
[5] University of Lyon, INSA Lyon, CNRS, LIRIS, Lyon, France
[6] University of Minho, Braga, Portugal
[7] Université de Bretagne Sud, HCTI, Morbihan, France
[8] Aix Marseille Univ, Université de Toulon, CNRS, LIS, Toulon, France

Abstract. Although citizens agree on the importance of objective scientific information, yet they tend to avoid scientific literature due to access restrictions, its complex language or their lack of prior background knowledge. Instead, they rely on shallow information on the web or social media often published for commercial or political incentives rather than the correctness and informational value. This paper presents an overview of the CLEF 2022 SimpleText track addressing the challenges of text simplification approaches in the context of promoting scientific information access, by providing appropriate data and benchmarks, and creating a community of IR and NLP researchers working together to resolve one of the greatest challenges of today. The track provides a corpus of scientific literature abstracts and popular science requests. It features three tasks. First, *content selection* (what is in, or out?) challenges systems to select passages to include in a simplified summary in response to a query. Second, *complexity spotting* (what is unclear?) given a passage and a query, aims to rank terms/concepts that are required to be explained for understanding this passage (definitions, context, applications). Third, *text simplification* (rewrite this!) given a query, asks to simplify passages from scientific abstracts while preserving the main content.

Keywords: Scientific text simplification · (Multi-document) summarization · Contextualization · Background knowledge · Scientific information distortion

1 Introduction

Scientific literacy is an important ability for people. It is one of the keys for critical thinking, objective decision-making and judgment of the validity and significance of findings and arguments, which allows discerning facts from fiction. Thus, having a basic scientific knowledge may also help maintain one's health, both physiological and

A. Barrón-Cedeño et al. (Eds.): CLEF 2022, LNCS 13390, pp. 470–494, 2022.
https://doi.org/10.1007/978-3-031-13643-6_28

mental. The COVID-19 pandemic provides a good example of such a matter. Understanding the issue itself, choosing to use or avoid particular treatment or prevention procedures can become crucial. However, the recent pandemic has also shown that simplification can be modulated by political needs and the scientific information can be distorted [14]. Thus, the evaluation of the alteration of scientific information during the simplification process is crucial but underrepresented in the state-of-the-art.

Digitization and open access have made scientific literature available to every citizen. While this is an important first step, there are several remaining barriers preventing laypersons to access the objective scientific knowledge in the literature. In particular, scientific texts are often hard to understand as they require solid background knowledge and use tricky terminology. Although there were some recent efforts on text simplification (e.g. [23]), removing such understanding barriers between scientific texts and general public in an automatic manner is still an open challenge. The CLEF 2022 SimpleText track[1] brings together researchers and practitioners working on the generation of simplified summaries of scientific texts. It is a new evaluation lab that follows up the SimpleText-2021 Workshop [11]. All perspectives on automatic science popularisation are welcome, including but not limited to: Natural Language Processing (NLP), Information Retrieval (IR), Linguistics, Scientific Journalism, etc.

SimpleText provides data and benchmarks for discussion of challenges of automatic text simplification by bringing in the following interconnected tasks:

Task 1: What is in (or out)? Select passages to include in a simplified summary, given a query.

Task 2: What is unclear? Given a passage and a query, rank terms/concepts that are required to be explained for understanding this passage (definitions, context, applications, ...).

Task 3: Rewrite this! Given a query, simplify passages from scientific abstracts.

Automatic scientific text simplification is a very ambitious problem which cannot be addressed by a simple solution, but we have isolated three clear challenges that need to be addressed to improve non-expert access to scientific literature. In order to simplify scientific texts, one has to (1) select the information to be included in a simplified summary, (2) decide whether the selected information is sufficient and comprehensible or provide some background knowledge if not, (3) improve the readability of the text [10]. Our lab is organised around this pipeline. Our test data was built accordingly as we asked to rank difficult terms (Task 2) and simplify sentences (Task 3) retrieved for Task 1 and we evaluated the results with regard to the queries from Task 1.

In the CLEF 2022 edition of SimpleText, a total of 62 teams registered for the SimpleText track. A total of 40 users downloaded data from the server. A total of 9 distinct teams submitted 24 runs, of which 10 runs were updated. The details of statistics on runs submitted for shared tasks are presented in Table 1.

This introduction is followed by Section 2 presenting a brief overview of related evaluation initiatives, related tasks and related approaches. The bulk of this paper presents the tasks with the datasets and evaluation metrics used, as well as the results of the participants, in three self-contained sections: Section 3 on the first task about content

---

[1] https://simpletext-project.com.

Table 1. CLEF 2022 SimpleText official run submission statistic

Team	Task 1	Task 2	Task 3	Total runs
UAms	2	1		3
NLP@IISERB	3 (3 updated)			3
SimpleScientificText		1 (1 updated)		1
aaac		1 (1 updated)		1
LEA_T5		1	1	2
PortLinguE			1 (1 updated)	1
CYUT Team2	1		1	2
HULAT-UC3M			10 (4 updated)	10
CLARA-HD			1	1
Total runs	6	4	14	24

selection, Section 4 on the second task about complexity spotting, and Section 5 on the third task about text simplification proper. We end with Section 6 discussing the results and findings, and lessons for the future.

2   Related Work

This section presents a brief overview of related evaluation initiatives, related tasks and related approaches.

In parallel with the CLEF SimpleText track, which was accepted in 2020, there have been a range of related initiatives on scholarly document processing at NLP conference. In 2020, Scholarly Document Processing[2] provided the shared tasks on

- CL-SciSumm: Scientific Document Summarization;
- CL-LaySumm:Lay Summary;
- LongSumm: Generating Long Summaries for Scientific Documents.

CL-SciSumm and LongSumm are focused on summarization task but no adaptation to general public is previewed. The CL-SciSumm'20 LaySummary [6] subtask asked to produce a scientific paper summary without technical jargon. However, terms are not often replaceable due to the risk of information distortion and these complex concepts should be explained to a reader.

In 2022 the Third Workshop on Scholarly Document Processing[3] hosted the following shared tasks which are related to our track although they don't tackle the simplification aspect:

- MSLR22: Multi-Document Summarization for Literature Reviews;
- DAGPap22: Detecting automatically generated scientific papers;
- LongSumm 2022: Generating Long Summaries for Scientific Documents;

---

[2] https://ornlcda.github.io/SDProc/sharedtasks.html.

[3] https://sdproc.org/2022/sharedtasks.html.

- SV-Ident 2022: Survey Variable Identification in Social Science Publications;
- Scholarly Knowledge Graph Generation;
- Multi Perspective Scientific Document Summarization.

As it turns out, the SimpleText tasks and SDProc tasks are complementary, and together build a larger community to work on this important problem.

Popular science articles are generally much shorter than scientific publications. Thus, summarization is a step to text simplification as it reduces the amount of information to be processed. However, information selection is understudied task in document simplification [41] as existing works mainly focus on word/phrase-level [24] or sentence-level simplifications [9]. However, the lack of background knowledge can become a barrier to reading comprehension and there is a knowledge threshold allowing reading comprehension [30]. Scientific text simplification presupposes the facilitation of readers' understanding of complex content by establishing links to basic lexicon while traditional methods of text simplification try to eliminate complex concepts and constructions [24]. SimpleText is not limited to a "Split and Rephrase" task [26] but also aims to provide a sufficient context to a scientific text. Entity linking could mitigate the background knowledge problem, by providing definitions, illustrations, examples, and related entities, but the existing entity linking datasets are focused on people, places, and organisation [19], while a non-expert reader of a scientific article needs assistance with new concepts and methods. INEX/CLEF'11–14 Tweet Contextualization [4] and CLEF'16–17 Cultural Microblog Contextualization [13] tracks aim to provide lacking background knowledge to a tweet. Besides completely different nature of tweets and popular science, this use case differs from the text simplification as this lack of background knowledge is due to the tweet length. In contrast to the Background Linking task at TREC'20 News Track [3], SimpleText focuses on (1) scientific text; (2) selection of notions to be explained; (3) helpfulness of the provided information rather than its relevance.

Large pre-trained AI models, like Jurassic-1 [20], Google's T5 [38], BERT or GPT-3 [5], outperformed other state-of-the-art models on several NLP tasks, including automatic summarization and text simplification [40], but their serious issues are (1) consistency and coherency (coreference errors) [35] and (2) limitation to short texts ($<$2k tokens) [39]. Simple Wikipedia based datasets could be useful to train AI models but (1) they are not scientific publications; (2) there is no direct correspondence between Wikipedia and Simple Wikipedia articles [14]. Another dataset was introduced at TAC 2014 Biomedical Summarization Track [1] with a goal to retrieve important aspects of a paper from the perspective of the community.

Automatic evaluation metrics have been designed to measure the results of text simplification: SARI [37] targets lexical complexity, while SAMSA estimates sentence structural complexity [32]. Standard evaluation measures (e.g. BLEU, ROUGE) are difficult to apply as one should consider the end user as well as source document content. Since traditional readability indices can be misleading [36], researchers proposed various approaches based on expert judgement [8], readability level [17], relevance judgement [7], crowd-sourcing [2], eye-tracking [18].

In contrast to that, we evaluate simplification in terms of lexical and syntax complexity combining with error analysis. As we demonstrated previously, scientific information is often distorted accidentally due to misunderstanding of terminology, omission of essential details, insertion of erroneous background etc. [14]. Information distortion analysis is close to scientific claim verification [25, 34] but fact checking is limited to search for relevant evidence and decide whether it supports the claim. Another close work is [31], where the TF-IDF cosine similarity between documents is computed on (1) a collection of abstracts of scientific papers from the Citation Network Dataset V1 AMINER [33] and (2) a set of articles from Huffington Post. However, this approach is not robust to lexical changes, which are crucial for text simplification. To the best of our knowledge, no other automatic nor semi-automatic method for information distortion analysis exists.

3   Task 1: What Is in (or Out)?

In this section, we discuss the first task about content selection (and *avoiding* complexity) from a corpus of scientific abstracts, addressing the task:

Select passages to include in a simplified summary, given a query.

The task aims at finding references in computer science that could be inserted as citations in original press articles of general audience for illustration, fact checking or actualization. For each of the selected references, more relevant sentences need to be extracted. These passages can be complex and require further simplification to be carried out in Tasks 2 and 3. Task 1 focuses on content retrieval.

3.1   Evaluation Framework

Corpus. As in 2021, we use the Citation Network Dataset: DBLP+Citation, ACM Citation network (12th version) [33] as source of scientific documents that can be used as reference passages [10]. It contains:

- 4,894,083 bibliographic references published before 2020;
- 4,232,520 abstracts in English;
- 3,058,315 authors with their affiliations;
- 45,565,790 ACM citations.

Textual content together with authorship can be extracted from this corpus. Although we manually preselected abstracts for topics, participants also have access to an Elastic Search index; this index is adequate to passage retrieval using BM25.

Additional datasets have been extracted to generate Latent Dirichlet Allocation models for query expansion or train Graph Neural Networks for citation recommendation as carried out in StellarGraph[4] for example. The shared datasets provide: document abstract content for LDA (Latent Dirichlet Allocation) or Word Embedding (WE); document authors for coauthoring analysis; citation relationship between documents for co-citation analysis; citations by author for author impact factor analysis. These extra datasets are intended to be used to select passages by authors who are experts on the topic (highly cited by the community).

---

[4] https://stellargraph.readthedocs.io/.

Table 2. SimpleText Task 1: Examples of topics and queries

Topic ID	Query ID	Title or Query
G12		*Patient data from GP surgeries sold to US companies*
	G12.1	`patient data`
G13		*Baffled by digital marketing? Find your way out of the maze*
	G13.1	`digital marketing`
	G13.2	`advertising`

Topics. Topics are a selection of 40 press articles: 20 from *The Guardian*,[5] a major international newspaper for a general audience with a tech section, and 20 from *Tech Xplore*[6] a website taking part in the Science X Network to provide a comprehensive coverage of engineering and technology advances. Each article was selected in the computer science field to be in accordance with the provided corpus. URLs to original articles, the title and textual content of each topic were provided to participants. Articles were enriched with queries manually extracted from their content to provide an indication of the essential technical concepts covered. We manually checked that each query allows participants to retrieve from the corpus at least 5 relevant passages that could be inserted as citations in the press article. The use of these queries were optional. Examples of topics and queries are given in Table 2.

Output Formats. Results had to be provided in a TREC style tabulated format (with a ".csv" extension). The following columns were required (including the first line):

run_id Run ID starting with team ID, followed by "task1" and run name
manual Whether the run is manual $\{0,1\}$
topic_id Topic ID
query_id Query ID used to retrieve the document (if one of the queries provided for the topic was used; 0 otherwise)
doc_id ID of the retrieved document (to be extracted from the JSON output)
passage Text of the selected passage (abstract)

For each topic, the maximum number of distinct DBLP references (_id json field) was 100 and the total length of passages was not to exceed 1,000 tokens. Table 3 shows an example of Task 1 output.

Evaluation Metrics. All passages retrieved from DBLP by participants are expected to have some overlap (lexical or semantic) with the article content. Passage relevance were evaluated through:

1. Lexical and semantic overlap of extracted passages with topic article content,
2. Manual assessment of a pool of passages.

---

[5] https://www.theguardian.com/science.
[6] https://techxplore.com/.

Table 3. SimpleText Task 1: Examples of output

Run	M/A	Topic	Query	Doc	Passage
ST1_task1_1	0	G01	G01.1	1564531496	A CDA is a mobile user device, similar to a Personal Digital Assistant (PDA). It supports the citizen when dealing with public authorities and proves his rights - if desired, even without revealing his identity.
ST1_task1_1	0	G01	G01.1	3000234933	People are becoming increasingly comfortable using Digital Assistants (DAs) to interact with services or connected objects
ST1_task1_1	0	G01	G01.2	1448624402	As extensive experimental research has shown individuals suffer from diverse biases in decision-making.

To build a pooled test collection, we first extracted all the article IDs ranked by the number of participants who used the article to select passages. From this extraction, we only kept articles chosen by at least two participants and gave a relevance score on a scale of 0 to 5:

0 for irrelevant articles;
1 for marginally relevant articles;
2 when the abstract is relevant with the query;
3 when the abstract and keywords are relevant with the query;
4 when the abstract and keywords are relevant with the query and the topic (title of the original article);
5 when the abstract and keywords are relevant with the query and the extended topic (content of the original article).

In order to speed up the judgment process, for this edition we only evaluated relevance at the article level, and not at the sentence level. The abstract was considered as relevant as soon it has a sentence useful to explain the title or the original article.

Among documents returned by at least three runs we found out:

– 14 Guardian topics with lightly relevant documents;
– 11 Guardian topics with highly relevant documents;
– 10 Tech topics with lightly relevant documents; and
– 9 Tech topics with highly relevant documents.

For the documents returned by two runs, we had a high number of 1 and 2 scores for the Guardian topics. As regards the Tech Xplore topics, which have more technical queries since they deal with more technical and specific areas, queries were less ambivalent and more in keeping with the content of DBLP corpus. This has resulted in usually higher relevant scores, with many articles retrieved by two participants having a score of 3. Globally, whether the query comes from the Guardian or Tech Xplore, human evaluators found abstracts, among the articles retrieved by the participants from DBLP, that really explain the article or have matters which should have been addressed in the

Table 4. SimpleText Task 1: Evaluation scores of official runs

Team	Score	#Docs	Doc Avg	#Queries	Query Avg	NDCG
CYUT	125	44	0.53	77	1.62	0.3322
UAMS-MF*	163	54	0.87	99	1.65	0.2761
UAMS	52	17	0.22	40	1.30	0.1048
NLP@IISERB	26	7	0.35	13	2.00	0.0290

* Manual run.

original article. Passages were often issued from publications that are more related to cognitive or information sciences than to technical fields, which shows that the DBLP corpus has expanded beyond computer science.

3.2 Results

A total of 3 teams submitted 6 runs: 4 automatic runs extracted 100 documents or abstracts per subquery, the CYUT automatic run extracted 5 sentences per subquery, and the manual extracted passages for a selection of subqueries.

We consider here the reduced pool of documents returned by at least two runs; there are 72 topics with judgments, with a mean of 6.7 and a median of 4 judged documents per topic. Since we have participants that focused on a short list of documents, we only report results computed at a depth of 5 returned documents. Table 4 shows cumulative (0–5) scores obtained by each run (*Score*), the number of returned documents with a score ≥ 1 (*#Docs*), the number of queries with at least one returned document (*#Queries*) and the average scores per document and query. We also provide NDCG@5 as the metrics used for official ranking on this task. These values show that the automatic run made by CYUT and the manual run significantly outperform other automatic runs in terms of selecting the abstracts with a high relevance.

4   Task 2: What Is Unclear?

In this section, we discuss the second task about complexity spotting in an extracted sentence from a scientific abstract, addressing the task:

> *Given a passage and a query, rank terms/concepts that are required to be explained for understanding this passage (definitions, context, applications etc.).*

The goal of this task is to decide which terms (up to 5) require explanation and contextualization to help a reader to understand a complex scientific text—for example, with regard to a query, terms that need to be contextualized (with a definition, example and/or use-case). For each passage, participants should provide a ranked list of difficult terms with corresponding scores on the scale 1–3 (3 to be the most difficult terms, while the meaning of terms scored 1 can be derived or guessed) and on the scale 1–5 (5 to be the most difficult terms). Passages (sentences) are considered to be independent, i.e. difficult term repetition was allowed.

4.1 Evaluation Framework

Train Dataset. For this task, data is two-fold: *Medicine* and *Computer Science*, as these two domains are the most popular on forums like ELI5 [12,29]. As in 2021, for *Computer Science*, we use scientific abstracts from the Citation Network Dataset: DBLP+Citation, ACM Citation network (12th version)[7] [10]. A master student in Technical Writing and Translation manually annotated each sentence by extracting difficult terms and attributing difficulty scores on a scale of 1–3 (3 to be the most difficult terms, while the meaning of terms scored 1 can be derived or guessed) and on a scale of 1–5 (5 to be the most difficult terms).

In 2022, we introduced new data based on Google Scholar and PubMed articles on muscle hypertrophy and health annotated by a master student in Technical Writing and Translation, specializing in these domains. The selected abstracts included the objectives of the study, the results and sometimes the methodology. The abstracts including only the topic of the study were excluded because of the lack of information. To avoid the curse of knowledge, another master student in Technical Writing and Translation not familiar with the domain was solicited for complexity spotting.

We provided 453 annotated examples in total.

Test Dataset. To construct the test data, we retrieved 116,763 sentences from the DBLP abstracts according to the queries from Task 1. We then manually evaluated 592 distinct sentences for 11 queries. For the query *Digital assistant* we took the first 1,000 sentences retrieved by ElasticSearch. We pool terms submitted by all participants for all these queries, representing a number of 4,167 distinct pairs *sentence-term* in total. We ensured that for each evaluated source sentence the pool contained the results of all participants. Statistics of the number of evaluated sentences per query for Task 2 are given in Table 5.

Input and Output Formats. The input for the train and the test data was provided in JSON and CSV formats with the following fields:

snt_id a unique passage (sentence) identifier.
source_snt passage text.
doc_id a unique source document identifier.
query_id a query ID.
query_text difficult terms should be extracted from sentences with regard to this query.

Input example (JSON format):

```
{"snt_id":"G06.2_2548923997_3", "source_snt":"These
  communication systems render self-driving vehicles
  vulnerable to many types of malicious attacks, such as Sybil
  attacks, Denial of Service (DoS), black hole, grey hole and
  wormhole attacks.", "doc_id":2548923997, "query_id":"G06.2",
  "query_text":"self driving"}
```

---

[7] https://www.aminer.org/citation.

Table 5. SimpleText Task 2: Statistics of the number of evaluated sentences per query

Query		# Sentences	# Sentence-term pairs
1	*guessing attack*	60	389
2	*end to end encryption*	55	390
3	*imbalanced data*	55	381
4	*distributed attack*	54	385
5	*genetic algorithm*	51	374
6	*quantum computing*	51	385
7	*qbit*	50	363
8	*side-channel attack*	49	340
9	*traffic optimization*	47	344
10	*quantum applications*	42	320
11	*cyber-security*	35	244
12	*conspiracy theories*	23	180
13	*crowsourcing*	15	104
14	*digital assistant*	5	32

Participants had to submit a list of terms to be contextualized in a JSON format or a tabulated file TSV (for manual runs) with the following fields:

run_id Run ID starting with (team_id)_(task_id)_(*name*).
manual Whether the run is manual $\{0, 1\}$.
snt_id a unique passage (sentence) identifier from the input file.
term Term or other phrase to be explained.
term_rank_snt term difficulty rank within the given sentence.
score_5 term difficulty score on the scale from 1 to 5 (5 to be the most difficult terms).
score_3 term difficulty score on the scale from 1 to 3 (3 to be the most difficult terms).

Output example (JSON format):

```
{"run_id":"NP_task_2_run1", "manual":1,
  ↪  "snt_id":"G06.2_2548923997_3", "term":"black hole attack",
  ↪  "term_rank_snt":1, "score_5":5, "score_3":3},
{"run_id":"NP_task_2_run1", "manual":1,
  ↪  "snt_id":"G06.2_2548923997_3", "term":"grey hole attack",
  ↪  "term_rank_snt":2, "score_5":5, "score_3":3},
{"run_id":"NP_task_2_run1", "manual":1,
  ↪  "snt_id":"G06.2_2548923997_3", "term":"Sybil attack",
  ↪  "term_rank_snt":3, "score_5":5, "score_3":3},
{"run_id":"NP_task_2_run1", "manual":1,
  ↪  "snt_id":"G06.2_2548923997_3", "term":"wormhole attack",
  ↪  "term_rank_snt":4, "score_5":5,"score_3":3},
{"run_id":"NP_task_2_run1", "manual":1,
  ↪  "snt_id":"G06.2_2548923997_3", "term":"Denial of service
  ↪  attack", "term_rank_snt":5, "score_5":4, "score_3":3}
```

Table 6. Examples of the term difficulty scale used for evaluation. Difficult terms are highlighted with the green color

Grade	Non-abbreviated (ordinary) term	Abbreviation
7	*external qubit* in "The qubit—qutrit pair acts as a closed system and one *external qubit* serve as the environment for the pair."	*XCSFHP* in "We compared *XCSFHP* to XCSF on several problems."
6	"This paper bring forward based on immune genetic algorithm to solve *man on board automated storage and retrieval system* optimized problem, immune genetic algorithm remains the characteristic which is not ..." "*Tile coding* is a well-known function approximator that has been successfully applied to many reinforcement learning tasks."	"*XCS* with computed prediction, namely XCSF, extends XCS by replacing the classifier prediction with a parametrized prediction function." "Side-channel attack (*SCA*) is a very efficient cryptanalysis technology to attack cryptographic devices."
5	"Experiment simulation result express: the result of *immune genetic algorithm* is better than traditional genetic algorithm in the circumstance of the same clusters and the same evolution generation."	"This paper presents a simple real-coded estimation of distribution algorithm (EDA) design using x-ary extended compact genetic algorithm (*XECGA*) and discretization methods."
4	"Immune genetic algorithm can shorten storage or retrieval distance in application, and enhance storage or *retrieval efficiency*." "*Deep learning* has become increasingly popular in both academic and industrial areas in the past years."	"This paper presents a simple real-coded estimation of distribution algorithm (*EDA*) design using x-ary extended compact genetic algorithm (XECGA) and discretization methods."
3	"The XECGA is then used to build the probabilistic model and to sample a new population based on the *probabilistic model*."	"We evaluate each measure's performance by *AUC* which is usually used for evaluation of imbalanced data classification."
2	"Experiment simulation result express: the result of immune genetic algorithm is better than traditional genetic algorithm in the circumstance of the same *clusters* and the same evolution generation." "Specifically, the real-valued *decision variables* are mapped to discrete symbols of user-specified cardinality using discretization methods."	*NIST* (The National Institute of Standards and Technology) in "Recently *NIST* has published the second draft document of recommendation for the entropy sources used for random bit generation."
1	"video labeling game is a *crowsourcing* tool to collect user-generated metadata for video clips." "On the other hand, a 3dimensional (3D) map, which is one of major themes in machine vision research, has been utilized as a simulation tool in city and *landscape planning*, and other engineering fields."	*2D* (2-dimensional), *3D* (3-dimensional) *maps* as in "The *3D maps* will give more intuitive information compared to conventional 2-dimensional (*2D*) ones."
0	"This *device* has two work modes: "native" and "remote"." "The proposed rECGA is *simple*, making it amenable for further empirical and theoretical analysis."	*et al.* (from latin "*et alii*" meaning "and others") in "However, Nam *et al.* pointed out..."

Table 7. SimpleText Task 2: Scale conversion rules

Term difficulty scale	0	1	2	3	4	5	6	7
7 point scale	0	1	2	3	4	5	6	7
⇒ 5 point scale	0	1		2		3	4		5
7 point scale	0	1	2	3	4	5	6	7
⇒ 3 point scale	0		1			2			3

Table 8. SimpleText Task 2: Examples of the annotation

Sentence	Term	Limits		Diffi-
		OK	Corrected	culty
This device has two work modes: 'native' and ' remote'.	remote	YES		1
This device has two work modes: 'native' and 'remote'.	work modes	YES		0
This device has two work modes: 'native' and 'remote'.	modes native	NO	work modes	0
This device has two work modes: 'native' and 'remote'.	device work	NO	device	0
This device has two work modes: 'native' and 'remote'.	native remote	NO	native	1

Evaluation Metrics. We evaluated terms according to:

– correctness of term limits;
– term difficulty score on the scale 1–3;
– term difficulty score on the scale 1–5.

For both scales of term difficulty, we used a converted scale 1–7. This scale 1–7 was chosen following the psycho-linguistic research of the perception and evaluation of lexical meanings performed by Osgood and his colleagues [27], in contrast to the psychometric Likert scale (1–5, Strongly disagree/Disagree/Neither agree nor disagree/Agree/Strongly agree), commonly used in the research that employs questionnaires [21]. In the classical version of the semantic differential technique, the scale shows the variety of the human perception of semantic nuances from negative (-3) to positive (+3) polarity where 0 marks the "norm" [27]. The scale 1–7 matches the Osgood's scale and seems more suitable to evaluate concepts and features avoiding associations with negative/positive assessment. Since the 1970s, the scale has been employed in various studies as an evaluation tool for qualitative features.

Table 6 provides examples of the used term difficulty scale. We separate the examples of abbreviations from non-abbreviated phrases/words.

We added 0 for terms that should not be explained at all and we converted the original scale 1–7 as presented in Table 7.

Table 8 provides some examples of the annotation for Task 2. *TERM* refers to the terms retrieved by participants, *Correct limits* is a binary category showing whether the retrieved terms is well limited, *Corrected* is an eventual correction of retrieved term limits, *Difficulty* is a term difficulty score in scale 1–7.

Table 9. SimpleText Task 2: Results for the official runs

	Total	Evaluated		Score_3		Score_5	
			+Limits		+Limits		+Limits
aaac	581,285	2,951	1,388	702	318	415	175
SimpleScientificText	63,027	298	262	48	44	47	42
UAms	263,022	1,315	1,175	105	69	60	49
lea_t5	23,331	5	4	0	0	0	0

Table 10. SimpleText Task 2: Results on a subset of 167 common sentences

	Total	Evaluated		Score_3		Score_5	
			+Limits		+Limits		+Limits
aaac	581,285	833	414	200	104	127	67
UAms	263,022	574	514	46	28	25	21
SimpleScientificText	63,027	208	188	33	32	32	29

4.2 Results

A total of 4 teams submitted runs, of which 2 runs were updated. The results are given in Tables 9 and 10. In both tables, we present results for correctly attributed scores regardless the correctness of term limits (*Score_3* and *Score_5*) and the number of correctly limited terms with correctly attributed scores (+ *Limits*). Table 9 provides the results on all sentences we evaluated. However, to have comparable results for partial runs we also report scores on a subset 167 common sentences in Table 10, although we were constrained to exclude the run *lea_t5* due to a very low number of evaluated sentences.

5   Task 3: Rewrite This!

In this section, we discuss the third task about text simplification proper, rewriting an extracted sentence from a scientific abstract, addressing the task:

Given a query, simplify passages from scientific abstracts.

The goal of this task is to provide a simplified version of text passages (sentences) with regard to a query. Participants were provided with queries and abstracts of scientific papers. The abstracts could be split into sentences. The simplified passages were evaluated manually in terms of the produced errors as follows.

5.1   Evaluation Framework

Train Dataset. As for *Task 2: What is unclear?*, we provided a parallel corpus of simplified sentences from two domains: *Medicine* and *Computer Science* (see Sect. 4.1). As previously, we use scientific abstracts from the DBLP Citation Network Dataset for

Table 11. SimpleText Task 3: Statistics of the number of evaluated sentences per query

Query		# Distinct source sentences	# Distinct simplified sentences
1	*digital assistant*	370	1,280
2	*conspiracy theories*	195	398
3	*end to end encryption*	55	102
4	*imbalanced data*	55	87
5	*genetic algorithm*	51	85
6	*quantum computing*	51	85
7	*qbit*	50	76
8	*quantum applications*	42	73
9	*cyber-security*	28	47
10	*fairness*	18	22
11	*crowsourcing*	14	21

Computer Science and Google Scholar and PubMed articles on muscle hypertrophy and health *Medicine* [10, 12].

Text passages issued from abstracts on computer science were simplified by either a master student in Technical Writing and Translation or a pair of experts: (1) a computer scientist and (2) a professional translator, English native speaker but not specialist in computer science [12]. Each passage was discussed and rewritten multiple times until it became clear for non-computer scientists. Medicine articles were annotated by a master student in Technical Writing and Translation specializing in this domain. Sentences were shortened, excluding every detail that was irrelevant or unnecessary to the comprehension of the study, and rephrased, using simpler vocabulary. If necessary, concepts were explained.

We provided 648 parallel sentences in total.

Test Dataset. We used the same 116,763 sentences retrieved by the ElasticSearch engine from the DBLP dataset according to the queries as for Task 2 (see Sect. 4.1). We manually evaluated 2,276 pairs of sentences for 11 queries. For the query *Digital assistant* we took the first 1,000 sentences retrieved by ElasticSearch. We pool source sentences coupled with their simplified versions submitted by all participants for all these queries. We ensured that for each evaluated source sentence the pool contained the results of all participants. The detailed statistics of the number of evaluated sentences per query for Task 3 are given in Table 11.

Input and Output Format. The input train and the test data were provided in JSON and CSV formats with the following fields:

snt_id a unique passage (sentence) identifier.
source_snt passage text.
doc_id a unique source document identifier.
query_id a query ID.
query_text simplification should be done with regard to this query.

484   L. Ermakova et al.

Input example (JSON format):

```
{"snt_id":"G11.1_2892036907_2", "source_snt":"With the ever
 ⌐ increasing number of unmanned aerial vehicles getting
 ⌐ involved in activities in the civilian and commercial
 ⌐ domain, there is an increased need for autonomy in these
 ⌐ systems too.", "doc_id":2892036907, "query_id":"G11.1",
 ⌐ "query_text":"drones"}
```

Participants were asked to provide a list of terms to be contextualized in a JSON format or a tabulated file TSV (for manual runs) with the following fields:

run_id Run ID starting with (team_id)_(task_3)_(name).
manual Whether the run is manual $\{0, 1\}$.
snt_id a unique passage (sentence) identifier from the input file.
simplified_snt Text of the simplified passage.

Output example (JSON format):

```
{"run_id":"BTU_task_3_run1", "manual":1,
 ⌐ "snt_id":"G11.1_2892036907_2", "simplified_snt":"Drones are
 ⌐ increasingly used in the civilian and commercial domain and
 ⌐ need to be autonomous."}
```

Evaluation Metrics. We filtered out the simplified sentences identical to the source ones and the truncated simplified sentences by keeping only passages matching the regular expression (valid snippets) .+[?.!"]'*$'.

Professional linguists manually annotated simplifications provided with regard to a query according to the following criteria. We evaluated binary errors:

– Incorrect syntax;
– Unresolved anaphora due to simplification;
– Unnecessary repetition/iteration (lexical overlap);
– Spelling, typographic or punctuation errors.

The lexical and syntax complexity of the produced simplifications were assessed on an absolute scale, value 1 referring to a simple output sentence regardless of the complexity of the source one, 7 corresponding to a complex one. Lexical complexity is mostly identical to that presented in Section 4.1.

We consider **syntax complexity** based on syntactic dependencies, their length and depth. The dependency trees reveal latent complications for reading and understanding text; thus, psycholinguists consider the syntactic dependencies to be a relevant tool to evaluate text readability [16]. The depth and length of the syntactic chains we interpret according to [16].

We evaluate **syntax complexity** as follows:

1. Simple sentence (without negation/passive voice): *Over Facebook, we find many interactions.*

2. Simple sentence with negation/passive voice (e.g. *Many interactions were found over Facebook*) or Simple sentences with syntactic constructions that show chains of dependency and shallow embedding depth (e.g. *Over Facebook, we find many interactions between public pages and both political wings.*)
3. Simple sentences with long chains of dependency and shallow embedding depth, with syntactic constructions like complex object, gerund construction, etc. (e.g. *Despite the enthusiastic rhetoric about the so-called collective intelligence, conspiracy theories have emerged.*) or Short complex or compound sentence (e.g. *We propose a novel approach that was used in terms of information theory.*)
4. Simple sentences with long chains of dependency and deep embedding depth, with syntactic constructions like complex object, gerund construction, etc. (e.g. *Over Facebook, we find many interactions between public pages for military and veterans, and both sides of the political spectrum*) or Complex or compound sentence that contains long chains of dependency and deep embedding depth;
5. Simple sentences with long chains of dependency and deep embedding depth, with several syntactic constructions like complex object, gerund construction, etc. or & Complex or compound sentence that contains long chains of dependency and deep embedding depth;
6. Complex or compound sentences that contain long chains of dependency and deep embedding depth along with complex object, gerund construction, etc. or Simple sentence that contains modifications, topicalization, parenthetical constructions: *Moreover, we measure the effect of 4709 evidently false information (satirical version of conspiracist stories) and 4502 debunking memes (information aiming at contrasting unsubstantiated rumors) on polarized users of conspiracy claims.*
7. Long complex or compound sentences that contain several clauses of different types, long chains of dependency and deep embedding depth along with complex object, gerund construction, etc.

We evaluate the information quality of the simplified snippet based on its content and readability. Transformation of information from the source snippet brings in omission of details, insertion of basic terms to explain particular terminology and complex concepts, reference to resources. Due to necessary insertions and references, the simplified snippets often contain more words and syntactic constructions as compared to their source. Nevertheless, the goal is to reduce lexical and syntax complexity in the extended simplified snippets. In case the simplified snippet lacks information mentioned in the source, we evaluate the degree of the information loss. Irrelevant insertions, iterations and wordy statements in the extended simplified snippet we consider as a misrepresentation or distortion of source information when a reader experiences difficulties in processing source content due to wordiness of the loosely structured simplified snippet.

We assessed the information loss severity during the simplification with regard to a given query on the scale from 1 to 7, where 1 corresponds to unimportant information loss while 7 refers to the most severe information distortion. We consider the information loss as a kind of information damage even if the information in the simplified text contains the information of the source passage but has some insertions, which impedes perception of the content.

We distinguish the following 11 types of misrepresentation of source information. Our classification leans on the error typology in machine translation [22,28].

The simplified snippet often combines several types of distortion, e.g. omission and ambiguity. Nevertheless, we observed many instances of small distortions that severely diminish the quality of the simplification; therefore, we need to explain each type providing the clear and transparent instances. Our evaluation of the value of the information distortion leans on the calculation of the information loss and assessment of the diminished readability of the simplified snippet that generates difficulties in text semantic processing by readers.

We distinguish the following types of information distortion:

1. **Style** (distortion severity 1)
   Source snippet: *In order to facilitate knowledge transfer between specialists and generalists and between experts and novices, and to promote interdisciplinary communication, there is a need to provide methods and tools for doing so.*
   Simplified snippet: *There is a need to provide methods and tools for doing so. In order to facilitate knowledge transfer between specialists and generalists and between experts and novices, we need to promote interdisciplinary communication. We need to make it easier for people to share their knowledge with each other.*
   Comment: Deviations from the style norms do not lead to information loss; however, they diminish the quality of text structure and affect readers' assessment of the text and its content.

2. **Insertion of unnecessary details with regard to a query** (distortion severity 1)
   Source snippet: *In the clinical setting, availability of needed information can be crucial during the decision-making process.*
   Simplified snippet: *availability of needed information can be crucial during the decision-making process. In the clinical setting, needed information is often difficult to come by. For confidential support call the Samaritans on 08457 90 90 90 or visit a local Samaritans branch, see* www.samaritans.org *for details.*
   Comment: The simplified snippet often contains more information than the source since the terminology is needed to be explained. An irrelevant insertion does not lead to the loss of information; however, it may bring in diminishing of the text readability and generate discomfort during text perception. The irrelevant reference to the support in the simplified snippet does not clarify the source. The source does not need any simplification.

3. **Redundancy** (without lexical overlap) (distortion severity 2)
   Source snippet: *The capability to get updated information and news is an important and decisive factor in business and finance.*
   Simplified snippet: *The capability to get updated information and news is an important and decisive factor in business and finance. The ability to get updates on the latest news is also an important factor in the success of a business or finance company. For more information, visit CNN.com/News.*
   Comment: Irrelevant insertions, iterations and wordy statements in the extended simplified snippet we consider as a misrepresentation or distortion of source information when a reader may misunderstand source content due to wordiness of the loosely structured simplified snippet.

4. **Insertion of false or unsupported information** (distortion severity 3)
   Source snippet: *The proposed method leads to not only faster running time but also efficient text localization.*

Simplified snippet: *The proposed method leads to not only faster running time but also efficient text localization. The proposed method is based on the idea that text should be written in a single block of text, rather than a series of blocks of text. The method is being developed by the University of California, San Diego.*

Comment: False and unsupported information is inserted in the simplified snippet because of external links of the source snippet to the open Web resources. False information confuses the readership, thus it is inappropriate in simplified texts.

5. **Omission of essential details with regard to a query** (distortion severity 4)

Source snippet: *In short, this thesis aims to repatriate young people's web practices from the sterile, positivist methods space of questionnaires and tests of digital literacy to social contexts of everyday life.*

Simplified snippet: *In short, this thesis aims to repatriate young people's web practices from the sterile, and tests of digital literacy.*

Comment: Simplified texts regularly omit details and minor concepts since the objective of the simplification is to facilitate perception and processing of the main idea and novel information of the source. Therefore, distinguishing the essential details and concepts from the minor ones plays a crucial role in the text simplification. Omission of the essential details leads to the lack of information and novelty of the message. The simplified snippet lacks the essential information about the direction where the young people's web practice needs to be repatriated (*to social contexts of everyday life*).

6. **Overgeneralization** (distortion severity 5)

Source snippet: *Online Social Networks explode with activity whenever a crisis event takes place.*

Simplified snippet: *It explodes when a crisis event takes place.*

Comment: In the simplified snippet the subject of the source snippet *Online Social Networks* is omitted being substituted by the pronoun *It*; also the essential detail *with activity* is omitted that brings in overgeneralization: the statement refers to all cases of the dysfunctions (server, browser, users' activity, moderators' restrictions, etc.)

7. **Oversimplification** (distortion severity 5)

Source snippet: *If we accept the current orthodoxy and then blame the Web we offer a technological determinist explanation of reality: technology produces misinformed populations.*

Simplified snippet: *If we accept the current orthodoxy and then blame the Web we have an explanation of reality.*

Comment: Oversimplification appears in the shortened simplified snippets when source utterance is transformed into a trivial statement or even a claim. The simplified snippet claims that *we have an explanation of reality* thanks to acceptance of the current orthodoxy and accusation of the Web. Meanwhile, the source snippet discusses the technological determinist explanation of reality. The omission of the essential details leads to the oversimplified statement that cannot explain the reality of the technological epoch.

8. **Topic shift** (distortion severity 5)

Source snippet: *global warming induced by chemtrails or the link between vaccines and autism - find on the Web a natural medium for their dissemination.*

Simplified snippet: *The link between vaccines and autism - can be found on the Web a natural medium for changing.*

Comment: Topic shift is revealed in a substitution of the source topic by omitting its part or selecting a wrong basic word to replace the peculiar term in the source. The source snippet lost the essential part of its topic (*global warming induced by chemtrails*) during the simplification process; moreover, the simplification resulted in the inappropriate syntactic structure of the snippet.

9. **Contra sense/contradiction** (distortion severity 6)

   Source snippet: *In this paper we discuss architectural design issues and trade-offs in connection with our experiences porting our agent-based platform, Opal, to the Sharp Zaurus personal digital assistant (PDA).*

   Simplified snippet: *The Sharp Zaurus is a personal digital assistant (PDA) developed by Sharp. It is based on the Opal agent-based platform. We discuss architectural design issues and trade-offs in connection with our experiences porting Opal to the Zaurus PDA.*

   Comment: Contradictions in simplified snippets appear due to elimination of essential concepts or interrelations among concepts, omission of significant details, and transformation of the semantic structure of the source snippet. The simplified snippet mentions agent-based platform Opal as the basis for the Sharp Zaurus, but at the same time claims that Opal was ported to the Sharp Zaurus. The source snippet *But the new phenomena, the non-agenda ownership, overcome any ideological influence, especially under the conditions of punishment mechanism applied to old politicians* lost its semantic structure since the concepts *ideological influence* and *punishment mechanism* were eliminated in the process of its simplification. Thus, the simplified snippet *But the new phenomena, the ownership of the non-agenda, had a lot of influence on old politicians* lacks any explanation how *the non-agenda ownership* is related to *old politicians* and why they are influence by *the new phenomena.*

10. **Ambiguity** (distortion severity 6)

    Source snippet: *The experimental results show that 3D maps with texture on mobile phone display size, and 3D maps without texture on PDA display size are superior to 2D maps in search time and error rate.*

    Simplified snippet: *3D maps with texture on mobile phone display size are superior to 2D maps in search time and error rate. The experimental results show that 3D maps without texture on PDA display size were superior to those with texture. The results were published in the journal 3D Maps.*

    Comment: Ambiguity presupposes that a statement has several equiprobable interpretations. The instance of the ambiguous simplified snippet above lacks a key to understand whether the 3D maps without texture outperform those with texture or not. Ambiguity often appears due to syntactic simplification of the source. In the source, the clause *changes in the strength of competition also reveal key asymmetrical differences* is replaced by shorter clause *but they do not have any biases* that produces ambiguity: whether *evidence* corresponds to reality or not. The source clarifies the differences between two political parties: *Though both Republicans and Democrats show evidence of implicit biases, changes in the strength of competition also reveal key asymmetrical differences* however, the simplified snippet doubts the reliability of the evidence: *Both Republicans and Democrats show evidence of biases, but they do not have any biases.* Readers of the simplified snippet are unable to resolve the ambiguity.

11. **Nonsense** (distortion severity 7)

Source snippet: *The large availability of user provided contents on online social media facilitates people aggregation around shared beliefs, interests, worldviews and narratives.*

Simplified snippet: *The large amount of user provided contents on online social media is called aggregation.*

Comment: The source snippet was transformed into a simple sentence. The transformation brings in erroneous usage of the word aggregation that leads to the loss of meaning of the whole sentence. Instead of the original statement about accessibility of the social or public media on the Web, which facilitates dissemination of fake news and rumors, the simplified snippet claims that there is an opportunity to find a resource to read about fake news and rumors.

The final ranking for Task 3 was done by the average harmonic mean of normalized opposite values of *Lexical Complexity (LC)*, *Syntactic Complexity (SC)* and *Distortion Level (DL)* as follows:

$$s_i = \frac{3}{\frac{7}{7-LC} + \frac{7}{7-SC} + \frac{7}{7-DL}} \tag{1}$$

$$Score = \frac{\sum_i \begin{cases} s_i, & \text{if } No\ Error \\ 0, & \text{otherwise} \end{cases}}{n} \tag{2}$$

In Eq. 2, variable n refers to the total number of judged snippets and *No Error* means that the snippet i does not have any of *Uncorrect syntax, Unresolved anaphora*, nor *Unnecessary repetition/iteration* error.

5.2 Results

A total of 5 different teams submitted 14 runs (5 runs were updated). Absolute number of errors and average *Lexical Complexity, Syntax Complexity* and *Information Loss* are provided in Tables 12 and 13. The final ranking for Task 3 is given in Table 14. We removed all runs with the 0 score.

Very interesting partial runs were provided by the HULAT-UC3M team as the generated simplifications provided the explanations of difficult terms. However, HULAT-UC3M's 8 runs over 10 were not in the pool with selected topics. Thus, we provided only automatic evaluation results. The HULAT-UC3M's runs provide clear evidence of the interconnection of tasks 2 and 3.

Table 12. SimpleText Task 3: General results of official runs

Run	Total	Unchanged	Truncated	Valid	Longer	Length Ratio	Evaluated	Uncorrect Syntax	Unresolved Anaphora	Minors	Syntax Complexity	Lexical Complexity	Information Loss
CLARA-HD	116,763	128	2,292	111,627	201	0.61	851	28	3	68	2.10	2.42	3.84
CYUT Team2	116,763	549	101,104	111,818	49	0.81	126	1		32	2.25	2.30	2.26
PortLinguE_full	116,763	42,189	852	111,589	3,217	0.92	564	7		5	2.94	3.06	1.50
PortLinguE_run1	1,000	359	7	970	30	0.93	80	1			3.63	3.57	2.27
lea_task3_t5	23,360	52	23,201	22,062	24	0.35	.	.	.	.	.	.	.
HULAT-UC3M01	1,000	.	13	973	968	2.46	95	10	1	20	4.69	3.69	2.20
HULAT-UC3M02	2,001	3	58	1,960	1,920	2.53	205	10	1	37	3.60	3.53	2.34
HULAT-UC3M03	1,000	2	13	958	966	2.53	.	.	.	.	.	.	.
HULAT-UC3M04	2,000	.	33	1,827	1,957	37	.	.	.	.	.	.	.
HULAT-UC3M05	2,000	.	56	1,921	1,918	2.38	.	.	.	.	.	.	.
HULAT-UC3M06	2,000	.	47	1,976	1,921	2.45	.	.	.	.	.	.	.
HULAT-UC3M07	1,000	.	56	970	972	2.43	.	.	.	.	.	.	.
HULAT-UC3M08	2,000	.	62	1,964	1,919	2.59	.	.	.	.	.	.	.
HULAT-UC3M09	2,000	.	170	1,964	1,904	2.15	.	.	.	.	.	.	.
HULAT-UC3M10	2,000	.	215	1,963	1,910	2.13	.	.	.	.	.	.	.

Table 13. SimpleText Task 3: Information distortion in evaluated runs

Run	Evaluated	Non-Sense	Contresens	Topic Shift	Wrong Synonym	Ambiguity	Omission Of Essential Details	Overgeneralization	Oversimplification	Unsupported Information	Unnecessary Details	Redundancy	Style
CLARA-HD	851	162	68	37	20	80	314	59	203	26	10	29	13
CYUT Team2	126	2	1	.	.	4	42	4	5	.	.	.	4
PortLinguE_full	564	9	3	4	3	19	94	9	13	2	2	5	1
PortLinguE_run1	80	.	.	1	.	.	27	5	2	.	.	.	.
lea_task3_t5	.	.	.	.	.	.	.	.	.	.	.	.	.
HULAT-UC3M01	95	1	7	2	.	5	2	.	1	5	38	36	.
HULAT-UC3M02	205	4	9	4	.	9	4	.	.	12	72	61	1

Table 14. SimpleText Task 3: Ranking of official submissions on combined score

Run	Score
PortLinguE_full	0.149
CYUT Team2	0.122
CLARA-HD	0.119

6   Conclusion and Future Work

We introduced the CLEF 2022 SimpleText track, containing three interconnected shared tasks on scientific text simplification. We pipelined the passages retrieved for Task 1 in order to rank difficult terms (Task 2) and simplify sentences (Task 3). We evaluated term difficulty and simplifications with regard to the queries from Task 1.

For Task 1, we created a large corpus of scientific abstracts, a set of popular science requests with detailed relevance judgments on the level of relevance of scientific abstracts to the request and broader context of a newspaper article on this topic. The abstracts of scientific papers retrieved for these requests were used in the follow up tasks. For Task 2 and 3, we created a corpus of sentences extracted from the abstracts of scientific publications, with manual annotations of term complexity (Task 2). In contrast to previous work, we evaluate simplification in terms of lexical and syntax complexity combining with error analysis. We introduced a new classification of information distortion types for automatic simplification and we annotated the collected simplifications according to this error classification (Task 3). Recent pandemics have shown that simplification can be modulated by political needs and the scientific information can be distorted. Thus, in contrast to previous work, we evaluated the simplifications in terms of information distortion.

For next year, we plan continue the Task 1 setup, but also refine the relevance judgments to sentence level, and provide additional evaluation measures of readability levels. We will extend Task 2 to provide a context to difficult terms and we will work on automatic metrics based on the insights we obtained this year. In particular, for Task 2, participants will be asked to provide context for difficult terms. This context should provide a definition and take into account ordinary readers' needs to associate their particular problems with the opportunities that science provides them to solve the problems [29]. This year, the HULAT-UC3M team submitted runs which combine tasks 2 and 3 which demonstrates strong interconnection of the tasks as often the terminology cannot be removed nor simplified but it needs to be explained to a reader. Finally, we plan to continue the Task 3 setup, continuing the detailed manual annotations of samples, but also working on automatic metrics that best reflect the insights of this year's analysis.

For details about this year's track and the approaches of individual teams we refer to the CLEF CEUR proceedings [15]. Further details about the lab can be found at the SimpleText website: http://simpletext-project.com. Please join us and help to make scientific results understandable!

Acknowledgment. We like to acknowledge the support of the Lab Chairs of CLEF 2022, Allan Hanbury and Martin Potthast, for their help and patience.Special thanks to the University Translation Office of the Université de Bretagne Occidentale, and to Nicolas Poinsu and Ludivine Grégoire for their major impact in the train data construction and Léa Talec-Bernard and Julien Boccou for their help in evaluation of participants' runs. We thank Josiane Mothe for reviewing papers. We also thank Alain Kerhervé, and the MaDICS (https://www.madics.fr/ateliers/simpletext/ research group.

References

1. Text Analysis Conference (TAC) 2014 Biomedical Summarization Track (2014). https://tac.nist.gov/2014/BiomedSumm/
2. Alva-Manchego, F., Martin, L., Bordes, A., Scarton, C., Sagot, B., Specia, L.: Asset: a dataset for tuning and evaluation of sentence simplification models with multiple rewriting transformations (2020). https://arxiv.org/abs/2005.00481
3. Anand Deshmukh, A., Sethi, U.: IR-BERT: leveraging BERT for semantic search in background linking for news articles 2007, July 2020. http://adsabs.harvard.edu/abs/2020arXiv200712603A
4. Bellot, P., Moriceau, V., Mothe, J., SanJuan, E., Tannier, X.: INEX tweet contextualization task: evaluation, results and lesson learned. Inf. Process. Manage. **52**(5), 801–819 (2016). https://doi.org/10.1016/j.ipm.2016.03.002
5. Brown, T.B., et al.: Language models are few-shot learners, July 2020. http://arxiv.org/abs/2005.14165
6. Chandrasekaran, M.K., Feigenblat, G., Hovy, E., Ravichander, A., Shmueli-Scheuer, M., de Waard, A.: Overview and insights from the shared tasks at scholarly document processing 2020: Cl-scisumm, laysumm and longsumm. In: Proceedings of the First Workshop on Scholarly Document Processing, pp. 214–224 (2020)
7. Cohan, A., Goharian, N.: Revisiting summarization evaluation for scientific articles, April 2016. http://arxiv.org/abs/1604.00400
8. De Clercq, O., Hoste, V., Desmet, B., van Oosten, P., De Cock, M., Macken, L.: Using the crowd for readability prediction. Nat. Lang. Eng. **20**(3), 293–325 (2014). http://dx.doi.org/10.1017/S1351324912000344. ISSN 1469–8110
9. Dong, Y., Li, Z., Rezagholizadeh, M., Cheung, J.C.K.: EditNTS: an neural programmer-interpreter model for sentence simplification through explicit editing. In: Proceedings of the 57th Annual Meeting of the ACL, Florence, Italy, pp. 3393–3402. ACL, July 2019. https://www.aclweb.org/anthology/P19-1331
10. Ermakova, L., et al.: Overview of SimpleText 2021 - CLEF workshop on text simplification for scientific information access. In: Candan, K.S., et al. (eds.) CLEF 2021. LNCS, vol. 12880, pp. 432–449. Springer, Cham (2021). https://doi.org/10.1007/978-3-030-85251-1_27
11. Ermakova, L., et al.: Text simplification for scientific information access: CLEF 2021 SimpleText workshop. In: Proceedings of Advances in Information Retrieval - 43nd European Conference on IR Research, ECIR 2021, Lucca, Italy, 28 March–1 April 2021 (2021)
12. Ermakova, L., et al.: Automatic simplification of scientific texts: SimpleText lab at CLEF-2022. In: Hagen, M., et al. (eds.) Advances in Information Retrieval, vol. 13186, pp. 364–373. Springer, Cham (2022). ISBN 978-3-030-99738-0 978-3-030-99739-7
13. Ermakova, L., Goeuriot, L., Mothe, J., Mulhem, P., Nie, J.-Y., SanJuan, E.: CLEF 2017 microblog cultural contextualization lab overview. In: Jones, G.J.F., et al. (eds.) CLEF 2017. LNCS, vol. 10456, pp. 304–314. Springer, Cham (2017). https://doi.org/10.1007/978-3-319-65813-1_27

14. Ermakova, L.N., Nurbakova, D., Ovchinnikova, I.: Covid or not Covid? Topic shift in information cascades on Twitter. In: Association for Computational Linguistics (ed.) 3rd International Workshop on Rumours and Deception in Social Media (RDSM) Collocated with COLING 2020, pp. 32–37. Proceedings of the 3rd International Workshop on Rumours and Deception in Social Media (RDSM), Barcelona, Spain, December 2020. https://hal.archives-ouvertes.fr/hal-03066857

15. Faggioli, G., Ferro, N., Hanbury, A., Potthast, M. (eds.): Proc. of the Working Notes of CLEF 2022: Conference and Labs of the Evaluation Forum. CEUR Workshop Proceedings (2022)

16. Futrell, R., et al.: The natural stories corpus: a reading-time corpus of English texts containing rare syntactic constructions. Lang. Resour. Eval. **55**(1), 63–77 (2021). https://doi.org/10.1007/s10579-020-09503-7. ISSN 1574-0218

17. Gala, N., François, T., Fairon, C.: Towards a French lexicon with difficulty measures: NLP helping to bridge the gap between traditional dictionaries and specialized lexicons. In: eLex-Electronic Lexicography (2013)

18. Grabar, N., Farce, E., Sparrow, L.: Study of readability of health documents with eye-tracking approaches. In: 1st Workshop on Automatic Text Adaptation (ATA) (2018)

19. Hoffart, J., et al.: Robust disambiguation of named entities in text. In: Proceedings of EMNLP 2011, pp. 782–792 (2011)

20. Lieber, O., Sharir, O., Lentz, B., Shoham, Y.: Jurassic-1: technical details and evaluation, p. 9 (2021)

21. Likert, R.: A technique for the measurement of attitudes. Arch. Psychol. **22**(140), 55 (1932)

22. Lommel, A., Görög, A., Melby, A., Uszkoreit, H., Burchardt, A., Popović, M.: Harmonised metric. Qual. Transl. **21**(QT21) (2015). https://www.qt21.eu/wp-content/uploads/2015/11/QT21-D3-1.pdf

23. Maddela, M., Alva-Manchego, F., Xu, W.: Controllable text simplification with explicit paraphrasing, April 2021. http://arxiv.org/abs/2010.11004

24. Maddela, M., Xu, W.: A word-complexity lexicon and a neural readability ranking model for lexical simplification. In: Proceedings of EMNLP 2018, Brussels, Belgium, pp. 3749–3760. ACL (2018). https://www.aclweb.org/anthology/D18-1410

25. Nakov, P., et al.: Automated fact-checking for assisting human fact-checkers, May 2021. http://arxiv.org/abs/2103.07769

26. Narayan, S., Gardent, C., Cohen, S.B., Shimorina, A.: Split and rephrase. In: Proceedings of EMNLP 2017, Copenhagen, Denmark, pp. 606–616. ACL, September 2017. https://www.aclweb.org/anthology/D17-1064

27. Osgood, C.E.: Semantic differential technique in the comparative study of cultures. Am. Anthropol. **66**(3), 171–200 (1964). https://onlinelibrary.wiley.com/doi/abs/10.1525/aa.1964.66.3.02a00880. ISSN 1548-1433

28. Ovchinnikova, I.: Impact of new technologies on the types of translation errors. In: CEUR Workshop Proceedings (2020)

29. Ovchinnikova, I., Nurbakova, D., Ermakova, L.: What science-related topics need to be popularized? A comparative study. In: Faggioli, G., Ferro, N., Joly, A., Maistro, M., Piroi, F. (eds.) Proceedings of the Working Notes of CLEF 2021 - Conference and Labs of the Evaluation Forum, Bucharest, Romania, 21–24 September 2021, vol. 2936, pp. 2242–2255. CEUR Workshop Proceedings (2021). http://ceur-ws.org/Vol-2936/paper-203.pdf

30. O'Reilly, T., Wang, Z., Sabatini, J.: How much knowledge is too little? When a lack of knowledge becomes a barrier to comprehension. Psychol. Sci., July 2019. https://journals.sagepub.com/doi/10.1177/0956797619862276

31. Pradeep, R., Ma, X., Nogueira, R., Lin, J.: Scientific claim verification with VerT5erini, October 2020. http://arxiv.org/abs/2010.11930

32. Sulem, E., Abend, O., Rappoport, A.: Simple and effective text simplification using semantic and neural methods. In: Proceedings of the 56th Annual Meeting of the ACL (Volume 1: Long Papers), Melbourne, Australia, pp. 162–173. ACL, July 2018. https://www.aclweb.org/anthology/P18-1016

33. Tang, J., Zhang, J., Yao, L., Li, J., Zhang, L., Su, Z.: ArnetMiner: extraction and mining of academic social networks. In: Proceeding of the 14th ACM SIGKDD International Conference on Knowledge Discovery and Data Mining - KDD 2008, Las Vegas, Nevada, USA, p. 990. ACM Press (2008). http://dl.acm.org/citation.cfm?doid=1401890.1402008. ISBN 978-1-60558-193-4

34. Wadden, D., et al.: Fact or fiction: verifying scientific claims, October 2020. http://arxiv.org/abs/2004.14974

35. Wang, W., Li, P., Zheng, H.T.: Consistency and coherency enhanced story generation, October 2020. http://arxiv.org/abs/2010.08822

36. Wubben, S., van den Bosch, A., Krahmer, E.: Sentence simplification by monolingual machine translation. In: Proceedings of the 50th Annual Meeting of the ACL (Volume 1: Long Papers), pp. 1015–1024 (2012)

37. Xu, W., Callison-Burch, C., Napoles, C.: Problems in current text simplification research: new data can help. Trans. ACL **3**, 283–297 (2015). https://www.mitpressjournals.org/doi/abs/10.1162/tacl_a_00139. ISSN 2307-387X

38. Xue, L., et al.: mT5: a massively multilingual pre-trained text-to-text transformer. In: Proceedings of the 2021 Conference of the North American Chapter of the ACL: Human Language Technologies, pp. 483–498. ACL, June 2021. https://aclanthology.org/2021.naacl-main.41

39. Yang, L., Zhang, M., Li, C., Bendersky, M., Najork, M.: Beyond 512 tokens: siamese multi-depth transformer-based hierarchical encoder for long-form document matching, April 2020. arXiv:2004.12297

40. Zhao, S., Meng, R., He, D., Saptono, A., Parmanto, B.: Integrating transformer and paraphrase rules for sentence simplification. In: Proceedings of EMNLP 2018, Brussels, Belgium, pp. 3164–3173. ACL, October 2018. https://www.aclweb.org/anthology/D18-1355

41. Zhong, Y., Jiang, C., Xu, W., Li, J.J.: Discourse level factors for sentence deletion in text simplification. In: Proceedings of the AAAI Conference on Artificial Intelligence, vol. 34, no. 05, pp. 9709–9716, April 2020. https://ojs.aaai.org/index.php/AAAI/article/view/6520. ISSN 2374-3468

Overview of the CLEF–2022 CheckThat! Lab on Fighting the COVID-19 Infodemic and Fake News Detection

Preslav Nakov[1]([✉]), Alberto Barrón-Cedeño[2], Giovanni da San Martino[3], Firoj Alam[4], Julia Maria Struß[5], Thomas Mandl[6], Rubén Míguez[7], Tommaso Caselli[8], Mucahid Kutlu[9], Wajdi Zaghouani[10], Chengkai Li[11], Shaden Shaar[12], Gautam Kishore Shahi[13], Hamdy Mubarak[4], Alex Nikolov[14], Nikolay Babulkov[15], Yavuz Selim Kartal[16], Michael Wiegand[17], Melanie Siegel[18], and Juliane Köhler[5]

[1] Mohamed Bin Zayed University of Artificial Intelligence, Abu Dhabi, UAE
preslav.nakov@mbzuai.ac.ae
[2] Università di Bologna, Forlì, Italy
a.barron@unibo.it
[3] University of Padova, Padova, Italy
dasan@math.unipd.it
[4] Qatar Computing Research Institute, HBKU, Ar-Rayyan, Qatar
{fialam,hmubarak}@hbku.edu.qa
[5] University of Applied Sciences Potsdam, Potsdam, Germany
{julia.struss,juliane.koehler}@fh-potsdam.de
[6] University of Hildesheim, Hildesheim, Germany
mandl@uni-hildesheim.de
[7] Newtral Media Audiovisual, Madrid, Spain
{ruben.miguez,javier.beltran}@newtral.es
[8] University of Groningen, Groningen, Netherlands
t.caselli@rug.nl
[9] TOBB University of Economics and Technology, Ankara, Turkey
m.kutlu@etu.edu.tr
[10] Hamad Bin Khalifa University, Ar-Rayyan, Qatar
wzaghouani@hbku.edu.qa
[11] University of Texas at Arlington, Arlington, USA
cli@uta.edu
[12] Cornell University, Ithaca, USA
ss2753@cornell.edu
[13] University of Duisburg-Essen, Duisburg, Germany
gautam.shahi@uni-due.de
[14] CheckStep, London, UK
alex.nikolov@checkstep.com
[15] Sofia University, Sofia, Bulgaria
[16] GESIS-Leibniz Institute for the Social Sciences, Mannheim, Germany
yavuzselim.kartal@gesis.org
[17] University of Klagenfurt, Klagenfurt, Austria
michael.wiegand@aau.at
[18] Darmstadt University of Applied Sciences, Darmstadt, Germany
melanie.siegel@h-da.de

Abstract. We describe the fifth edition of the `CheckThat!` lab, part of the 2022 Conference and Labs of the Evaluation Forum (CLEF). The lab evaluates technology supporting tasks related to factuality in multiple languages: Arabic, Bulgarian, Dutch, English, German, Spanish, and Turkish. Task 1 asks to identify relevant claims in tweets in terms of check-worthiness, verifiability, harmfullness, and attention-worthiness. Task 2 asks to detect previously fact-checked claims that could be relevant to fact-check a new claim. It targets both tweets and political debates/speeches. Task 3 asks to predict the veracity of the main claim in a news article. CheckThat! was the most popular lab at CLEF-2022 in terms of team registrations: 137 teams. More than one-third (37%) of them actually participated: 18, 7, and 26 teams submitted 210, 37, and 126 official runs for tasks 1, 2, and 3, respectively.

Keywords: Fact-Checking · Disinformation · Misinformation · Check-Worthiness · Verified Claim Retrieval · Fake News · COVID-19

1 Introduction

The mission of the `CheckThat!` lab is to foster the development of technology to assist in the process of fact-checking claims made in political debates, social media posts and news articles. The five editions of the lab have been held in 2018–2022, targeting various natural language processing and information retrieval tasks related to factuality [11,12,24,25,61,62,65,66]. The aim is to develop systems that can be useful as supportive technology for investigative journalism, as they could provide help and guidance, thus saving time [30,35,37,42,63,76,97]. For example, a system could automatically identify check-worthy claims, make sure they have not been fact-checked already by a reputable fact-checking organization, and then present them to a journalist for further analysis in a ranked list [83]. In addition, we can develop systems to identify whether documents are potentially *useful* for human fact-checkers to verify a claim [63,100], and it could also estimate a *veracity score* supported by evidence to increase the journalist's understanding and trust in the system's decision [82].

`CheckThat!` at CLEF 2022 is the fifth edition of the lab [61], and aims to foster the technology on three timely problems in multiple languages: Arabic, Bulgarian, Dutch, English, German, Spanish, and Turkish. Task 1 asks to detect relevant tweets: check-worthy, verifiable, harmful, and attention-worthy. Task 2 aims at detecting previously fact-checked claims in tweets, political debates and speeches. Task 3 focuses on checking the veracity of news articles.

2 Previously on `CheckThat!`

The tasks in the current edition of `CheckThat!` are a follow up or reformulations of those from 2021 [27,65,66], where the focus was on (*i*) *tweets*, (*ii*) *political debates and speeches*, and (*iii*) *news articles*. It featured five languages: Arabic, Bulgarian, English, Spanish, and Turkish. Next, we include a brief overview of the most successful approaches explored in the tasks of that edition.

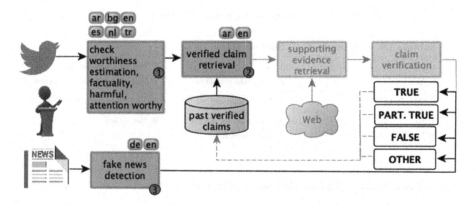

Fig. 1. The full verification pipeline. The 2022 lab covers three tasks from that pipeline: (*i*) check-worthiness estimation, (*ii*) verified claim retrieval, and (*iii*) fake news detection. The gray tasks were addressed in previous editions of the lab [12,25].

Task 1$_{2021}$. *Determine whether a piece of text is worth fact-checking* [85]. The most successful submissions used BERT, AraBERT, and RoBERTa [85,99], and some systems used WordNet [101] and LIWC [80].

Task 2$_{2021}$. *Given a check-worthy claim in the form of a tweet, and a set of previously fact-checked claims, rank these previously fact-checked claims in order of their usefulness to fact-check that new claim* [84]. The most successful approaches were based on AraBERT, RoBERTa, and Sentence-BERT [18,55,74].

Task 3$_{2021}$. *Given the text and the title of a news article, determine whether the main claim it makes is true, partially true, false, or other. Also, identify the domain of the article: health, crime, climate, elections, or education* [88]. The task was offered in English. The most successful pre-trained language model was RoBERTa [9,20,44]. Ensembles were also popular, with components using BERT [44] and LSTMs [20,44].

Previous editions of the lab had targeted other tasks of the verification pipeline (cf. Fig. 1) on different kinds of texts. **The 2020 edition** featured three main tasks: detecting previously fact-checked claims, evidence retrieval, and actual fact-checking of claims [11,12]. The major focus was on Twitter. **The 2019 edition** covered the various modules necessary to verify a claim: from check-worthiness, to ranking and classification of evidence in the form of Web pages, to actual fact-checking of claims against specific text snippets [24,25]. **The 2018 edition** of the lab focused on check-worthiness and fact-checking of claims in political debates [62].

3   Description of the Tasks

The 2022 edition of the `CheckThat!` lab is organized around three tasks, each of which in turn has several subtasks. Figure 1 shows the full `CheckThat!` verification pipeline, and the three tasks we target this year are highlighted.

Table 1. Class labels for Subtasks 1A, 1B, 1C, and 1D.

Subtask 1A	Subtask 1C	Subtask 1D	
1. No	1. No	1. No	6. Yes, contains advice
2. Yes	2. Yes	2. Yes, asks question	7. Yes, discusses action taken
Subtask 1B		3. Yes, blame authorities	8. Yes, discusses cure
1. No		4. Yes, calls for action	9. Yes, other
2. Yes		5. Yes, Harmful	

3.1 Task 1: Identifying Relevant Claims in Tweets

The aim of Task 1 is to determine whether a claim in a tweet is worth fact-checking. In order to do that, we either resort to the judgments of professional fact-checkers or we ask human annotators to answer several auxiliary questions [3,4], such as "does it contain a verifiable factual claim?", "is it harmful?" and "is it of general interest?", before deciding on the final check-worthiness label. Tasks 1A to 1C are all binary problems and the models are expected to establish whether a tweet is relevant according to different criteria. Task 1D is a multi-class problem. Table 1 shows the class labels for all four subtasks. Regarding languages, Arabic, Bulgarian, Dutch, English, and Turkish are present in all four subtasks, whereas Spanish is included only in Subtask 1A. The participants were free to work on any language(s) of their choice, and they could also use multilingual approaches that make use of all datasets for training.

Subtask 1A: Check-worthiness of tweets. Given a tweet, predict whether it is worth fact-checking.

Subtask 1B: Verifiable factual claims detection. Given a tweet, predict whether it contains a verifiable claim or not.

Subtask 1C: Harmful tweet detection. Given a tweet, predict whether it is harmful to the society.

Subtask 1D: Attention-worthy tweet detection. Given a tweet, predict whether it should get the attention of policy makers and why. Table 1 shows the nine classes. More details of the label definitions can be found in [4].

3.2 Task 2: Detecting Previously Fact-Checked Claims

Given a check-worthy claim, and a set of previously-checked claims, determine whether the claim has been previously fact-checked with respect to a collection of fact-checked claims. Both subtasks are ranking problems, where systems are asked to produce a list of top-n candidates. Subtask 2A focuses on tweets and was offered in both Arabic and English. Subtask 2B focuses on political debates and speeches and was given only in English.

Subtask 2A: Detect previously fact-checked claims from tweets. Given a tweet, detect whether the claim it makes was previously fact-checked with respect to a collection of previously fact-checked claims. This is a ranking task, where the systems were asked to produce a list of top-n candidates.

Subtask 2B: Detect previously fact-checked claims in political debates or speeches. Within the context of a political debate or a speech, detect whether a claim has been previously fact-checked with respect to a collection of previously fact-checked claims.

3.3   Task 3: Fake News Detection

Task 3 asks to predict the veracity news articles and is designed as a multi-class classification problem. This task was offered as a monolingual task in English and as a cross-lingual task for English and German (English training data, German test data). The idea of the latter is to use the English data and cross-language representation learning (e.g., [19,71]) to build a classification model for the German language as well.

Task 3: Multi-class fake news detection of news articles. Given the text of a news article, determine whether the claims made in the article are *true*, *partially true*, *false*, or *other* (e.g., claims in dispute).

4   Datasets

Here, we briefly describe the datasets for each of the three tasks. For more details, refer to the task description papers for Task 1 [60], Task 2 [64], and Task 3 [43].

4.1   Task 1: Identifying Relevant Claims in Tweets

For all **1A, 1B, 1C**, and **1D** subtasks and all languages, but Spanish, we used the dataset reported in [4]. The dataset is developed based on a multi-question annotation schema and annotated tweets for Arabic, Bulgarian, Dutch, and English [3]. Following the same annotation schema, a Turkish dataset has also been produced. The dataset reported in [4] comes with a training, development, and test set. For the shared task, we provided the test set as a dev-test set to enable the participants to validate their systems internally, while they can use the dev set for parameter tuning. For each language and subtask, we have annotated new instances, using three or four annotators per instance. Class labels have been assigned by majority voting and disagreements have been solved by a consolidator.

For Spanish, the tweets were manually annotated by journalists from Newtral—a Spanish fact-checking organization—and came from the Twitter accounts of 300 Spanish politicians. The Spanish collection is the largest one compared to the other languages ; more than three times the second largest dataset: 14,990 vs 4,121 for Arabic. However, Spanish is only available for Subtask **1A**.

Table 2 summarizes the data available for each subtask and each language.

Table 2. Task 1 (Identifying Relevant Claims in Tweets): Statistics about the CT–CWT–22 corpus for all six languages. The bottom part of the table shows the main topics covered.

Subtask	Partition	AR	BG	NL	EN	ES	TR	Total
	Train	2,513	1,871	923	2,122	4,990	2,417	14,836
	Dev	235	177	72	195	2,500	222	3,401
1A	Dev-Test	691	519	252	574	2,500	660	5,196
	Test	682	130	666	149	5,000	303	6,930
	Total	4,121	2,697	1,913	3,040	14,990	3,602	
	Train	3,631	2,710	1,950	3,324		2,417	14,032
	Dev	339	251	181	307		222	1,300
1B	Dev-Test	996	736	534	911		660	3,837
	Test	1,248	329	1,358	251		512	3,698
	Total	6,214	4,026	4,023	4,793		3,811	
	Train	3,624	2,708	1,946	3,323		2,417	14,018
	Dev	336	250	179	307		222	1,294
1C	Dev-Test	994	735	531	910		660	3,830
	Test	1,201	325	1,360	251		512	3,649
	Total	6,155	4,018	4,016	4,791		3,811	
	Train	3,621	2,710	1,949	3,321		1,904	13,505
	Dev	338	251	179	306		178	1,252
1D	Dev-Test	995	736	533	909		533	3,706
	Test	1,186	329	1,356	251		465	3,587
	Total	6,140	4,026	4,017	4,787		3,080	
Main topics								
COVID-19		■	■	■	■	■	■	
Politics							■	■

4.2 Task 2: Detecting Previously Fact-Checked Claims

Subtask 2A: Detecting previously fact-checked claims from tweets. For English, we have 1,610 annotated tweets, each matching a single claim in a set of 13,835 verified claims from Snopes. For Arabic, we have 858 tweets, matching 1,089 verified claims (some tweets match more than one verified claim) in a collection of 30,379 previously fact-checked claims. The latter include 5,921 Arabic claims from AraFacts [5] and 24,408 English claims from ClaimsKG [91], translated to Arabic using the Google Translate API.[1]

---

[1] http://cloud.google.com/translate.

Table 3. Task 2: Statistics about the CT–VCR–22 corpus, including the number of *Input–VerClaim* pairs and the number of *VerClaim* claims to match the input claim against.

Partition	2A-Arabic	2A-English	2B-English
Input Claims	**908**	**1,610**	**752**
Training	512	999	472
Development	85	200	119
Dev-Test	261	202	78
Test	50	209	83
Input-VerClaims pairs	**1,089**	**1,610**	**869**
Training	602	999	562
Development	102	200	139
Dev-Test	335	202	103
Test	50	209	65
Verified claims (to match against)	**30,379**	**13,835**	**20,771**

Subtask 2B: Detecting previously fact-checked claims in political debates/speeches. We have 752 claims from political debates [83], matched against 869 verified claims (some input claims match more than one verified claim) in a collection of 20,771 verified claims in PolitiFact.

Table 3 shows statistics about the CT–VCR–22 corpus for Task 2, including both subtasks and languages. CT–VCR–22 stands for `CheckThat!` verified claim retrieval 2022. *Input–VerClaim* pairs represent input claims with their corresponding verified claims by a fact-checking source. The input for subtask 2A (2B) is a tweet (sentence from a political debate or a speech). More details about the corpus construction can be found in [84].

4.3 Task 3: Fake News Detection

For the creation of the data for Task 3 the AMUSED framework [86] was followed. The starting point for the data collection was finding suitable fact-checking organizations and their websites. On those websites, the authors of the individual articles discuss and rate the truthfulness of claims that are made in different sources. We scraped the links to those sources as well as the judgment about the claim made on the fact-checking sites. To ensure that only news articles are in the corpus, automatic filtering was applied. Thus, all links leading to a social media platform or a non-textual document (e.g., image, video) were deleted. Furthermore, the remaining links were manually checked. During this step, in addition to deleting non-relevant URLs, we examined, if the links actually (still) lead to the claim source and if the document still existed in its original form. Following those quality evaluations, we scraped the title and the full text for each of the remaining articles.

Table 4. Task 3: Statistics about the number of documents and class distribution for the CT-FAN-22 corpus for English and German fake news detection.

Class	EN Training	EN Dev.	EN Test	DE Test
False	465	113	315	191
True	142	69	210	243
Partially false	217	141	56	97
Other	76	41	31	55
Total	900	364	612	586

Task 3 was offered in English and as a cross-language task in German. As training material, we only provided the English data of last year's iteration. Thus, 900 English news articles for training and 364 articles as development set were given to the participants. Those documents were collected from a total of 15 fact-checking websites (e.g., PolitiFact) [88]. Because the German task was intended as a cross-language classification problem, no German training data was necessary. The training data contained an ID for each article as unique identifier, the title of the given target article as well as its full-text, and finally, a label stating the truthfulness of the article. We took the labels from the judgment on the fact-checking sites. Yet, each fact-checking site had their own label inventory if any at all, such as *incorrect, inaccurate,* or *misinformation* for *false.* Therefore, we merged the labels with a similar meaning according to [87], leading to the following four classes: *true, false, partially false* (meaning any mix of false and true information, such as mostly true or mostly false), and *other.*

As test data, we collected 612 English and 586 German articles from a total of 20 fact-checking websites (14 for the English data and 7 for the German data; the AFP website was consulted for both languages). We did not provide any other information (e.g., a link to the article, a publication date, eventual tags, authors, location of publication, etc.). An overview of the different datsets can be found in Table 4. Both training and test data set are available on Zenodo[2].

5   Evaluation

We used different official evaluation metrics, depending on the nature of the tasks at hand and the involved datasets.

Task 1 and Task 3 included both binary and multi-class classification subtasks. For **Subtasks 1A and 1C**, we used the F_1-measure with respect to the positive class (yes), to account for class imbalance. For **Subtask 1B**, we used accuracy, as the data is fairly balanced. For **Subtask 1D**, we used weighted-F_1, as there are multiple classes and we wanted them appropriately weighted.

Task 2 included ranking subtasks. The official measure for both **Subtasks 2A and 2B** was mean-average precision at 5 (MAP@5); these are the same evaluation measures as in the 2021 edition of the CheckThat! lab.

For **Task 3**, we used macro F_1-measure, as in the previous iteration of the task.

---

[2] https://zenodo.org/record/6555293.

6   Results for Task 1: Identifying Relevant Claims in Tweets

Below, we report the evaluation results for Task 1 and its four subtasks for all six languages.

6.1   Task 1A. Check-Worthiness Estimation

A total of 20 teams took part in this task, with English, Bulgarian, and Dutch being the most popular languages. Two teams (TOBB ETU [26] and NUS-IDS [57]) participated in five languages out of six. For all six languages, we had a monolingual random baseline. Table 5 shows the performance of the official submissions—the last valid blind submission by each team—on the test set, in addition to the random baseline. The table shows the runs ranked on the basis of the official F_1 with respect to the positive class and includes all six languages.

Table 5. Task 1A: Check-Worthiness estimation, results for the official submissions in all six languages. F1 with respect to the positive class. Baseline is the random baseline.

Team	F1	Team	F1	Team	F1
Arabic		**English**		**Spanish**	
1. NUS-IDS [57]	0.628	1. AI Rational [77]	0.698	1. NUS-IDS [57]	0.571
2. TOBB ETU [26]	0.495	2. Zorros [16]	0.667	2. PoliMi-FlatEarthers [2]	0.323
3. iCompass [13]	0.462	3. PoliMi-FlatEarthers [2]	0.626	3. Z-Index [90]	0.303
4. Baseline	0.347	4. TOBB ETU [26]	0.561	4. Baseline	0.139
5. PoliMi-FlatEarthers [2]	0.321	5. Fraunhofer SIT [29]	0.552	**Turkish**	
Bulgarian		6. RUB-DFL [39]	0.525	1. RUB-DFL [39]	0.801
1. NUS-IDS [57]	0.617	7. hinokicrum*	0.522	2. AI Rational [77]	0.789
2. TOBB ETU [26]	0.542	8. NUS-IDS [57]	0.519	3. ARC-NLP [93]	0.760
3. AI Rational [77]	0.483	9. TonyTTTTT	0.500	4. TOBB ETU [26]	0.729
4. Baseline	0.434	10. Asatya [50]	0.500	5. Baseline	0.496
5. PoliMi-FlatEarthers [2]	0.341	11. VTU_BGM [41]	0.482		
6. pogs2022*	0.000	12. Z-Index [90]	0.478		
Dutch		13. NLP&IR@UNED*	0.469		
1. NUS-IDS [57]	0.642	14. Baseline	0.253		
2. AI Rational [77]	0.620				
3. TOBB ETU [26]	0.534				
4. PoliMi-FlatEarthers [2]	0.532				
5. Z-Index [90]	0.497				
6. Baseline	0.451				

*No working note submitted.

Arabic. Four teams participated for Arabic, submitting a total of 12 runs. All participating teams fine-tuned existing pre-trained models, such as BERT, AraBERT, GPT-3 and mT5 models. The top performing system, NUS-IDS [57], used mT5 model, which is a multilingual sequence-to-sequence transformer pre-trained on the mC4 corpus covering 101 languages. They performed both data augmentation and preprocessing. The second best system, TOBB ETU [26], used fine-tuned AraBERT.

Bulgarian. Five teams took part for Bulgarian, submitting a total of six runs. Once again **NUS-IDS** [57] was the top-ranked team, followed by Team **TOBB ETU** [26]. BERT, RoBERTa, DistilBERT and the common pretrained models have been used by all participating systems. Several systems also used data augmentation and standard preprocessing.

Dutch. Five teams participated for Dutch, submitting a total of 11 runs. Team **NUS-IDS** [57] also ranked first, followed by Team **AI Rational** [77] is the second-best system. Across different teams, BERT is the most commonly used pre-trained model. Other pre-trained models include RoBERTa, DistilBERT, and GPT-3. Data augmentation and standard preprocessing have also been used for Dutch.

English. A total of 13 teams took part in task 1A for English, with a total of 59 runs. The top-ranked team was **AI Rational** [77], and they fine-tuned several pre-trained transformers models such as DistilBERT, BERT, RoBERTa. For the system submission they used RoBERTa-large. The second best system—Team **Zorros** [16]—also used BERT and RoBERTa with an ensemble approach.

Spanish. Three teams took part for Spanish, with a total of eight runs. Team **NUS-IDS** [57] is the top-ranked team. Team **PoliMi-FlatEarthers** [2] is second, with a system based on a GPT-3 pretrained model.

Turkish. Four teams participated for Turkish, submitting a total of five runs. All participants used BERT-based models and GPT-3. The top ranked team is **RUB-DFL** [39], which used BERT-based models and LIWC features. The runner up team **AI Rational** applied standard pre-processing and data augmentation with back translation.

6.2   Subtask 1B: Verifiable Factual Claims Detection

Thirteen teams took part in Subtask 1B, with English, Bulgarian and Arabic being the most popular languages. Team **TOBB ETU** [26] participated in all five languages. Team **AI Rational** participated in four languages. Table 6 shows the performance of the official submissions on the test set including the random baseline. The table shows the runs ranked on the basis of the official *accuracy* measure in all five languages for this subtask.

Arabic. Three teams participated in the Arabic factual claim detection subtask, submitting a total of seven runs. The system of team **TOBB ETU** [26] ranked best for this subtask, which uses a four-layer feed-forward network with Manifold Mixup regularization and BERT embeddings.

Bulgarian. Two teams submitted three runs: Team **AI Rational** [77] tops the ranking, followed by Team **TOBB ETU** [26]. AI Rational used XLM-RoBERTa with data augmentation while TOBB ETU used fine-tuned RoBERTa.

Dutch. As for Bulgarian, two teams submitted three runs. Team **AI Rational** [77] and **TOBB ETU** [26] ranked as the first and second systems. Similar approaches (i.e., BERT, RoBERTa, and DistilBERT) have been used.

Table 6. Task 1B: Verifiable Factual Claims Detection, results for the official submissions in all five languages.

Team	Acc	Team	Acc	Team	Acc
Arabic		**English**		**Turkish**	
1. TOBB ETU [26]	0.570	1. PoliMi-FlatEarthers [2]	0.761	1. RUB-DFL [39]	0.801
2. Baseline	0.531	2. Asatya [50]	0.749	3. AI Rational [77]	0.789
3. claeser*	0.454	3. NLP&IR@UNED*	0.725	3. ARC-NLP [93]	0.760
4. pogs2022*	0.454	4. AI Rational [77]	0.713	4. TOBB ETU [26]	0.729
Bulgarian		5. Zorros [16]	0.709	5. Baseline	0.496
1. AI Rational [77]	0.839	6. RUB-DFL [39]	0.709		
2. TOBB ETU [26]	0.742	7. VTU_BGM [41]	0.709		
3. Baseline	0.535	8. hinokicrum*	0.665		
Dutch		9. TOBB ETU [26]	0.641		
1. AI Rational [77]	0.736	10. Baseline	0.494		
2. TOBB ETU [26]	0.658				
3. Baseline	0.521				

*No working note submitted.

English. Nine teams participated with 21 runs. Team **PoliMi-FlatEarthers** [2] ranked as the best system and **Asatya** [50] as the second. The top-performing system used GPT-3, whereas other teams used BERT, RoBERTa, and Distil-BERT as pretrained models for fine-tuning.

Turkish. Four teams participated, submitting five runs. The top-ranked team is **RUB-DFL** [39], which used RoBERTa, Electra, and BERTurk pre-trained models. The second-best team is **AI Rational** [77], which used BERT, RoBERTa, and DistilBERT.

6.3   Subtask 1C: Harmful Tweet Detection

Thirteen teams participated in Subtask 1C, with English and Turkish being the most popular languages. Teams TOBB ETU [26] and AI Rational [77] participated in five and four languages, respectively. Table 7 shows the performance of the official submissions on the test set, together with the random baseline. The table shows the runs ranked based on the official *F1 with respect to positive class* for five languages.

Table 7. Task 1C: Harmful Tweet Detection, results for the official submissions in all five languages.

Team	F1	Team	F1	Team	F1
Arabic		**English**		**Turkish**	
1. iCompass [13]	0.557	1. Zorros [16]	0.397	1. ARC-NLP [93]	0.366
2. TOBB ETU [26]	0.268	2. AI Rational [77]	0.361	2. RUB-DFL [39]	0.353
3. Baseline	0.118	3. Asatya [50]	0.361	3. AI Rational [77]	0.346
Bulgarian		4. NLP&IR@UNED*	0.347	4. TOBB ETU [26]	0.262
1. AI Rational [77]	0.286	5. TOBB ETU [26]	0.329	5. Baseline	0.061
2. TOBB ETU [26]	0.054	6. ARC-NLP [93]	0.300		
3. Baseline	0.000	7. hinokicrum*	0.281		
Dutch		8. COURAGE [47]	0.280		
1. TOBB ETU [26]	0.358	9. RUB-DFL [39]	0.273		
2. AI Rational [77]	0.147	10. PoliMi-FlatEarthers [2]	0.270		
3. Baseline	0.114	11. Baseline	0.200		
		12 VTU_BGM [41]	0.000		

*No working note submitted.

Arabic. Two teams participated, submitting a total of 12 runs. Team **iCompass** [13] is the best system, followed by Team **TOBB ETU** [26]. iCompass finetuned the AraBERT and ARBERT pre-trained language models.

Bulgarian. Two teams participated, submitting 4 runs. Team **AI Rational** [77] ranked as the best system using XLM-RoBERTa while the second best system **TOBB ETU** [26] fine-tuned RoBERTa. Both teams applied data augmentation via back-translation.

Dutch. Two teams participated with 3 runs. Team **TOBB ETU** [26] ranked on top and **AI Rational** [77] ranked second. For this subtask, AI Rational used XLM-RoBERTa without data-augmentation while TOBB ETU fine-tuned BERT and applied data-augmentation via back-translation.

English. A total of 11 teams participated with 17 submissions. Team **Zorros** [16] ranked as the best system, using an ensemble of five transformer-based models. Team **ARC-NLP** [93] ranked second. Besides transformer-based models across all approaches, some teams have also used data augmentation.

Turkish. Four teams participated with five runs submitted. Team **ARC-NLP** [93] ranked as the best system by approaching harmful detection as a contradiction detection problem. They first extracted facts related to the COVID-19 pandemic from reliable sources, and then associated tweets with facts based on their textual similarity. Next, they fine-tuned BERTurk using fact and tweet pairs as data instances. The second best system is by Team **RUB-DFL** [39], which fine-tuned ConvBert with standard pre-processing.

6.4   Subtask 1D: Attention-Worthy Tweet Detection

Seven teams participated in subtask 1D, with English being the most popular language. As for subtask 1C, teams TOBB ETU [26] and AI Rational [77] participated in five and four languages, respectively. Table 8 shows the performance of the official submissions on the test, together with the random baseline. The ranking is based on the official *weighted F1*.

Arabic. Only one team participated. The random baseline outperformed feed-forward network with BERT embeddings and Manifold Mixup regularization proposed by team **TOBB ETU** [26].

Bulgarian. Two teams participated, submitting 4 runs. Team **AI Rational** [77] ranked on top whereas **TOBB ETU** [26] arrived second. While AI Rational used the same transformer-based model in Subtask 1C, TOBB ETU utilized a manifold mixup approach.

Dutch. Two teams participated, making three runs. As for Bulgarian, teams **AI Rational** [77] and **TOBB ETU** [26] ranked first and second.

English. Six teams participated with a total of 14 runs. Team **Zorros** [16] ranked first, by fine-tuning a COVID Twitter BERT pre-trained model. The random baseline ranked second.

Table 8. Task 1D: Attention-Worthy Tweet Detection, results for the official submissions in all five languages. Performance is reported as weighted F1.

Team	F1	Team	F1	Team	F1
Arabic		**English**		**Turkish**	
1. Baseline	0.206	1. Zorros [16]	0.725	1. AI Rational [77]	0.895
2. TOBB ETU [26]	0.184	2. Baseline	0.695	2. Baseline	0.853
Bulgarian		3. AI Rational [77]	0.684	3. TOBB ETU [26]	0.806
1. AI Rational [77]	0.915	4. TOBB ETU [26]	0.670	**Dutch**	
2. TOBB ETU [26]	0.877	5. NLP&IR@UNED*	0.650	1. AI Rational [77]	0.715
3. Baseline	0.875	6. hinokicrum*	0.643	2. TOBB ETU [26]	0.694
		7. PoliMi-FlatEarthers [2]	0.636	3. Baseline	0.641

*No working note submitted.

Turkish. Two teams participated, with three runs. Team **AI Rational** [77] ranked on top, followed by a random baseline.

7   Results for Task 2: Verified Claim Retrieval

Six teams took part in Task 2. Subtask 2A was more popular than subtask 2B. Only team **SimBa** took part in both subtasks, whereas team **BigIR** was the only one that participated in both languages.

7.1   Subtask 2A: Detecting Previously Fact-Checked Claims in Tweets

Table 9 shows the official results for Task 2A English for all participated teams. We do not report results for Arabic as the scores are zero for both random baseline and the submitted system.

Arabic. Team **bigIR** submitted a run for this subtask. They used AraBERT to rerank a list of candidates retrieved by a BM25 model. Their approach consists of three main steps such as preprocessing, retrieving an initial list using BM25 and finally reranking the initial list using an AraBERT-based model.

As with the random baseline, since the system did not match any input with the verified claims, the performance end up being 0.0.

English. Six teams participated, submitting a total of thirty-two runs. All teams improved over the random baseline. Team **RIET Lab** [54] submitted the top run, based on a sentence transformer (sentence-t5) for candidate selection and a generative model (gpt-neo [14]) for re-ranking. Team **AI Rational** ranked second, using a pretrained SBERT, ElasticSearch, and an SVM.

Table 9. Task 2A and 2B: Official evaluation results, in terms of MRR, MAP@k, and Precision@k. The teams are ranked by the official evaluation measure: MAP@5. Here, *Baseline* refers to the random baseline.

Team	MRR	MAP				Precision		
		@1	@3	@5	@10	@3	@5	@10
Task 2A: English								
1. RIET Lab [54]	0.957	0.943	0.955	0.956	0.956	0.322	0.194	0.098
2. AI Rational	0.922	0.904	0.919	0.922	0.922	0.313	0.190	0.095
3. BigIR [51]	0.923	0.900	0.921	0.921	0.921	0.316	0.189	0.095
4. SimBa [38]	0.907	0.876	0.905	0.907	0.907	0.314	0.190	0.095
5. motlogelwan*	0.878	0.833	0.870	0.873	0.876	0.306	0.187	0.095
6. Fraunhofer SIT [28]	0.624	0.557	0.601	0.610	0.617	0.221	0.141	0.075
Task 2B: English								
SimBa [38]	0.475	0.408	0.446	0.459	0.459	0.190	0.126	0.063

7.2   Subtask 2B: Detecting Previously Fact-Checked Claims in Political Debates and Speeches

Table 9 shows the official results for Task 2B, which was offered in English only. The table does not report the random baseline results as scores are zero for all metrics.

Team **SimBa** [38] submitted a total of four runs. They computed different kinds of similarities between input claims and verified claims, including the cosine between sentence embeddings and different lexical similarity metrics. They made use of a blocking approach to filter dissimilar pairs that can easily be excluded based on sentence-embedding-based similarity scores, training and applying their classifier only to distinguish between harder cases.

8   Results for Task 3: Fake News Detection

In this section, we present the results of the evaluation for Task 3 and for each of the two languages, English (monolingual subtask) and German (cross-language subtask). Each team could submit up to 200 runs. Yet, only the last submission was taken into account for the evaluation. In total, there were 32 teams submitting runs for the English and 14 for the German task. Runs which were either incorrectly formatted or consisted of incomplete files were rejected, resulting in 25 and 8 runs for the English and German subtasks, respectively.

As in the 2021 edition [88], most experiments involved deep learning models (16 teams), especially applications of BERT (12 teams), RoBERTa (6 teams) or other BERT variations (8 teams) and one of the publicly available BERT language models. However, almost as many teams (14 teams) experimented with feature-based supervised-learning approaches as well. Examples are SVMs (10 teams), logistic regression (9 teams), random forests (8 teams) and naïve bayes (7 teams). Yet, the majority merely fine-tuned a pre-trained language model and only very few experimented with other approaches.

English. Last year, the best submission made extensive use of external data resources [88]. This year, in total, 12 teams worked with additional English, and one team with additional German training data that was not provided by the organizers of this task. The best submission for the monolingual subtask was by team **iCompass** (macro-averaged F_1: 0.339). They applied *bert-base-uncased* and fine-tuned their model. They also experimented with RoBERTa for which they got worse results. No additional external resources were employed in the final classifier.

The second-best submission, by team **NLP&IR@UNED** (macro-averaged F_1: 0.332), made use of an ensemble of classifiers. It was built out of a Funnel Transformer and a Feed Forward Neural Network. The features were extracted by the *LIWC* text analysis tool.

Overall, all teams had a macro-averaged F_1 score lower than 0.5. Table 10 shows a complete overview of the teams and their results. The baseline system [79], a standard bert-base-cased model from HuggingFace, was made available to the participants at the beginning of the lab cycle.

German. Eight teams attempted to solve the second subtask, which was the English–German cross-language setting. Team **ur-iw-hn** was the team with the most successful submission (macro-averaged F_1: 0.290). They translated the first 5,000 tokens of an article from the German test data using the service

Table 10. Task 3 English: Official evaluation results for English Fake News Detection ranked by the macro-F_1 score, including the F_1 scores for individual classes and the overall accuracy

Team	True	False	Partially False	Other	Accuracy	Macro-F1
1 iCompass [89]	0.383	0.721	0.173	0.080	0.547	0.339
2 NLP&IR@UNED [52]	0.446	0.729	0.097	0.057	0.541	0.332
3 Awakened [95]	0.328	0.744	0.185	0.035	0.531	0.323
4 UNED	0.346	0.725	0.191	0.000	0.544	0.315
Baseline	0.244	0.701	0.157	0.144	0.480	0.312
5 NLytics [75]	0.339	0.707	0.184	0.000	0.513	0.308
6 SCUoL [6]	0.377	0.709	0.133	0.000	0.526	0.305
7 NITK-IT_NLP [34]	0.325	0.734	0.133	0.000	0.536	0.298
8 CIC [7]	0.111	0.682	0.215	0.136	0.475	0.286
9 ur-iw-hnt [94]	0.290	0.733	0.110	0.000	0.533	0.283
10 BUM [46]	0.207	0.694	0.140	0.063	0.472	0.276
11 boby232	0.255	0.676	0.126	0.045	0.475	0.275
12 HBDCI [17]	0.177	0.708	0.209	0.000	0.508	0.273
13 DIU_SpeedOut	0.195	0.706	0.182	0.000	0.521	0.271
14 DIU_Carbine	0.192	0.626	0.157	0.056	0.472	0.258
15 CODE [15]	0.126	0.662	0.203	0.029	0.444	0.255
16 MNB	0.160	0.701	0.142	0.000	0.507	0.251
17 subMNB	0.160	0.701	0.142	0.000	0.507	0.251
18 FoSIL [48]	0.141	0.670	0.169	0.022	0.462	0.251
19 TextMinor [45]	0.250	0.555	0.086	0.048	0.377	0.235
20 DLRG	0.009	0.694	0.092	0.000	0.513	0.199
21 DIU_Phoenix	0.420	0.040	0.092	0.000	0.278	0.159
22 AIT_FHSTP [78]	0.280	0.146	0.154	0.039	0.199	0.155
23 DIU_SilentKillers	0.407	0.070	0.135	0.000	0.260	0.153
24 DIU_Fire71	0.430	0.006	0.094	0.000	0.275	0.133
25 AI Rational	0.296	0.000	0.196	0.090	0.098	0.117

of Google Translate. They applied an extractive summarization techniques and a $BERT_{Large}$ model for the multi-class classification.

Team **NITK-IT_NLP**, which was the first runner up, divided the news article into windows of 500 tokens. Those windows were shifted over the text to avoid losing context. They experimented with different transformer models, with an $mDeBERTa$ model yielding the best results. Table 11 shows the individual results of all eight submissions. Again, the baseline [79] (macro-averaged F_1 score 0.242) results are listed in the table as well. The baseline translated the German articles into English to classify them in accordance to the monolingual subtask.

Table 11. Task 3 German: Official evaluation results for German Fake News Detection ranked by the macro-F_1 score, including the F_1 scores for individual classes and the overall accuracy

Team	True	False	Partially False	Other	Accuracy	Macro-F1
1 ur-iw-hnt [94]	0.401	0.536	0.189	0.033	0.427	0.290
Baseline	0.405	0.328	0.029	0.204	0.280	0.242
2 NITK-IT_NLP [34]	0.268	0.490	0.077	0.063	0.362	0.225
3 UNED	0.298	0.166	0.210	0.162	0.213	0.209
4 AIT_FHSTP [78]	0.378	0.168	0.151	0.081	0.254	0.195
5 Awakened [95]	0.098	0.452	0.194	0.000	0.283	0.186
6 CIC [7]	0.000	0.449	0.240	0.000	0.282	0.172
7 NoFake	0.000	0.492	0.000	0.000	0.326	0.123
8 AI Rational	0.268	0.000	0.166	0.122	0.114	0.111

9   Related Work

There has been a significant number of work on detecting fake news, identifying factuality/credibility of a claim appearing in different sources [8, 10, 40, 49, 68, 69, 73, 102]. Typical sources include news article, social media (e.g., Facebook status, tweets, WhatsApp messages, posts in different forums). Major research attention has been paid to the social media [63, 81]. Within the realm of misinformation and disinformation there are a number of research areas such as identifying the checkworthiness of a claim [74, 85], claim detection [30, 35–37], fact-checked claims [32, 83, 96] etc.

Shared tasks has also been organized in the last few years, which are similar to CheckThat! . Such initiatives include SemEval on determining rumour veracity and support for rumours [22, 31], on stance detection [58], on fact-checking in community question answering forums [56], on propaganda detection [21, 23], and on semantic textual similarity [1, 67]. It is also related to the FEVER task [92] on fact extraction and verification, Fake News Challenge [33], and the FakeNews task at MediaEval [72], fact verification and evidence finding for tabular data [98], detecting and rating humor and offense [53], toxic span detection [70], and multimodal fake news detection [59].

10   Conclusion and Future Work

We have presented the 2022 edition of the CheckThat! lab, which was again the most popular lab regarding the number of registrations, with a total of 137 registered teams.

Task 1 asked to identify relevant claims in tweets in terms of check-worthiness, verifiability, harmfulness, and attention-worthiness. Task 2 asked to detect previously fact-checked claims that could be relevant to fact-check a new claim. Task 3 asked to predict the veracity of the main claim in a news article. As in CheckThat! 2021, BERT and BERT-derived transformers were at the core of the majority of the explored approaches (other transformers explored were GPT-3 and sentence-t5). Back-translation was a popular data augmentation strategy. Regarding Task 1, the use of the mT5 transformer outperformed all other participants in four out of six languages for subtask 1A. The most successful model for subtask 1B approached harmful detection as a contradiction problem. Addressing the retrieval Task 2 with the sentence-t5 transformer and gpt-neo resulted in the best performance, whereas search engines ran short. As for Task 3, the most successful approaches fine-tuned a BERT-based model (which also represented the baseline) and feature-based approaches ran short. The cross-language nature of this task was addressed by machine translating German instances into English.

The approaches to all CheckThat! 2022 tasks reflect convergence toward the fine-tuning of transformers. In the future, we are considering targeting other tasks which could play a relevant role in the analysis of journalistic and social media posts, besides the explicit factuality decision. We are considering both coverage bias in the news and subjectivity, among others.

Acknowledgments. Part of this research is carried out under the Tanbih mega-project, developed at the Qatar Computing Research Institute, HBKU, which aims to limit the impact of "fake news", propaganda, and media bias, thus promoting digital literacy and critical thinking.

Part of this work has been funded by the German Federal Ministry of Education and Research (BMBF) under the grant no. 01FP20031J. The responsibility for the contents of this publication lies with the authors.

References

1. Agirre, E., et al.: SemEval-2016 task 1: semantic textual similarity, monolingual and cross-lingual evaluation. In: Proceedings of the 10th International Workshop on Semantic Evaluation, SemEval 2016. pp. 497–511 (2016)
2. Agrestia, S., Hashemianb, A.S., Carmanc, M.J.: PoliMi-FlatEarthers at Check-That! 2022: GPT-3 applied to claim detection. In: Working Notes of CLEF 2022 - Conference and Labs of the Evaluation Forum, CLEF 202022, Bologna, Italy (2022)
3. Alam, F., et al.: Fighting the COVID-19 infodemic in social media: a holistic perspective and a call to arms. In: Proceedings of the International AAAI Conference on Web and Social Media, ICWSM 2021, pp. 913–922 (2021)
4. Alam, F., et al.: Fighting the COVID-19 infodemic: modeling the perspective of journalists, fact-checkers, social media platforms, policy makers, and the society. In: Findings of EMNLP 2021, pp. 611–649 (2021)
5. Ali, Z.S., Mansour, W., Elsayed, T., Al-Ali, A.: Arafacts: the first large Arabic dataset of naturally occurring claims. In: Proceedings of the Sixth Arabic Natural Language Processing Workshop, pp. 231–236 (2021)

6. Althabiti, S., Alsalka, M.A., Atwell, E.: SCUoL at CheckThat! 2022: fake news detection using transformer-based models. In: Working Notes of CLEF 2022 - Conference and Labs of the Evaluation Forum, CLEF 2022, Bologna, Italy (2022)
7. Arif, M., et al.: CIC at CheckThat! 2022: multi-class and cross-lingual fake news detection. In: Working Notes of CLEF 2022 - Conference and Labs of the Evaluation Forum, CLEF 2022, Bologna, Italy (2022)
8. Ba, M.L., Berti-Equille, L., Shah, K., Hammady, H.M.: VERA: a platform for veracity estimation over web data. In: Proceedings of the 25th International Conference on World Wide Web, WWW 2016, pp. 159–162 (2016)
9. Balouchzahi, F., Shashirekha, H., Sidorov, G.: MUCIC at CheckThat! 2021: FaDofake news detection and domain identification using transformers ensembling. In: Faggioli, G., Ferro, N., Joly, A., Maistro, M., Piroi, F. (eds.) CLEF 2021 Working Notes. Working Notes of CLEF 2021-Conference and Labs of the Evaluation Forum, pp. 455–464 (2021)
10. Baly, R., et al.: What was written vs. who read it: news media profiling using text analysis and social media context. In: Proceedings of the 58th Annual Meeting of the Association for Computational Linguistics, ACL 2020, pp. 3364–3374 (2020)
11. Barrón-Cedeño, A., et al.: **CheckThat**! at CLEF 2020: enabling the automatic identification and verification of claims in social media. In: Jose, J.M., Yilmaz, E., Magalhães, J., Castells, P., Ferro, N., Silva, M.J., Martins, F. (eds.) ECIR 2020. LNCS, vol. 12036, pp. 499–507. Springer, Cham (2020). https://doi.org/10.1007/978-3-030-45442-5_65
12. Barrón-Cedeño, A., et al.: Overview of CheckThat! 2020: automatic identification and verification of claims in social media. In: Arampatzis, A., et al. (eds.) CLEF 2020. LNCS, vol. 12260, pp. 215–236. Springer, Cham (2020). https://doi.org/10.1007/978-3-030-58219-7_17
13. Bilel, T., Mohamed Aziz, B.N., Haddad, H.: iCompass at CheckThat! 2022: ARBERT and AraBERT for Arabic checkworthy tweet identification. In: Working Notes of CLEF 2022 - Conference and Labs of the Evaluation Forum, CLEF 2022, Bologna, Italy (2022)
14. Black, S., Gao, L., Wang, P., Leahy, C., Biderman, S.: GPT-neo: large scale autoregressive language modeling with mesh-tensorflow (2021). https://doi.org/10.5281/zenodo.5297715
15. Blanc, O., Pritzkau, A., Schade, U., Geierhos, M.: CODE at CheckThat! 2022: multi-class fake news detection of news articles with BERT. In: Working Notes of CLEF 2022 - Conference and Labs of the Evaluation Forum, CLEF 2022, Bologna, Italy (2022)
16. Buliga Nicu, R.M.: Zorros at CheckThat! 2022: ensemble model for identifying relevant claims in tweets. In: Working Notes of CLEF 2022 - Conference and Labs of the Evaluation Forum, CLEF 2022, Bologna, Italy (2022)
17. Capetillo, C.P., Lecuona-Gómez, D., Gómez-Adorn, H., Arroyo-Fernández, I., Neri-Chávez, J.: HBDCI at CheckThat! 2022: fake news detection using a combination of stylometric features and deep learning. In: Working Notes of CLEF 2022 - Conference and Labs of the Evaluation Forum, CLEF 2022, Bologna, Italy (2022)
18. Chernyavskiy, A., Ilvovsky, D., Nakov, P.: Aschern at CLEF CheckThat! 2021: Lambda-calculus of fact-checked claims. In: Faggioli, G., Ferro, N., Joly, A., Maistro, M., Piroi, F. (eds.) CLEF 2021 Working Notes. Working Notes of CLEF 2021-Conference and Labs of the Evaluation Forum (2021)

19. Conneau, A., et al.: Unsupervised cross-lingual representation learning at scale. In: Proceedings of the Annual Meeting of the Association for Computational Linguistics (ACL), pp. 8440–8451 (2020). https://doi.org/10.18653/v1/2020.acl-main.747

20. Cusmuliuc, C.G., Amarandei, M.A., Pelin, I., Cociorva, V.I., Iftene, A.: UAICS at CheckThat! 2021: fake news detection. In: Faggioli, G., Ferro, N., Joly, A., Maistro, M., Piroi, F. (eds.) CLEF 2021 Working Notes. Working Notes of CLEF 2021-Conference and Labs of the Evaluation Forum (2021)

21. Da San Martino, G., Barrón-Cedeno, A., Wachsmuth, H., Petrov, R., Nakov, P.: SemEval-2020 task 11: detection of propaganda techniques in news articles. In: Proceedings of the 14th Workshop on Semantic Evaluation, SemEval 2020, pp. 1377–1414 (2020)

22. Derczynski, L., Bontcheva, K., Liakata, M., Procter, R., Hoi, G.W.S., Zubiaga, A.: SemEval-2017 task 8: RumourEval: determining rumour veracity and support for rumours. In: Proceedings of the 11th International Workshop on Semantic Evaluation, SemEval 2017, pp. 69–76 (2017)

23. Dimitrov, D., et al.: SemEval-2021 task 6: detection of persuasion techniques in texts and images. In: Proceedings of the International Workshop on Semantic Evaluation, SemEval 2021, pp. 70–98 (2021)

24. Elsayed, T., et al.: CheckThat! at CLEF 2019: automatic identification and verification of claims. In: Azzopardi, L., Stein, B., Fuhr, N., Mayr, P., Hauff, C., Hiemstra, D. (eds.) Advances in Information Retrieval, pp. 309–315. Springer International Publishing, Cham (2019). https://doi.org/10.1007/978-3-030-15719-7_41

25. Elsayed, T., et al.: Overview of the CLEF-2019 CheckThat!: automatic identification and verification of claims. In: Experimental IR Meets Multilinguality, Multimodality, and Interaction, pp. 301–321. LNCS (2019)

26. Eyuboglu, A.B., Arslan, M.B., Sonmezer, E., Kutlu, M.: TOBB ETU at CheckThat! 2022: detecting attention-worthy and harmful tweets and check-worthy claims. In: Working Notes of CLEF 2022 - Conference and Labs of the Evaluation Forum, CLEF 2022, Bologna, Italy (2022)

27. Faggioli, G., Ferro, N., Joly, A., Maistro, M., Piroi, F. (eds.): CLEF 2021 Working Notes. Working Notes of CLEF 2021-Conference and Labs of the Evaluation Forum (2021)

28. Frick, R.A., Vogel, I.: Fraunhofer SIT at CheckThat! 2022: ensemble similarity estimation for finding previously fact-checked claims. In: Working Notes of CLEF 2022 - Conference and Labs of the Evaluation Forum, CLEF 2022, Bologna, Italy (2022)

29. Frick, R.A., Vogel, I., Nunes Grieser, I.: Fraunhofer SIT at CheckThat! 2022: semi-supervised ensemble classification for detecting check-worthy tweets. In: Working Notes of CLEF 2022 - Conference and Labs of the Evaluation Forum, CLEF 2022, Bologna, Italy (2022)

30. Gencheva, P., Nakov, P., Màrquez, L., Barrón-Cedeño, A., Koychev, I.: A context-aware approach for detecting worth-checking claims in political debates. In: Proceedings of the International Conference Recent Advances in Natural Language Processing, RANLP 2017, pp. 267–276 (2017)

31. Gorrell, G., et al.: SemEval-2019 task 7: RumourEval, determining rumour veracity and support for rumours. In: Proceedings of the 13th International Workshop on Semantic Evaluation, SemEval 2019, pp. 845–854 (2019)

32. Guo, Z., Schlichtkrull, M., Vlachos, A.: A survey on automated fact-checking. Trans. Assoc. Comput. Linguist. **10**, 178–206 (2022)
33. Hanselowski, A., et al.: A retrospective analysis of the fake news challenge stance-detection task. In: Proceedings of the 27th International Conference on Computational Linguistics, COLING 2018, pp. 1859–1874 (2018)
34. Hariharan, R.L., Anand Kumar, M.: NITK-IT_NLP at CheckThat! 2022: window based approach for fake news detection using transformers. In: Working Notes of CLEF 2022 - Conference and Labs of the Evaluation Forum, CLEF 2022, Bologna, Italy (2022)
35. Hassan, N., Li, C., Tremayne, M.: Detecting check-worthy factual claims in presidential debates. In: Proceedings of the 24th ACM International on Conference on Information and Knowledge Management, CIKM 2015, pp. 1835–1838 (2015)
36. Hassan, N., Tremayne, M., Arslan, F., Li, C.: Comparing automated factual claim detection against judgments of journalism organizations. In: Computation+Journalism Symposium, pp. 1–5 (2016)
37. Hassan, N., et al.: ClaimBuster: the first-ever end-to-end fact-checking system. Proc. VLDB Endowment **10**(12), 1945–1948 (2017)
38. Hövelmeyer, A., Boland, K., Dietze, S.: SimBa at CheckThat! 2022: lexical and semantic similarity based detection of verified claims in an unsupervised and supervised way. In: Working Notes of CLEF 2022 - Conference and Labs of the Evaluation Forum, CLEF 2022, Bologna, Italy (2022)
39. Hüsünbeyi, Z.M., Deck, O., Scheffler, T.: RUB-DFL at CheckThat! 2022: transformer models and linguistic features for identifying relevant claims. In: Working Notes of CLEF 2022 - Conference and Labs of the Evaluation Forum, CLEF 2022, Bologna, Italy (2022)
40. Karadzhov, G., Nakov, P., Màrquez, L., Barrón-Cedeño, A., Koychev, I.: Fully automated fact checking using external sources. In: Proceedings of the International Conference Recent Advances in Natural Language Processing, RANLP 2017, pp. 344–353 (2017)
41. Kavatagi, S., Rachh, R., Mulimani, M.: VTU_BGM at Check That! 2022: an autoregressive encoding model for verifying check-worthy claims. In: Working Notes of CLEF 2022 - Conference and Labs of the Evaluation Forum, CLEF 2022, Bologna, Italy (2022)
42. Kazemi, A., Garimella, K., Shahi, G.K., Gaffney, D., Hale, S.A.: Research note: Tiplines to uncover misinformation on encrypted platforms: a case study of the 2019 Indian general election on whatsapp. Harvard Kennedy School Misinformation Review (2022)
43. Köhler, J., et al.: Overview of the CLEF-2022 CheckThat! lab task 3 on fake news detection. In: Working Notes of CLEF 2022–Conference and Labs of the Evaluation Forum, CLEF 2022, Bologna, Italy (2022)
44. Kovachevich, N.: BERT fine-tuning approach to CLEF CheckThat! fake news detection. In: Faggioli, G., Ferro, N., Joly, A., Maistro, M., Piroi, F. (eds.) CLEF 2021 Working Notes. Working Notes of CLEF 2021-Conference and Labs of the Evaluation Forum (2021)
45. Kumar, S., Kumar, G., Singh, S.R.: TextMinor at CheckThat! 2022: fake news article detection using RoBERT. In: Working Notes of CLEF 2022 - Conference and Labs of the Evaluation Forum, CLEF 2022, Bologna, Italy (2022)
46. La Barbera, D., Roitero, K., Mackenzie, J., Damiano, S., Demartini, G., Mizzaro, S.: BUM at CheckThat! 2022: a composite deep learning approach to fake news detection using evidence retrieval. In: Working Notes of CLEF 2022 - Conference and Labs of the Evaluation Forum, CLEF 2022, Bologna, Italy (2022)

47. Lomonaco, F., Donabauer, G., Siino, M.: COURAGE at CheckThat! 2022: harmful tweet detection using graph neural networks and ELECTRA. In: Working Notes of CLEF 2022 - Conference and Labs of the Evaluation Forum, CLEF 2022, Bologna, Italy (2022)

48. Ludwig, A., Felser, J., Xi, J., Labudde, D., Spranger, M.: FoSIL at CheckThat! 2022: using human behaviour-based optimization for text classification. In: Working Notes of CLEF 2022 - Conference and Labs of the Evaluation Forum, CLEF 2022, Bologna, Italy (2022)

49. Ma, J., Gao, W., Mitra, P., Kwon, S., Jansen, B.J., Wong, K.F., Cha, M.: Detecting rumors from microblogs with recurrent neural networks. In: Proceedings of the International Joint Conference on Artificial Intelligence, IJCAI 2016, pp. 3818–3824 (2016)

50. Manan Suri, P.K., Dudeja, S.: Asatya at CheckThat! 2022: multimodal BERT for identifying claims in tweets. In: Working Notes of CLEF 2022 - Conference and Labs of the Evaluation Forum, CLEF 2022, Bologna, Italy (2022)

51. Mansour, W., Elsayed, T., Al-Ali, A.: Did i see it before? detecting previously-checked claims over twitter. In: European Conference on Information Retrieval. pp. 367–381 Springer (2022)

52. Martinez-Rico, J.R., Martinez-Romo, J., Araujo, L.: NLP&IRUNED at CheckThat! 2022: ensemble of classifiers for fake news detection. In: Working Notes of CLEF 2022 - Conference and Labs of the Evaluation Forum, CLEF 2022, Bologna, Italy (2022)

53. Meaney, J., Wilson, S., Chiruzzo, L., Lopez, A., Magdy, W.: Semeval 2021 task 7: hahackathon, detecting and rating humor and offense. In: Proceedings of the 15th International Workshop on Semantic Evaluation (SemEval-2021), pp. 105–119 (2021)

54. Michael Shliselberg, S.D.H.: RIET Lab at CheckThat! 2022: improving decoder based re-ranking for claim matching. In: Working Notes of CLEF 2022 - Conference and Labs of the Evaluation Forum, CLEF 2022, Bologna, Italy (2022)

55. Mihaylova, S., Borisova, I., Chemishanov, D., Hadzhitsanev, P., Hardalov, M., Nakov, P.: DIPS at CheckThat! 2021: verified claim retrieval. In: Faggioli, G., Ferro, N., Joly, A., Maistro, M., Piroi, F. (eds.) CLEF 2021 Working Notes. Working Notes of CLEF 2021-Conference and Labs of the Evaluation Forum (2021)

56. Mihaylova, T., Karadzhov, G., Atanasova, P., Baly, R., Mohtarami, M., Nakov, P.: SemEval-2019 task 8: Fact checking in community question answering forums. In: Proceedings of the 13th International Workshop on Semantic Evaluation, SemEval 2019, pp. 860–869 (2019)

57. Mingzhe, D., Sujatha Das Gollapalli, S.K.N.: NUS-IDS at CheckThat! 2022: identifying check-worthiness of tweets using CheckthaT5. In: Working Notes of CLEF 2022 - Conference and Labs of the Evaluation Forum, CLEF 2022, Bologna, Italy (2022)

58. Mohammad, S., Kiritchenko, S., Sobhani, P., Zhu, X., Cherry, C.: SemEval-2016 task 6: Detecting stance in tweets. In: Proceedings of the 10th International Workshop on Semantic Evaluation, SemEval 2016, pp. 31–41 (2016)

59. Nakamura, K., Levy, S., Wang, W.Y.: r/fakeddit: a new multimodal benchmark dataset for fine-grained fake news detection. arXiv preprint arXiv:1911.03854 (2019)

60. Nakov, P., et al.: Overview of the CLEF-2022 CheckThat! lab task 1 on identifying relevant claims in tweets. In: Working Notes of CLEF 2022–Conference and Labs of the Evaluation Forum, CLEF 2022, Bologna, Italy (2022)

61. Nakov, P., et al.: The CLEF-2022 CheckThat! Lab on fighting the covid-19 infodemic and fake news detection. In: Hagen, M., et al. (eds.) Advances in Information Retrieval, pp. 416–428. Springer, Cham (2022). https://doi.org/10.1007/978-3-030-99739-7_52

62. Nakov, P., et al.: Overview of the CLEF-2018 lab on automatic identification and verification of claims in political debates. In: Working Notes of CLEF 2018 - Conference and Labs of the Evaluation Forum, CLEF 2018 (2018)

63. Nakov, P., et al.: Automated fact-checking for assisting human fact-checkers. In: Proceedings of the 30th International Joint Conference on Artificial Intelligence, IJCAI 2021, pp. 4551–4558 (2021)

64. Nakov, P., Da San Martino, G., Alam, F., Shaar, S., Mubarak, H., Babulkov, N.: Overview of the CLEF-2022 CheckThat! lab task 2 on detecting previously fact-checked claims. In: Working Notes of CLEF 2022–Conference and Labs of the Evaluation Forum, CLEF 2022, Bologna, Italy (2022)

65. Nakov, P., et al.: The CLEF-2021 CheckThat! lab on detecting check-worthy claims, previously fact-checked claims, and fake news. In: Hiemstra, D., Moens, M.-F., Mothe, J., Perego, R., Potthast, M., Sebastiani, F. (eds.) ECIR 2021. LNCS, vol. 12657, pp. 639–649. Springer, Cham (2021). https://doi.org/10.1007/978-3-030-72240-1_75

66. Nakov, P., et al.: Overview of the CLEF–2021 CheckThat! lab on detecting check-worthy claims, previously fact-checked claims, and fake news. In: Candan, S., et al. (eds.) CLEF 2021. LNCS, vol. 12880, pp. 264–291. Springer, Cham (2021). https://doi.org/10.1007/978-3-030-85251-1_19

67. Nakov, P., et al.: SemEval-2016 Task 3: community question answering. In: Proceedings of the 10th International Workshop on Semantic Evaluation, SemEval 2015, pp. 525–545 (2016)

68. Nguyen, V.H., Sugiyama, K., Nakov, P., Kan, M.Y.: FANG: leveraging social context for fake news detection using graph representation. In: Proceedings of the 29th ACM International Conference on Information & Knowledge Management, CIKM 2020, pp. 1165–1174 (2020)

69. Oshikawa, R., Qian, J., Wang, W.Y.: A survey on natural language processing for fake news detection. In: Proceedings of the 12th Language Resources and Evaluation Conference, LREC 2020, pp. 6086–6093 (2020)

70. Pavlopoulos, J., Sorensen, J., Laugier, L., Androutsopoulos, I.: Semeval-2021 task 5: toxic spans detection. In: Proceedings of the 15th International Workshop on Semantic Evaluation (SemEval-2021), pp. 59–69 (2021)

71. Pires, T., Schlinger, E., Garrette, D.: How Multilingual is Multilingual BERT. In: Proceedings of the Annual Meeting of the Association for Computational Linguistics (ACL), Florence, Italy, pp. 4996–5001 (2019). https://doi.org/10.18653/v1/P19-1493

72. Pogorelov, K., et al.: FakeNews: corona virus and 5G conspiracy task at MediaEval 2020. In: Proceedings of the MediaEval 2020 Workshop, MediaEval 2020 (2020)

73. Popat, K., Mukherjee, S., Strötgen, J., Weikum, G.: Credibility assessment of textual claims on the web. In: Proceedings of the 25th ACM International Con-

ference on Information and Knowledge Management, CIKM 2016, pp. 2173–2178 (2016)

74. Pritzkau, A.: NLytics at CheckThat! 2021: check-worthiness estimation as a regression problem on transformers. In: Faggioli, G., Ferro, N., Joly, A., Maistro, M., Piroi, F. (eds.) CLEF 2021 Working Notes. Working Notes of CLEF 2021-Conference and Labs of the Evaluation Forum (2021)

75. Pritzkau, A., Blanc, O., Geierhos, M., Schade, U.: NLytics at CheckThat! 2022: hierarchical multi-class fake news detection of news articles exploiting the topic structure. In: Working Notes of CLEF 2022 - Conference and Labs of the Evaluation Forum, CLEF 2022, Bologna, Italy (2022)

76. Röchert, D., Shahi, G.K., Neubaum, G., Ross, B., Stieglitz, S.: The networked context of covid-19 misinformation: informational homogeneity on youtube at the beginning of the pandemic. Online Social Netw. Media **26**, 100164 (2021)

77. Savchev, A.: AI Rational at CheckThat! 2022: using transformer models for tweet classification. In: Working Notes of CLEF 2022 - Conference and Labs of the Evaluation Forum, CLEF 2022, Bologna, Italy (2022)

78. Schütz, M., et al.: AIT FHSTP at CheckThat! 2022: cross-lingual fake news detection with a large pre-trained transformer. In: Working Notes of CLEF 2022 - Conference and Labs of the Evaluation Forum, CLEF 2022, Bologna, Italy (2022)

79. Schütz, M., Siegel, M.: Baseline for clef2022 - checkthat! lab task 3 (2022). https://doi.org/10.5281/zenodo.6362498

80. Sepúlveda-Torres, R., Saquete, E.: GPLSI team at CLEF CheckThat! 2021: fine-tuning BETO and RoBERTa. In: Faggioli, G., Ferro, N., Joly, A., Maistro, M., Piroi, F. (eds.) CLEF 2021 Working Notes. Working Notes of CLEF 2021-Conference and Labs of the Evaluation Forum (2021)

81. Shaar, S., Alam, F., Da San Martino, G., Nakov, P.: The role of context in detecting previously fact-checked claims. Arxiv:2104.07423 (2021)

82. Shaar, S., Alam, F., Martino, G.D.S., Nakov, P.: Assisting the human fact-checkers: detecting all previously fact-checked claims in a document. arXiv preprint arXiv:2109.07410 (2021)

83. Shaar, S., Babulkov, N., Da San Martino, G., Nakov, P.: That is a known lie: Detecting previously fact-checked claims. In: Proceedings of the 58th Annual Meeting of the Association for Computational Linguistics, ACL 2020, pp. 3607–3618 (2020)

84. Shaar, S., et al.: Overview of the CLEF-2021 CheckThat! lab task 2 on detecting previously fact-checked claims in tweets and political debates. In: Faggioli, G., Ferro, N., Joly, A., Maistro, M., Piroi, F. (eds.) CLEF 2021 Working Notes. Working Notes of CLEF 2021-Conference and Labs of the Evaluation Forum (2021)

85. Shaar, S., et al.: Overview of the CLEF-2021 CheckThat! lab task 1 on check-worthiness estimation in tweets and political debates. In: Faggioli, G., Ferro, N., Joly, A., Maistro, M., Piroi, F. (eds.) CLEF 2021 Working Notes. Working Notes of CLEF 2021-Conference and Labs of the Evaluation Forum (2021)

86. Shahi, G.K.: AMUSED: an annotation framework of multi-modal social media data. arXiv:2010.00502 (2020)

87. Shahi, G.K., Dirkson, A., Majchrzak, T.A.: An exploratory study of COVID-19 misinformation on Twitter. Online Social Netw. Media **22**, 100104 (2021)

88. Shahi, G.K., Struß, J.M., Mandl, T.: Overview of the CLEF-2021 CheckThat! lab: Task 3 on fake news detection. In: Faggioli, G., Ferro, N., Joly, A., Maistro, M., Piroi, F. (eds.) CLEF 2021 Working Notes. Working Notes of CLEF 2021-Conference and Labs of the Evaluation Forum (2021)

89. Taboubi, B., Nessir, M.A.B., Haddad, H.: iCompass at CheckThat! 2022: combining deep language models for fake news detection. In: Working Notes of CLEF 2022 - Conference and Labs of the Evaluation Forum, CLEF 2022, Bologna, Italy (2022)

90. Tarannum, P., Md. Arid, H., Alam, F., Noori, S.R.H.: Z-Index at CheckThat! Lab 2022: check-worthiness identification on tweet text. In: Working Notes of CLEF 2022 - Conference and Labs of the Evaluation Forum, CLEF 2022, Bologna, Italy (2022)

91. Tchechmedjiev, A., et al.: ClaimsKG: a knowledge graph of fact-checked claims. In: Ghidini, C., et al. (eds.) ISWC 2019. LNCS, vol. 11779, pp. 309–324. Springer, Cham (2019). https://doi.org/10.1007/978-3-030-30796-7_20

92. Thorne, J., Vlachos, A., Christodoulopoulos, C., Mittal, A.: FEVER: a large-scale dataset for fact extraction and VERification. In: Proceedings of the Conference of the North American Chapter of the Association for Computational Linguistics: Human Language Technologies, NAACL 2018, pp. 809–819 (2018)

93. Toraman, C., Ozcelik, O., Şahinuç, F., Sahin, U.: ARC-NLP at CheckThat! 2022: contradiction for harmful tweet detection. In: Working Notes of CLEF 2022 - Conference and Labs of the Evaluation Forum, CLEF 2022, Bologna, Italy (2022)

94. Tran, H.N., Kruschwitz, U.: ur-iw-hnt at CheckThat! 2022: cross-lingual text summarization for fake news detection. In: Working Notes of CLEF 2022 - Conference and Labs of the Evaluation Forum, CLEF 2022, Bologna, Italy (2022)

95. Truică C.O., Apostol, E.S., Paschke, A.: Awakened at CheckThat! 2022: fake news detection using BiLSTM and sentence transformer. In: Working Notes of CLEF 2022 - Conference and Labs of the Evaluation Forum, CLEF 2022, Bologna, Italy (2022)

96. Varma, H., Jain, A., Ratadiya, P., Rathi, A.: Attestable at semeval-2021 task 9: extending statement verification with tables for unknown class, and semantic evidence finding. In: Proceedings of the 15th International Workshop on Semantic Evaluation (SemEval-2021), pp. 1276–1282 (2021)

97. Vasileva, S., Atanasova, P., Màrquez, L., Barrón-Cedeño, A., Nakov, P.: It takes nine to smell a rat: neural multi-task learning for check-worthiness prediction. In: Proceedings of the International Conference on Recent Advances in Natural Language Processing, RANLP 2019, pp. 1229–1239 (2019)

98. Wang, N.X., Mahajan, D., Danilevsky, M., Rosenthal, S.: Semeval-2021 task 9: fact verification and evidence finding for tabular data in scientific documents (sem-tab-facts). In: Proceedings of the 15th International Workshop on Semantic Evaluation (SemEval-2021), pp. 317–326 (2021)

99. Williams, E., Rodrigues, P., Tran, S.: Accenture at CheckThat! 2021: interesting claim identification and ranking with contextually sensitive lexical training data augmentation. In: Faggioli, G., Ferro, N., Joly, A., Maistro, M., Piroi, F. (eds.) CLEF 2021 Working Notes. Working Notes of CLEF 2021-Conference and Labs of the Evaluation Forum (2021)

100. Yasser, K., Kutlu, M., Elsayed, T.: Re-ranking web search results for better fact-checking: a preliminary study. In: Proceedings of the 27th ACM International Conference on Information and Knowledge Management, pp. 1783–1786 (2018)

101. Zhou, X., Wu, B., Fung, P.: Fight for 4230 at CLEF CheckThat! 2021: domain-specific preprocessing and pretrained model for ranking claims by check-worthiness. In: Faggioli, G., Ferro, N., Joly, A., Maistro, M., Piroi, F. (eds.) CLEF 2021 Working Notes. Working Notes of CLEF 2021-Conference and Labs of the Evaluation Forum (2021)
102. Zubiaga, A., Liakata, M., Procter, R., Hoi, G.W.S., Tolmie, P.: Analysing how people orient to and spread rumours in social media by looking at conversational threads. PLoS ONE **11**(3), e0150989 (2016)

Overview of ChEMU 2022 Evaluation Campaign: Information Extraction in Chemical Patents

Yuan Li[1], Biaoyan Fang[1], Jiayuan He[1,4], Hiyori Yoshikawa[1,5],
Saber A. Akhondi[2], Christian Druckenbrodt[3], Camilo Thorne[3], Zubair Afzal[2],
Zenan Zhai[1], Timothy Baldwin[1], and Karin Verspoor[1,4(✉)] ⓘ

[1] The University of Melbourne, Melbourne, Australia
`karin.verspoor@rmit.edu.au`
[2] Elsevier BV, Amsterdam, The Netherlands
[3] Elsevier Information Systems GmbH, Frankfurt, Germany
[4] RMIT University, Melbourne, Australia
[5] Fujitsu Limited, Tokyo, Japan

Abstract. In this paper, we provide an overview of the Cheminformatics Elsevier Melbourne University (ChEMU) evaluation lab 2022, part of the Conference and Labs of the Evaluation Forum 2022 (CLEF 2022). The ChEMU campaign focuses on information extraction tasks over chemical reactions in patents. The ChEMU 2020 lab provided two information extraction tasks, named entity recognition and event extraction. The ChEMU 2021 lab introduced one more task, anaphora resolution. This year, we re-run all the three tasks with new test data. Together, the tasks support comprehensive automatic chemical patent analysis. Herein, we describe the resources created for these tasks and the evaluation methodology adopted. We also provide a brief summary of the methods employed by participants of this lab and the results obtained across 22 runs from 3 teams, finding that several submissions achieve better results than the baseline methods prepared by the organizers.

Keywords: Chemical patents · Text mining · Information Extraction

1 Introduction

The discovery of new chemical compounds is a key driver of the chemistry and pharmaceutical industries, *inter alia*. Patents serve as a critical source of information about new chemical compounds, providing timely and comprehensive information about new chemical compounds [4,6,36]. Despite the significant commercial and research value of the information in patents, manual effort is still the primary mechanism for extracting and organizing this information. This is costly, considering the large volume of patents available [16,31]. Development of automatic natural language processing (NLP) systems for chemical patents, which aim to convert text corpora into structured knowledge about chemical compounds, has become a focus of recent research [12,20].

The ChEMU campaign focuses on information extraction tasks over chemical reactions in patents. The ChEMU2020 lab [12,33] provided two information extraction tasks, named entity recognition (NER) and event extraction (EE). The ChEMU 2021 lab [11,26] introduced one more task, anaphora resolution (AR). This year, we re-run all the three tasks with new test sets. Together, the tasks support comprehensive automatic chemical patent analysis.

In collaboration with chemical domain experts, we have built upon the datasets used in ChEMU 2020/2021 (1500 snippets) and prepared 500 snippets from selected chemical patents that specifically target all three tasks. For the NER and the EE tasks, three chemical experts were hired to manually annotate the corpus, labeling named entities and event steps in these text segments. Two of them reviewed all text segments independently and the third annotator acted as an adjudicator who resolved their disagreements and merged their annotations into the final gold-standard corpus. For the AR task, two chemical experts, a PhD candidate and a final year bachelor student in Chemistry were hired to annotate the same set of snippets. The dataset was first annotated by the two annotators individually, and then their annotations were compared and combined by an adjudicator.

The ChEMU2022 lab has received considerable interest, attracting 54 registrants. Specifically, we received 8 runs from 3 teams in the NER task, 11 runs from 3 teams in the EE task, and 3 runs from 1 team in the AR task, respectively. Several submissions achieved exciting results, with a few of them outperforming baseline models significantly.

The rest of the paper is structured as follows. We first discuss related work and shared tasks in Sect. 2 and introduce the corpus we created for use in the lab in Sect. 3. Then we give an overview of the tasks in Sect. 4 and detail the valuation framework of ChEMU in Sect. 5 including the evaluation methods and baseline models. We present the evaluation results in Sect. 6 and finally conclude this paper in Sect. 7.

2   Related Work

To assess and advance the natural language processing (NLP) techniques in the biochemical domain, many shared tasks/labs have been organized, including n2c2[1], TREC[2], BioCreative[3], BioNLP[4], and CLEF workshops[5]. These shared tasks have covered a range of benchmark text mining tasks: information retrieval, such as document retrieval (CLEF eHealth 2014 [18]) and text classification (CoNLL 2010 [9]); word semantics, such as named entity recognition (BioCreative II [30] Task 1) and mention normalization (BioCreative III [5,28] Gene Normalization Task); relation semantics, such as event extraction (GENIA Event

---

[1] https://n2c2.dbmi.hms.harvard.edu/.
[2] https://trec.nist.gov/.
[3] https://biocreative.bioinformatics.udel.edu/.
[4] https://2019.bionlp-ost.org/.
[5] https://sites.google.com/site/clefehealth/.

Extraction [19]) and interaction extraction (Drug-Drug Interaction [14]); and high-level applications, such as question answering (Semantic QA [37]) and document summarization (Biomed-Summ [17]).

Nevertheless, most of these shared tasks/labs did not focus on the domain of chemical patents. These shared tasks mainly focused on the text mining over biomedical texts (e.g., scientific literature, such as PubMed abstracts) or clinical data (e.g., clinical health records). Text mining techniques that are developed for biomedical or biochemical texts, such as scientific journals and clinical records may not be effective for chemical patents. This is because their purpose is distinct-chemical patents are written for protection of intellectual property related to chemical compounds-and their content has different scope and characteristics, including variations in linguistic structures. Thus, it is critical to develop text mining techniques that are tailored for chemical patents.

Only two shared tasks have previously considered chemical patents. TREC 2009 [29] provided a chemical information retrieval track for the tasks of ad hoc retrieval of chemical patents and prior art search. However, this track differs significantly from the subtasks in our ChEMU lab: it addresses document-level retrieval and relevance to queries instead of considering the detailed content of each document. The ChemDNER-patents task [23] at the BioCreative V workshop was the task that is most similar with ours. It aimed at detection of chemical compounds and genes/proteins in patent text. However, the ChemDNER-patents task only considered entity detection within patent abstracts while we consider data extracted from the full texts of patents. Moreover, our definition of chemical compound entities is much richer as our label set defines not only that a chemical or drug compound is mentioned, but also what its specific role is with respect to the chemical reaction that it is related to in the description, e.g., starting material, catalyst, or product.

The ChEMU labs also contribute new corpus on chemical text mining for the research community[6]. Most existing benchmark datasets for biochemical text mining focus on biomedical texts, i.e., texts that consider the interaction of chemicals with molecular biology or human disease. CHEMProt [21] consists of 1,820 PubMed[7] abstracts with chemical-protein interactions, DDI extraction 2013 corpus [14] is a collection of 792 texts selected from the DrugBank database[8] and other 233 PubMed abstracts, and BC5CDR is a collection of 1,500 PubMed titles and abstracts selected from the CTD-Pfizer corpus, just to give a few examples.

The number of public datasets that focus on the chemistry domain is limited. Further, several existing chemical datasets are based on structured/semi-structured texts rather than free, natural language, texts. For example, the ZINC 15 250 k corpus[9] is a collection of 250,000 molecules with their Simplified Molecular Input Line Entry System (SMILES) strings. The Tox21 dataset contains

---

[6] https://chemu.eng.unimelb.edu.au/.

[7] https://pubmed.ncbi.nlm.nih.gov/.

[8] https://go.drugbank.com/.

[9] https://github.com/aspuru-guzik-group/chemical_vae/tree/master/models/zinc.

roughly 7,000 molecules and typical 120 characteristics, such as atomic number, aromicity, donor status. There are two datasets that are constructed from free patent texts: (1) the dataset released by the ChemDNER patents task and (2) the dataset created by Akhondi et al. [2]. However, these two datasets only contain entity annotations. Our chemical reaction corpus is further enriched by the relations between the annotated entities.

Despite the limited number of shared tasks on chemical patent mining, there is an increasing interest in developing information extraction models for patents in general research communities [4,38,41]. Various text mining techniques have been proposed for information extraction over chemical patents [22], addressing fundamental NLP tasks, such as named entity recognition and relation extraction [3,38,39,42]. Early techniques for chemical text mining, such as dictionary-based methods [3,15,35] and grammar-based methods [1,27,32], heavily rely on expert knowledge in the chemical domain. Recently, machine learning-based techniques have reported state-of-the-art effectiveness in chemical text mining [13,42]. However, such techniques require a large amount of annotated text data, which still remains limited. Thus, ChEMU lab 2020 was hosted to provide an opportunity for NLP experts to develop information extraction systems over chemical patents. The new ChEMU reaction corpus was also made publicly available to all researchers as an important benchmark dataset for future research in this domain [40].

3   The ChEMU Chemical Reaction Corpus

In this section, we explain how the dataset is created for our shared tasks. The complete annotation guidelines are made available on our website[10].

3.1   Data Selection

The ChEMU chemical reaction corpus was built with the aid of Elsevier Reaxys® database[11]. Reaxys® is a rich information resource for chemical reactions, which contains detailed descriptions of chemical reactions that are extracted via an "excerption" process, i.e., manual selection of information from literature sources, such as patents and scientific publications.

In ChEMU 2020, we selected 180 English patents from the European Patent Office and the United States Patent and Trademark Office, for which information had been included in the Reaxys database. From these patents, 1500 text segments were sampled from chemical reaction descriptions pre-identified by expert domain annotators, available as a product of the process used to populate information in Reaxys® . We refer to each text segment as a patent "snippet" and use the two expressions interchangeably in the remainder of this paper. The 1500

---

[10] http://chemu2022.eng.unimelb.edu.au/.

[11] Reaxys® Copyright ©2022 Elsevier Life Sciences IP Limited except certain content provided by third parties. Reaxys is a trademark of Elsevier Life Sciences IP Limited, used under license. https://www.reaxys.com.

[Step 4] Synthesis of N-((5-(hydrazinecarbonyl)pyridin-2-yl)methyl)-1-methyl-N-phenylpiperidine-4-carboxamide Methyl 6-((1-methyl-N-phenylpiperidine-4-carboxamido)methyl)nicotinate (0.120 g, 0.327 mmol), synthesized in step 3, and hydrazine monohydrate (0.079 mL, 1.633 mmol) were dissolved in ethanol (10 mL) at room temperature, and the solution was heated under reflux for 12 hours, and then cooled to room temperature to terminate the reaction. The reaction mixture was concentrated under reduced pressure to remove the solvent, and the concentrate was purified by column chromatography (SiO2, 4 g cartridge; methanol/dichloromethane = from 5% to 30%) and concentrated to give the title compound (0.115 g, 95.8%) as a foam solid.

Fig. 1. An example of one patent snippet in ChEMU chemical reaction corpus.

snippets were annotated for the named entity recognition (NER) and the event extraction (EE) tasks. In ChEMU 2021, we annotated the same 1500 snippets for the anaphora resolution (AR) task. In ChEMU 2022, we further collect 500 snippets from the selected patents and annotate them for all three tasks.

We present an example of a patent snippet in Fig. 1. This snippet describes the synthesis of a particular chemical compound, N-((5-(hydrazinecarbonyl)pyridin-2-yl)methyl)-1-methyl-N-phenylpiperidine-4-carboxamide. The synthesis process consists of an ordered sequence of reaction steps:

1. dissolving the chemical compound synthesized in step 3 and hydrazine monohydrate in ethanol;
2. heating the solution under reflux;
3. cooling the solution to room temperature;
4. concentrating the cooled mixture under reduced pressure;
5. purification of the concentrate by column chromatography;
6. concentration of the purified product to get the title compound.

Our shared tasks aim at extraction of chemical reactions from chemical patents, e.g., extracting the above synthesis steps given the patent snippet in Fig. 1. To achieve this goal, it is crucial for us to first identify the entities that are involved in these reaction steps (e.g., hydrazine monohydrate and ethanol) and then determine the relations between the involved entities (e.g., hydrazine monohydrate is dissolved in ethanol).

Furthermore, our shared tasks also aim at resolving the reference in the chemical reactions. For example, the *solution* in the second step refers to the title compound (0.120 g, 0.327 mmol), hydrazine monohydrate (0.079 mL, 1.633 mmol), and ethanol (10 mL).

3.2   Annotation Guidelines

NER Annotations. Four categories of entities are annotated over the corpus: (1) chemical compounds that are involved in a chemical reaction; (2) conditions under which a chemical reaction is carried out; (3) yields obtained for the final chemical product; and (4) example labels that are associated with reaction specifications. Ten labels are further defined under the above four categories. We define five different roles that a chemical compound can play within a chemical reaction, corresponding to five labels under this category: STARTING MATERIAL, REAGENT CATALYST, REACTION PRODUCT, SOLVENT, and OTHER COMPOUND. We also define two labels under the category of conditions: TIME and TEMPERATURE; and two labels under the category of yields: YIELD PERCENT and YIELD OTHER.

The definitions of all resultant labels are summarized as follows:

1. *Reaction product*: A substance that is formed during a chemical reaction.
2. *Starting material*: A substance that is consumed in the course of a chemical reaction providing atoms to products.
3. *Reagent catalyst*: A compound added to a system to cause or help with a chemical reaction. Compounds like catalysts, bases to remove protons or acids to add protons must be also annotated with this tag.
4. *Solvent*: A chemical entity that dissolves a solute resulting in a solution.
5. *Other compound*: Other chemical compounds that are not the products, starting materials, reagents, catalysts and solvents.
6. *Example label*: A label associated with a reaction specification.
7. *Temperature*: The temperature at which the reaction was carried out.
8. *Time*: The reaction time of the reaction.
9. *Yield percent*: Yield given in percent values.
10. *Yield other*: Yields provided in other units than %.

EE Annotations. A chemical reaction process is usually a sequence of steps, and these steps can be categorized into two types: (1) reaction steps, i.e., the steps required to convert the starting materials to the target reaction product; and (2) work-up steps, i.e., the manipulations required to purify or isolate a chemical product. For example, in Fig. 1, the step of heating the solution under reflux for 12 h is a reaction step while the step of cooling it to room temperature is a work-up step.

We define two types of trigger words: WORKUP which refers to an event step where a chemical compound is isolated/purified, and REACTION STEP which refers to an event step that is involved in the conversion from a starting material to an end product. When labelling event arguments, we adapt semantic argument role labels Arg1 and ArgM from the Proposition Bank to label the relations between the trigger words and other arguments. Specifically, the label Arg1 refers to the relation between an event trigger word and a chemical compound. Here, Arg1 represents argument roles of being causally affected by another participant in the event. ArgM represents adjunct roles with respect to an event, used to label the relation between a trigger word and a temperature, time or yield entity. The definitions of trigger word types and relation types are summarized as follows:

1. *Workup*: An event step which is a manipulation required to isolate and purify the product of a chemical reaction.
2. *Reaction step*: An event within which starting materials are converted into the product.
3. *Arg1*: The relation between an event trigger word and a chemical compound.
4. *ArgM*: The relation between an event trigger word and a temperature, time, or yield entity.

AR Annotations - Mentions. We aim to capture anaphora in chemical patents, with a focus on identifying chemical compounds during the reaction process. Consistent with other anaphora corpora [7,10,34], only mentions that are involved in referring relationships (as defined in Sect. 3.2) and related to chemical compounds are annotated. The mention types that are considered for anaphora annotation are listed below.

1. *Chemical names*: the formal name of chemical compounds.
2. *Identifiers*: identifiers or labels that uniquely represent chemical compounds which occur earlier in the text.
3. *Phrases and noun types*: pronouns that refer to a previously mentioned chemical compounds, e.g. *they* or *it*, and definite and indefinite noun phrases that refer to chemical compounds, e.g. *the solvent, the title compound, the mixture*, and *a white solid, a crude product*.

It should be noted that verbs (e.g. *mix, purify, distil*) and descriptions that refer to events (e.g. *the same process, step 5*) are not annotated in this corpus.

Unlike many annotation schemes, our annotation allows discontinuous mentions. For example, the underlined spans of the fragment *114 mg of 4-((4aS,7aS)-6-benzyloctahydro-1-pyrrolo[3,4-b]pyridine-1-yl)-7H-pyrrolo[2,3-d]pyrimidine* was obtained *with a yield of about 99.1%* are treated as a single discontinuous mention. This introduces further complexity into the task and helps to capture more comprehensive anaphora phenomena.

There are some differences in the definitions of entities for the NER task and the AR task. For the NER task, entity annotations identify chemical compounds (i.e. REACTION_PRODUCT, STARTING_MATERIAL, REAGENT_CATALYST, SOLVENT, and OTHER COMPOUND), reaction conditions (i.e. TIME, TEMPERATURE), quantity information (i.e. YIELD_PERCENT, YIELD_OTHER), and example labels (i.e. EXAMPLE_LABEL). There is overlap with our definition of mention for the labels relating to chemical compounds. However, in AR annotation, chemical names are annotated along with additional quantity information, as we consider this information to be an integral part of the chemical compound description. Furthermore, the original entity annotations do not include generic expressions that corefer with chemical compounds such as *the mixture, the organic layer*, or *the filtrate*, and neither do they include equipment descriptions.

AR Annotations - Relation. Anaphora resolution subsumes both coreference and bridging. In the context of chemical patents, we define four sub-types of bridging, incorporating generic and chemical knowledge.

1. *Coreference*: two expressions/mentions that refer to the same entity.
2. *Bridging*:
   (a) *Transformed*: two chemical compound entities that are initially based on the same chemical components and have undergone possible changes through various conditions (e.g., pH and temperature).
   (b) *Reaction-associated*: the relationship between a chemical compound and its immediate sources via a mixing process. The immediate sources do need to be reagents, but they need to end up in the corresponding product. The source compounds retain their original chemical structure.
   (c) *Work-up*: the relationship between chemical compounds that were used for isolation or purification purposes, and their corresponding output products.
   (d) *Contained*: the association holding between chemical compounds and the related equipment in which they are placed. The direction of the relation is from the related equipment to the previous chemical compound.

A referring mention which cannot be interpreted on its own, or an indirect mention, is called an *anaphor*, and the mention which it refers back to is called the *antecedent*. In relation annotation, we preserve the direction of the anaphoric relation, from the anaphor to the antecedent. Following similar assumptions in recent work, we restrict annotations to cases where the antecedent appears earlier in the text than the anaphor.

3.3   Annotation Process

To facilitate the annotation process, a silver standard set was first prepared based on information captured in the Elsevier Reaxys® database. The extracted records from Reaxys® are linked to the IDs of their source patents. However, the precise locations of the key entity and relation information in these records in source patents are needed to construct the gold-standard corpus. The silver-standard dataset was prepared by automatically mapping elements of the records in the Reaxys® database to the source patents from which the records were extracted. This mapping process was performed by scanning patent texts and searching for excerpted entity mentions.

For the NER and the EE tasks, three chemical experts were hired to prepare the gold standard corpus. They manually reviewed all texts and pre-annotations in the silver-standard dataset to add or correct precise locations of the relevant entities and relations in the texts, according to annotation guidelines in Sect. 3.2. Two of the experts first independently reviewed and updated the silver standard annotations. Then, a third chemical expert served as an adjudicator who resolved their disagreements to produce the final gold-standard corpus. For the AR task, one of the chemical experts who had annotated for the anaphora resolution task in ChEMU 2021 was hired to annotate the same set of snippets.

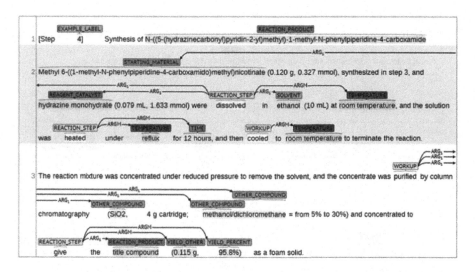

Fig. 2. Visualization of the annotations in the snippet in Fig. 1 for the NER and the EE tasks.

The annotation process was conducted using the BRAT annotation tool[12], which is an interactive web-based tool for adding annotations to input texts. Continuing with the example snippet shown in Fig. 1, a visualization of the snippet after annotation is presented in Fig. 2 for the NER and the EE tasks, and Fig. 3 for the AR task.

3.4 Data Partitions

We combine the training/development/test sets for ChEMU 2020/2021 (1500 snippets) and use it as the training set for ChEMU 2022. The 500 new snippets that we annotated for ChEMU 2022 are used as the test set.

In ChEMU 2020 and 2021, the evaluation results of all submissions to the test set were only available when the shared tasks ended. This year, we run all shared tasks in a Kaggle-style where the test set (500 snippets) is randomly partitioned into two splits public/private with a ratio of 30%/70%, and the participants will get immediate feedback on the public test set (150 snippets) after making a submission, while the evaluation results on the private test set (350 snippets) remain secret until the end of the shared tasks. Note that the participants are not aware of the specific split of public and private test sets.

4   Task Definition

The three tasks, named entity recognition, event extraction, and anaphora resolution, are all snippet-level tasks since they only consider entities or relations

---

[12] https://brat.nlplab.org/.

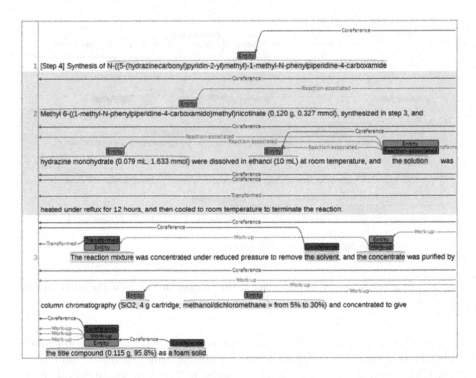

Fig. 3. Visualization of the annotations in the snippet in Fig. 1 for the AR task.

between them within a few consecutive sentences. In our ChEMU corpus, every snippet has been annotated for all three tasks, which opens the opportunity to explore multi-task learning since the input data is the same for all three tasks, as illustrated in Table 1.

4.1 Task 1: Named Entity Recognition

In order to understand and extract a chemical reaction from natural language texts, the first essential step is to identify the entities that are involved in the chemical reaction. The first task aims to accomplish this step by identifying the ten types of entities described in Sect. 3.2. The task requires the detection of the entity names in patent snippets and the assignment of correct labels to the detected entities. For example, given a detected chemical compound, the task requires the identification of both its text span and its specific type according to the role in which it plays within a chemical reaction description.

4.2 Task 2: Event Extraction

A chemical reaction usually consists of an ordered sequence of event steps that transforms a starting product to an end product, such as the six reaction steps

Table 1. Illustration of three tasks performed on the same snippet (NER, EE, and AR).

Raw text	The title compound was used without purification (1.180 g, 95.2%) as yellow solid.
NER	The **title compound** was used without purification (**1.180 g**, **95.2%**) as yellow solid REACTION_PRODUCT: **title compound** YIELD_OTHER: **1.180 g** YIELD_PERCENT: **95.2%**
EE	The **title compound** was *used* without purification (**1.180 g**, **95.2%**) as yellow solid REACTION_STEP: *used* → REACTION_PRODUCT: **title compound** REACTION_STEP: *used* → YIELD_OTHER: **1.180 g** REACTION_STEP: *used* → YIELD_PERCENT: **95.2%**
AR	**The title compound** was used without purification (**1.180 g**, **95.2%**) as *yellow solid* COREFERENCE: *yellow solid* → **The title compound (1.180 g, 95.2%)**

in the synthesis process of the chemical compound described in the example in Fig. 1. The event extraction task (Task 2) targets identifying these event steps. Similarly to conventional event extraction problems, the EE task involves three subtasks: event trigger word detection, event typing and argument prediction. First, it requires the detection of event trigger words and assignment of correct labels for the trigger words. Second, it requires the determination of argument entities that are associated with the trigger words, i.e., which entities identified in the NER task participate in event or reaction steps. This is done by labelling the connections between event trigger words and their arguments. Given an event trigger word e and a set S of arguments that participate in e, the EE task requires the creation of $|S|$ relation entries connecting e to an argument entity in S. Here, $|S|$ represents the cardinality of the set S. Finally, this task requires the assignment of correct relation type labels (Arg1 or ArgM) to each of the detected relations.

4.3 Task 3: Anaphora Resolution

This task requires the resolution of anaphoric dependencies between expressions in chemical patents. The participants are required to find five types of anaphoric relationships in chemical patents, i.e. coreference, reaction-associated, work-up, contained, and transform.

Taking the text snippet in Fig. 4 as an example, several anaphoric relationships can be extracted from it. [**The mixture**]₄ and [**the mixture**]₃ refer to the same "mixture" and thus, form a coreference relationship. The two expressions [**The mixture**]₁ and [**the mixture**]₂ are initially based on the same chemical components but the property of [**the mixture**]₂ changes after the "stir" and "cool" action. Thus, the two expressions should be linked as "Transformed". The expression [**The mixture**]₁ comes from mixing the chemical compounds prior to it, e.g., [**water (4.9 ml)**] . Thus, the two expressions are linked as "Reaction-associated". The expression [**The combined organic layer**] comes from the

[Acetic acid (9.8 ml)] and [water (4.9 ml)] were added to [the solution] in [a flask]. [The mixture]$_1$ was stirred for 3 hrs at 50°C and then cooled to 0°C . 2N-sodium hydroxide aqueous solution was added to [the mixture]$_2$ until the pH of [the mixture]$_3$ became 9. [The mixture]$_4$ was extracted with [ethyl acetate] for 3 times. [The combined organic layer] was washed with water and saturated aqueous sodium chloride.

ID	Relation type	Anaphor	Antecedent
AR1	Coreference	[The mixture]$_4$	[the mixture]$_3$
AR2	Transformed	[the mixture]$_2$	[The mixture]$_1$
AR3	Reaction_associated	[The mixture]$_1$	[water (4.9 ml)]
AR4	Work-up	[The combined organic layer]	[ethyl acetate]
AR5	Contained	[a flask]	[the solution]

Fig. 4. Text snippet containing a chemical reaction, with its anaphoric relationships. The expressions that are involved are highlighted in **bold**. In the cases where several expressions have identical text form, subscripts are added according to their order of appearance.

extraction of [ethyl acetate] . Thus, they are linked as "Work-up". Finally, the expression [the solution] is contained by the entity [a flask] , and the two are linked as "Contained".

5   Evaluation Framework

5.1   Evaluation Methods

We use BRATEval[13] to evaluate all the runs that we receive. Three metrics are used to evaluate the performance of all the submissions: Precision, Recall, and F_1 score. We use two difference matching criteria, exact matching and relaxed matching (approximate matching), as in some practical applications it also makes sense to understand if the model can identify the *approximate* region of mentions.

Formally, let $E = (ET, A, B)$ denote an entity where ET is the type of E, A and B are the beginning position (inclusive) and end position (exclusive) of the text span of E. Then two entities E_1 and E_2 are exactly matched ($E_1 = E_2$), if $ET_1 = ET_2$, $A_1 = A_2$, and $B_1 = B_2$. While two entities E_1 and E_2 are approximately matched ($E_1 \approx E_2$) if $ET_1 = ET_2$, $A_2 < B_1$, and $A_1 < B_2$, i.e. the two spans $[A_1, B_1)$ and $[A_2, B_2)$ overlaps.

Furthermore, let $R = (RT, E^{ana}, E^{ant})$ be a relation where RT is the type of R, E^{ana} the anaphor of R, E^{ant} the antecedent of R. Then R_1 and R_2 are exactly matched ($R_1 = R_2$) if $RT_1 = RT_2$, $E_1^{ana} = E_2^{ana}$, and $E_1^{ant} = E_2^{ant}$. While R_1

---

[13] https://bitbucket.org/nicta_biomed/brateval/src/master/.

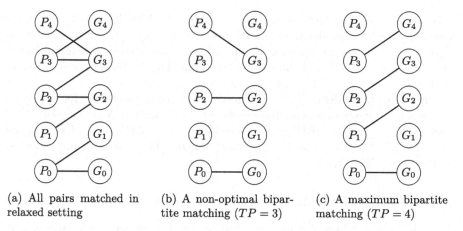

(a) All pairs matched in relaxed setting

(b) A non-optimal bipartite matching ($TP = 3$)

(c) A maximum bipartite matching ($TP = 4$)

Fig. 5. An example matching graph and two bipartite matching for it.

and R_2 are approximately matched ($R_1 \approx R_2$) if $RT_1 = RT_2$, $E_1^{ana} \approx E_2^{ana}$, and $E_1^{ant} \approx E_2^{ant}$.

In summary, we require strict type match in both exact and relaxed matching, but are lenient in span matching.

Exact Matching. With the above definitions, the metrics for exact matching can be easily calculated. The true positives (TP) are exact matching pairs found in gold relations and predicted relations. Then false positives (FP) are the predicted relations that don't have a match, i.e. $FP = \#pred - TP$, where $\#pred$ is the number of predicted relations. Similarly, false negatives FN are the gold relations that are not matched by any predicted relations, i.e. $FN = \#gold - TP$ where $\#gold$ is the number of gold relations. Finally Precision $P = TP/(TP + FP)$, Recall $R = TP/(TP + FN)$, and $F_1 = 2/(1/P + 1/R)$.

Relaxed Matching. Unlike exact matching, relaxed matching is not well-defined and metrics in this setting have more than one way to calculate, therefore we need to clearly define all the metrics.

Let consider an example shown in Fig. 5a where nodes $\{P_i\}_{i=1}^5$ are predicted relations, $\{G_i\}_{i=1}^5$ are gold relations, and every edge between a P node and a G node means they are approximately matched. At first glance, one may think that $FN = FP = 0$ because every gold relation has at least a match and so does every predicted relation. However, it is impossible to find 5 true positive pairs from this graph without using one node more than once. Therefore, if $FN = FP = 0$, then $FN + TP \neq \#gold = 5$ and $FP + TP \neq \#pred = 5$, which is inconsistent with the formulas in exact setting.

So, instead of defining FN as the number of gold relations that don't have a match, we just define $FN = \#gold - TP$. Similarly FP is defined as $\#pred - TP$.

Then the problem remained is how to calculate TP. Actually, finding true positive pairs can be considered as bipartite matching. Figure 5b shows a matching with $TP = 3$ but is not optimal. Figure 5c shows one possible maximum bipartite matching with $TP = 4$. Another optimal matching is replacing edge $P_0 - G_0$ with $P_0 - G_1$.

In summary, we define TP as the maximum bipartite matching for the graph constructed by all approximately matched pairs, then $FN = \#gold - TP$ and $FP = \#pred - TP$, finally Precision $P = TP/(TP + FP)$, Recall $R = TP/(TP + FN)$, and $F_1 = 2/(1/P + 1/R)$. This has been implemented in the latest BRATEval.

5.2 Coreference Linkings in Anaphora Resolution Task

We consider two types of coreference linking, i.e. (1) surface coreference linking and (2) atomic coreference linking, due to the existence of *transitive coreference relationships*. By transitive coreference relationships we mean multi-hop coreference such as a link from an expression T1 to T3 via an intermediate expression T2, viz., "T1→T2→T3". Surface coreference linking will restrict attention to one-hop relationships, viz., to: "T1→T2" and "T2→T3". Whereas atomic coreference linking will tackle coreference between an anaphoric expression and its first antecedent, i.e. intermediate antecedents will be collapsed. Thus, these two links will be used for the above example, "T1→T3" and "T2→T3". Note that we only consider transitive linking in coreference relationships.

Note that {T1→T2,T2→T3} infers {T1→T3,T2→T3}, but the reverse is not true. This leads to a problem about how to score a prediction {T1→T3,T2→T3}, when the gold relation is {T1→T2,T2→T3}. Both T1→T3 and T2→T3 are true, but some information is missing here.

Our solution is to first expand both the prediction set and gold set where all valid relations that can be inferred will be generated and added to the set, and then to evaluate the two sets normally. In the above example, the gold set will be expanded to {T1→T2,T2→T3,T1→T3}, and then the result is $TP = 2$, $FN = 1$. Likewise, when evaluate {T1→T4,T2→T4,T3→T4} against {T1→T2,T2→T3,T3→T4}, the gold set will be expanded into 6 relations, while the prediction set won't be expanded as no new relation can be inferred. So the evaluation result will be $TP = 3$, $FN = 3$. One may worry that if there is a chain of length n then its expanded set will be in $O(n^2)$, when n is large, this local evaluation result will have too much influence on the overall result. But we find in practice that coreference chains are relatively short, with 3 or 4 being the most typical lengths, so it is unlikely to be a big issue.

5.3 Baselines

5.4 NER and EE Baseline

We use a joint model for recognizing named entities and classifying relations between them. The model first processes the input snippet using a BERT model

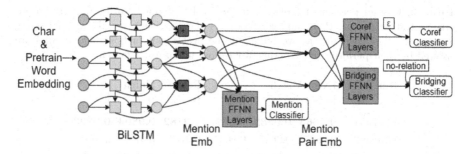

Fig. 6. The architecture of our baseline model for Task 3: Anaphora Resolution. This figure is taken from [8].

to obtain the contextualized word representations. We adopt the BIO tagging schema for training the NER classifier which classifies every word into entity tags. Then a list of identified entities is created based on the output of the NER classifier. For each entity in the list, the contextualized word representations are max pooled to obtain the representation for the entity. Then the model enumerates all possible pairs of entities and provides them to a relation classifier which classifies every pair of entities by concatenating the representations of both entities.

AR Baseline Our baseline model adopts an end-to-end architecture for coreference resolution [24,25], as depicted in Fig. 6. Following the methods presented in [8], we use GloVe embeddings and a character-level CNN as input to a BiLSTM to obtain contextualized word representations. Then all possible spans are enumerated and fed to a mention classifier which detects if the input is a mention. Based on the same mention representations, pairs of mentions are fed to a coreference classifier and a bridging classifier, where the coreference classifier does binary classification and the bridging one classifies pairs into 4 bridging relation types and a special class for no relation. Training is done jointly with all losses added together.

6 Results and Discussions

A total of 54 participants registered on our submission website for the shared tasks. Among them, we finally received 22 submissions from 3 teams on the test set. The 3 teams are LG AI Research (LG), Hokkaido University (HUKB), and Virginia Commonwealth University (VCU). In this section, we report their results along with the performance of our baseline systems.

6.1 Task 1: Named Entity Recognition

We report the overall performance of all runs in Table 2. The baseline achieves 0.9367 in F1-score under exact-match. Four runs outperform the baseline in

Table 2. Overall performance of all runs in Task 1 Named Entity Recognition on private test set. Here, P, R, and F represents the Precision, Recall, and F1-score, respectively. For each metric, we highlight the best result in bold. The results are ordered by their performance in terms of F1-score under exact-match.

Run	Exact-Match			Relaxed-Match		
	P	R	F	P	R	F
LG-run1	**0.9663**	**0.9683**	**0.9673**	**0.9782**	**0.9803**	**0.9793**
LG-run2	0.9627	0.9655	0.9641	0.9758	0.9787	0.9772
LG-run3	0.9628	0.9652	0.964	0.9758	0.9782	0.977
HUKB	0.9401	0.9422	0.9412	0.9561	0.9583	0.9572
Baseline	0.947	0.9267	0.9367	0.964	0.9432	0.9535
VCU-run1	0.7335	0.8072	0.7686	0.8345	0.9185	0.8745
VCU-run2	0.734	0.7501	0.742	0.8802	0.8996	0.8898
VCU-run3	0.695	0.7869	0.7381	0.7944	0.8994	0.8436
VCU-run4	0.7263	0.7501	0.738	0.8726	0.9012	0.8867

terms of F1-score under exact-match. The best run was submitted by team LG AI Research, achieving a high F1-score of 0.9673. The F1-scores for submissions from team VCU in relaxed match are 10%-15% higher than those in exact-match. This difference between exact-match and relaxed-match may be related to the long text spans of chemical compounds, which is one of the main challenges in NER tasks in the domain of chemical documents.

6.2 Task 2: Event Extraction

The overall performance of all runs is summarized in Table 3 in terms of Precision, Recall, and F1-score under both exact-match and relaxed-match. The rankings of different systems are almost fully consistent across all metrics. Our baseline obtains 0.9088 F1-score under exact-match and the best run is from team LG AI Research which archieves 0.9199 F1-score under exact-match. The performance gap between our baseline and the best run indicates the difficulty of the event extraction task comparing to the NER task. We also notice that recall scores of most runs are consistently lower than their precision scores, which may reveal that the task of identifying a relation from a chemical patent is harder than the task of typing an identified relation.

6.3 Task 3: Anaphora Resolution

The evaluation results of all submission to the anaphora resolution task are shown in Table 4. The first run from the Hokkaido University team achieves an F1-score of 0.7085 in exact-match, outperforming our baseline which gets 0.687. The lead of the best run is even larger in relaxed matching, with an F1-score of 0.7893, about 6 points higher than our baseline. This shows the potential of the

Table 3. Overall performance of all runs in Task 2 Event Extraction on private test set. Here, P, R, and F represents the Precision, Recall, and F1-score, respectively. For each metric, we highlight the best result in bold. The results are ordered by their performance in terms of F1-score under exact-match.

Run	Exact-Match			Relaxed-Match		
	P	R	F	P	R	F
LG-run1	**0.9258**	0.9141	**0.9199**	0.9403	0.9284	0.9343
LG-run2	0.9251	**0.9147**	0.9198	**0.9416**	**0.9309**	**0.9362**
LG-run3	0.9241	0.9129	0.9185	0.9403	0.929	0.9346
LG-run4	0.9234	0.9135	0.9184	0.9398	0.9298	0.9348
LG-run5	**0.9258**	0.907	0.9163	0.942	0.9229	0.9323
Baseline	0.9087	0.9089	0.9088	0.9244	0.9246	0.9245
HUKB	0.9058	0.8685	0.8868	0.9222	0.8842	0.9028
VCU-run1	0.8249	0.6831	0.7473	0.8771	0.7264	0.7946
VCU-run2	0.826	0.6776	0.7445	0.8775	0.7199	0.7909
VCU-run3	0.7533	0.6883	0.7193	0.8015	0.7323	0.7653
VCU-run4	0.64	0.6238	0.6318	0.703	0.6852	0.694
VCU-run5	0.2675	0.6263	0.3749	0.3075	0.7199	0.4309

Table 4. Overall performance of all runs in Task 3 Anaphora Resolution on private test set. Here, P, R, and F represents the Precision, Recall, and F1-score, respectively. For each metric, we highlight the best result in bold. The results are ordered by their performance in terms of F1-score under exact-match.

Run	Exact-Match			Relaxed-Match		
	P	R	F	P	R	F
HUKB-run1	0.6876	**0.7307**	**0.7085**	0.766	**0.814**	**0.7893**
HUKB-run2	0.729	0.6838	0.7057	0.8107	0.7604	0.7848
HUKB-run3	0.7393	0.6616	0.6983	**0.8222**	0.7358	0.7766
Baseline	**0.7418**	0.6398	0.687	0.7867	0.6784	0.7286

model built by the Hokkaido University team and indicates that the performance in exact matching may be further boosted if the boundary errors of their model could be corrected in a post-processing step. Our baseline has higher precision in the exact setting, which indicates that our model is more conservative and could possibly be enhanced by making more aggressive predictions to improve recall.

7   Conclusions

This paper presents a general overview of the activities and outcomes of the ChEMU 2022 evaluation lab. As the third instance of our ChEMU lab

series, ChEMU 2022 targets three tasks focusing on chemical reaction information extraction from chemical patents. The evaluation result includes different
approaches to tackling the shared task, with several submissions outperforming
our baseline methods. We look forward to fruitful discussion and deeper understanding of the methodological details of these submissions at the workshop.

Acknowledgements. We are grateful for the detailed excerption and annotation work
of the domain experts that support Reaxys, and the support of Ivan Krstic, Director
of Chemistry Solutions at Elsevier. Funding for the ChEMU project is provided by
an Australian Research Council Linkage Project, project number LP160101469, and
Elsevier. We acknowledge the support of annotators for the anaphora resolution task,
Dr. Sacha Novakovic and Colleen Hui Shiuan Yeow at the University of Melbourne.

References

1. Akhondi, S.A., Hettne, K.M., Van Der Horst, E., Van Mulligen, E.M., Kors, J.A.:
   Recognition of chemical entities: combining dictionary-based and grammar-based
   approaches. J. Cheminformatics **7**(1), 1–11 (2015)
2. Akhondi, S.A., et al.: Annotated chemical patent corpus: a gold standard for text
   mining. PLoS One **9**(9), e107477 (2014)
3. Akhondi, S.A., et al.: Chemical entity recognition in patents by combining
   dictionary-based and statistical approaches. Database, **2016** (2016)
4. Akhondi, S.A., et al.: Automatic identification of relevant chemical compounds
   from patents. Database, **2019** (2019)
5. Arighi, C.N., et al.: Overview of the biocreative III workshop. BMC Bioinform.
   12(8), 1–9 (2011). https://doi.org/10.1186/1471-2105-12-S8-S1
6. Bregonje, M.: Patents: a unique source for scientific technical information in chemistry related industry? World Pat. Inf. **27**(4), 309–315 (2005)
7. Cohen, K.B., et al.: Coreference annotation and resolution in the colorado richly
   annotated full text (CRAFT) corpus of biomedical journal articles. BMC Bioinform. **18**(1), 372:1–372:14 (2017). https://doi.org/10.1186/s12859-017-1775-9
8. Fang, B., Druckenbrodt, C., Akhondi, S.A., He, J., Baldwin, T., Verspoor, K.:
   ChEMU-Ref: a corpus for modeling anaphora resolution in the chemical domain.
   In: Proceedings of the 16th Conference of the European Chapter of the Association
   for Computational Linguistics. Association for Computational Linguistics (2021)
9. Farkas, R., Vincze, V., Móra, G., Csirik, J., Szarvas, G.: The conll-2010 shared task:
   learning to detect hedges and their scope in natural language text. In: Proceedings of the Fourteenth Conference on Computational Natural Language Learning-
   Shared Task, pp. 1–12 (2010)
10. Ghaddar, A., Langlais, P.: Wikicoref: an english coreference-annotated corpus of
    wikipedia articles. In: Calzolari, N. (eds.) Proceedings of the Tenth International
    Conference on Language Resources and Evaluation LREC 2016, Portorož, Slove-
    nia, 23–28 May 2016. European Language Resources Association (ELRA) (2016).
    http://www.lrec-conf.org/proceedings/lrec2016/summaries/192.html
11. He, J., et al.: ChEMU 2021: reaction reference resolution and anaphora resolu-
    tion in chemical patents. In: Hiemstra, D., Moens, M.-F., Mothe, J., Perego, R.,
    Potthast, M., Sebastiani, F. (eds.) ECIR 2021. LNCS, vol. 12657, pp. 608–615.
    Springer, Cham (2021). https://doi.org/10.1007/978-3-030-72240-1_71

12. He, J., et al.: ChEMU 2020: natural language processing methods are effective for information extraction from chemical patents. Front. Res. Metrics Anal. **6**, 654438 (2021)
13. Hemati, W., Mehler, A.: Lstmvoter: chemical named entity recognition using a conglomerate of sequence labeling tools. J. Cheminformatics **11**(1), 1–7 (2019)
14. Herrero-Zazo, M., Segura-Bedmar, I., Martínez, P., Declerck, T.: The ddi corpus: an annotated corpus with pharmacological substances and drug-drug interactions. J. Biomed. Inform. **46**(5), 914–920 (2013)
15. Hettne, K.M., et al.: A dictionary to identify small molecules and drugs in free text. Bioinformatics **25**(22), 2983–2991 (2009)
16. Hu, M., Cinciruk, D., Walsh, J.M.: Improving automated patent claim parsing: dataset, system, and experiments. arXiv preprint arXiv:1605.01744 (2016)
17. Jaidka, K., Chandrasekaran, M.K., Rustagi, S., Kan, M.Y.: Overview of the cl-scisumm 2016 shared task. In: Proceedings of the Joint Workshop on Bibliometric-Enhanced Information Retrieval and Natural Language Processing for Digital Libraries (BIRNDL), pp. 93–102 (2016)
18. Kelly, L., et al.: Overview of the ShARe/CLEF ehealth evaluation lab 2014. In: Kanoulas, E. (ed.) CLEF 2014. LNCS, vol. 8685, pp. 172–191. Springer, Cham (2014). https://doi.org/10.1007/978-3-319-11382-1_17
19. Kim, J.D., Wang, Y., Yasunori, Y.: The genia event extraction shared task, 2013 edition-overview. In: Proceedings of the BioNLP Shared Task 2013 Workshop, pp. 8–15 (2013)
20. Krallinger, M., Leitner, F., Rabal, O., Vazquez, M., Oyarzabal, J., Valencia, A.: CHEMDNER: the drugs and chemical names extraction challenge. J. Cheminformatics **7**(1), 1–11 (2015)
21. Krallinger, M., et al.: Overview of the biocreative vi chemical-protein interaction track. In: Proceedings of the Sixth Biocreative Challenge Evaluation Workshop, vol. 1, pp. 141–146 (2017)
22. Krallinger, M., Rabal, O., Lourenco, A., Oyarzabal, J., Valencia, A.: Information retrieval and text mining technologies for chemistry. Chem. Rev. **117**(12), 7673–7761 (2017)
23. Krallinger, M., et al.: Overview of the chemdner patents task. In: Proceedings of the Fifth BioCreative Challenge Evaluation Workshop, pp. 63–75 (2015)
24. Lee, K., He, L., Lewis, M., Zettlemoyer, L.: End-to-end neural coreference resolution. In: Palmer, M., Hwa, R., Riedel, S. (eds.) Proceedings of the 2017 Conference on Empirical Methods in Natural Language Processing, EMNLP 2017, Copenhagen, Denmark, 9–11 September 2017. pp. 188–197. Association for Computational Linguistics (2017). https://doi.org/10.18653/v1/d17-1018
25. Lee, K., He, L., Zettlemoyer, L.: Higher-order coreference resolution with coarse-to-fine inference. In: Walker, M.A., Ji, H., Stent, A. (eds.) Proceedings of the 2018 Conference of the North American Chapter of the Association for Computational Linguistics: Human Language Technologies, NAACL-HLT, New Orleans, Louisiana, USA, 1–6 June 2018, Volume 2 (Short Papers), pp. 687–692. Association for Computational Linguistics (2018). https://doi.org/10.18653/v1/n18-2108
26. Li, Y., et al.: Overview of ChEMU 2021: reaction reference resolution and anaphora resolution in chemical patents. In: Candan, K.S. (ed.) CLEF 2021. LNCS, vol. 12880, pp. 292–307. Springer, Cham (2021). https://doi.org/10.1007/978-3-030-85251-1_20
27. Liu, H., Christiansen, T., Baumgartner, W.A., Verspoor, K.: Biolemmatizer: a lemmatization tool for morphological processing of biomedical text. J. Biomed. Semant. **3**(1), 1–29 (2012)

28. Lu, Z., et al.: The gene normalization task in biocreative III. BMC Bioinf. **12**(8), 1–19 (2011). https://doi.org/10.1186/1471-2105-12-S8-S2
29. Lupu, M., et al.: Overview of the trec 2011 chemical ir track. In: TREC (2011)
30. MorganMorgan, A., et al.: Overview of biocreative II gene normalization. Genome Biol. **9**, S3 (2008). https://doi.org/10.1186/gb-2008-9-s2-s3
31. Muresan, S., et al.: Making every SAR point count: the development of chemistry connect for the large-scale integration of structure and bioactivity data. Drug Discov. Today **16**(23–24), 1019–1030 (2011)
32. Narayanaswamy, M., Ravikumar, K., Vijay-Shanker, K.: A biological named entity recognizer. In: Biocomputing 2003, pp. 427–438. World Scientific (2002)
33. Nguyen, D.Q., et al.: ChEMU: named entity recognition and event extraction of chemical reactions from patents. In: Jose, J.M. (ed.) ECIR 2020. LNCS, vol. 12036, pp. 572–579. Springer, Cham (2020). https://doi.org/10.1007/978-3-030-45442-5_74
34. Pradhan, S., Moschitti, A., Xue, N., Uryupina, O., Zhang, Y.: Conll-2012 shared task: Modeling multilingual unrestricted coreference in ontonotes. In: Pradhan, S., Moschitti, A., Xue, N. (eds.) Joint Conference on Empirical Methods in Natural Language Processing and Computational Natural Language Learning - Proceedings of the Shared Task: Modeling Multilingual Unrestricted Coreference in OntoNotes, EMNLP-CoNLL 2012, 13 July 2012, Jeju Island, Korea, pp. 1–40. ACL (2012). https://www.aclweb.org/anthology/W12-4501/
35. Rebholz-Schuhmann, D., Kirsch, H., Arregui, M., Gaudan, S., Riethoven, M., Stoehr, P.: Ebimed-text crunching to gather facts for proteins from medline. Bioinformatics **23**(2), e237–e244 (2007)
36. Senger, S., Bartek, L., Papadatos, G., Gaulton, A.: Managing expectations: assessment of chemistry databases generated by automated extraction of chemical structures from patents. J. Cheminformatics **7**(1), 1–12 (2015). https://doi.org/10.1186/s13321-015-0097-z
37. Tsatsaronis, G., et al.: Bioasq: a challenge on large-scale biomedical semantic indexing and question answering. In: AAAI Fall Symposium: Information Retrieval and Knowledge Discovery in Biomedical Text. Citeseer (2012)
38. Tseng, Y.H., Lin, C.J., Lin, Y.I.: Text mining techniques for patent analysis. Inf. Process. Manag. **43**(5), 1216–1247 (2007)
39. Vazquez, M., Krallinger, M., Leitner, F., Valencia, A.: Text mining for drugs and chemical compounds: methods, tools and applications. Mol. Inf. **30**(6–7), 506–519 (2011)
40. Verspoor, K., et al.: Chemu dataset for information extraction from chemical patents. Mendeley Data **2**, 10–17632 (2020)
41. Yoshikawa, H., et al.: Detecting chemical reactions in patents. In: Proceeding 17th Annual Workshop of the Australasian Language Technology Association, ALTA 2019, Sydney, Australia, 4–6 December 2019, pp. 100–110 (2019)
42. Zhai, Z., et al.: Improving chemical named entity recognition in patents with contextualized word embeddings. arXiv preprint arXiv:1907.02679 (2019)

Overview of the ImageCLEF 2022: Multimedia Retrieval in Medical, Social Media and Nature Applications

Bogdan Ionescu[1]([✉]), Henning Müller[2], Renaud Péteri[3], Johannes Rückert[4],
Asma Ben Abacha[5], Alba G. Seco de Herrera[6], Christoph M. Friedrich[4],
Louise Bloch[4], Raphael Brüngel[4], Ahmad Idrissi-Yaghir[4], Henning Schäfer[4],
Serge Kozlovski[7], Yashin Dicente Cid[8], Vassili Kovalev[7], Liviu-Daniel Ştefan[1],
Mihai Gabriel Constantin[1], Mihai Dogariu[1], Adrian Popescu[9],
Jérôme Deshayes-Chossart[9], Hugo Schindler[9], Jon Chamberlain[6],
Antonio Campello[10], and Adrian Clark[6]

[1] Politehnica University of Bucharest, Bucharest, Romania
bogdan.ionescu@upb.ro
[2] University of Applied Sciences Western Switzerland (HES-SO), Sierre, Switzerland
[3] University of La Rochelle, La Rochelle, France
[4] University of Applied Sciences and Arts Dortmund, Dortmund, Germany
[5] Microsoft, Redmond, WA, USA
[6] University of Essex, Colchester, UK
[7] Institute for Informatics, Minsk, Belarus
[8] University of Warwick, Coventry, UK
[9] Université Paris-Saclay, CEA, LIST, 91120 Palaiseau, France
[10] Wellcome Trust, London, UK

Abstract. This paper presents an overview of the ImageCLEF 2022 lab that was organized as part of the Conference and Labs of the Evaluation Forum – CLEF Labs 2022. ImageCLEF is an ongoing evaluation initiative (first run in 2003) that promotes the evaluation of technologies for annotation, indexing and retrieval of visual data with the aim of providing information access to large collections of images in various usage scenarios and domains. In 2022, the 20th edition of ImageCLEF runs four main tasks: (i) a *medical* task that groups two previous tasks, i.e., caption analysis and tuberculosis prediction, (ii) a *social media* aware task on estimating potential real-life effects of online image sharing, (iii) a *nature* coral task about segmenting and labeling collections of coral reef images, and (iv) a new *fusion* task addressing the design of late fusion schemes for boosting the performance, with two real-world applications: image search diversification (retrieval) and prediction of visual interestingness (regression). The benchmark campaign received the participation of over 25 groups submitting more than 258 runs.

Keywords: Medical image classification · medical image caption analysis · tuberculosis prediction · coral image segmentation and classification · prediction of effects of online image sharing · late fusion for search diversification and interestingness prediction · ImageCLEF lab

© The Author(s), under exclusive license to Springer Nature Switzerland AG 2022
A. Barrón-Cedeño et al. (Eds.): CLEF 2022, LNCS 13390, pp. 541–564, 2022.
https://doi.org/10.1007/978-3-031-13643-6_31

1   Introduction

ImageCLEF[1] is the image retrieval and classification lab of the CLEF (Conference and Labs of the Evaluation Forum) conference. ImageCLEF has started in 2003 with only four participants [8]. It increased its impact with the addition of medical tasks in 2004 [7], attracting over 20 participants already in the second year. An overview of ten years of the medical tasks can be found in [22]. It continued the ascending trend, reaching over 200 participants in 2019 and over 110 in 2020 despite the outbreak of the covid-19 pandemic. The tasks have changed much over the years but the general objective has always been the same, i.e., *to combine text and visual data to retrieve and classify visual information.* Tasks have evolved from more general object classification and retrieval to many specific application domains, e.g., nature, security, medical, Internet. A detailed analysis of several tasks and the creation of the data sets can be found in [29]. ImageCLEF has shown to have an important impact over the years, already detailed in 2010 [39,40].

Since 2018, ImageCLEF uses the crowdAI platform, now migrated to AIcrowd[2] from 2020, to distribute the data and receive the submitted results. The system allows having an online leader board and gives the possibility to keep data sets accessible beyond competition, including a continuous submission of runs and addition to the leader board. Over the years, ImageCLEF and also CLEF have shown a strong scholarly impact that was analyzed in [39,40]. For instance, the term "ImageCLEF" returns on Google Scholar[3] over 5,990 article results (search on June 13th, 2022). This underlines the importance of evaluation campaigns for disseminating best scientific practices. We introduce here the four tasks that were run in the 2022 edition[4], namely: ImageCLEFmedical, ImageCLEFfusion, ImageCLEFaware, and ImageCLEFcoral.

2   Overview of Tasks and Participation

ImageCLEF 2022 consists of four main tasks with the objective of covering a *diverse range* of multimedia retrieval applications, namely: *medicine, social media and Internet,* and *nature* applications. It followed the 2019 tradition [20] of diversifying the use cases [5,11,23,34,36]. The 2022 tasks are presented as follows:

– **ImageCLEFmedical.** Medical tasks have been part of ImageCLEF every year since 2004. In 2018, all but one task were medical, but little interaction happened between the medical tasks. For this reason, starting with 2019, the medical tasks were focused towards one specific problem but combined as a single task with several subtasks. This allows exploring synergies between the domains (Fig. 1):

---

[1] http://www.imageclef.org/.
[2] https://www.aicrowd.com/.
[3] https://scholar.google.com/.
[4] https://www.imageclef.org/2022/.

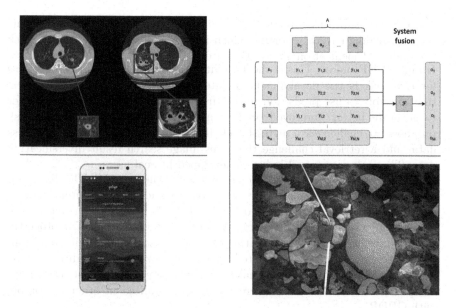

Fig. 1. Sample images from (left to right, top to bottom): ImageCLEFmedical tuberculosis prediction, ImageCLEFfusion with late fusion scheme, ImageCLEFaware with estimating potential real-life effects of online image sharing mobile application, and ImageCLEFcoral with segmenting and labeling collections of coral reef images.

- *Caption*: This is the 6th edition of the task in this format, however, it is based on previous medical tasks. The task is once again running with both the "concept detection" and "caption prediction" subtasks [36], after the former was brought back last year based on the lessons learned in previous editions [14,17,18,30–32]. The "caption prediction" subtask focuses on composing coherent captions for the entirety of a radiology image, while the "concept detection" subtask focuses on identifying the presence of relevant concepts in the same corpus of radiology images. After a smaller data set of manually annotated radiology images was used last year, the 2022 edition once again uses a larger dataset based on ROCO data [33], which was already used in 2020 and 2019.

- *Tuberculosis*: This is the 6th edition of the task. The main objective is to provide an automatic CT-based evaluation of tuberculosis (TB) patients. This is done by detecting and assessing visual TB-related findings based on the automatic analysis of lung CT scans. Caverns are one of the finding types which need specific attention. Even after successful treatment which fulfills the existing criteria of recovery caverns may still contain colonies of Mycobacterium Tuberculosis that could lead to unpredictable disease relapse. Therefore, finding and describing caverns helps to evaluate the quality of the treatment and plan recovery and control routines after the active treatment phase. In this year's edition, participants need to solve two subtasks - the first one is cavern detection, and the second one is providing cavern reporting which includes three binary labels: "Thick walls", "Calcification", and "Foci around" [23].

– **ImageCLEFfusion**. This is the 1st edition of the task. The main objective for this task is the development of late fusion or ensembling approaches, that are able to use prediction results from pre-computed inducers in order to generate better, improved prediction outputs. This edition of the task proposes two challenges: a regression challenge that uses media interestingness data, and a retrieval challenge that uses image search result diversification data. The task uses actual inducers developed by real users.

– **ImageCLEFaware**. This was the 2nd edition of the task [24]. The disclosure of personal data is done in a particular context and users are often unaware that their data can be reused in other contexts. It is thus important to give feedback to users about the effects of personal data sharing. The objective was to automatically provide a rating of a visual user profile in different real-life situations. The dataset created specifically for the 2021 edition of the task was expanded in order to make the evaluation more robust. Data were sampled from YFCC100 and were further anonymized in order to comply with GPDR.

Table 1. Key figures regarding participation in ImageCLEF 2022.

Task	Groups that submitted results	Submitted runs	Submitted working notes
Caption	12	157	13
Tuberculosis	6	43	5
Fusion	5	39	4
Aware	3	9	2
Coral	2	11	2
Overall	28	258	26

– **ImageCLEFcoral**. The 4th edition of the task follows the directions of previous years [3,4,6]. The task consists on two subtasks which aim to automatically segment and label with types of benthic substrate a collection of coral reef images. The first subtask uses bounding boxes to annotate the images while the second subtask segment the images pixel-wise using polygons. As in the third edition, in 2022 [5] the training and test data form the complete set of images required to form a 3D reconstruction of the environment.

To participate in the evaluation campaign, the research groups had to register by following the instructions on the ImageCLEF 2022 web page[5]. To ease the overall management of the campaign, in 2022 the challenge was organized through the AIcrowd platform[6]. To actually get access to the data sets, the

---

[5] https://www.imageclef.org/2022/.
[6] https://www.aicrowd.com/.

participants were required to submit a signed End User Agreement (EUA). Table 1 summarizes the participation in ImageCLEF 2022, indicated both per task and for the overall lab. The table also shows the number of groups that submitted runs and the ones that submitted a working notes paper describing the techniques used. Teams were allowed to register for participating in several tasks.

After a decrease in participation in 2016, the participation increased in 2017 and 2018, and increased again in 2019. In 2018, 31 teams completed the tasks and 28 working notes papers were received. In 2019, 63 teams completed the tasks and 50 working notes papers were retrieved. In 2020, 40 teams completed the tasks and submitted working notes papers. In 2021, 42 teams completed the tasks and we received 30 working notes papers. In 2022, 28 teams completed the tasks and we received 26 working notes papers. We can observe a drop in participation compared to 2019 and also 2021. The 2022 edition marks the end of the pandemic. Also, one of the medical tasks, i.e., the visual question answering, was not organized this year. Nevertheless, the number of submitted runs is similar to 2021 regardless the fact that less teams submitted, namely 258 vs. 256. Teams were more involved in finding solutions. Overall, even in its 20th anniversary, ImageCLEF continues to provide a strong evaluation benchmark.

In the following sections, we present the tasks. Only a short overview is reported, including general objectives, description of the tasks and data sets, and a short summary of the results. A detailed review of the received submissions for each task is provided with the task overview working notes: Caption [36], Tuberculosis [23], Fusion [11], Aware [34], and Coral [5].

3   The Caption Task

The caption task was first proposed as part of the ImageCLEFmedical [18] in 2016 aiming to extract the most relevant information from medical images. Hence, the task was created to condense visual information into textual descriptions. In 2017 and 2018 [14,17] the ImageCLEFcaption task comprised two subtasks: concept detection and caption prediction. In 2019 [31] and 2020 [32], the task concentrated on extracting Unified Medical Language System® (UMLS) Concept Unique Identifiers (CUIs) [1] from radiology images. In 2021 [30], both subtasks, concept detection and caption prediction, were running again due to participants demands. The focus in 2021 was on making the task more realistic by using fewer images which were all manually annotated by medical doctors. For the 2022 ImageCLEFmedical Caption task [36], both subtasks are continued albeit with an extended version of the ROCO data set used for both subtasks, which was already used in 2020 and 2019.

3.1   Task Setup

The ImageCLEFmedical Caption 2022 [36] follows the format of the previous ImageCLEFmedical caption tasks. In 2022, the overall task comprises two subtasks: "Concept Detection" and "Caption prediction". The concept detection subtask focuses on predicting Unified Medical Language System® (UMLS) Concept Unique Identifiers (CUIs) [1] based on the visual image representation in a given image. The caption prediction subtask focuses composing coherent captions for the entirety of the images.

The detected concepts are evaluated using the balanced precision and recall trade-off in terms of F1-scores, as in previous years. This year, a secondary F1-score based on manually curated concepts regarding image modality and x-ray anatomy was introduced. The predicted captions are evaluated using the BLEU score independent from the first subtask and designed to be robust to variability in style and wording. In addition to the BLEU score, a secondary ROUGE score was provided. After the submission period ended, a number of additional scores were calculated and published: METEOR, CIDEr, SPICE, and BERTScore.

3.2 Data Set

In 2022, an extended subset of the ROCO [33] data set is used for both subtasks, which originates from biomedical articles of the PMC Open Access Subset[7] [35] and was extended with new images added since the last time the data set was updated. In the previous edition, in an attempt to make the task more realistic, the data set contained a smaller number of real radiology images annotated by medical doctors which resulted in high-quality concepts. Additional data of similar quality is hard to acquire and so it was decided to return to the data set already used in 2020 and 2019. From the captions, UMLS® concepts were generated and concepts regarding anatomy and image modality were manually validated for all images.

Following this approach, we provided new training, validation, and test sets for both tasks:

- *Training set* including 83,275 radiology images and associated captions and concepts.
- *Validation set* including 7,645 radiology images and associated captions and concepts.
- *Test set* including 7,645 radiology images.

3.3 Participating Groups and Submitted Runs

In the sixth edition of the ImageCLEFmedical Caption task, 20 teams registered and signed the End-User-Agreement that is needed to download the development data. 12 teams submitted 157 runs for evaluation (all 12 teams submitted working notes) attracting more attention than in 2021. Each of the groups was allowed a maximum of 10 graded runs per sub task. 11 teams participated in the concept detection subtask this year, 3 of those teams also participated in 2021. 10 teams submitted runs to the caption prediction subtask, 4 of those teams also participated in 2021. Overall, 9 teams participated in both subtasks, two teams

[7] https://www.ncbi.nlm.nih.gov/pmc/tools/openftlist/.

Table 2. Performance of the participating teams in the ImageCLEFmedical 2022 concept detection subtask. The best run per team is selected. Teams with previous participation in 2021 are marked with an asterisk.

Team	Institution	F1-Score
AUEB-NLP-Group*	Department of Informatics, Athens University of Economics and Business, Athens, Greece	0.4511
CMRE-UoG (fdallaserra)	Canon Medical Research Europe, Edinburgh, UK and University of Glasgow, Glasgow, UK	0.4505
CSIRO*	Australian e-Health Research Centre, Commonwealth Scientific and Industrial Research Organisation, Herston, Queensland, Australia and CSIRO Data61, Imaging and Computer Vision Group, Pullenvale, Queensland, Australia and Queensland University of Technology, Brisbane, Queensland, Australia	0.4471
eecs-kth	KTH Royal Institute of Technology, Stockholm, Sweden	0.4360
vcmi	University of Porto, Porto, Portugal and INESC TEC, Porto, Portugal	0.4329
PoliMi-ImageClef	Politecnico di Milano, Milan, Italy	0.4320
SSNSheerinKavitha	Department of CSE, Sri Sivasubramaniya Nadar College of Engineering, India	0.4184
IUST_NLPLAB	School of Computer Engineering, Iran University of Science and Technology, Tehran, Islamic Republic Of Iran	0.3981
Morgan_CS	Morgan State University, Baltimore, MD, USA	0.3520
kdelab*	KDE Laboratory, Department of Computer Science and Engineering, Toyohashi University of Technology, Aichi, Japan	0.3104
SDVA-UCSD	San Diego VA HCS, San Diego, CA, USA	0.3079

participated only in the concept detection subtask and one team participated only in the caption prediction subtask.

In the concept detection subtasks, the groups used primarily multi-label classification systems and image retrieval systems, much like in the 2021 challenge. Multi-label classification systems outperformed retrieval-based systems for most of the teams who experimented with both, and while the winner was a multi-label classification approach, the second placing team with an F1-score only 0.0006

Table 3. Performance of the participating teams in the ImageCLEFmedical 2022 caption prediction subtask. The best run per team is selected. Teams with previous participation in 2021 are marked with an asterisk.

Team	Institution	BLEU Score
IUST_NLPLAB	School of Computer Engineering, Iran University of Science and Technology, Tehran, Islamic Republic Of Iran	0.4828
AUEB-NLP-Group*	Department of Informatics, Athens University of Economics and Business, Athens, Greece	0.3221
CSIRO*	Australian e-Health Research Centre, Commonwealth Scientific and Industrial Research Organisation, Herston, Queensland, Australia and CSIRO Data61, Imaging and Computer Vision Group, Pullenvale, Queensland, Australia and Queensland University of Technology, Brisbane, Queensland, Australia	0.3114
vcmi	University of Porto, Porto, Portugal and INESC TEC, Porto, Portugal	0.3058
eecs-kth	KTH Royal Institute of Technology, Stockholm, Sweden	0.2917
CMRE-UoG (fdallaserra)	Canon Medical Research Europe, Edinburgh, UK and University of Glasgow, Glasgow, UK	0.2913
kdelab*	KDE Laboratory, Department of Computer Science and Engineering, Toyohashi University of Technology, Aichi, Japan	0.2782
Morgan_CS	Morgan State University, Baltimore, MD, USA	0.2549
MAI_ImageSem*	Institute of Medical Information and Library, Chinese Academy of Medical Sciences and Peking Union Medical College, Beijing, China	0.2211
SSNSheerinKavitha	Department of CSE, Sri Sivasubramaniya Nadar College of Engineering, India	0.1595

less than the winning team, used an image retrieval system based on the winning approach from last year.

In the caption prediction subtask, most teams experimented with Transformer-based architectures and image retrieval systems. Only one team used a multi-label classification approach, and it achieved by far the best BLEU score. However, it did not score as well on most of the other employed metrics. Transfer

Learning has frequently been used for pre-training, from a variety of different data sets.

To get a better overview of the submitted runs, the primary scores of the best results for each team are presented in Tables 2 and 3.

3.4 Results

This year's models for concept detection do not show an increased F1-score compared to last year, however due to the much larger data set and number of concepts used in this year's challenge, this is not surprising. Comparing it to the 2020 results, where a data set of similar size was used, the F1-scores show a clear improvement. There are no radically new approaches used in this year's concept detection subtask, but the teams experimented with, optimised and re-combined many different existing techniques and created competitive solutions using both multi-label classification systems and image retrieval systems.

Similar to the concept detection, the BLEU scores in the caption prediction subtask are overall lower compared to last year, which can be explained by the larger data set and more varied captions. Since there was no caption prediction subtask running in 2020, no comparable scores for a similar data set exist. An in-depth analysis is presented in [36].

3.5 Lessons Learned and Next Steps

This year's caption task of ImageCLEFmedical once again ran with both subtasks, concept detection and caption prediction. It returned to a larger, ROCO-based data set for both challenges after a smaller, manually annotated data set was used last year. It attracted 12 teams who submitted 157 runs overall, a stronger participation compared to last year. Some changes were introduced for the scores, with a secondary F1-score related to manually curated concepts for image modality and x-ray anatomy added to the concept detection, and several new scores added to the caption prediction subtask which was appreciated by the teams as it highlighted the difficulty of evaluating caption similarity and showed that models performing worse on the BLEU score could perform better in several of the other metrics instead.

With the bigger data set, most teams were more successful in training multi-label classification models compared to image retrieval models for the concept detection. For the caption prediction, most teams used Transformer-based models, but the winning models in terms of the BLEU score was a multi-label classification model.

For next year's ImageCLEFmedical Caption challenge, some possible improvements include adding more manually validated concepts like increased anatomical coverage and directionality information, reducing recurring captions, more fine-grained CUI filters, improving the caption pre-processing, and using a different primary score for the caption prediction challenge, since the BLEU score has some disadvantages which were highlighted by this year's caption prediction results. It will also be important to make sure that no models are used that were pre-trained on PubMedCentral data, since these models will already have seen the original captions.

4   The Tuberculosis Task

Tuberculosis (TB) is a bacterial infection caused by a germ called Mycobacterium tuberculosis. More than a century after its discovery, the disease remains a persistent threat and one of the top 10 causes of death worldwide according to the WHO [41]. The bacteria usually attack the lungs and generally TB can be cured with antibiotics. However, the different types of TB require different treatments, and therefore detection of the specific case characteristics is an important real-world task.

In the previous editions of this task, the setup evolved from year to year. In the first two editions [14,16] participants had to detect Multi-drug resistant patients (MDR subtask) and classify the TB type (TBT subtask) both based only on the CT image. After 2 editions it was concluded to drop the MDR subtask because it seemed impossible to solve based only on the image, and the TBT subtask was also suspended because of a very little improvement in the results between the 1st and the 2nd editions. At the same time, most of the participants obtained good results in the severity scoring (SVR) subtask introduced in 2018. In the 3d edition Tuberculosis task [15] was restructured to allow usage of the uniform dataset, and included two subtasks - a continued Severity Score (SVR) prediction subtask and a new subtask based on providing an automatic report (CT Report) on the TB case. In the 4th edition [25], the SVR subtask was dropped and the automated CT report generation task was modified to be lung-based rather than CT-based. In the 5th edition [24], the task organizers have decided to discontinue the CTR task and brought back to life the Tuberculosis Type classification task from the 1st and 2nd ImageCLEFmed Tuberculosis editions to check if recent Machine Learning and Deep Learning methods allow improving previous rather low results.

In this year's edition [23] the task was dedicated to the caverns detection and report, which were split into two subtasks. The first subtask (Caverns Detection) focused on detection, i.e., participants must detect lung cavern regions in lung CT images associated with lung tuberculosis. The problem is important because even after successful treatment which fulfills the existing criteria of recovery the caverns may still contain colonies of Mycobacterium Tuberculosis that could lead to unpredictable disease relapse. The second subtask (Caverns Report) was the classification of caverns. Participants must predict 4 binary features of caverns suggested by experienced radiologists.

4.1   Task Setup

In this task, participants had to automatically detect lung cavern regions in lung CT images associated with lung tuberculosis in the first subtask, and predict 3 binary features of caverns suggested by experienced radiologists. So the first subtask was a 3D object detection task, and the second one was a multi-label classification problem.

4.2 Data Set

In this edition, separate data sets were provided for each subtask. The Caverns Detection dataset contained 559 train and 140 test cases, while the Caverns Report data included just 60 train and 16 test cases due to labelled data scarcity. Each CT image corresponded to one unique patient. For all patients, we provided 3D CT images with a slice size of 512×512 pixels and a variable number of slices (the median number was 128). All train CTs for both subtasks were accompanied by caverns area bounding boxes (if any), and labelling of caverns was provided for Cavern Report subtask. Since bounding boxes were provided for all CTs, participants were welcomed to use data from one subtask for the another.

Same as in the previous year, for all patients we provided two versions of automatically extracted masks of the lungs obtained using the methods described in [13, 27].

4.3 Participating Groups and Submitted Runs

In 2022, 6 groups from 5 countries submitted at least one run. 4 group participated in each task, and 2 groups participated in both tasks. Similar to the previous editions, each group could submit up to 10 runs. 43 scored runs were submitted in total (17 for Caverns Detection and 26 for Caverns Report).

All groups used 2D or/and 3D CNNs in both tasks. For the Caverns Detection subtask one group tried both 3D approach using customized 3D Retina U-Net based model and projection-wise 2D approach using YOLO v5 detection netwoks; another group reported 2D slice-wise approach using the YOLO v3. For the Caverns Report subtask three participants reported usage of 3D-only approach, two of them utilized custom 3D CNN, and one used ResNet34 with convolutional block attention model (CBAM); one group used slice-wise approach, but in addition to 2D CNN (EffcientNet, DenseNet) also used SRGAN for data preprocessing. The majority of participants used transfer learning techniques where possible, and executed some pre-processing steps, such as resizing, grouping, normalization, slice filtering etc.

4.4 Results

The Caverns Detection task was scored using the mean average precision at the different intersection over union (IoU) thresholds score. The Caverns Report task was evaluated as a multi-label classification problem and scored using mean AUC as primary score and minimum AUC as secondary score. Tables 4 and 5 shows the final results for each group's best run in both tasks. More detailed results, including metric description and other performance measures, are presented in the overview article [23].

Table 4. Results obtained by the participants of the Caverns Detection task. Only the best run of each participant is reported.

Group name	Institution	map_iou
CSIRO	Australian e-Health Research Centre, Commonwealth Scientific and Industrial Research Organisation, Herston, Queensland, Australia and CSIRO Data61, Imaging and Computer Vision Group, Pullenvale, Queensland, Australia and Queensland University of Technology, Brisbane, Queensland, Australia	0.504
SenticLab.UAIC	Alexandru Ioan Cuza University of Iasi, Romania	0.295
KDE-lab	KDE Laboratory, Department of Computer Science and Engineering, Toyohashi University of Technology, Aichi, Japan	0.185
SDVA-UCSD	San Diego VA HCS, San Diego, CA, USA	0.000

Table 5. Results obtained by the participants of the Caverns Report task. Only the best run of each participant is reported.

Group name	Institution	Mean AUC	Min AUC
SDVA-UCSD	San Diego VA HCS, San Diego, CA, USA	0.687	0.513
KDE-lab	KDE Laboratory, Department of Computer Science and Engineering, Toyohashi University of Technology, Aichi, Japan	0.658	0.317
KL_BP_SSN	Sri Sivasubramaniya Nadar college of Engineering, Chennai, India	0.536	0.413
SSN_Dheepak_Kavitha	SSN College of Engineering, Chennai, India	0.461	0.256

4.5   Lessons Learned and Next Steps

The results obtained in the task cannot be compared to the previous editions, since it's the first appearance of caverns-dedicated tasks. Furthermore, this is the first time for the TB task when we switched from classification problems to the detection problem.

The best result of Caverns Detection subtask was achieved by the CSIRO group using a custom neural network with 3D Retina U-Net-based architecture in a combination with developed plane-based bounding box merging postprocessing routine. The second-ranked SenticLab.UAIC group used nodule detection CNN. The 3rd ranked KDE-lab group used slice-wise analysis with YOLO v3 CNN.

The best result of the Caverns Report subtask was achieved by the SDVA-UCSD group using the 3D CBAM Resnet model and a semi-supervised training strategy which allowed involving data set provided in the detection task. The second-ranked KDE-lab group used slice-wise analysis using pre-trained 2D CNN (EffcientNet, DenseNet) and also used resolution increase using SRGAN as a preprocessing step. The 3rd ranked SSN_Dheepak_Kavitha group used custom 3D CNN.

Results analysis shows, that the best scores are reasonably high for both subtasks, and the top score for the Caverns Detection is better than we expected taking into account the complexity of the 3D detection problem. Based on the participants' approach analysis we can note that both winning solutions used advantages of volumetric analysis to the contrary of previous task editions, where projection-based approaches were more effective. As a result, we can conclude that despite a rather low number of participants this year, we see interesting approaches with quite a high score, so in general, the task is successful and its outcome is informative and useful.

Possible updates for future editions of caverns-related TB task should consider: (i) extending data set sizes and labels count for caverns report; (ii) switching from detection to segmentation problem.

5   The Fusion Task

The generalization ability and performance of machine learning models show signs of reaching a plateau in many domains, where the performance improvements over the years are not significant. Therefore, exploring the performance and optimizing the efficiency of machine learning methods is important for real-world applications as they can only use limited, noisy data. In this context, fusion methods are gaining popularity by harnessing the complementary knowledge of multiple base models to build more robust and accurate models compared with single models.

Several challenges must be explored by the participants in this task, such as *diversity*, which refers to a set of classifiers that, given the same instance, output different predictions; *voting mechanism*, which regulate how individual outputs from the base models are used during prediction; *dependency*, which refers to the way a base model affects the construction of the next model in the fusion chain; *cardinality*, which refers to the number of individual base models that form the ensemble – one needs to find a balance, as diversity may be reduced if too many models are incorporated in the fusion; the *learning mode* of the base models, which is the property that balance the classifiers' ability to adapt properly to new, previously unseen, data while at the same time retaining the previously learned knowledge.

5.1 Task Setup

This first edition of the ImageCLEFfusion task [11] consists of two challenges: a regression challenge involving media interestingness (ImageCLEFfusion-int) for which we provide output data from 29 base models, and a retrieval challenge involving result diversification (ImageCLEFfusion-div) for which we provide outputs data from 56 inducers. Participants were asked to develop late fusion learning strategies based on the outputs of the inducers associated with the media samples for each of the subtask. Evaluation was performed using MAP@10 for the ImageCLEFfusion-int task, and F1@20 and Cluster Recall@20 for ImageCLEFfusion-div task. Participants were invited to submit for either or both tasks.

5.2 Data Set

The ImageCLEFfusion-int task uses data from the Interestingness10k dataset [10], specifically, the image-based prediction data associated with the 2017 MediaEval Predicting Media Interestingness task [12]. We provide prediction outputs from the 29 systems submitted during this benchmarking task, dividing the available data into 1877 data samples for the training of the proposed fusion systems and 558 for testing.

On the other hand, the ImageCLEFfusion-div task uses data associated with the e Retrieving Diverse Social Images dataset [21], specifically the DIV150Multi challenge [19]. The retrieval outputs provided from the 56 systems are divided into 60 queries for the training data and 63 queries for the testing set.

In both training sets, we provide the inducer outputs, the necessary scripts for metric computation, the performance for each of the inducers according to the official metrics, and ground truth data. For the testing sets we only provide the inducer outputs. It is also important to note that participants were not allowed to use external inducers, being limited only to the ones we provide, as our intention is to have a fair assessment of the performance of the late fusion approach in itself, without changing the inducer set.

Table 6. Participation in the ImageCLEF-int 2022 task: the best score from all runs for each team.

Team	#Runs	MAP@10
AIMultimediaLab	5	0.2192
ssn_it	1	0.1106
UECORK	8	0.1097

5.3 Participating Groups and Submitted Runs

Three teams submitted runs for each task, while only one team participated in both tasks. A total of 14 runs are submitted for the interestingness task, while the diversification task is more successful, with 25 submitted runs.

Table 7. Participation in the ImageCLEF-div 2022 task: the best score from all runs for each team.

Team	#Runs	F1@20	CR@20
AIMultimediaLab	5	0.6216	0.4916
klssncse	10	0.5634	0.4414
shreya_sriram	10	0.5604	0.4373

The analysis of the submitted methods shows two important types of approaches proposed by the participants for this task. The first significant type is based on weighting the inducer output by implementing several different techniques. For example, one group used a simple grid search based on the performance of inducers on the training set, where higher weights are assigned to better-performing inducers. Other weighted approaches use a learning method for determining the optimal inducer weights, including learning methods based on Genetic Algorithms, Particle Swarm Optimization, and Trust Region Constrained Optimization.

The second type of approach is based on passing the inducer prediction outputs through a learning mechanism that provides the final fusion results, thus learning the way inducers interact for a given sample. In this category, some participants proposed implementing sets of traditional learning methods like kNN, Classification and Regression Trees, or SVR, while others chose neural networks as the base for the fusion engine, including approaches based on DeepFusion, MLP models, and Keras Regressors. Finally, it is worth noting that one team proposed a method where the output of several DNN-based fusion engines is passed through a final stage represented by a voting regressor.

5.4   Results

The results are presented in Table 6 for the interestingness task, and Table 7 for the diversification task. In both tasks, the best performance is achieved with a DeepFusion type approach [9], submitted by the AIMultimediaLab team. The best performance for the ImageCLEF-int task is a MAP@10 value of 0.2192, while for the ImageCLEF-div task a F1@20 of 0.6216 and a CR@20 of 0.4916 is achieved.

Overall, while results for the diversification task seem to be higher than those recorded in the interestingness task, it is important to note that the improvement over the provided inducers is greater for the interestingness task. Specifically, the improvement in the interestingness task over the average inducer performance, which is a MAP@10 value of 0.0946 is 131%. For the diversification task, the average inducer performance is an F1@20 value of 0.5313, thus the submitted systems show an improvement of almost 17%. While this may be the result of greater initial performance on the diversification task, it is also worth to note that the degree of complexity associated with the diversification task and its inducer outputs is greater.

5.5 Lessons Learned and Next Steps

The results presented this year are encouraging, especially considering the fact that all teams performed above the performance of the average inducers. A large variety of approaches, ranging from simple statistical methods to more complex approaches that require learning inducer interactions, like SVMs, classification and regression trees, and deep neural networks.

For the next edition of this task, we believe it is very important to continue with these two datasets, as this will allow us to study the year-to-year improvement of the proposed fusion techniques. Also, we will study the possibility of adding another dataset, that will target another complex type of machine learning task, whether it is a multi-class classification task, or multi-label classification.

6 The Aware Task

Social networks engage the users to share their personal data in order to interact with other users. The context of the sharing is chosen by the users but they do not have control on further data use. These data are automatically aggregated into profiles which are exploited by social networks to propose personalized advertising/services to users. Depending on their visibility, data can be also consulted by other entities to make decisions which have a high impact on the user's life. It is thus important to give users feedback about the potential real-life effects of their personal data sharing.

We designed a task focused on the automatic rating of visual user profile in four impactful situations. Each profile includes 100 photos and its appeal is manually evaluated via crowdsourcing. Participants are asked to provide automatic visual profile ratings obtained by using a training set which includes visual- and situation-related information. These ratings are then ranked and compared to manual ones in order to assess the feasibility of providing automatic feedback related to the effects of personal photos sharing. Three teams submitted results for this second edition of the task.

6.1 Task Setup

This is the second edition of the task and consists of one challenge. Participants are provided with automatic object detections for the images and with object ratings per situation. Then, the objective is to propose a ranking of user profiles which is as close as possible to the crowdsourced one. Data were split into 600/200/200 profiles for training/validation and test. The Pearson correlation coefficient between manual and automatic profile rankings was used to evaluate the quality of proposed runs. The final scores were calculated by averaging correlations obtained for individual situations.

6.2   Data Set

A data set of 1,000 user profiles with 100 photos per profile was created and annotated with an "appeal" score for four real-life situations via crowdsourcing. The modeled situations are demands for: a bank credit, an accommodation, a job as an IT engineer, a job as a waiter. Participants to the experiment were asked to provide a global rating of each profile in each situation modeled using a 7-points Likert scale ranging from "strongly unappealing" to "strongly appealing". The averaged "appeal" score was used to create a ground truth composed of ranked users in each modeled situation. User profiles are created by repurposing a subset of the YFCC100M dataset [38].

Situations are modeled by crowdsourcing visual objects ratings. Similar to profile crowdsourcing, object ratings are collected for each situation using a 7-points Likert scale with ratings between -3 (strongly negative influence) to $+3$ (strongly positive influence). The averaged rating is computed and provided to participants. A Faster R-CNN object detector was trained in order to detect objects in images. The detection dataset combines objects from OpenImages [26], ImageNet [37] and COCO [28]. Only objects with at least one non-zero situation rating were kept. All objects detected in the 100 images of a profile were provided to participants, along with the detection probability and the associated bounding box. Given a situation, the combination of the ratings of objects and of their automatic detection enables the automatic computation of a profile score.

Given the personal nature of the included profiles, the dataset was anonymized in order to comply with GDPR. Participants did not have access to the images, and the user IDs and the object names were hashed.

6.3   Participating Groups and Submitted Runs

Three teams registered for the task this year, all from the SSN College of Engineering, Chennai, India. All three teams submitted a total of nine runs.

6.4   Results

The participants tested a range of techniques to rate and rank user profiles, notably: random forest regressors, extra tree regressors and dense neural networks. Attention was also given to the preprocessing step in order to make the most of the available training data, with different runs using various combinations of object detections, confidence scores, object counts, and/or bounding boxes. The best reported Pearson correlation is 0.544, and was obtained with random forest regressor. The best score reported this year is similar to the one from 2021.

6.5   Lessons Learned and Next Steps

The participation this year was better than last year, but still low. The interaction with participants was smooth, and there were no problems with the dataset

Table 8. Results of the Aware 2022 task.

Team	#Runs	Pearson
SSNCSE_KS_NA_AKR_CB	5	0.544
JBTTM	2	0.139
ssnce-cse-JT	2	0.0

usage. The availability of a larger dataset allowed the use of different learning techniques, including deep learning ones. The scores reported by participants are interesting, but the task is far from being solved (Table 8).

For the next edition of the task, we will continue the extension of the dataset to make it more robust and timely. Focus will be put on: (1) further increase the number of user profiles, and (2) use large-scale object detection methods, such as Detic [42], to provide finer-grained profiles.

7   The Coral Task

Marine ecosystem monitoring is a key priority for evaluating ecosystem conditions [2]. Despite a wide range of monitoring programs for tropical coral reefs, there is still a crucial need to establish an effective monitoring process. This process can be made by collecting 3D visual data using autonomous underwater vehicles. The ImageCLEFcoral task organisers have developed a novel multi-camera system that allows large amounts of imagery to be captured by a SCUBA diver or autonomous underwater vehicle in a single dive which will provide useful information for both annotation and further study of the coral.

Previous editions of ImageCLEFcoral in 2019 [3] and 2020 [4] have shown improvements in task performance and promising results on cross-learning between images from geographical regions. The 3rd edition [6] increased the complexity of the task and size of data available to participants through supplemental data, resulting in lower performance than previous years. As with the 3rd edition, in 2022 [5], the training and test data form a complete set of images required to form 3D reconstructions of the marine environment.

7.1   Task Setup

In 2022, the ImageCLEFcoral task followed the format of previous editions [3, 4,6]. In 2021 participants were again asked to devise and implement algorithms for automatically annotating regions in a collection of images containing several types of benthic substrate, such as hard coral or sponge. As in previous editions, 2022 comprised two sub-tasks: "Coral reef image annotation and localisation" and "Coral reef image pixel-wise parsing" subtasks. The "Coral reef image annotation and localisation" subtask uses bounding boxes, with sides parallel to the edges of the image, for the annotation of regions in a collection of images containing several types of benthic substrates. The "Coral reef image pixel-wise parsing" subtask uses a series of boundary image coordinates which form a single polygon around each identified region in the coral reef images; this has been dubbed *pixel-wise parsing* (these polygons should not have self-intersections). Participants were invited to make submissions for either or both tasks.

Algorithmic performance is evaluated on the unseen test data using the popular intersection over union metric from the PASCAL VOC[8] exercise. This computes the area of intersection of the output of an algorithm and the corresponding ground truth, normalising that by the area of their union to ensure its maximum value is bounded.

7.2   Data Set

As in previous editions, the data for this ImageCLEFcoral task originates from a growing, large-scale collection of images taken from coral reefs around the world as part of a coral reef monitoring project with the Marine Technology Research Unit at the University of Essex. The images contain annotations of the following 13 types of substrates: Hard Coral - Branching, Hard Coral - Submassive, Hard Coral - Boulder, Hard Coral - Encrusting, Hard Coral - Table, Hard Coral - Foliose, Hard Coral - Mushroom, Soft Coral, Soft Coral - Gorgonian, Sponge, Sponge - Barrel, Fire Coral - Millepora and Algae - Macro or Leaves.

In addition, participants are encouraged to use the publicly available NOAA NCEI data[9] and/or CoralNet[10] to train their approaches. They were also encouraged to explore novel probabilistic computer vision techniques based around image overlap and transposition of data points.

7.3   Participating Groups and Submitted Runs

In 2022, 6 teams registered, of which 2 teams submitted 11 valid runs. Teams were limited to submit 10 runs per subtask. To get a better overview of the submitted runs, the best results for each team are presented in Table 9. Unfortunately, there were no participants to the "Coral reef image pixel-wise parsing" subtask this year. An in-depth analysis is presented in [5].

Table 9. Coral reef image annotation and localisation performance in terms of $MAP0.5IoU$. The best run per team is selected.

Run id	Team	$MAP0.5IoU$	$MAP0.0IoU$
183919	HHU	0.396	0.752
185373	UTK	0.003	0.327

---

[8] http://host.robots.ox.ac.uk/pascal/VOC/.

[9] https://www.ncei.noaa.gov/.

[10] https://coralnet.ucsd.edu/.

7.4 Results

The results from the "Coral reef image annotation and localisation" achieved better higher than in the 2021 edition although they are not directly comparable since the data has been updated in 2022. There was no participation in the "Coral reef image pixel-wise parsing", which is a more complicated task while closer to the real-word problem. More detailed analysis of the results is presented in [6].

7.5 Lessons Learned

As with the 3rd edition, the training and test data formed a complete set of images required to form 3D reconstructions of the marine environment. Unfortunately, no participant has explored yet computer vision techniques based around image overlap and transposition of data points. Therefore, we can still unlock the true potential of the dataset to provide meaningful insights for the analysis of the coral reefs.

8 Conclusion

This paper presents a general overview of the activities and outcomes of the ImageCLEF 2022 evaluation campaign. Four tasks were organised, covering challenges in the medical domain (caption analysis, tuberculosis prediction), social networks and Internet (analysis of the real-life effects of personal data sharing, fusion techniques for retrieval and interestingness prediction), and nature (segmenting and labeling collections of coral images). 28 teams completed the tasks and submitted over 258 runs.

As anticipated already, most of the proposed solutions evolved around state-of-the-art deep neural network architectures. In ImageCLEFcaption, with the bigger data set, most teams were more successful in training multi-label classification models compared to image retrieval models for the concept detection. For the caption prediction, most teams used Transformer-based models, but the winning models in terms of the BLEU score was a multi-label classification model. In ImageCLEFtuberculosis, the results cannot be compared to the previous editions, since it's the first time appearance of caverns-dedicated tasks. Furthermore, this is the first time when we switched from classification problems to the detection problem. The best result was achieved using a custom neural network with 3D Retina U-Net-based architecture in a combination with developed plane-based bounding box merging postprocessing routine. In ImageCLEFfusion, although in its first edition, the results are encouraging. All teams performed above the performance of the average inducer. A large variety of approaches, ranging from simple statistical methods to more complex approaches that require learning inducer interactions, like SVMs, classification and regression trees, and deep neural networks, were explored. In ImageCLEFaware, the participation was better than last year, but still low. The availability of a larger dataset allowed the use of different learning techniques, including deep learning ones. The scores reported by participants are interesting, but the task is far from being solved. In ImageCLEFcoral, the training and test data formed a complete set of images required to form 3D reconstructions of the marine environment. Unfortunately, no participant has explored yet computer vision techniques based around image overlap and transposition of data points. Therefore, we can still unlock the true potential of the dataset to provide meaningful insights for the analysis of the coral reefs.

ImageCLEF 2022 brought again together an interesting mix of tasks and approaches and we are looking forward to the fruitful discussions at the CLEF 2022 workshop.

Acknowledgements. The ImageCLEFaware and ImageCLEFfusion tasks were supported under the H2020 AI4Media "A European Excellence Centre for Media, Society and Democracy" project, contract #951911. The work of Louise Bloch and Raphael Brüngel was partially funded by a PhD grant from the University of Applied Sciences and Arts Dortmund (FH Dortmund), Germany. The work of Ahmad Idrissi-Yaghir and Henning Schäfer was funded by a PhD grant from the DFG Research Training Group 2535 Knowledge- and data-based personalisation of medicine at the point of care (WisPerMed).

References

1. Bodenreider, O.: The Unified Medical Language System (UMLS): integrating biomedical terminology. Nucleic Acids Res. **32**(Database-Issue), 267–270 (2004). https://doi.org/10.1093/nar/gkh061
2. Carrillo-García, D.M., Kolb, M.: Indicator framework for monitoring ecosystem integrity of coral reefs in the Western Caribbean. Ocean Sci. J. 1–24 (2022)
3. Chamberlain, J., Campello, A., Wright, J.P., Clift, L.G., Clark, A., García Seco de Herrera, A.: Overview of ImageCLEFcoral 2019 task. In: CLEF 2019 Working Notes. CEUR Workshop Proceedings. CEUR-WS.org (2019)
4. Chamberlain, J., Campello, A., Wright, J.P., Clift, L.G., Clark, A., García Seco de Herrera, A.: Overview of the ImageCLEFcoral 2020 task: automated coral reef image annotation. In: CLEF 2020 Working Notes. CEUR Workshop Proceedings. CEUR-WS.org (2020)
5. Chamberlain, J., García Seco de Herrera, A., Campello, A., Clark, A.: ImageCLEFcoral task: coral reef image annotation and localisation. In: Experimental IR Meets Multilinguality, Multimodality, and Interaction. Proceedings of the 13th International Conference of the CLEF Association (CLEF 2022). LNCS. Springer, Cham (2022)
6. Chamberlain, J., García Seco de Herrera, A., Campello, A., Clark, A., Oliver, T.A., Moustahfid, H.: Overview of the ImageCLEFcoral 2021 task: coral reef image annotation of a 3D environment. In: CLEF 2021 Working Notes. CEUR Workshop Proceedings, Bucharest, Romania, 21–24 September 2021. CEUR-WS.org (2021)
7. Clough, P., Müller, H., Sanderson, M.: The CLEF 2004 cross-language image retrieval track. In: Peters, C., Clough, P., Gonzalo, J., Jones, G.J.F., Kluck, M., Magnini, B. (eds.) CLEF 2004. LNCS, vol. 3491, pp. 597–613. Springer, Heidelberg (2005). https://doi.org/10.1007/11519645_59
8. Clough, P., Sanderson, M.: The CLEF 2003 cross language image retrieval task. In: Proceedings of the Cross Language Evaluation Forum (CLEF 2003) (2004)

9. Constantin, M.G., Ştefan, L.-D., Ionescu, B.: DeepFusion: deep ensembles for domain independent system fusion. In: Lokoč, J., et al. (eds.) MMM 2021. LNCS, vol. 12572, pp. 240–252. Springer, Cham (2021). https://doi.org/10.1007/978-3-030-67832-6_20

10. Constantin, M.G., Ştefan, L.D., Ionescu, B., Duong, N.Q., Demarty, C.H., Sjöberg, M.: Visual interestingness prediction: a benchmark framework and literature review. Int. J. Comput. Vis. **129**(5), 1526–1550 (2021)

11. Ştefan, L.D., Constantin, M.G., Dogariu, M., Ionescu, B.: Overview of Image-CLEFfusion 2022 task - ensembling methods for media interestingness prediction and result diversification. In: CLEF 2022 Working Notes. CEUR Workshop Proceedings, Bologna, Italy, 5–8 September 2022. CEUR-WS.org (2022)

12. Demarty, C.H., Sjöberg, M., Ionescu, B., Do, T.T., Gygli, M., Duong, N.: Mediaeval 2017 predicting media interestingness task. In: MediaEval Workshop (2017)

13. Dicente Cid, Y., Jimenez-del-Toro, O., Depeursinge, A., Müller, H.: Efficient and fully automatic segmentation of the lungs in CT volumes. In: Goksel, O., Jimenez-del-Toro, O., Foncubierta-Rodriguez, A., Müller, H. (eds.) Proceedings of the VIS-CERAL Challenge at ISBI, vol. 1390, pp. 31–35. CEUR Workshop Proceedings, April 2015

14. Dicente Cid, Y., Kalinovsky, A., Liauchuk, V., Kovalev, V., Müller, H.: Overview of ImageCLEFtuberculosis 2017 - predicting tuberculosis type and drug resistances. In: CLEF 2017 Working Notes. CEUR Workshop Proceedings, Dublin, Ireland, 11–14 September 2017. CEUR-WS.org (2017). http://ceur-ws.org

15. Dicente Cid, Y., Liauchuk, V., Klimuk, D., Tarasau, A., Kovalev, V., Müller, H.: Overview of ImageCLEFtuberculosis 2019 - automatic CT-based report generation and tuberculosis severity assessment. In: CLEF 2019 Working Notes. CEUR Workshop Proceedings, Lugano, Switzerland, 9–12 September 2019. CEUR-WS.org (2019). http://ceur-ws.org

16. Dicente Cid, Y., Liauchuk, V., Kovalev, V., Müller, H.: Overview of ImageCLEF-tuberculosis 2018 - detecting multi-drug resistance, classifying tuberculosis type, and assessing severity score. In: CLEF 2018 Working Notes. CEUR Workshop Proceedings, Avignon, France, 10–14 September 2018. CEUR-WS.org (2018). http://ceur-ws.org

17. García Seco de Herrera, A., Eickhoff, C., Andrearczyk, V., Müller, H.: Overview of the ImageCLEF 2018 caption prediction tasks. In: CLEF 2018 Working Notes. CEUR Workshop Proceedings, Avignon, France, 10–14 September 2018. CEUR-WS.org (2018). http://ceur-ws.org

18. García Seco de Herrera, A., Schaer, R., Bromuri, S., Müller, H.: Overview of the ImageCLEF 2016 medical task. In: Working Notes of CLEF 2016 (Cross Language Evaluation Forum), September 2016

19. Ionescu, B., Gînscă, A.L., Boteanu, B., Lupu, M., Popescu, A., Müller, H.: Div150multi: a social image retrieval result diversification dataset with multi-topic queries. In: Proceedings of the 7th International Conference on Multimedia Systems, pp. 1–6 (2016)

20. Ionescu, B., et al.: ImageCLEF 2019: multimedia retrieval in medicine, lifelogging, security and nature. In: Crestani, F., et al. (eds.) CLEF 2019. LNCS, vol. 11696, pp. 358–386. Springer, Cham (2019). https://doi.org/10.1007/978-3-030-28577-7_28

21. Ionescu, B., Rohm, M., Boteanu, B., Gînscă, A.L., Lupu, M., Müller, H.: Bench-marking image retrieval diversification techniques for social media. IEEE Trans. Multimed. **23**, 677–691 (2020)

22. Kalpathy-Cramer, J., García Seco de Herrera, A., Demner-Fushman, D., Antani, S., Bedrick, S., Müller, H.: Evaluating performance of biomedical image retrieval systems: overview of the medical image retrieval task at ImageCLEF 2004–2014. Comput. Med. Imaging Graph. **39**, 55–61 (2015)

23. Kozlovski, S., Dicente Cid, Y., Kovalev, V., Müller, H.: Overview of ImageCLEFtuberculosis 2022 - CT-based caverns detection and report. In: CLEF 2022 Working Notes. CEUR Workshop Proceedings, Bologna, Italy, 5–8 September 2022. CEUR-WS.org (2022). http://ceur-ws.org

24. Kozlovski, S., Liauchuk, V., Dicente Cid, Y., Kovalev, V., Müller, H.: Overview of ImageCLEFtuberculosis 2021 - CT-based tuberculosis type classification. In: CLEF 2021 Working Notes. CEUR Workshop Proceedings, Bucharest, Romania, 21–24 September 2021. CEUR-WS.org (2021). http://ceur-ws.org

25. Kozlovski, S., Liauchuk, V., Dicente Cid, Y., Tarasau, A., Kovalev, V., Müller, H.: Overview of ImageCLEFtuberculosis 2020 - automatic CT-based report generation. In: CLEF 2020 Working Notes. CEUR Workshop Proceedings, Thessaloniki, Greece, 22–25 September 2020. CEUR-WS.org (2020). http://ceur-ws.org

26. Kuznetsova, A., et al.: The open images dataset V4: unified image classification, object detection, and visual relationship detection at scale. CoRR abs/1811.00982 (2018). http://arxiv.org/abs/1811.00982

27. Liauchuk, V., Kovalev, V.: ImageCLEF 2017: supervoxels and co-occurrence for tuberculosis CT image classification. In: CLEF 2017 Working Notes. CEUR Workshop Proceedings, Dublin, Ireland, 11–14 September 2017. CEUR-WS.org (2017). http://ceur-ws.org

28. Lin, T.-Y., et al.: Microsoft COCO: common objects in context. In: Fleet, D., Pajdla, T., Schiele, B., Tuytelaars, T. (eds.) ECCV 2014. LNCS, vol. 8693, pp. 740–755. Springer, Cham (2014). https://doi.org/10.1007/978-3-319-10602-1_48

29. Müller, H., Clough, P., Deselaers, T., Caputo, B. (eds.): ImageCLEF - Experimental Evaluation in Visual Information Retrieval. The Springer International Series On Information Retrieval, vol. 32. Springer, Heidelberg (2010). https://doi.org/10.1007/978-3-642-15181-1

30. Pelka, O., Ben Abacha, A., García Seco de Herrera, A., Jacutprakart, J., Friedrich, C.M., Müller, H.: Overview of the ImageCLEFmed 2021 concept & caption prediction task. In: CLEF 2021 Working Notes. CEUR Workshop Proceedings, Bucharest, Romania, 21–24 September 2021, pp. 1101–1112. CEUR-WS.org (2021)

31. Pelka, O., Friedrich, C.M., García Seco de Herrera, A., Müller, H.: Overview of the ImageCLEFmed 2019 concept prediction task. In: CLEF 2019 Working Notes. CEUR Workshop Proceedings, Lugano, Switzerland, 09–12 September 2019. CEUR-WS.org (2019). http://ceur-ws.org

32. Pelka, O., Friedrich, C.M., García Seco de Herrera, A., Müller, H.: Overview of the ImageCLEFmed 2020 concept prediction task: medical image understanding. In: CLEF 2020 Working Notes. CEUR Workshop Proceedings, Thessaloniki, Greece, 22–25 September 2020. CEUR-WS.org (2020)

33. Pelka, O., Koitka, S., Rückert, J., Nensa, F., Friedrich, C.M.: Radiology Objects in COntext (ROCO): a multimodal image dataset. In: Stoyanov, D., et al. (eds.) LABELS/CVII/STENT -2018. LNCS, vol. 11043, pp. 180–189. Springer, Cham (2018). https://doi.org/10.1007/978-3-030-01364-6_20

34. Popescu, A., Deshayes-Chossart, J., Schindler, H., Ionescu, B.: Overview of the ImageCLEF 2022 aware task. In: Experimental IR Meets Multilinguality, Multimodality, and Interaction. Proceedings of the 13th International Conference of the CLEF Association (CLEF 2022). LNCS. Springer, Cham (2022)

35. Roberts, R.J.: PubMed central: the GenBank of the published literature. Proc. Natl. Acad. Sci. U.S.A. **98**(2), 381–382 (2001). https://doi.org/10.1073/pnas.98.2. 381
36. Rückert, J., et al.: Overview of ImageCLEFmedical 2022 - caption prediction and concept detection. In: CLEF 2022 Working Notes. CEUR Workshop Proceedings, Bologna, Italy, 5–8 September 2022. CEUR-WS.org (2022)
37. Russakovsky, O., et al.: ImageNet large scale visual recognition challenge. Int. J. Comput. Vis. **115**(3), 211–252 (2015)
38. Thomee, B., et al.: YFCC100M: the new data in multimedia research. Commun. ACM **59**(2), 64–73 (2016)
39. Tsikrika, T., de Herrera, A.G.S., Müller, H.: Assessing the scholarly impact of ImageCLEF. In: Forner, P., Gonzalo, J., Kekäläinen, J., Lalmas, M., de Rijke, M. (eds.) CLEF 2011. LNCS, vol. 6941, pp. 95–106. Springer, Heidelberg (2011). https://doi.org/10.1007/978-3-642-23708-9_12
40. Tsikrika, T., Larsen, B., Müller, H., Endrulis, S., Rahm, E.: The scholarly impact of CLEF (2000–2009). In: Forner, P., Müller, H., Paredes, R., Rosso, P., Stein, B. (eds.) CLEF 2013. LNCS, vol. 8138, pp. 1–12. Springer, Heidelberg (2013). https://doi.org/10.1007/978-3-642-40802-1_1
41. World Health Organization, et al.: Global tuberculosis report 2019 (2019)
42. Zhou, X., Girdhar, R., Joulin, A., Krähenbühl, P., Misra, I.: Detecting twenty-thousand classes using image-level supervision. arXiv preprint arXiv:2201.02605 (2022)

Author Index

Printed in the United States
by Baker & Taylor Publisher Services